Algebra Know-It-ALL

About the Author

Stan Gibilisco is an electronics engineer, researcher, and mathematician who has authored a number of titles for the McGraw-Hill *Demystified* series, along with more than 30 other books and dozens of magazine articles. His work has been published in several languages.

Algebra Know-It-ALL

Beginner to Advanced, and Everything in Between

Stan Gibilisco

New York Chicago San Francisco Lisbon London Madrid
Mexico City Milan New Delhi San Juan Seoul
Singapore Sydney Toronto

The McGraw-Hill Companies

Library of Congress Cataloging-in-Publication Data

Gibilisco, Stan.
Algebra know-it-all : beginner to advanced, and everything in between / Stan Gibilisco.—1st ed.
p. cm.
ISBN 978-0-07-154617-1 (alk. paper)
1. Algebra. I. Title.
QA155.G52 2008
512—dc22 2008012916

1 2 3 4 5 6 7 8 9 0 DOC/DOC 0 1 4 3 2 1 0 9 8

ISBN 978-0-07-154617-1
MHID 0-07-154617-0

Sponsoring Editor
Judy Bass

Production Supervisor
Pamela A. Pelton

Editing Supervisor
Stephen M. Smith

Project Manager
Vasundhara Sawhney, International Typesetting and Composition

Copy Editor
Priyanka Sinha, International Typesetting and Composition

Proofreaders
Nigel O'Brien and Sanjukta Chandra, International Typesetting and Composition

Art Director, Cover
Jeff Weeks

Composition
International Typesetting and Composition

Printed and bound by RR Donnelley.

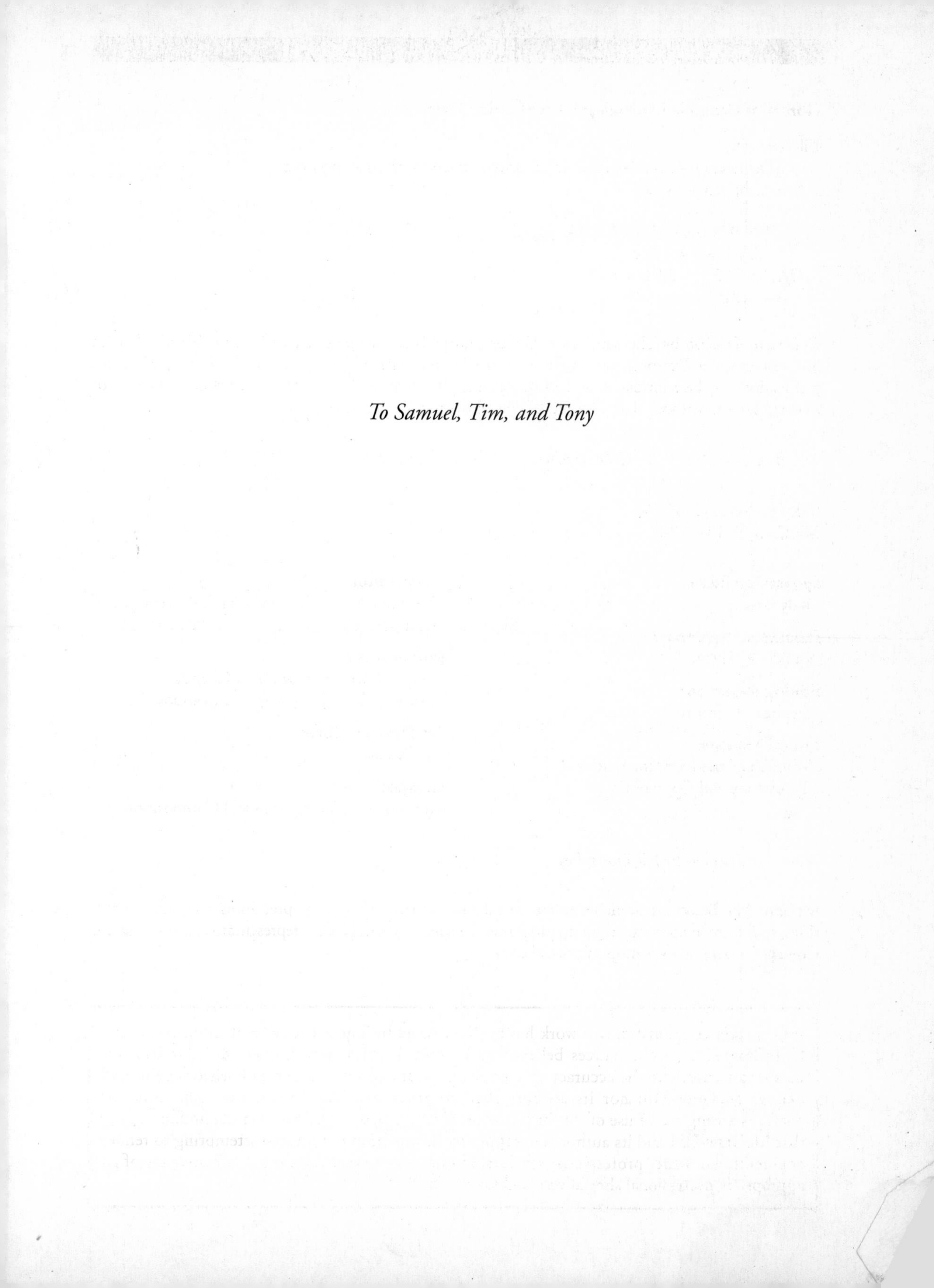

To Samuel, Tim, and Tony

Contents

Part 2 Linear Equations and Relations

Preface

If you want to improve your understanding of algebra, then this book is for you. It can supplement standard texts at the middle-school and high-school levels. It can also serve as a self-teaching or home-schooling supplement. The essential prerequisite is a solid background in arithmetic. It will help if you've had some pre-algebra as well.

This book contains three major sections. Part 1 involves numbers, sets, arithmetic operations, and basic equations. Part 2 is devoted to first-degree equations, relations, functions, and systems of linear equations. Part 3 deals with quadratic, cubic, and higher-degree equations, and introduces you to logarithms, exponentials, and systems of nonlinear equations.

Chapters 1 through 9, 11 through 19, and 21 through 29 end with practice exercises. You may (and should) refer to the text as you solve these problems. Worked-out solutions appear in Apps. A, B, and C. Often, these solutions do not represent the only way a problem can be figured out. Feel free to try alternatives!

Chapters 10, 20, and 30 contain question-and-answer sets that finish up Parts 1, 2, and 3, respectively. These chapters will help you review the material. A multiple-choice final exam concludes the course. Don't refer to the text while taking the exam. The questions in the exam are more general (and less time consuming) than the practice exercises at the ends of the chapters. The final exam is designed to test your grasp of the concepts, not to see how well you can execute calculations. The correct answers are listed in App. D.

In my opinion, middle-school and high-school students aren't sufficiently challenged in mathematics these days. I think that most textbooks place too much importance on "churning out answers," and often fail to explain how and why you get those answers. I wrote this book to address these problems. The presentation sometimes gets theoretical, but I've tried to introduce the language gently so you won't get lost in a wilderness of jargon. Many of the examples and problems are easy, some take work, and a few are designed to make you think hard.

If you complete one chapter per week, you'll get through this course in a school year. But don't hurry. Proceed at your own pace. When you've finished this book, I highly recommend McGraw-Hill's *Algebra Demystified* and *College Algebra Demystified*, both by Rhonda Huettenmueller, for further study.

Stan Gibilisco

Acknowledgment

I extend thanks to my nephew Tony Boutelle. He spent many hours helping me proofread the manuscript, and he offered insights and suggestions from the viewpoint of the intended audience.

PART 1

Numbers, Sets, and Operations

CHAPTER

1

Counting Methods

Algebra is a science of numbers. To work with numbers, you need symbols to represent them. The way these symbols relate to actual quantities is called a *numeration system*. In this chapter, you'll learn about numeration systems for whole-unit quantities such as 4, 8, 1,509, or 1,580,675. Fractions, negative numbers, and more exotic numbers will come up later.

Fingers and Sticks

Throughout history, most cultures developed numeration systems based on the number of fingers and thumbs on human hands. The word *digit* derives from the Latin word for "finger." This is no accident. Fingers are convenient for counting, at least when the numbers are small!

Number or numeral?

The words *number* and *numeral* are often used as if they mean the same thing. But they're different. A number is an abstraction. You can't see or feel a number. A numeral is a tangible object, or a group of objects, that represents a number. Suppose you buy a loaf of bread cut into eighteen slices. You can consider the whole sliced-up loaf as a numeral that represents the number eighteen, and each slice as a digit in that numeral. You can't eat the number eighteen, but you can eat the bread.

In this chapter, when we write about numbers as quantities, let's write them out fully in words, like eighteen or forty-five or three hundred twenty-one. When we want to write down a numeral, it's all right to put down 18 or 45 or 321, but we have to be careful about this sort of thing. When you see a large quantity written out in full here, keep this in mind: It means we're dealing with a number, not a numeral.

Figuring with fingers

Imagine it's the afternoon of the twenty-fourth day of July. You have a doctor's appointment for the afternoon of the sixth of August. How many days away is your appointment?

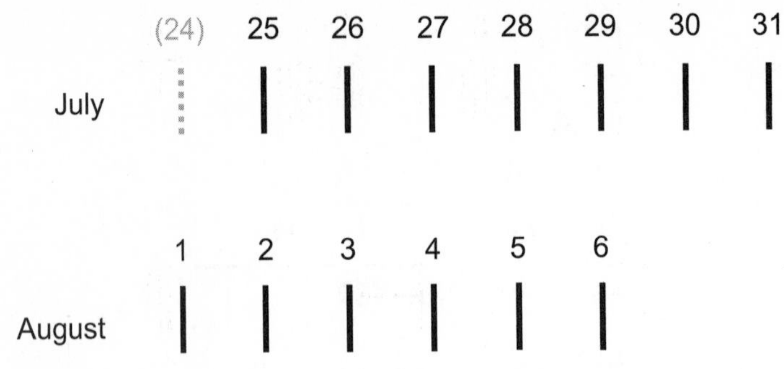

Figure 1-1 How many days pass from the afternoon of July 24 until the afternoon of August 6? You can make marks on a piece of paper and then count them to figure out the answer.

A calculator won't work very well to solve this problem. Try it and see! You can't get the right answer by any straightforward arithmetic operation on twenty-four and six. If you attack this problem as I would, you'll count out loud starting with tomorrow, July twenty-fifth (under your breath): "twenty-five, twenty-six, twenty-seven, twenty-eight, twenty-nine, thirty, thirty-one, one, two, three, four, five, six!" While jabbering away, I would use my fingers to count along or make "hash marks" on a piece of paper (Fig. 1-1). You might use a calendar and point to the days one at a time as you count them out. However you do it, you'll come up with thirteen days if you get it right. But be careful! This sort of problem is easy to mess up.

Don't be embarrassed if you find yourself figuring out simple problems like this using your fingers or other convenient objects. You're making sure that you get the right answer by using numerals to represent the numbers. Numerals are tailor-made for solving number problems because they make abstract things easy to envision.

Toothpicks on the table

Everyone has used "hash marks" to tally up small numbers. You can represent one item by a single mark and five items by four marks with a long slash. You might use objects such as toothpicks to create numerals in a system that expands on this idea, as shown in Fig. 1-2. You can represent ten by making a capital letter T with two toothpicks. You can represent fifty by using three toothpicks to make a capital letter F. You can represent a hundred by making a capital letter H with three toothpicks. This lets you express rather large numbers such as seventy-four or two hundred fifty-three without having to buy several boxes of toothpicks and spend a lot of time laying them down.

In this system, any particular arrangement of sticks is a numeral. You can keep going this way, running an F and H together to create a symbol that represents five hundred. You can run a T and an H together to make a symbol that represents a thousand. How about ten thousand? You could stick another T onto the left-hand end of the symbol for a thousand, or you could run two letters H together to indicate that it's a hundred hundred! Use your imagination. That's what mathematicians did when they invented numeration systems in centuries long past.

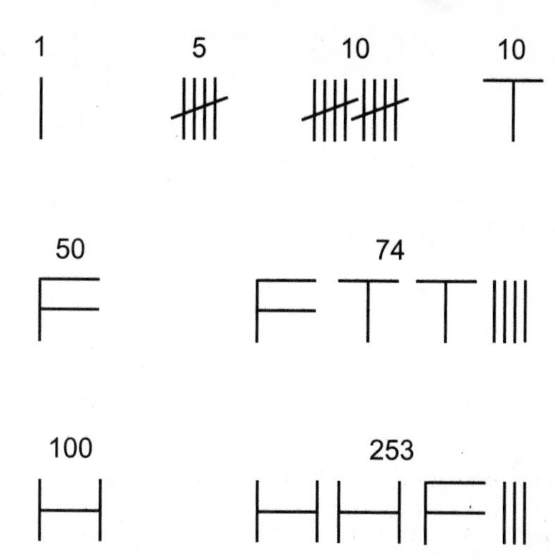

Figure 1-2 Toothpick-numeral equivalents of some numbers. In this system, most numbers can be represented by more than one numeral. But there is always a "best numeral" that uses the smallest possible number of toothpicks.

Are you confused?

If the toothpick numeral system puzzles you, don't feel bad. It's awkward. It's impractical for expressing gigantic numbers. People aren't used to counting in blocks of five or fifty or five hundred. It's easier to go straight from blocks of one to blocks of ten, and then from ten to a hundred, then to a thousand, then to ten thousand, and so on. But using blocks of five, fifty, five hundred and so on, in addition to the traditional multiples of ten, conserves toothpicks.

Here's a challenge!

Using toothpick numerals represent the number seven hundred seventy-seven in two different ways. Make sure one of your arrangements is the most "elegant" possible way to represent seven hundred seventy-seven, meaning that it uses the smallest possible number of toothpicks.

Solution

Figure 1-3 shows two ways you can represent this number. In order to represent five hundred, you build the F and the H together so they're a single connected pattern of sticks. The arrangement on top is the most "elegant" possible numeral.

You can represent seven hundred seventy-seven in more ways than just the two shown here. You can make numerals that are far more "inelegant" than the bottom arrangement. The worst possible approach is to lay down seven hundred seventy-seven toothpicks side-by-side.

Figure 1-3 Two different ways of expressing seven hundred seventy-seven in toothpick numerals. The top method is preferred because it is more "elegant."

Roman Numerals

The toothpick numeration system just described bears a resemblance to another system that was actually used in much of the world until a few centuries ago: the *Roman numeration system,* more often called *Roman numerals.*

Basic symbols

In Roman numerals, a quantity of one is represented by a capital letter I. A quantity of five is represented by a capital V. A quantity of ten is denoted as a capital X, fifty is a capital L, a hundred is a capital C, five hundred is a capital D, and a thousand is usually represented by a capital M. (Sometimes K is used instead.)

So far, this looks like a refinement of the toothpick numeration scheme. But there are some subtle differences. You don't always write the symbols in straightforward order from left to right, as you lay down the sticks in the toothpick system. There are exceptions, intended to save symbols.

Arranging the symbols

The people who designed the Roman system did not like to put down more than three identical symbols in a row. Instead of putting four identical symbols one after another, the writer would jump up to the next higher symbol and then put the next lower one to its left, indicating that the smaller quantity should be taken away from the larger.

For example, instead of IIII (four ones) to represent four, you would write IV (five with one taken away). Instead of XXXX (four tens) to represent forty, you'd write XL (fifty with ten taken away). Instead of MDXXXX to represent one thousand nine hundred, you'd write MCM (a thousand and then another thousand with a hundred taken away).

What about zero?

By now you must be thinking, "No wonder people got away from Roman numerals, let alone hash marks. They're confusing!" But that's not the only trouble with the Roman numeral system or the toothpick numeral system we made up earlier. There's a more serious issue. Neither of these schemes give you any way to express the quantity zero. This might not seem important at first thought. Why make a big fuss over a symbol that represents nothing?

Sometimes the best way to see why something is important is to try to get along without it. When you start adding and subtracting, and especially when you start multiplying and dividing, it's almost impossible to get along without zero. In a computer, the numeral 0 is one of only two possible digits (the other being 1) for building large numerals. In accounting, the presence or absence of a single 0 on a piece of paper can represent the difference between the price of a car and the price of a house.

Are you confused?

Let's write down all the *counting numbers* from one to twenty-one as Roman numerals. This will give you a "feel" for how the symbols are arranged to represent adding-on or taking-away of quantities.

The first three are easy: the symbol I means one, II means two, and III means three. Then for four, we write IV, meaning that one is taken away from five. Proceeding, V means five, VI means six, VII means seven, and VIII means eight. To represent nine, we write IX, meaning that one is taken away from ten. Then going on, X means ten, XI means eleven, XII means twelve, and XIII means thirteen. Now for fourteen, we write XIV, which means ten with four more added on. Then XV means fifteen, XVI means sixteen, XVII means seventeen, and XVIII means eighteen. For nineteen, we write XIX, which means ten with nine more added on. Continuing, we have XX that stands for twenty, and XXI to represent twenty-one.

Here's a challenge!

Write down some Roman numerals in a table as follows. In the first column, put down the equivalents of one to nine in steps of one. In a second column, put down the equivalents of ten to ninety in steps of ten. In a third column, put down the equivalents of one hundred to nine hundred in steps of one hundred. In a fourth column, put down the equivalents of nine hundred ten to nine hundred ninety in steps of ten. In a fifth column, put down the equivalents of nine hundred ninety-one to nine hundred ninety-nine in steps of one.

Solution

Refer to Table 1-1. The first column is farthest to the left, and the fifth column is farthest to the right. For increasing values in each column, read downward. "Normal" numerals are shown along with their Roman equivalents for clarification.

Table 1-1. Some examples of Roman numerals. From this progression, you should be able to see how the system works for fairly large numbers. You should also begin to understand why mathematicians abandoned this system centuries ago.

1 = I	10 = X	100 = C	910 = CMX	991 = CMXCI
2 = II	20 = XX	200 = CC	920 = CMXX	992 = CMXCII
3 = III	30 = XXX	300 = CCC	930 = CMXXX	993 = CMXCIII
4 = IV	40 = XL	400 = CD	940 = CMXL	994 = CMXCIV
5 = V	50 = L	500 = D	950 = CML	995 = CMXCV
6 = VI	60 = LX	600 = DC	960 = CMLX	996 = CMXCVI
7 = VII	70 = LXX	700 = DCC	970 = CMLXX	997 = CMXCVII
8 = VIII	80 = LXXX	800 = DCCC	980 = CMLXXX	998 = CMXCVIII
9 = IX	90 = XC	900 = CM	990 = CMXC	999 = CMXCIX

Hindu-Arabic Numerals

The numeration system we use today was invented in the seventh century by mathematicians in Southern Asia. During the next two or three hundred years, invaders from the Middle East picked it up. Good ideas have a way of catching on, even with invading armies! Eventually, most of the civilized world adopted the *Hindu-Arabic numeration system.* The "Hindu" part of the name comes from India, and the "Arabic" part from the Middle East. You will often hear this scheme called simply *Arabic numerals.*

The idea of "place"

In an Arabic numeral, every digit represents a quantity ranging from nothing to nine. These digits are the familiar 0, 1, 2, 3, 4, 5, 6, 7, 8, and 9. The original Hindu inventors of the system came up with an interesting way of expressing numbers larger than nine. They gave each digit more or less "weight" or value, depending on where it was written in relation to other digits in the same numeral. The idea was that every digit in a numeral should have ten times the value of the digit (if any) to its right. When building up the numeric representation for a large number, there would occasionally be no need for a digit in a particular place, but a definite need for one on either side of it. That's where the digit 0 became useful.

Zero as a "placeholder"

Figure 1-4 shows an example of a numeral that represents a large number. Note that the digit 0, also called a *cipher,* is just as important as any other digit. The quantity shown is

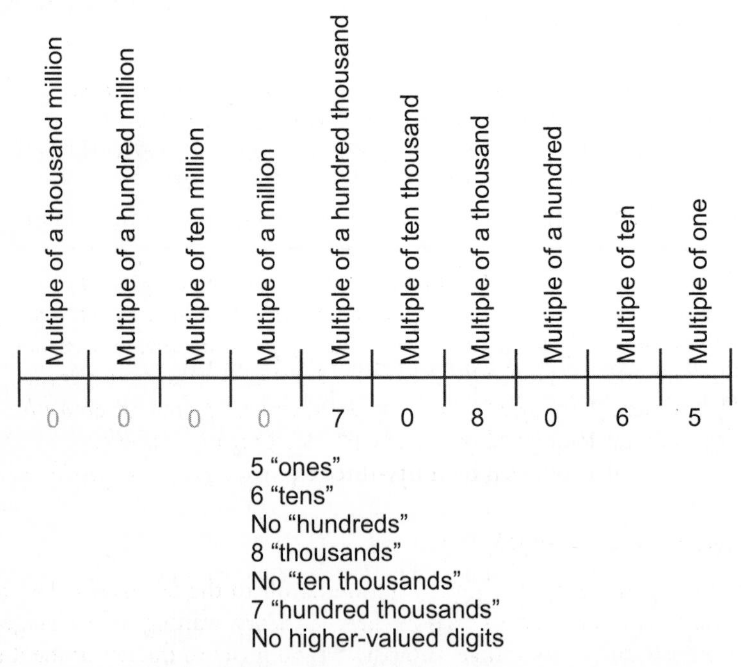

Figure 1-4 In the Hindu-Arabic numeration system, large numbers are represented by building up numerals digit-by-digit from right to left, giving each succeeding digit ten times the value of the digit to its right.

seven hundred eight thousand sixty-five. (Some people would call it seven hundred *and* eight thousand *and* sixty-five.) It's customary to place a comma or space after every third digit as you proceed from right to left in a multi-digit numeral like this. Once you get to a certain nonzero digit as you work your way from right to left, all the digits farther to the left are understood to be ciphers.

Every digit 0 "inside" a numeral serves as a placeholder, making it clear what the values of digits to its left should be. All those ciphers to the left of the last nonzero digit are insignificant in most situations, and it is unusual to see any of them written down. But once in a while you might find it helpful to insert one or more of them during a calculation.

Counting vs. whole numbers

Let's make sure we understand the difference between a counting number and a *whole number.* Usage varies depending on which text you happen to read. For our purposes, the counting numbers go as one, two, three, four, five, and so on. They can be defined with the Roman numeration system, and can also be defined with the toothpick system we "invented" here. We'll define the whole numbers as zero, one, two, three, four, five, and so on. The only difference is whether we start at one or zero.

Names for some huge numbers

People who used Roman numerals hardly ever had to work with numbers much larger than a thousand. But in today's scientific world, we deal with numbers that make a thousand seem tiny by comparison. Here are some of the names for numbers that are represented as a 1 followed by multiples of three ciphers:

- The numeral 1 followed by three ciphers represents a *thousand.*
- The numeral 1 followed by six ciphers represents a *million.*
- The numeral 1 followed by nine ciphers represents a *billion* in the United States or a *thousand million* in England.
- The numeral 1 followed by twelve ciphers represents a *trillion* in the United States or a *billion* in England.
- The numeral 1 followed by fifteen ciphers represents a *quadrillion.*
- The numeral 1 followed by eighteen ciphers represents a *quintillion*
- The numeral 1 followed by twenty-one ciphers represents a *sextillion.*
- The numeral 1 followed by twenty-four ciphers represents a *septillion.*
- The numeral 1 followed by twenty-seven ciphers represents an *octillion.*
- The numeral 1 followed by thirty ciphers represents a *nonillion.*
- The numeral 1 followed by thirty-three ciphers represents a *decillion.*

How many numbers exist?

Envision an endless string of ciphers continuing off to the left in Fig. 1-4, all of them gray (just to remind you that each of them is there in theory, waiting to be changed to some other digit if you need to express a huge number). If you travel to the left of the digit 7 in Fig. 1-4 by dozens of places, passing through 0 after 0, and then change one of those ciphers to the digit 1, the value of the represented number increases fantastically. This is an example of the power of the Arabic numeration system. A simple change in a numeral can make a big difference in the number it represents.

Another interesting property of the Arabic system is the fact that there is no limit to how large a numeral you can represent. Even if a string of digits is hundreds of miles long, even if it circles the earth, even if it goes from the earth to the moon—all you have to do is put a nonzero digit on the left or any digit on the right, and you get the representation for a larger whole number. Mathematicians use the term *finite* to describe anything that ends somewhere. No matter how large a whole number you want to express, the Arabic system lets you do it in a finite number of digits, and every single one of those digits is from the basic set of 0 through 9. You don't have to keep inventing new symbols when numbers get arbitrarily large, as people did when the Roman system ruled.

Every imaginable number can be represented as an Arabic numeral that contains a finite number of digits. But there is no limit to the number of whole numbers you can denote that way. The group, or *set*, of all whole numbers is said to be *infinite* (not finite). That means there is no largest whole number.

What about "infinity"?

That elusive thing we call "infinity" is entirely different from any whole number, or any other sort of number people usually imagine. Mathematicians have found more than one type of "infinity"! Depending on the context, "infinity" can be represented by a *lemniscate* (∞), the small Greek letter *omega* (ω), or the capital Hebrew letter *aleph* ($\aleph$) with a numeric subscript that defines its "density."

Are you confused?

Do you still wonder why the digit 0 is needed? After all, it represents "nothing." Why bother with commas or spaces, either?

The quick answer to these questions is that the digit 0 and the comma (or space) are *not* actually needed in order to write numerals. The original inventors of the Arabic system put down a dot or a tiny circle instead of the full-size digit 0. But the cipher and the comma (or space) make errors a lot less likely.

Here's a challenge!

Imagine a whole number represented by a certain string of digits in the Arabic system. How can you change the Arabic numeral to make the number a hundred times as large, no matter what the digits happen to be?

Solution

You can make any counting numeral stand for a number a hundred times as large by attaching two ciphers to its right-hand end. Try it with a few numerals and see. Don't forget to include the commas where they belong! For example:

- 700 represents a quantity that's a hundred times as large as 7.
- 1,400 represents a quantity that's a hundred times as large as 14.
- 78,900 represents a quantity that's a hundred times as large as 789.
- 1,400,000 represents a quantity that's a hundred times as large as 14,000.

The Counting Base

The *radix* or *base* of a numeration system is the number of single-digit symbols it has. The *radix-ten* system, also called *base-ten* or the *decimal numeration system*, therefore has ten symbols, not counting commas (or decimal points, which we'll get into later). But there are systems that use bases other than ten, and that have more or less than ten symbols to represent the digits. In this section we'll look at some of them. You can get a good "mental workout" by playing with these! But they're more than mind games. The base-two and base-sixteen systems, in particular, are commonly used in computer science.

A subtle distinction

Doesn't 5 always mean the quantity five, 8 always mean the quantity eight, 10 always mean the quantity ten, and 16 always mean the quantity sixteen? Not necessarily! It's true in base-ten, but it is not necessarily true in other bases.

- Here are five pound signs: #####
- Here are eight pound signs: ########
- Here are ten pound signs: ##########
- Here are twelve pound signs: ############
- Here are sixteen pound signs: ################

In the *base-eight numeration system*, the total number of pound signs in the second line in the above list would be written as 10, the third line as 12, the fourth line as 14, and the last line as 20. In the *base-sixteen numeration system*, the total number of pound signs in the third line would be written as the letter A, the fourth line as C, and the last line as 10. (If you're confused right now, just hold on a couple of minutes!)

When the expression for a number is a spelled-out word like "eighteen" or "forty-five" or "three hundred twenty-one," we mean the actual quantity, regardless of the radix. If I write, "There are forty-five apples in this basket," it is absolutely clear what I mean. But if I write, "There are 45 apples in this basket," you must know the radix to be sure of how many apples the basket contains.

The decimal system

As you count upward from zero in the base-ten system, imagine proceeding clockwise around the face of a ten-hour clock as shown in Fig. 1-5A. When you have completed the first revolution, place a digit 1 to the left of the 0 and then go around again, keeping the 1 in the tens place. When you have completed the second revolution, change the tens digit to 2. You can keep going this way until you have completed the tenth revolution in which you have a 9 in the tens place. Then you must change the tens digit back to 0 and place a 1 in the hundreds place.

The Roman system

The Roman numeration scheme can be considered as a base-five system, at least when you start counting in it. Imagine a five-hour clock such as the one shown in Fig. 1-5B. You start with I (which stands for the number one), not with 0. You can complete one revolution and go through part of the second and the system works well.

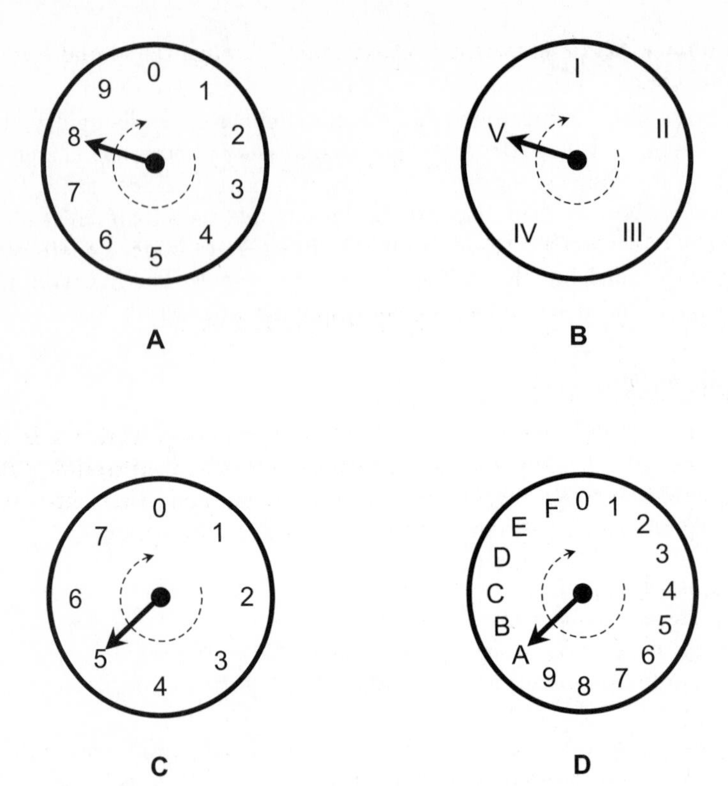

Figure 1-5 Clock-like representations of digits in base-ten or decimal (**A**), Roman base-five (**B**), base-eight or octal (**C**), and base-sixteen or hexadecimal (**D**). As you count, proceed clockwise.

After the first revolution, you keep the V and then start adding symbols to its right: VI, VII, VIII. But when you get past VIII (which stands for the number eight), a problem occurs. The number nine is not represented as VIV, although technically it could be. It's written as IX, but X is not on the clock face. The orderliness of this system falls apart before you even get twice around!

The octal system

Now imagine an eight-hour clock as shown in Fig. 1-5C. This shows how the base-eight or *octal numeration system* works. Use the same upward-counting scheme as you did with the ten-hour clock. But skip the digits 8 and 9. They do not exist in this system. When you finish the first revolution and are ready to start the second, place a 1 to the left of the digits shown, so you count

... 5, 6, 7, 10, 11, 12, ...

The string of three dots is called an *ellipsis.* It indicates that a pattern continues for a while, or perhaps even forever, saving you from having to do a lot of symbol scribbling. (You'll see this notation often in mathematics.) Continuing through the second revolution and into the third, you count

... 15, 16, 17, 20, 21, 22, ...

When you finish up the eighth revolution and enter the ninth, you count

... 75, 76, 77, 100, 101, 102, ...

The hexadecimal system

Let's invent one more strange clock. This one has sixteen hours, as shown in Fig. 1-5D. You can see from this drawing how the base-sixteen or *hexadecimal numeration system* works. Use the same upward-counting scheme as you did with the ten-hour and eight-hour clocks. There are six new digits here, in addition to the digits in the base-ten system:

- A stands for ten
- B stands for eleven
- C stands for twelve
- D stands for thirteen
- E stands for fourteen
- F stands for fifteen

When you finish the first revolution and move into the second, place a 1 to the left of the digits shown. You count

... 8, 9, A, B, C, D, E, F, 10, 11, 12, 13, ...

Continuing through the second revolution and into the third, you count

... 18, 19, 1A, 1B, 1C, 1D, 1E, 1F, 20, 21, 22, 23, ...

When you complete the tenth revolution and move into the eleventh, you count

... 98, 99, 9A, 9B, 9C, 9D, 9E, 9F, A0, A1, A2, A3, ...

It goes on like this with B, C, D, E, and F in the sixteens place. Then you get to the end of the sixteenth revolution and move into the seventeenth, like this:

F8, F9, FA, FB, FC, FD, FE, FF, 100, 101, 102, 103, ...

Get the idea?

The binary system

When engineers began to design electronic calculators and computers in the twentieth century, they wanted a way to count up to large numbers using only two digits, one to represent the "off" condition of an electrical switch and the other to represent the "on" condition. These two states can also be represented as "false/true," "no/yes," "low/high," "negative/positive," or as the numerals 0 and 1. The result is a base-two or *binary numeration system.*

Figure 1-6 shows how numerals in the binary system are put together. Instead of going up by multiples of ten, eight, or sixteen, you double the value of each digit as you move one place to the left. Numerals in the binary system are longer than numerals in the other systems, but binary numerals can be easily represented by the states of simple, high-speed electronic switches.

Every binary numeral has a unique equivalent in the decimal system, and vice versa. When you use a computer or calculator and punch in a series of decimal digits, the machine converts it into a binary numeral, performs whatever calculations or operations you demand,

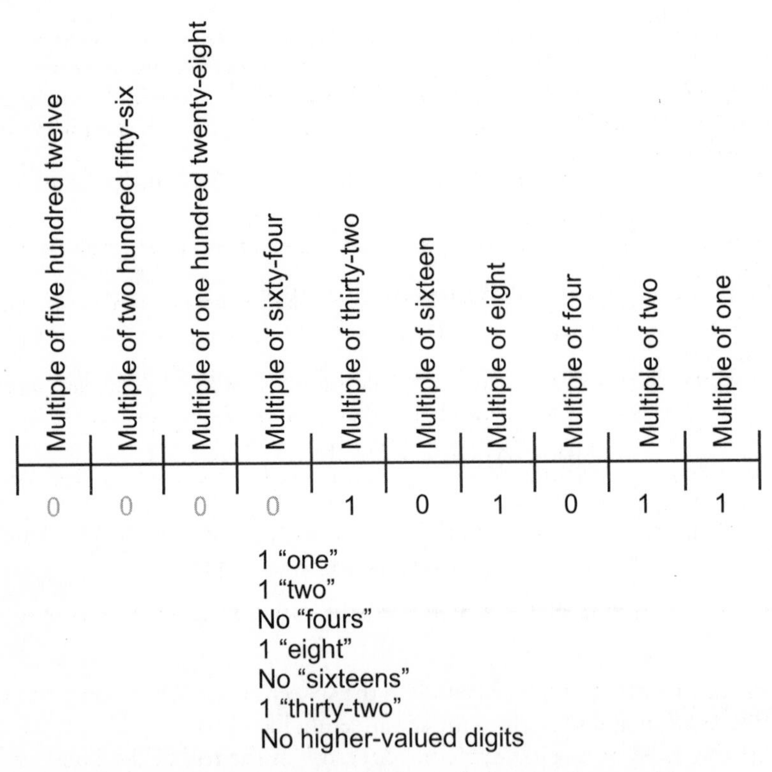

Figure 1-6 In the binary system, large numbers are represented by building up numerals digit-by-digit from right to left, giving each succeeding digit twice the value of the digit to its right. Note the absence of commas in this system.

converts the result back to a decimal numeral, and then displays that numeral for you. All of the conversions and calculations, all of the electronic switching actions and manipulations take place out of your sight, at incredible speed.

Are you confused?

Table 1-2 compares numerical values in the base-ten, base-two, base-eight, and base-sixteen systems from zero to sixty-four. From this table, you should be able to figure out (with a little bit of thought and scribbling) how to convert larger decimal numerals to any of the other forms. Fortunately, there are plenty of computer programs and Web sites that will do such conversions for you up to millions, billions, and trillions!

Here's a challenge!

Convert the hexadecimal numeral 2D03 to decimal form. Don't use a computer or go on the Internet to find a Web site that will do it for you. Grind it out the long way.

Solution

To solve this, you need to know the place values. The digit farthest to the right represents a multiple of one (that is, just itself). The next digit to the left represents a multiple of sixteen. After that comes a multiple of two hundred fifty-six (or sixteen times sixteen). Then comes a multiple of four thousand ninety-six (sixteen times two hundred fifty-six). Note that D represents thirteen. Thinking in the decimal system, you can figure it out as follows.

- In the ones place you have 3, so you start out with that
- In the sixteens place you have 0, so you must add zero times sixteen, which is zero, to what you have so far
- In the two hundred fifty-sixes place you have D which means thirteen, so you must add thirteen times two hundred fifty-six, which is three thousand three hundred twenty-eight, to what you have so far
- In the four thousand ninety-sixes place you have 2, so you must add two times four thousand ninety-six, which is eight thousand one hundred ninety-two, to what you have so far

Because there are no digits to the left of the 2, you are finished at this point. The final result, expressed as a sum in decimal numerals, is

$$3 + 0 + 3{,}328 + 8{,}192 = 11{,}523$$

One more thing ...

Are you getting tired of reading numbers as words? In the rest of this book, we'll be dealing in the decimal system exclusively. So we'll start using numerals to represent specific quantities most of the time. We won't have to worry about ambiguity that could result from an alternative radix such as eight or sixteen. Numerals will also come in handy when numbers get large or "messy." That's one of the reasons why numerals were invented!

Table 1-2. The conventional (or decimal) numerals 0 through 64, along with their binary, octal, and hexadecimal equivalents.

Decimal	Binary	Octal	Hexadecimal
0	0	0	0
1	1	1	1
2	10	2	2
3	11	3	3
4	100	4	4
5	101	5	5
6	110	6	6
7	111	7	7
8	1000	10	8
9	1001	11	9
10	1010	12	A
11	1011	13	B
12	1100	14	C
13	1101	15	D
14	1110	16	E
15	1111	17	F
16	10000	20	10
17	10001	21	11
18	10010	22	12
19	10011	23	13
20	10100	24	14
21	10101	25	15
22	10110	26	16
23	10111	27	17
24	11000	30	18
25	11001	31	19
26	11010	32	1A
27	11011	33	1B
28	11100	34	1C
29	11101	35	1D
30	11110	36	1E
31	11111	37	1F
32	100000	40	20
33	100001	41	21
34	100010	42	22
35	100011	43	23
36	100100	44	24
37	100101	45	25
38	100110	46	26
39	100111	47	27
40	101000	50	28
41	101001	51	29

(*Continued*)

Table 1-2. The conventional (or decimal) numerals 0 through 64, along with their binary, octal, and hexadecimal equivalents. (*Continued*)

Decimal	Binary	Octal	Hexadecimal
42	101010	52	2A
43	101011	53	2B
44	101100	54	2C
45	101101	55	2D
46	101110	56	2E
47	101111	57	2F
48	110000	60	30
49	110001	61	31
50	110010	62	32
51	110011	63	33
52	110100	64	34
53	110101	65	35
54	110110	66	36
55	110111	67	37
56	111000	70	38
57	111001	71	39
58	111010	72	3A
59	111011	73	3B
60	111100	74	3C
61	111101	75	3D
62	111110	76	3E
63	111111	77	3F
64	1000000	100	40

Practice Exercises

This is an open-book quiz. You may (and should) refer to the text as you solve these problems. Don't hurry! You'll find worked-out answers in App. A. The solutions in the appendix may not represent the only way a problem can be figured out. If you think you can solve a particular problem in a quicker or better way than you see there, by all means try it!

1. How many days pass in a given place between noon local time on June 24 and noon local time of October 2 of any given year?

2. Convert the following decimal numerals to Roman numerals.

   (a) 200

   (b) 201

   (c) 209

   (d) 210

3. Convert the following Roman numerals to decimal numerals.
   (a) MMXX
   (b) MMXIX
   (c) MMIX
   (d) MMVI
4. Write down the number three hundred two trillion, seventy billion, one hundred forty-nine million, six thousand, one hundred ten as a decimal numeral. Include commas where appropriate. Consider a billion as a thousand million, and a trillion as a million million.
5. How many ciphers could you add to the left of the digit 3 in the decimal numeral in the situation of Problem 4 without changing the value of the number it represents?
6. How can you make the number represented by the numeral in the answer to Problem 4 ten times as large? A hundred times as large? A thousand times as large?
7. How can you write out the final answer to Problem 6 as a number in words rather than as a numeral in digits?
8. What numeral in the decimal (base-ten) system represents the same quantity that the binary numeral shown in Fig. 1-6 represents?
9. What numeral in the octal (base-eight) system represents the same quantity that the binary numeral shown in Fig. 1-6 represents?
10. What numeral in the hexadecimal (base-sixteen) decimal system represents the same quantity that the binary numeral shown in Fig. 1-6 represents?

CHAPTER

2

The Language of Sets

Before going farther with numbers, you should be familiar with sets and the symbols that describe their behavior. Sets are important in all branches of mathematics, including algebra. Put on your "abstract thinking cap"!

The Concept of a Set

A *set* is a collection or group of things called *elements* or *members*. An element of a set can be anything you can imagine, even another set. Sets, like numbers, are abstractions. If you have a set of a dozen eggs, you have something more than just the eggs. You have the fact that those eggs are all in the same group. Maybe you plan to use them to "rustle up" flapjacks for your ranch hands. Maybe your sister wants to try to hatch chickens from them.

To belong, or not to belong

If you want to call some entity x an element of set A, then you write

$$x \in A$$

The "lazy pitchfork" symbol means "is an element of." You can also say that x belongs to set A, or that x is in set A. If some other entity y is *not* an element of set A, then you can write that as

$$y \notin A$$

An element is a "smallest possible piece" that can exist in any set. You can't break an element down into anything smaller and have it remain a legitimate element of the original set. This little notion becomes important whenever you have a set that contains another set as one of its elements.

Listing the elements

When the elements of a set are listed, the list is enclosed in "curly brackets," usually called *braces*. The order of the list does not matter. Repetition doesn't matter either. The following sets are all the same:

$$\{1, 2, 3\}$$
$$\{3, 2, 1\}$$
$$\{1, 3, 3, 2, 1\}$$
$$\{1, 2, 3, 1, 2, 3, 1, 2, 3, \ldots\}$$

The ellipsis (string of three dots) means that the list goes on forever in the pattern shown. In this case, it's around and around in an endless cycle.

Now look at this example of a set with five elements:

$$S = \{2, 4, 6, 8, 10\}$$

Are the elements of this set S meant to be numbers or numerals? That depends on the context. Usually, when you see a set with numerals in it like this, the author means to define the set containing the numbers that those numerals represent.

Here's another example of a set with five elements:

$$P = \{\text{Mercury, Venus, Earth, Mars, Jupiter}\}$$

You're entitled to assume that the elements of this set are the first five planets in our solar system, not the words representing them.

The empty set

A set can exist even if there are no elements in it. This is called the *empty set* or the *null set*. It can be symbolized by writing two braces facing each other with a space between, like this:

$$\{\ \}$$

Another way to write it is to draw a circle and run a forward slash through it, like this:

$$\varnothing$$

Let's use the circle-slash symbol in the rest of this chapter, and anywhere else in this book the null set happens to come up.

You might ask, "How can a set have no elements? That would be like a club with no members!" Well, so be it, then! If all the members of the Pingoville Ping-Pong Club quit today and no new members join, the club still exists if it has a charter and by laws. The set of members of the Pingoville Ping-Pong Club might be empty, but it's a legitimate set as long as someone says the club exists.

Finite or infinite?

Sets can be categorized as either *finite* or *infinite*. When a set is finite, you can name all of its elements if you have enough time. This includes the null set. You can say "This set has no elements," and you've named all the elements of the null set. When a set is infinite, you can't name all of its elements, no matter how much time you have.

Even if a set is infinite, you might be able to write an "implied list" that reveals exactly what all of its elements are. Consider this:

$$W = \{0, 1, 2, 3, 4, 5, ...\}$$

This is the set of whole numbers as it is usually defined in mathematics. You know whether or not something is an element of set W, even if it is not shown above, and even if you could not reach it if you started to scribble down the list right now and kept at it for days. You can tell right away which of the following numbers are elements of W, and which are not:

12

1/2

23

100/3

78,883,505

356.75

90,120,801,000,000,000

−65,457,333

The first, third, fifth, and seventh numbers are elements of W, but the second, fourth, sixth, and eighth numbers are not.

Some infinite sets cannot be totally defined by means of any list, even an "implied list"! You'll learn about this type of set in Chap. 9.

Sets within sets

A set can be an element of another set. Remember again, anything can be a member of a set! You can have sets that get confusing when listed. Here are some examples, in increasing order of strangeness:

$$\{1, 2, 3, 4, 5\}$$

$$\{1, 2, \{3, 4, 5\}\}$$

$$\{1, \{2, \{3 ,4, 5\}\}\}$$

$$\{1, \{2, \{3, \{4, 5\}\}\}\}$$

$$\{1, \{2, \{3, \{4, \{5\}\}\}\}\}$$

An "inner" or "member" set can sometimes have more elements than the set to which it belongs. Here is an example:

$$\{1, 2, \{3, 4, 5, 6, 7, 8\}\}$$

Here, the main set has three elements, one of which is a set with six elements.

Are you confused?

Do you still wonder what makes a bunch of things a set? If you have a basket full of apples and you call it a set, is it still a set when you dump the apples onto the ground? Were those same apples elements of a set before they were picked? Questions like this can drive you crazy if you let them. A collection of things is a set if you decide to call it a set. It's that simple.

As you go along in this course, you'll eventually see how sets are used in algebra. Here's an easy example. What number, when multiplied by itself, gives you 4? The obvious answer is 2. But −2 will also work, because "minus times minus equals plus." In ordinary mathematics, a number can't have more than one value. But two or more numbers can be elements of a set. A mathematician would say that the *solution set* to this problem is $\{-2, 2\}$.

Here's a riddle!

You might wonder if a set can be an element of itself. At first, it is tempting to say "No, that's impossible. It would be like saying the Pingoville Ping-Pong Club is one of its own members. The elements are the Ping-Pong players, not the club."

But wait! What about the set of all abstract ideas? That's an abstract idea. So a set *can* be a member of itself. This is a strange scenario because it doesn't fit into the "real world." In a way, it's just a riddle. Nevertheless, riddles of this sort sometimes open the door to important mathematical discoveries.

Here's a challenge!

Define the set of all the positive and negative whole numbers in the form of an "implied list" of numerals. Make up the list so that, if someone picks a positive or negative number, no matter how big or small it might be, you can easily tell whether or not it is in the set by looking at the list.

Solution

You can do this in at least two ways. You can start with zero and then list the numerals for the positive and negative whole numbers alternately:

$$\{0, 1, -1, 2, -2, 3, -3, 4, -4, \ldots\}$$

You can also make the list open at both ends, implying unlimited "travel" to the left as well as to the right:

$$\{\ldots, -4, -3, -2, -1, 0, 1, 2, 3, 4, \ldots\}$$

How Sets Relate

Now let's see how sets can be broken down, compared, and combined. Pictures can do the work of thousands of words here.

Venn diagrams

One of the most useful illustrations for describing relationships among sets is a *Venn diagram,* in which sets are shown as groups of points or as geometric figures. Figure 2-1 is an example. The

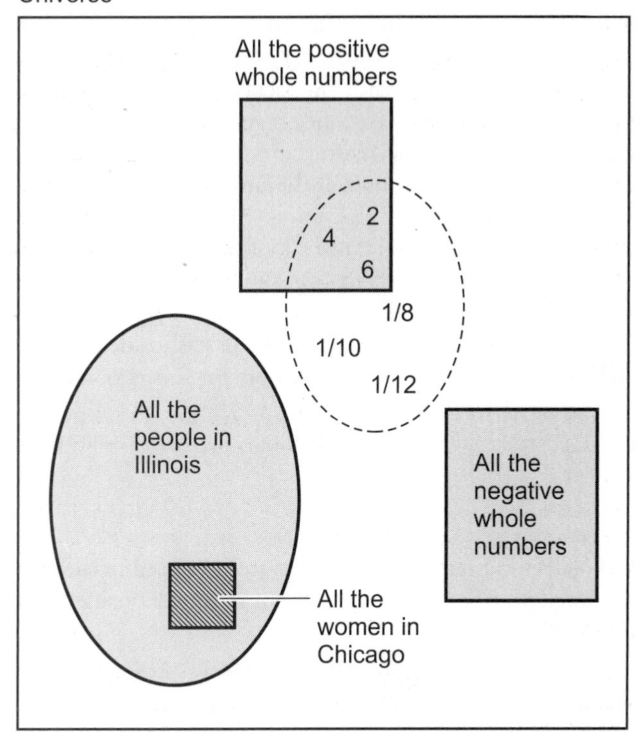

Figure 2-1 A Venn diagram showing the set of all sets (the universe) along with a few specific sets within it.

large, heavy rectangle represents the set of all things that can exist, whether real or imaginary (and that includes all possible sets). This "emperor of sets" is called the *universal set* or the *universe.*

In Fig. 2-1, three of the sets shown inside the universe are finite and two are infinite. Note how the objects overlap or are contained within one another or are entirely separate. This is important, because it describes the various ways sets can relate to each other. You can see how this works by examining the diagram carefully.

All the women in Chicago are people in Illinois, but there are plenty of people in Illinois who aren't women in Chicago. The numbers 2, 4, and 6 are positive whole numbers, but there are lots of positive whole numbers different from 2, 4, or 6. The sets of positive and negative whole numbers are entirely separate, even though both sets are infinite. None of the positive or negative whole numbers is a person in Illinois, and no person in Illinois is number (except according to the government, maybe).

Subsets

When all the elements of a set are also contained in a second set, the first set is called a *subset* of the second. If you have two sets A and B, and every element of A is also an element of B, then A is a subset of B. That fact can be written

$$A \subseteq B$$

Figure 2-1 shows that the set of all the women in Chicago is a subset of the set of all the people in Illinois. That is expressed by a hatched square inside a shaded oval. Figure 2-1 also shows that the set {2, 4, 6} is a subset of the set of positive whole numbers. That is expressed by placing the numerals 2, 4, and 6 inside the rectangle representing the positive whole numbers. All five of the figures inside the large, heavy rectangle of Fig. 2-1 represent subsets of the universe. Any set you can imagine, no matter how large, small, or strange it might be, and no matter if it is finite or infinite, is a subset of the universe. Technically, a set is always a subset of itself.

Often, a subset represents only part, not all, of the main set. Then the smaller set is called a *proper subset* of the larger one. In the situation shown by Fig. 2-1, the set of all the women in Chicago is a proper subset of the set of all the people in Illinois. The set {2, 4, 6} is a proper subset of the set of positive whole numbers. All five of the sets inside the main rectangle are proper subsets of the universe. When a certain set C is a proper subset of another set D, we write

$$C \subset D$$

Congruent sets

Once in a while, you'll come across two sets that are expressed in different ways, but they turn out to be exactly the same when you look at them closely. Consider these two sets:

$$E = \{1, 2, 3, 4, 5, ...\}$$
$$F = \{7/7, 14/7, 21/7, 28/7, 35/7, ...\}$$

At first glance, these two sets look completely different. But if you think of their elements as numbers (not as symbols representing numbers) the way a mathematician would regard them, you can see that they're really the same set. You know this because

$$7/7 = 1$$
$$14/7 = 2$$
$$21/7 = 3$$
$$28/7 = 4$$
$$35/7 = 5$$
$$\downarrow$$

and so on, forever

Every element in set E has exactly one "mate" in set F, and every element in set F has exactly one "mate" in set E. In a situation like this, the elements of the two sets exist in a *one-to-one correspondence.*

When two sets have elements that are identical, and all the elements in one set can be paired off one-to-one with all the elements in the other, they are said to be *congruent sets.* Sometimes they're called *equal sets* or *coincident sets.* In the above situation, we can write

$$E = F$$

Once in a while, you'll see a three-barred equals sign to indicate that two sets are congruent. In this case, we would write

$$E \equiv F$$

Disjoint sets

When two sets are completely different, having no elements in common whatsoever, then they are called *disjoint sets*. Here is an example of two disjoint sets of numbers:

$$G = \{1, 2, 3, 4\}$$

$$H = \{5, 6, 7, 8\}$$

Both of these sets are finite. But infinite sets can also be disjoint. Take the set of all the even whole numbers and all the odd whole numbers:

$$W_{\text{even}} = \{0, 2, 4, 6, 8, ...\}$$

$$W_{\text{odd}} = \{1, 3, 5, 7, 9, ...\}$$

No matter how far out along the list for W_{even} you go, you'll never find any element that is also in W_{odd}. No matter how far out along the list for W_{odd} you go, you'll never find any element that is also in W_{even}. We won't try to prove this right now, but you should not have any trouble sensing that it's a fact. Sometimes the mind's eye can see forever!

Figure 2-2 is a Venn diagram showing two sets, J and K, with no elements in common. You can imagine J as the set of all the points on or inside the circle and K as the set of all the points on or inside the oval. Sets J and K are disjoint. When you have two disjoint sets, neither of them is a subset of the other.

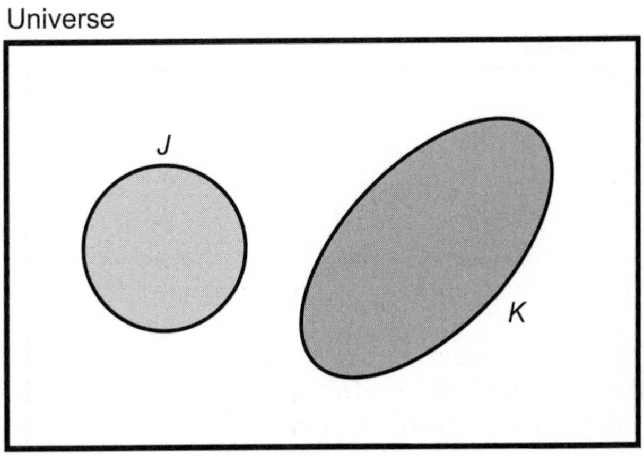

Figure 2-2 Two disjoint sets, J and K. They have no elements in common.

Overlapping sets

When two sets have at least one element in common, they are called *overlapping sets*. In formal texts you might see them called *nondisjoint sets*. Congruent sets overlap in the strongest possible sense, because they have all their elements in common. More often, two overlapping sets share some, but not all, of their elements. Here are two sets of numbers that overlap with one element in common:

$$L = \{2, 3, 4, 5, 6\}$$
$$M = \{6, 7, 8, 9, 10\}$$

Here is a pair of sets that overlap more:

$$P = \{21, 23, 25, 27, 29, 31, 33\}$$
$$Q = \{25, 27, 29, 31, 33, 35, 37\}$$

Technically, these sets overlap too:

$$R = \{11, 12, 13, 14, 15, 16, 17, 18, 19\}$$
$$S = \{12, 13, 14\}$$

Here, you can see that S is a subset of R. In fact, S is a proper subset of R. Now, let's look at a pair of infinite sets that overlap with four elements in common:

$$W_{3-} = \{..., -5, -4, -3, -2, -1, 0, 1, 2, 3\}$$
$$W_{0+} = \{0, 1, 2, 3, 4, 5, ...\}$$

The notation W_{3-} (read "W sub three-minus") means the set of all positive or negative whole numbers starting at 3 and decreasing, one by one, without end. The notation W_{0+} (read "W sub zero-plus") means the set of whole numbers starting with 0 and increasing, one by one, without end. That's the set of whole numbers as it is usually defined.

Figure 2-3 is a Venn diagram that shows two sets, T and U, with some elements in common, so they overlap. You can imagine T as the set of all the points on or inside the circle, and

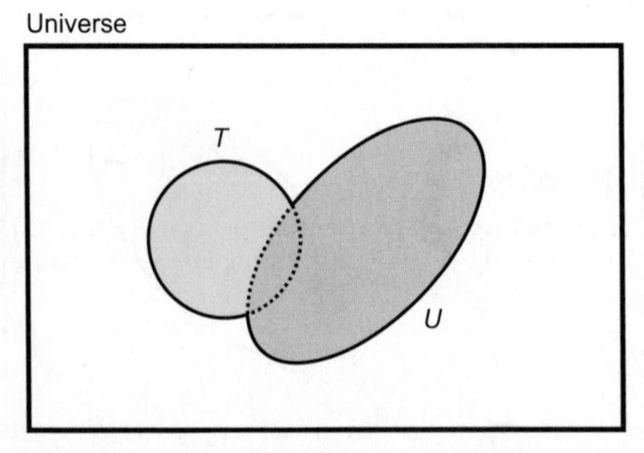

Figure 2-3 Two overlapping sets, T and U. They have some elements in common.

U as the set of all the points on or inside the oval. When you have two overlapping sets, one of them can be a subset of the other, but this does not have to be the case. It is clearly not true of the two sets T and U in Fig. 2-3. Neither of these sets is a subset of the other, because both have some elements all their own.

Set Intersection

The *intersection* of two sets is made up of all elements that belong to both of the sets. When you have two sets, say V and W, their intersection is also a set, and it is written $V \cap W$. The upside-down U-like symbol is read "intersect," so you would say "V intersect W."

Intersection of two congruent sets

When two nonempty sets are congruent, their intersection is the set of all elements in either set. You can write it like this, for any nonempty sets X and Y,

$$\text{If } X = Y$$

$$\text{then}$$

$$X \cap Y = X$$

$$\text{and}$$

$$X \cap Y = Y$$

But really, you're dealing with only one set here, not two! So you could just as well write

$$X \cap X = X$$

This also holds true for the null set:

$$\varnothing \cap \varnothing = \varnothing$$

Intersection with the null set

The intersection of the null set with any nonempty set gives you the null set. This fact is not so trivial. You might have to think awhile to fully understand it. For any nonempty set V, you can write

$$V \cap \varnothing = \varnothing$$

Remember, any element in the intersection of two sets has to belong to both of those sets. But nothing can belong to a set that contains no elements! Therefore, nothing can belong to the intersection of the null set with any other set.

Intersection of two nonempty disjoint sets

When two nonempty sets are disjoint, they have no elements in common, so it's impossible for anything to belong to them both. The intersection of two disjoint sets is always the null set. It doesn't matter how big or small the sets are. Remember the sets of even and odd whole numbers, W_{even} and W_{odd}? They're both infinite, but

$$W_{even} \cap W_{odd} = \varnothing$$

Intersection of two overlapping sets

When two sets overlap, their intersection contains at least one element. There is no limit to how many elements the intersection of two sets can have. The only requirement is that every element in the intersection set must belong to both of the original sets.

Let's look at the examples of overlapping sets you saw a little while ago, and figure out the intersection sets. First, examine these

$$L = \{2, 3, 4, 5, 6\}$$
$$M = \{6, 7, 8, 9, 10\}$$

Here, the intersection set contains one element:

$$L \cap M = \{6\}$$

That means the set containing the number 6, not just the number 6 itself. Now look at these:

$$P = \{21, 23, 25, 27, 29, 31, 33\}$$
$$Q = \{25, 27, 29, 31, 33, 35, 37\}$$

The intersection set in this case contains five elements:

$$P \cap Q = \{25, 27, 29, 31, 33\}$$

Now check these sets out:

$$R = \{11, 12, 13, 14, 15, 16, 17, 18, 19\}$$
$$S = \{12, 13, 14\}$$

In this situation, $S \subset R$, so the intersection set is the same as S. We can write that down as follows:

$$R \cap S = S$$
$$= \{12, 13, 14\}$$

How about the set W_{3-} of all positive, negative, or zero whole numbers less than or equal to 3, and the set W_{0+} of all the nonnegative whole numbers?

$$W_{3-} = \{..., -5, -4, -3, -2, -1, 0, 1, 2, 3\}$$
$$W_{0+} = \{0, 1, 2, 3, 4, 5, ...\}$$

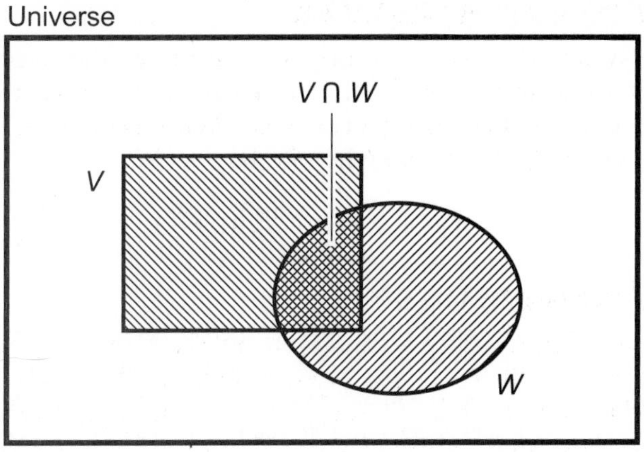

Figure 2-4 Two overlapping sets, V and W. Their intersection is shown by the double-hatched region.

Here, the intersection set has four elements:

$$W_{3-} \cap W_{0+} = \{0, 1, 2, 3\}$$

Figure 2-4 is a Venn diagram that shows two overlapping sets. Think of V as the rectangle and everything inside it. Imagine W as the oval and everything inside it. The two regions are hatched diagonally, but in different directions. The intersection $V \cap W$ shows up as a double-hatched region.

Are you confused?

Go back and look again at Fig. 2-1. You can see that the set of all women in Chicago (call it C_w) is a proper subset of the set of all people in the state of Illinois (call it I_p). You would write down this fact as follows:

$$C_w \subset I_p$$

The diagram also makes it clear that the intersection of set C_w with set I_p is just the set C_w. In order to be in both sets, a person must be a woman in Chicago, that is, an element of C_w. Here's how you would write that

$$C_w \cap I_p = C_w$$

You can always draw a Venn diagram if it will help you understand how sets are related.

Here's a challenge!

Find two sets of whole numbers that overlap, with neither set being a subset of the other, and whose intersection set contains infinitely many elements.

Solution

There are countless examples of set pairs like this. Let's look at the set of all positive whole numbers divisible by 4 without a remainder. (When there is no remainder, a quotient comes out as a whole number.) Name this set W_{4d}. Similarly, name the set of all positive whole numbers divisible by 6 without a remainder W_{6d}. Then

$$W_{4d} = \{0, 4, 8, 12, 16, 20, 24, 28, 32, 36, ...\}$$
$$W_{6d} = \{0, 6, 12, 18, 24, 30, 36, 42, 48, ...\}$$

Both of these sets have infinitely many elements. They overlap, because they share certain elements. But neither is a subset of the other, because they both have some elements all their own. Their intersection is the set of elements divisible by both 4 and 6. Let's call it W_{4d6d}. If you're willing to write out both of the above lists up to all values less than or equal to 100, you will see that

$$\begin{aligned} W_{4d} \cap W_{6d} &= W_{4d6d} \\ &= \{0, 12, 24, 36, 48, 60, 72, 84, 96, ...\} \end{aligned}$$

This is an infinite set, and it happens to be the set of all positive whole numbers divisible by 12 without a remainder (call it W_{12d}). We can write

$$W_{4d} \cap W_{6d} = W_{12d}$$

Set Union

The *union* of two sets contains all of the elements that belong to one set or the other, or both. When you have two sets, say X and Y, their union is also a set, written $X \cup Y$. The U-like symbol is read "union," so you would say "X union Y."

Union of two congruent sets

When two nonempty sets are congruent, their union is the set of all elements in either set. For any nonempty sets X and Y,

$$\text{If } X = Y$$

then

$$X \cup Y = X$$

and

$$X \cup Y = Y$$

But you're really dealing with only one set here, so you could just as well write

$$X \cup X = X$$

And for the null set

$$\varnothing \cup \varnothing = \varnothing$$

When two sets are congruent, their union is the same as their intersection. This might seem trivial right now, but there are situations where it's not clear that two sets are congruent. In cases like that, you can compare the union with the intersection as a sort of congruence test. If the union and intersection turn out identical, then you know the two sets in question are congruent.

Union with the null set

The union of the null set with any nonempty set gives you that nonempty set. For any nonempty set X, you can write

$$X \cup \varnothing = X$$

Remember, any element in the union of two sets only has to belong to one of them.

Union of two disjoint sets

When two nonempty sets are disjoint, they have no elements in common, but their union always contains some elements. Consider again the sets of even and odd whole numbers, W_{even} and W_{odd}. Their union is the set of all the whole numbers. So

$$W_{even} \cup W_{odd} = \{0, 1, 2, 3, 4, 5, ...\}$$

Union of two overlapping sets

Again, let's look at the same examples of overlapping sets we checked out when we worked with intersection. First

$$L = \{2, 3, 4, 5, 6\}$$
$$M = \{6, 7, 8, 9, 10\}$$

The union set here contains nine elements:

$$L \cup M = \{2, 3, 4, 5, 6, 7, 8, 9, 10\}$$

The number 6 appears in both sets, but we count it only once in the union. (An element can only "belong to a set once.") Now look at these:

$$P = \{21, 23, 25, 27, 29, 31, 33\}$$
$$Q = \{25, 27, 29, 31, 33, 35, 37\}$$

The union set in this case is

$$P \cup Q = \{21, 23, 25, ..., 33, 35, 37\}$$

That's all the odd whole numbers between, and including, 21 and 37. We count the duplicate elements 25 through 33 only once. Now look at these:

$$R = \{11, 12, 13, 14, 15, 16, 17, 18, 19\}$$
$$S = \{12, 13, 14\}$$

In this situation, $S \subset R$, so the union set is the same as R. We can write that down this way:

$$\begin{aligned} R \cup S &= R \\ &= \{11, 12, 13, 14, 15, 16, 17, 18, 19\} \end{aligned}$$

We count the elements 12, 13, and 14 only once. Now these:

$$W_{3-} = \{..., -5, -4, -3, -2, -1, 0, 1, 2, 3\}$$
$$W_{0+} = \{0, 1, 2, 3, 4, 5, ...\}$$

Here, the union set consists of all the positive and negative whole numbers, along with zero. Let's write that set as $W_{0\pm}$ (read "W sub zero plus-or-minus"). Then

$$\begin{aligned} W_{3-} \cup W_{0+} &= W_{0\pm} \\ &= \{..., -5, -4, -3, -2, -1, 0, 1, 2, 3, 4, 5, ...\} \end{aligned}$$

The elements 0, 1, 2, and 3 are counted only once. This set $W_{0\pm}$ is usually called the set of *integers*. We'll work more with integers in the chapters to come.

Figure 2-5 is a Venn diagram that shows two overlapping sets. Think of X as the rectangle and everything inside it. Imagine Y as the oval and everything inside it. The union of the sets, $X \cup Y$, is shown by the entire shaded region inside the outer solid line. Part of that line is the

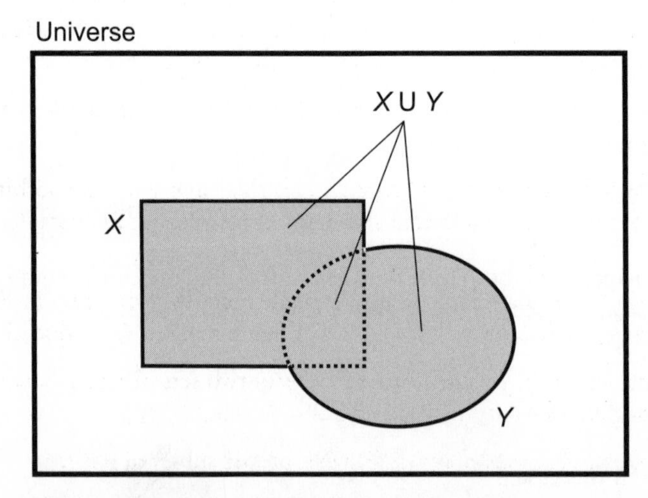

Figure 2-5 Two overlapping sets, X and Y. Their union is shown by the entire shaded region.

outside of the rectangle and part of it is the outside of the oval. Any element inside the region bounded by the dashed line is counted only once.

Are you confused?

Once more, go back and look at Fig. 2-1, again noting that the set of all the women in Chicago is a proper subset of the set of all the people in Illinois, that is, $C_w \subset I_p$. The diagram also makes it plain that the union of C_w with I_p is just I_p. To be in one set or the other (or both), a person only has to be a resident of Illinois, that is, an element of I_p. It's not necessary to be a woman, and it's not necessary to be in Chicago. Here's how you would write that:

$$C_w \cup I_p = I_p$$

Here's a challenge!

Can you find two sets of whole numbers, with one of them infinite, but such that their union contains only a finite number of elements?

Solution

Don't think about this for too long. You'll never find two such sets! An element in the union of two sets only has to belong to one of the sets. If a set has infinitely many elements, then the union of that set with any other set—even the null set—must have infinitely many elements as well.

Practice Exercises

This is an open-book quiz. You may (and should) refer to the text as you solve these problems. Don't hurry! You'll find worked-out answers in App. A. The solutions in the appendix may not represent the only way a problem can be figured out. If you think you can solve a particular problem in a quicker or better way than you see there, by all means try it!

1. Is there any set that is a subset of every other set? If so, what is it? If such a set can't exist, why not?
2. Continuing with the theme of Problem 1, is there a way to take nothing and build up an unlimited number of different sets from it? If so, show an example. If not, explain why not.
3. What set does the small, dark-shaded triangle marked *P* represent in Fig. 2-6? What set does the dark-shaded, irregular, four-sided figure marked *Q* represent?
4. If you consider all the possible intersections of two sets in Fig. 2-6, which of those intersection sets are empty?
5. Is the universal set a subset of itself? Is it a proper subset of itself?
6. Give an example of two sets, both with infinitely many elements, but such that one is a proper subset of the other.

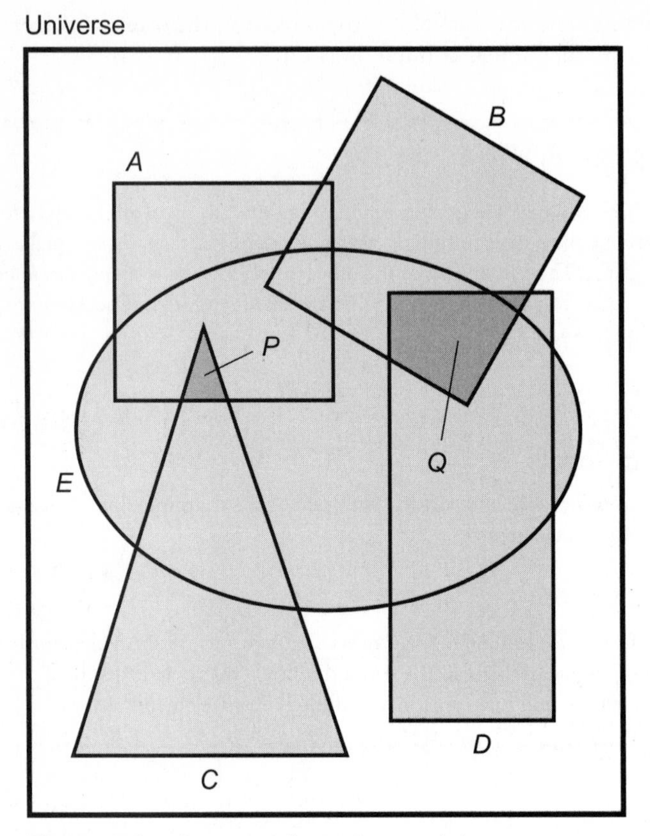

Figure 2-6 Illustration for Probs. 3 and 4.

7. What is the intersection of these two sets? What is their union?

$$A = \{1, 1/2, 1/3, 1/4, 1/5, 1/6, ...\}$$
$$G = \{1, 1/2, 1/4, 1/8, 1/16, 1/32, ...\}$$

In set A, the denominator of the fraction increases by 1 as you go down the list. In set G, the denominator doubles as you go down the list. All the numerators in both sets are equal to 1.

8. List all the subsets of {1, 2, 3}. Here's a hint: Whenever you want to find all the subsets of a small set like this, first list its individual elements. Then make up every possible set that contains at least one of those elements. Finally, be sure to include the empty set, which is a subset of any other set.

9. List all the subsets of {1, {2, 3}}. Be careful. The hint given with Problem 8 is important here.

10. List all the subsets of {1, {2, {3}}}. Be extra careful! The hint given with Problem 8 is even more important here.

CHAPTER

3

Natural Numbers and Integers

The whole numbers starting with 0 and counting upward are usually called the *natural numbers.* Some mathematicians don't call 0 a natural number. It's a little like the dispute among astronomers over whether Pluto should be called a planet, or whether empty space should be called a part of nature. In this book, we'll call 0 a natural number.

How Natural Numbers are Made

In Chap. 2, you saw how we can build sets from nothing. A similar scheme can be used to generate the natural numbers. From the natural numbers, we can create fractions, square roots, and all the other kinds of numbers.

The starting point: 0

In order to build anything, we need a foundation. The natural numbers start from 0 and increase one by one. Zero is an excellent place to begin the number-building process. If you think of the natural numbers as evenly spaced points along a straight *ray* or *half-line,* 0 is at the very beginning. The empty set is a good way to define 0. So let's agree that the number 0 is the set containing no elements, and illustrate it as a point at the left-hand end of an infinitely long half-line (Fig. 3-1).

How 1 is defined

We can define the number 1 as the set containing the number 0, so it has one element. That makes it different from 0, but doesn't require that we invent anything new. Because the number 0 is the null set, the number 1 can also be imagined as the set containing the null set (Fig. 3-2). We now have three different ways we can write the number 1 in terms of other things we've already defined:

$$1 = \{0\}$$
$$1 = \{\varnothing\}$$
$$1 = \{\{\ \}\}$$

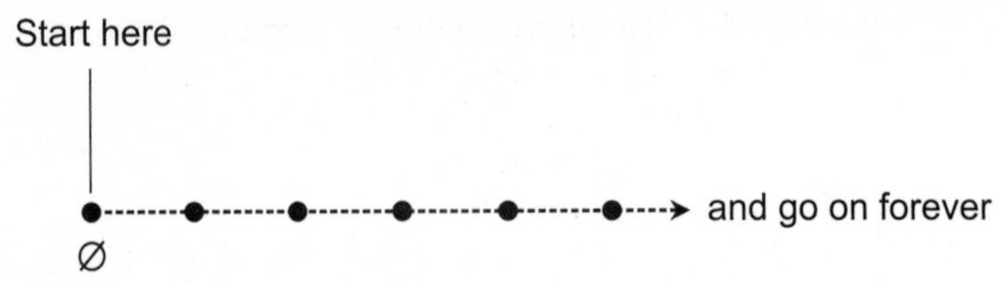

Figure 3-1 The number 0 can be defined as the null set. We can show how it starts to generate natural numbers by placing it at the beginning of an endless string of points.

Building new numbers

Let's make up a rule that we can use to generate new natural numbers, one after another. Suppose we've built a certain natural number. Call it n. If we want to create the next higher natural number, $n + 1$, we can take all the natural numbers up to and including n, make them into elements of a set, and then call that set the new number. A mathematician would write it like this:

$$n + 1 = \{0, 1, 2, 3, ..., n\}$$

If we have any particular natural number p, we can define it on the basis of all the natural numbers less than itself, like this:

$$p = \{0, 1, 2, 3, 4, ..., p - 1\}$$

If that's a little too abstract for you, look at Fig. 3-3. This drawing shows the first six natural numbers (0, 1, 2, 3, 4, and 5) as they are built up and assigned to a vertical stack of points that ascends upward as far as we care to go.

This is a nifty scheme! A natural number p is a set containing p elements. In theory, those elements could be anything, such as apples, stars, atoms, or people. But it's convenient to use all the natural numbers less than p as those elements. That allows us to build the set of natural numbers, one on top of the other, like stacking coins. Just as each coin in the stack rests on all the coins below itself, every element p in the set of natural numbers "rests on" all the natural numbers "below" itself. Once this process is defined, it sets in motion a mathematical "chain reaction" that never ends.

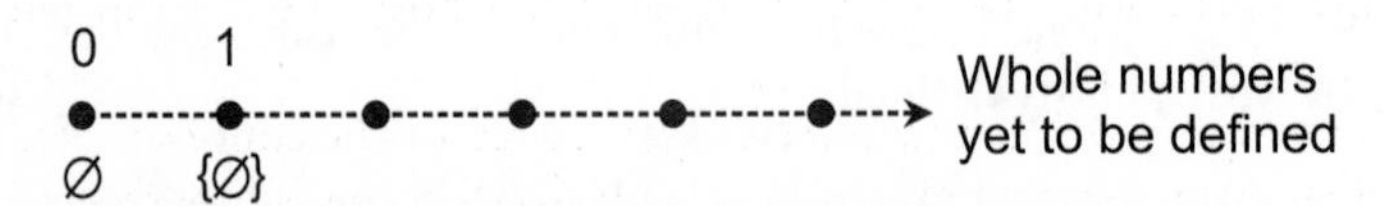

Figure 3-2 The number 1 is the set containing the null set, which is the same as the set containing the number 0. It has one element.

Figure 3-3 The natural numbers can be built up, each one "on top" of its predecessors.

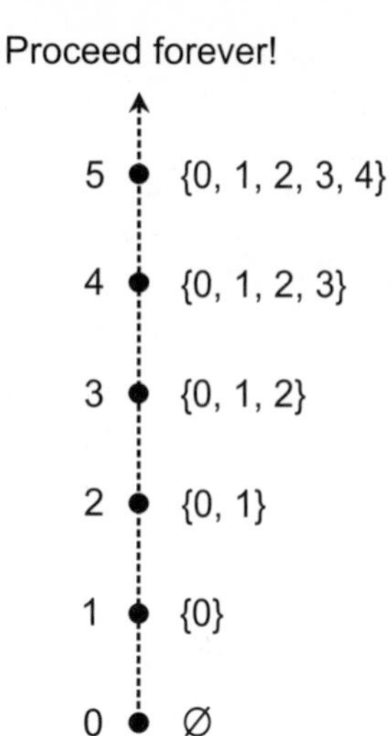

A definition for infinity

When we write out any particular natural number p in the form of a set of smaller numbers, that set has exactly p elements in it:

$$0 = \varnothing \text{ which has no elements}$$
$$1 = \{0\} \text{ which has one element}$$
$$2 = \{0, 1\} \text{ which has two elements}$$
$$3 = \{0, 1, 2\} \text{ which has three elements}$$
$$\downarrow$$
$$p = \{0, 1, 2, 3, ..., p - 1\} \text{ which has } p \text{ elements}$$
$$\downarrow$$
and so on, forever

Now think of the set of all natural numbers. This set, which we called W (for whole numbers) in the last chapter, is symbolized N. We can write

$$N = \{0, 1, 2, 3, ...\}$$

Think about this set N for a minute. What *number* does it represent, according to the scheme we've just invented for building natural numbers? Can the set N represent any number at all? We can never finish writing the list of its elements. But the mere fact that we can't write the whole list doesn't mean that the set itself does not exist. We can imagine it, so in the mathematical world, it exists!

Mathematicians define the number represented by the entire set N as a form of "infinity" and denote it using the last letter in the Greek alphabet, *omega*, in lowercase (ω). "Omega" is a traditional expression for "the end of all things." In formal terms, ω is called an *infinite ordinal* or *transfinite ordinal*, and it has some strange properties. There are infinitely many infinite ordinals! We won't delve into their properties here, but if you're interested in learning more about them, there are plenty of Web sites you can explore. Go to your favorite search engine and enter "infinite ordinal" or "transfinite ordinal" in the phrase-search mode.

Are you confused?

In the solution to Prob. 2 in Chap. 2, we saw how sets can be built on each other by "tacking on braces" to the null set:

$$\varnothing$$
$$\{\varnothing\}$$
$$\{\{\varnothing\}\}$$
$$\{\{\{\varnothing\}\}\}$$
$$\{\{\{\{\varnothing\}\}\}\}$$
$$\{\{\{\{\{\varnothing\}\}\}\}\}$$
$$\downarrow$$

and so on, forever

Why can't we do that to construct the natural numbers? Well, we could, perhaps. But this method doesn't increase the number of elements in each set as the defined number gets bigger. Each of the above sets, except for the first, contains one element. To make a good definition of the natural numbers, number theorists want to have the sets contain more and more elements as the defined numbers get larger.

Here's a challenge!

Write down the natural numbers 0 through 4 purely in terms of braces and the null set symbol, showing how each number can be assembled from "multiple nothings."

Solution

Start with 0, which is equal to $\varnothing$. It helps if you write the numbers as sets of smaller numbers and then break those expressions down:

$$0 = \varnothing$$
$$1 = \{0\} = \{\varnothing\}$$
$$2 = \{0, 1\} = \{\varnothing, \{\varnothing\}\}$$
$$3 = \{0, 1, 2\} = \{\varnothing, \{\varnothing\}, \{\varnothing, \{\varnothing\}\}\}$$
$$4 = \{0, 1, 2, 3\} = \{\varnothing, \{\varnothing\}, \{\varnothing, \{\varnothing\}\}, \{\varnothing, \{\varnothing\}, \{\varnothing, \{\varnothing\}\}\}\}$$

Special Natural Numbers

You can classify natural numbers in various ways, just as you can classify people according to blood type, postal zone of residence, country of residence, or even (maybe someday) planet of residence. Here are a few of the most well-known types of natural numbers.

Even numbers

An *even natural number* is a whole number whose numeral ends in 0, 2, 4, 6, or 8. If you multiply every number in the set N by 2, you get the set N_{even} of all the even natural numbers. This is the familiar set

$$N_{even} = \{0, 2, 4, 6, 8, 10, ...\}$$
$$= \{0\times2, 1\times2, 2\times2, 3\times2, 4\times2, 5\times2, ...\}$$

Now do the same thing, but backwards. If you divide every number in the set N_{even} by 2, you get the set N of all natural numbers:

$$N = \{0, 1, 2, 3, 4, 5, ...\}$$
$$= \{0/2, 2/2, 4/2, 6/2, 8/2, 10/2, ...\}$$

Odd numbers

An *odd natural number* is a whole number whose numeral ends in 1, 3, 5, 7, or 9. If you multiply every number in the set N by 2 and then add 1 to the result, you get the set N_{odd} of all the odd natural numbers:

$$N_{odd} = \{1, 3, 5, 7, 9, 11, ...\}$$
$$= \{(0\times2)+1, (1\times2)+1, (2\times2)+1, (3\times2)+1, (4\times2)+1, (5\times2)+1, ...\}$$

Now do the same thing, but backwards. If you subtract 1 from every number in the set N_{odd} and then divide the result by 2, you get the set N of all natural numbers:

$$N = \{0, 1, 2, 3, 4, 5, ...\}$$
$$= \{(1-1)/2, (3-1)/2, (5-1)/2, (7-1)/2, (9-1)/2, (11-1)/2 ...\}$$

The union of the set of all the even natural numbers and the set of all the odd natural numbers is the entire set of natural numbers. You might find this mouthful of words easier to understand if you write it in symbols:

$$N_{even} \cup N_{odd} = N$$

This means that you can pick any natural number, as large as you want, and it will always be either even or odd. But you'll never see a natural number that is both even and odd.

Factors

Whenever you multiply two natural numbers together, you get another natural number. This is obvious, even trivial, when one of the numbers is 0 or 1. Any number times 0 is equal to 0, and any number times 1 is equal to itself. When you start working with the other numbers, things get more interesting.

Let's pick a natural number, preferably a fairly large one. How about 99? How can we break this number down into a product of other natural numbers besides itself and 1? It's easy to see that 33 and 3 will work:

$$33 \times 3 = 99$$

We can't break the number 3 down into a product of natural numbers other than itself and 1. How about 33? This can break down into the product of 11 and 3:

$$33 = 11 \times 3$$

We can't break 11 down into a product of natural numbers other than itself and 1. Now we have 99 as a product of "unbreakables":

$$99 = 11 \times 3 \times 3$$

Whenever you have a natural number expressed as a product of other numbers, those other numbers are called *factors*. The process of breaking a number down into a product of other numbers is called *factorization* or *factoring*.

Prime and composite numbers

An "unbreakable" natural number is called a *prime number*, or simply a *prime*. It's a natural number larger than 1 that can only be factored into a product of itself and 1. Table 3-1 lists the first 24 prime numbers. If this doesn't go high enough for you, there are plenty of lists of primes on the Internet.

Any nonprime natural number can be factored into a product of two or more primes. The numbers in such a product are called the *prime factors*, and the whole product is called a *composite number*. When you want to find the prime factors of a large natural number, you can get some help from a calculator that has a *square root* key. The square root of a number is a smaller number, not always whole, that gives you the original number when multiplied by itself.

Here's how the process goes. First, use a calculator to find the square root of the number you want to factor. Once you have done that, "chop off" any nonzero digits that might appear after the decimal point, so you get a whole number. Then add 1 to that number. Call this new whole number s. Now divide the original number by all the primes (referring to Table 3-1) less than or equal to s, starting with the largest prime and working your way down. If you ever get a whole-number quotient as you go through this process, then you know that the divisor and

Table 3-1. The first 24 prime numbers. The number 1 is not considered prime. Any natural number, no matter how large, can be factored into a product of primes.

Order	Prime	Order	Prime	Order	Prime
1st	2	9th	23	17th	59
2nd	3	10th	29	18th	61
3rd	5	11th	31	19th	67
4th	7	12th	37	20th	71
5th	11	13th	41	21st	73
6th	13	14th	43	22nd	79
7th	17	15th	47	23rd	83
8th	19	16th	53	24th	89

the quotient are both factors of the original number. Sometimes the quotient will be prime, and sometimes it won't be. If it isn't prime, then it can be factored down further. Don't stop dividing the original by primes until you get all the way down to 2.

Once you've found all the primes smaller than *s* that divide your original number without giving you a remainder, you have found the prime factors of your original number. Some product of these will give you that original number. You might find that one of the primes is a factor twice, three times, or more. For example:

$$99 = 11 \times 3 \times 3$$

$$297 = 11 \times 3 \times 3 \times 3$$

$$891 = 11 \times 3 \times 3 \times 3 \times 3$$

$$2{,}673 = 11 \times 3 \times 3 \times 3 \times 3 \times 3$$

Perfect squares

Whenever you multiply a number by itself, the process is called *squaring*, and the product is called a *square*. If you multiply a natural number by itself, you get a *perfect square*. If you take the square root of a perfect square, you always get a natural number.

Perfect squares, unlike primes, are easy to find if you have a calculator. Table 3-2 lists the first 24 perfect squares, with 0 and 1 included.

Are you confused?

If you see a large natural number, you can sometimes tell right away that it's not prime. If its numeral ends in an even number, you know that one of its factors is 2, so it can be factored into a pair of natural numbers other than 1 and itself.

Sometimes, numbers seem at first as if they ought to be prime, but it turns out that they are not. A good example is 39. It can be factored into 13 × 3. Another is 51, which can be factored into 17 × 3. Still another is 57, which can be factored into 19 × 3.

Table 3-2. The first 24 perfect squares. The numbers 0 and 1 are included here. When you take the square root of a perfect square, you always get a natural number. Note that we start with the "0th" rather than the "1st" in order here. That way, the order agrees with the number squared.

Order	Square	Order	Square	Order	Square
0th	0	8th	64	16th	256
1st	1	9th	81	17th	289
2nd	4	10th	100	18th	324
3rd	9	11th	121	19th	361
4th	16	12th	144	20th	400
5th	25	13th	169	21st	441
6th	36	14th	196	22nd	484
7th	49	15th	225	23rd	529

The best way to find out whether or not a large odd number is prime is to try to factor it into primes. If the only factors you get are itself and 1 (i.e., if you can't factor it into primes), then your number is prime. There are some other techniques you can use determine when a number is not prime, such as the "divisibility" tricks you'll see later in this chapter.

Here's a challenge!

When an even number is multiplied by 7, the result always even. Show why this is true.

Solution

For the first few even natural numbers, multiplication by 7 always gives you an even number. Here are the examples for all the single-digit even numbers:

$$0 \times 7 = 0$$
$$2 \times 7 = 14$$
$$4 \times 7 = 28$$
$$6 \times 7 = 42$$
$$8 \times 7 = 56$$

You can prove that multiplying any even number by 7 always gives you an even number if you realize that the *last digit* of an even number is always even. Think of an even number *p*—any even number. This number *p*, however large it might be, must look like one of the following:

______0
______2
______4
______6
______8

where the long underscore represents any string of digits you want to put there. Now think of "long multiplication" by 7. Remember how you arrange the numerals on the paper and then do the calculations. You always start out by multiplying the last digits of the two numbers together, getting the last digit of the product. The even number on top, which you are multiplying by the number on the bottom, must end in 0, 2, 4, 6, or 8. If the number on the bottom is 7, then the last digit in the product must be 0, 4 (the second digit in 14), 8 (the second digit in 28), 2 (the second digit in 42), or 6 (the second digit in 56) respectively. The product of any even number and 7 is therefore always even.

Natural Number Nontrivia

Here are some interesting facts about natural numbers. I was about to call them "trivia," but after thinking about it for awhile, I decided that the ones involving primes are not trivial at all!

Divisibility

If you want to know whether or not a large number can be divided by a single-digit number without leaving a remainder, there are some handy little tricks you can use. You can use a calculator to see immediately whether or not any number is "cleanly" divisible by any other, but the following rules can be interesting anyway.

- A natural number is divisible by 2 without a remainder if it is even.
- A natural number is divisible by 3 without a remainder if the sum of the digits in its numeral is a natural-number multiple of 3.
- A natural number is divisible by 5 without a remainder if its numeral ends in either 0 or 5.
- A natural number is divisible by 9 without a remainder if the sum of the digits in its numeral is a natural-number multiple of 9.
- A natural number is divisible by 10 without a remainder if its numeral ends in 0.

You can combine these tricks and get the following facts:

- A natural number is divisible by 4 without a remainder you get an even number after dividing it by 2.
- A natural number is divisible by 6 without a remainder if it is even and the sum of its digits is a natural-number multiple of 3.
- A natural number is divisible by 8 without a remainder if you can divide it by 2 and get an even number, and then divide that number by 2 again and get an even number.

There aren't any convenient tricks, other than using a calculator or performing "long division," to find out if a natural number is divisible by 7 without leaving a remainder.

Is there a largest prime?

Now that you know what a prime number is, and you know that any nonprime natural number can be broken down into a product of primes, you might ask, "Is there a largest prime?" The answer is "No." Here's why. You might have to read the following explanation two or three times to completely understand it. Try to follow it step-by-step. If you can accept each step of this argument one at a time, that's good enough. The fact that there is no such thing as a largest prime is one of the most important facts, or *theorems*, that have ever been proven in mathematics.

Let's start by imagining that there actually is a largest prime number. Then we'll prove that this assumption cannot be true by "painting ourselves into a corner" where we end up with something ridiculous. Now that we have decided there is a largest prime, suppose we give it a name. How about p? Theoretically, we can list the entire set of prime numbers (call it P). It might take mountains of paper and centuries of time, but if there is a largest prime, we can eventually write all of the primes. We can describe the set P in shorthand like this:

$$P = \{2, 3, 5, 7, 11, 13, ..., p\}$$

Suppose that we multiply all of these primes together. We get a composite number, because it is a product of primes. No doubt, this number is huge—larger than any calculator can display—but it will be finite. Let's call it y. What if we add 1 to y, getting a number even larger than the product of all the primes? If you call that new number z, you can express it like this:

$$z = y + 1$$
$$= (2 \times 3 \times 5 \times 7 \times 11 \times 13 \times \ldots \times p) + 1$$

Now we know that z has to be larger than p, because z is 1 more than, say, $2 \times p$ or $3 \times p$ or $5 \times p$ or $7 \times p$. But there's something else interesting about z. If we divide z by any prime number, we always get a remainder of 1. That's because if we divide y by any prime, there's no remainder, and z is exactly 1 more than y.

We know that z can't be prime, because we've already determined that z is bigger than p, and we have already assumed that p is the largest prime. So z is composite. Because z is composite, it must be divisible without a remainder by at least one prime, that is, one element of set P. But wait! We just figured out a minute ago that if we divide z by any element of P, we get a remainder of 1. Therefore, z can't be composite. But it can't be prime either. But every natural number larger than 1 is either prime or composite! But ... but ... but ... we are trapped!

There's only one way out of this situation. Our original assumption, that there is a largest prime number, must be false. *Reductio ad absurdum*, which we first encountered in the solution to Prob. 5 at the end of Chap. 2, comes to the rescue again!

How many primes?

The discovery that there is no largest prime number leads us straightaway into another important truth: there are infinitely many prime numbers. When a mathematician proves a major theorem like the one we just explored, and then some other fact follows on its heels, that secondary fact is called a *corollary*.

Let's start out by assuming that the number of primes is finite, and load up our *reductio ad absurdum* "cannon" for another shot. This time it's going to be easy. If the number of primes is finite, we can list them all. That means one of them has to be larger than all the others, so it is the largest prime. But we just discovered that there is no largest prime. Contradiction! The number of primes can't be finite, so it must be infinite.

Are you confused?

When you have found the prime factors for a composite number, you can write the product out in any order. But unlike a listing of the elements of a set, in which you are allowed to list any element only once, you must be sure to include all the occurrences of a prime factor if it appears more than once. Take this example:

$$6{,}615 = 3 \times 3 \times 3 \times 5 \times 7 \times 7$$

You will get into trouble if you say, "The set of prime factors of 6,615 is {3, 5, 7}." How do you know whether a given factor occurs once, twice, three times, or more?

Some people get around this issue by putting a little superscript called an *exponent* after a number in the set to indicate that it occurs more than once as a factor. They write that the set of prime factors of 6,615 is

$\{3^3, 5, 7^2\}$. That's okay if you can remember that 3^3 does not literally mean 27 in this context, and 7^2 does not literally mean 49. (Neither 27 nor 49 are prime!)

The clearest way to express the prime factors of a composite number is to write out the product, listing each factor as many times as it "deserves," and using multiplication symbols between them. You can arrange the product in any order, but it helps if you start with the smallest factor and go up, or start with the largest factor and go down.

There is only one way to factor a composite number into a product of primes. This fact is called the *Fundamental Theorem of Arithmetic.*

Here's a challenge!

The numbers 2 and 3 are both prime, and they are also consecutive whole numbers. Are there any other examples of two consecutive whole numbers that are both prime?

Solution

No, none of the even numbers larger than 2 (i.e., 4, 6, 8, 10, etc.) are prime. We can factor 2 out of any such number, and we always get a natural number bigger than 1. By elimination, then, all the primes larger than 2 are odd. If we take any of these primes and then find the next natural number, we are adding 1 to an odd number. That always produces an even number, which we have just seen can't be prime.

The conclusion: when we come across any prime number larger than 2, the next consecutive natural number is always composite.

The Integers

Centuries ago, negative numbers weren't taken seriously. How could you have less than none of anything? When the set of all negative whole numbers was finally joined together with the set of natural numbers, the result became known as the set of *integers.* That set is symbolized *Z*, like this:

$$Z = \{\ldots, -3, -2, -1, 0, 1, 2, 3, \ldots\}$$

Negative numbers

What is a negative number, exactly? That question is deeper than it seems at first thought. We can start to answer it by creating situations where negative numbers are really useful.

In the United States, most nonscientific people use the *Fahrenheit temperature scale*, where 32 degrees represents the freezing point of water. Scientists, and people outside the United States, use the *Celsius temperature scale*, where 0 degrees represents the freezing point of water. In either system, temperatures often get colder than 0 degrees. Then people start calling temperatures negative.

Here's another real-life situation where negative numbers come in handy. These days, nearly everyone has a credit card. When you first get the card, it has a balance of 0. That means you haven't put any money in the bank that gave you the card, but you don't owe the bank any money, either. What if you buy some items at the local department store, "charging" up a balance of $49? How much money is in the account now? If you think of it as the bank's account, they have a claim to \$49 of your money. If you think of it as your account, you're \$49 dollars in debt. You have, in a sense, negative \$49. If you go to another store and charge \$10 more, you'll end up with negative \$59. In theory, there is no limit to how *large negatively* your account, in dollars, can become. (In practice, the bank will put a limit on it.)

Negative whole numbers are denoted by putting a minus sign in front of a natural number. The exception is 0, where a negative sign doesn't change the meaning. "Negative 0" is the same thing as "positive 0" in ordinary mathematics. In the credit-card situation just described, you start out with \$0 and then go to –\$49, then to –\$59. The same thing can happen with temperature. If it was 0 degrees yesterday afternoon and then the temperature fell by 10 degrees overnight, it was –10 degrees in the morning.

A "number reflector"

We've already shown how the natural numbers can be generated from sets. How can we add the negative natural numbers to the "normal" or positive ones, making sure to include 0 so we get the entire set of integers?

We can take two natural-number rays (or half-lines), put minus signs in front of all the numbers on one of the rays, and then stick the rays together end-to-end so "positive 0" and "negative 0" are on top of each other. Figure 3-4 shows how this works. You might think of the

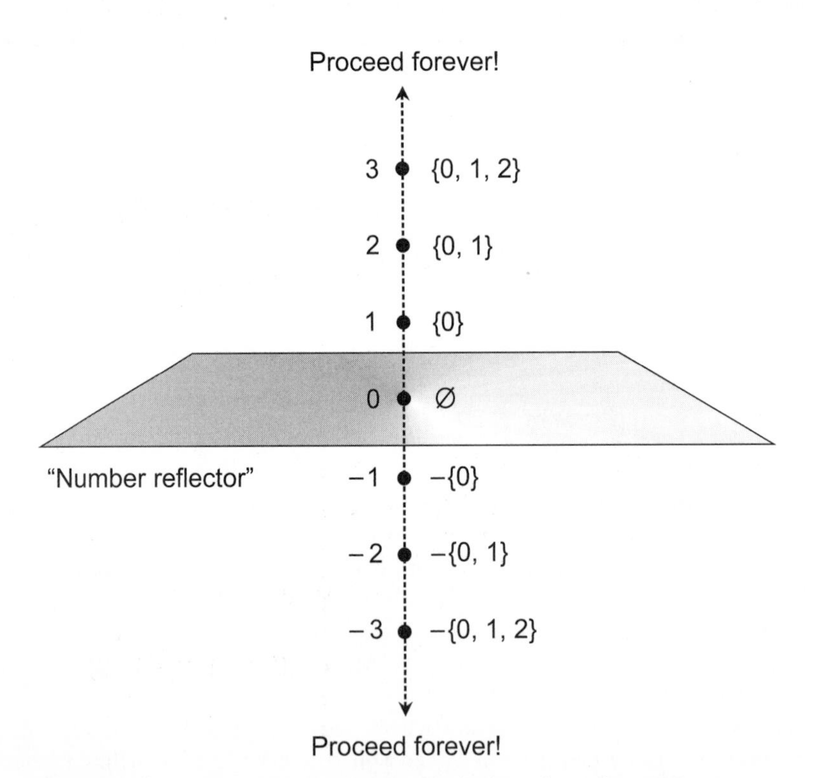

Figure 3-4 The negative numbers can be built up from the positive ones by inventing an imaginary "number reflector" that reverses the "sense" of every natural number and gives it a "twin."

natural numbers as being attached to a ray that stands straight up above the "number reflector," and their negatives as being attached to a ray that dangles straight down.

This is a fine way to imagine the integers, but in mathematical terms, it is a little "impure." In order to define the negative numbers this way, we have to come up with new gimmicks that we did not need to define the natural numbers. A pure mathematician would demand some way to define all the integers, positive, negative, and 0, using only the idea of a set and nothing else. We can define the entire set of integers in the same way as we defined the set of natural numbers. Put your "abstract thinking cap" on again (if it isn't glued to your head by now), and keep in mind that what you're about to read does not represent the only way the set of integers can be defined in a "pure" way.

Building the integers

The natural numbers have a clear starting point, which is 0. But the integers go on forever in two directions. At least, that's the impression you'll get if you look at Fig. 3-4. How can you start moving along a line that goes in two directions, and cover every point on it? You have to pick one direction or the other, right?

Wrong! In the real world that might be true, but in the "mathematical cosmos" we have powers that ordinary mortals lack.

Take a look at Fig. 3-5. Instead of hopping from 0 to 1, and then from 1 to 2, and then from 2 to 3, always moving in the same direction, suppose you hop alternately back and forth. Start at 0, then move up one unit to 1. Then go down two units to −1. Then go up three units

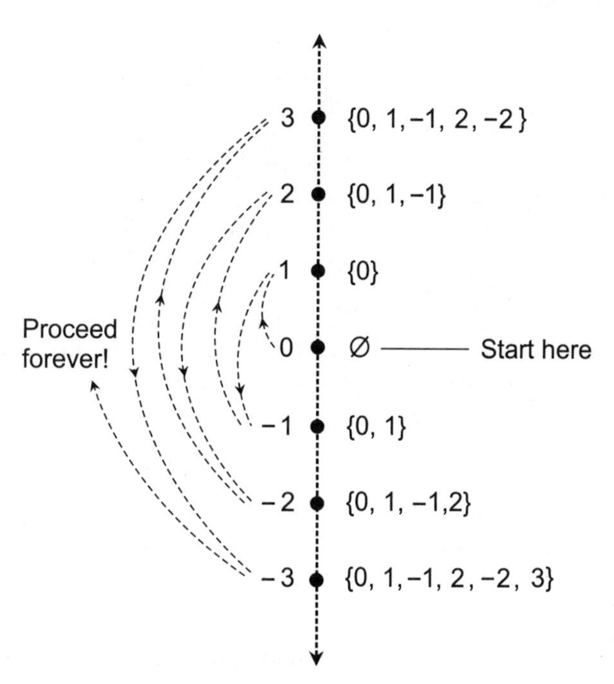

Figure 3-5 Here's a way to generate the set of integers with a scheme similar to the one we used to build up the set of natural numbers.

to 2, down four units to −2, up five units to 3, down six units to −3, and so on. Keep hopping alternately up and down, making your hop one unit longer every time. In Fig. 3-5, the integers themselves are shown to the left side of the vertical line, and their equivalents, built up as sets of previously defined integers, are shown on the right side. Pick any integer, positive or negative, as big or small as you want. You'll eventually reach it if you make enough hops.

The next time you are at a party with a bunch of mathematics lovers and somebody asks you, "What is the number −2, really?" you can say, "Well, that can be debated. But if you like, we can define it as the set containing 0, 1, −1, and 2." That should get you a raised eyebrow. If you want to bring down the house, you can go to an old-fashioned chalk blackboard (every good mathematics party has one, right?) and scribble out the following to make your point:

$$0 = \varnothing$$
$$1 = \{0\} = \{\varnothing\}$$
$$-1 = \{0, 1\} = \{\varnothing, \{\varnothing\}\}$$
$$2 = \{0, 1, -1\} = \{\varnothing, \{\varnothing\}, \{\varnothing,\{\varnothing\}\}\}$$
$$-2 = \{0, 1, -1, 2\} = \{\varnothing, \{\varnothing\}, \{\varnothing, \{\varnothing\}\},\{\varnothing, \{\varnothing\}, \{\varnothing, \{\varnothing\}\}\}\}$$
$$\downarrow$$

and so on, forever

Are you confused?

The integers can get confusing when you compare values. If you draw a number line and represent the integers as points on it, such as is done in Figs. 3-4 or 3-5, what does it mean if one number is "larger" or "smaller" than another? How about the expressions "less than" or "greater than"?

A mathematician will tell you that the integers get smaller as you move downward in Figs. 3-4 or 3-5, and larger as you go upward. For example, −5 is smaller (or less) than −2, and any negative integer is smaller (or less) than any natural number. Conversely, −2 is larger (or greater) than −5, and any natural number is larger (or greater) than any negative integer. But that can begin to seem strange if you think about it awhile. How can −158 be "smaller" than −12? If you find yourself in debt by $158, isn't it a bigger problem than if you are in debt by $12?

In the literal sense, −158 is indeed smaller (or less) than −12, just as −158° is colder than −12°. In fact, the integer −158 is less than −12 or −32 or −157. But −158 is *larger negatively* than −12 or −32 or −157. To avoid confusion when comparing numbers, the best policy is to be careful with your choice of words. Figure 3-6 should clear up any lingering uncertainty you might have about this.

Here's a challenge!

If we allow all negatives of primes (i.e., −2, −3, −5, −7, −11, −13, −17, −19, ...) to be called prime, does that make all the nonprime negative numbers composite?

Solution

Let's keep the traditional definition of *composite number*: a product of two or more primes. Now imagine that we have some positive composite number. It is therefore a product of primes that are all positive. If we make one of those primes negative, we get the negative of that composite number. For example:

$$100 = 5 \times 5 \times 2 \times 2$$

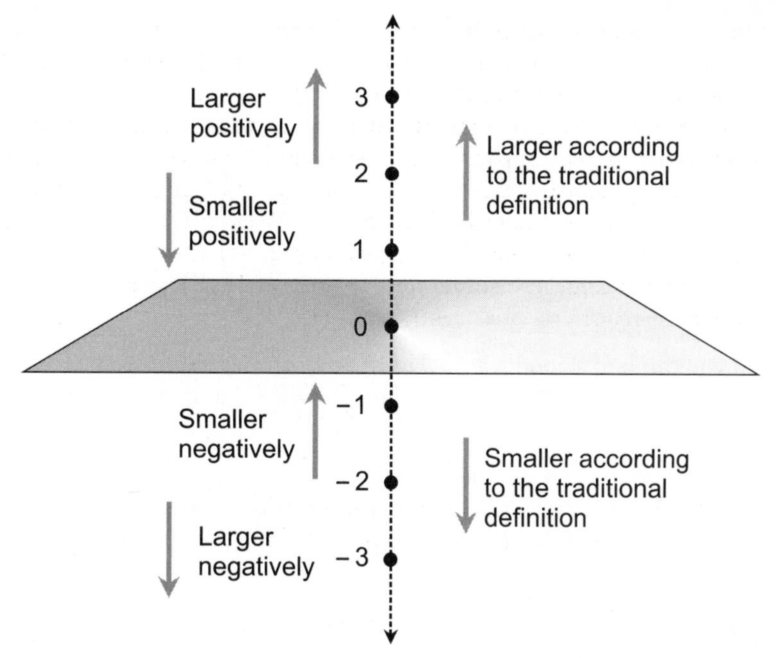

Figure 3-6 This drawing, and careful choice of words, can help you avoid confusion when comparing the values of integers.

If we remember the basic multiplication sign rules, we can see that

$$-100 = -5 \times 5 \times 2 \times 2$$

This same technique can be applied to any negative nonprime number smaller than –3 to show that it's composite! We have to be sure that "negative primes" are allowed in the mathematical system we're dealing with. According to the traditional definition, all the primes are natural numbers larger than 1, so this trick won't work.

Practice Exercises

This is an open-book quiz. You may (and should) refer to the text as you solve these problems. Don't hurry! You'll find worked-out answers in App. A. The solutions in the appendix may not represent the only way a problem can be figured out. If you think you can solve a particular problem in a quicker or better way than you see there, by all means try it!

1. If the number 0 is the set containing nothing, then what number does nothing represent?
2. The number 3 is odd. If a number n is divisible by 3 without a remainder, does that mean n must be odd?
3. When an odd number is multiplied by 3, is the result always odd? If so, demonstrate why. If not, show a *counterexample* (a situation where an odd number is multiplied by 3 to get an even number).

4. Find out whether or not 901 is a prime number.
5. Break down 1,081 into a product of primes.
6. Break down 841 into a product of primes. What interesting property does this number have?
7. Break down 2,197 into a product of primes. What interesting property does this number have?
8. Are any negative integers composite if we insist on using the traditional definition of a prime number?
9. Can you think of a good reason why the natural numbers 0 and 1 are not defined as prime? Here's a hint: It should *never* be possible for a number to be both prime and composite.
10. Show how the natural numbers can be paired off one-to-one with the integers. Here's a hint: Use Fig. 3-5 with its pattern of dashed, arrowed guidelines to create an "implied one-ended list" of the integers that captures them all.

CHAPTER

4

Addition and Subtraction

Let's take a close look at the processes, also called *operations*, known as *addition* and *subtraction.* Much of this material will seem like a review of arithmetic to you, but you'll need to know it "forward and backward" to work with the algebra to come later.

Moving Up and Down

Adding a number to another, or subtracting a number from another, are sophisticated ways of counting. When you do these operations with integers, it's like moving up or down, point-by-point, on a vertical number line of the sort you saw in the last chapter.

Absolute value

Imagine the "number reflector" from Chap. 3 as a flat plane perpendicular to the number line and passing through 0, as shown in Fig. 4-1. Every number is a certain distance above or below the "number reflector." The distance of an integer from the "number reflector" is called the *absolute* value of the integer.

To denote absolute value, you enclose a numeral or expression between vertical lines. The absolute value of 2 is written $|2|$, and the absolute value of -3 is written $|-3|$. If you have any quantity, no matter how complicated, you can always indicate its absolute value by putting vertical lines on either side of the set of symbols that represents it.

There's no such thing as negative absolute value, because it is an expression of distance *without taking direction into account.* To say that a number has an absolute value of -3 is like saying that your house is -3 miles from your cousin's house on the other side of town. That's nonsense! You can talk about direction as well as distance. Then negative values are possible. When you move or travel over a certain distance in a certain direction, it is called *displacement,* and it can be positive or negative. It can even go off in other directions, such as west, or straight up, or toward the sun, or toward your cousin's house.

The absolute value of any natural number (or *nonnegative integer*) is equal to that natural number. The absolute value of any negative integer is its "image" in the "number reflector." If

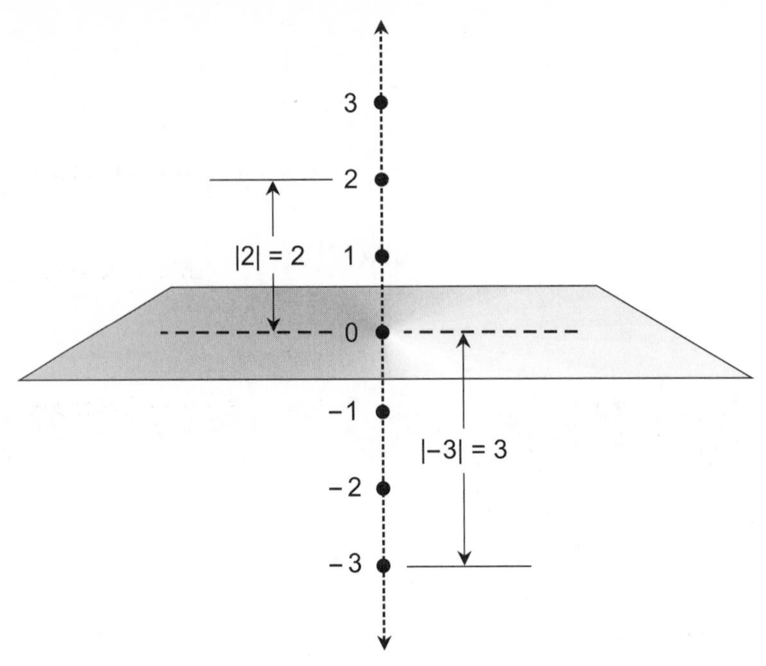

Figure 4-1 The absolute value of a number is its distance from 0 along the number line. The direction (up or down, positive or negative) doesn't matter. That is why absolute values can never be negative.

you have a numeral that represents a negative integer, you can get the numeral representing its absolute value by removing the minus sign.

Meet the variables!

When you want to talk about how numbers relate to each other but don't want to specify any particular numbers, you can use *variables* instead. For variables representing integers, mathematicians most often use small, italic letters from a through q. When you see something like $a + b = c$, you know you are supposed to add a quantity a to another quantity b to get a third quantity c. You don't have to know what the actual numbers are, but only that they are related in a certain way. The term *variable* means that a quantity doesn't have any fixed value; it can vary.

To add, move upward

Now let's get back to displacement. We'll go up and down here, because we've already illustrated the number line in a vertical sense. Think of upward distances as positive displacements, and downward distances as negative displacements. If we have an integer a and we want to add another integer b to it, we first find the point on the number line representing a. Then we move up b units. That will get us to the point representing $a + b$.

As an example, suppose $a = -3$ and $b = 2$. We start at the point for -3 and move up 2 units. That gets us to the point for $-3 + 2$. It happens to be -1, as shown on the left side of Fig. 4-2.

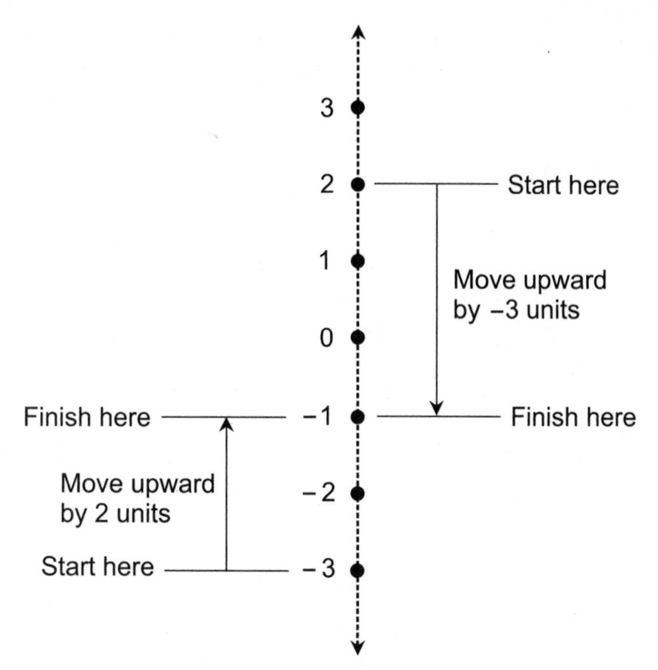

Figure 4-2 On the left, we add $-3 + 2$. On the right, we add $2 + (-3)$. When we move negatively upward, we move downward by the equivalent distance.

What if we reverse the order of this sum? We start with 2 and travel upward −3 units. We're talking about displacement here, not simple distance, so negatives can make sense! An upward displacement of −3 units is the same as a downward displacement of 3 units. This process is shown on the right in Fig. 4-2. When we add the integers −3 and 2 in either order as shown, we end up at the same point, which corresponds to −1. We have now analyzed these two facts:

$$-3 + 2 = -1$$

and

$$2 + (-3) = -1$$

To subtract, move downward

If you think it's ridiculous to imagine downward movement as negative upward movement, you're right, except for one little catch. You are going to come up with situations in mathematics where you'll get a negative quantity for an answer to a problem, and it won't seem to make sense. Suppose you fly a rush-hour traffic observation helicopter for your local TV station. You see that a rain shower has caused an existing jam in the westbound traffic on Boxelder Bug

Boulevard to be displaced by −4 miles. This means the jam has been displaced to the east by 4 miles. The jam has moved in the opposite direction from the flow of traffic!

Look again at the right-hand side of Fig. 4-2. You add a negative number to some other quantity. Adding a negative number is the same thing as subtracting the absolute value of that number. Using variables, you can write that statement as

$$a + (-b) = a - |-b|$$

which means the same thing as

$$a + (-b) = a - b$$

If you have an integer a and you want to subtract another integer b from it, first find the point on the number line representing a. Then travel down b units. That will get you to the point representing $a - b$.

Are you confused?

Have you been wondering why negative numbers always have a minus sign in front of them, but positive numbers don't have a plus sign? Is there something technically wrong with including a plus sign so people know when a number is positive? Why leave any doubt?

That's a good question. The answer is that there's no need for a plus sign in a positive number. A number is always assumed to be positive *unless* there's a minus sign in front of it to indicate that it's negative. It is a *mathematical convention.* (In this context, "convention" means "custom" or "way of doing things," and not "a huge gathering of people.") If you think about this for a little while, it makes sense. Why write +3 + (+7) when you can write 3 + 7?

Here's a challenge!

In terms of the integer line, express the fact that when you subtract −5 from −3, you get 2. Write it down in the simplest possible form.

Solution

When you subtract a negative number, you move negatively downward, meaning that you actually travel upward. Figure 4-3 is a number-line drawing that shows how this works. You start at −3 and move down −5 units, which means you really move up 5 units. You finish at the point corresponding to 2. You can write

$$-3 - (-5) = 2$$

In its simplest possible form, this fact is written

$$-3 + 5 = 2$$

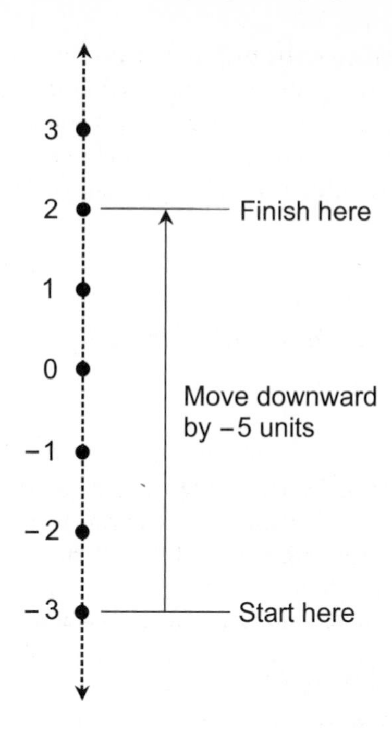

Figure 4-3 Here, we start with −3 and then subtract −5, ending up with 2. When we go negatively downward, we go upward by the equivalent distance.

Identity, Grouping, and Signs

Let's review the mechanics of addition and subtraction, and make sure you understand how the signs work. Then we'll proceed to two important rules, called *laws,* that govern addition and subtraction. These laws apply not only to the integers, but to all quantities, expressions, and numbers you'll encounter in algebra.

The identity element

There is one special integer that you can add to, or subtract from, any integer, and it will never change the value. That integer is 0. Because adding 0 to a number gives you the same number again, 0 is called the *additive identity element.* You can also call it the "subtractive identity element," although that term is informal. The word "identity" means that you always get an output value identical to the input value. For any integer a, the following two statements are always true:

$$a + 0 = 0$$

and

$$a - 0 = 0$$

Grouping with parentheses

A couple of minutes ago, you saw some expressions containing parentheses around a negative number or quantity. Here's one of those expressions, with the right-hand side changed to a variable:

$$-3 - (-5) = a$$

These parentheses indicate that you should consider -5 as "negative 5," not "the subtraction of 5." If you don't write the parentheses, the above expression is

$$-3 - -5 = a$$

This is an example of improper *mathematical grammar*. It's a problem because it's ambiguous. People can't be sure what it means. Should you subtract 5 from -3? That would give you -8, not 2. Should you subtract 5 from -3 twice? That would give you -13. Does the double minus sign mean something different from plain old subtraction or addition?

Pay attention to parentheses when you see them. When you write an expression, be sure to include parentheses when you need them. You can add extra parentheses to an expression as long as you don't change the meaning. Always be sure the number of opening parentheses in an expression is the same as the number of closing parentheses.

Signs in addition and subtraction

Here's a summary of how addition and subtraction work for negative integers as well as for positive ones:

- When you add a positive, the result grows larger.
- When you subtract a positive, the result grows smaller.
- When you add a negative, the result grows smaller; it's like subtracting a positive.
- When you subtract a negative, the result grows larger; it's like adding a positive.

You might want to memorize two general facts. For any two integers a and b,

$$a + (-b) = a - b$$

and

$$a - (-b) = a + b$$

Are you confused?

Here's a real-life situation where the idea of subtracting a negative number makes pretty good sense. Suppose the people in your town are trying to reduce their driving because of high gasoline prices. They've started a support group. People go to meetings once a month to share information about how much less they've driven in the past month compared with the month before that. One after another, the members of the group tell their stories. William says, "I drove 45 fewer miles." Anna says, "I cut back by 65 miles." Maria says, "I drove 200 miles less last month!" Then it comes to you. You have not done so well. The group leader asks, "How much did you cut back? Go on, don't be afraid to tell us." You blush, clear your throat, hold your head up, and declare, "I reduced my driving last month by negative 80 miles."

Here's a challenge!

Start with the integer 5, add −3 to it, then subtract −6 from that, then subtract 10 from that, then add 14 to that, and finally subtract −21 from that. What's the result?

Solution

We can break this down step-by-step, paying careful attention to signs and using parentheses when we need them. Here we go:

$$5 + (-3) = 5 - 3 = 2$$

$$2 - (-6) = 2 + 6 = 8$$

$$8 - 10 = -2$$

$$-2 + 14 = 12$$

$$12 - (-21) = 12 + 21 = 33$$

The Commutative Law for Addition

In basic arithmetic, you learned that you can add a long string of numbers backward or forward, and it doesn't matter. Good accountants take advantage of this when checking their work. They'll add up a column of numbers from top to bottom, then again from bottom to top, just to be sure they have done the arithmetic correctly.

It works when you add

The fact that you can add two integers in either order and get the same result is called the *commutative law for addition*. It means you can *commute* (interchange) the two numbers you're adding, called the *addends*, and get the same *sum* either way. In formal terms, a mathematician would say that for any two integers a and b,

$$a + b = b + a$$

This works whether the numbers are positive, negative, or 0. It also works if there are three, four, five, or more numbers in a sum, as long as the number of addends is not infinite.

It fails when you subtract

In subtraction, the order *does* matter. It's easy to find an example that shows why. Consider this:

$$3 - 5 = -2$$

but

$$5 - 3 = 2$$

In formal terms we would say that for any two integers a and b, it is *not* always true that

$$a - b = b - a$$

In fact, it is *almost never* true. It only works if a and b happen to be the same.

Turning it inside-out

We can use a trick that will make the commutative law "sort of work" with subtraction. This trick is often used by accountants who must work with long columns of *credits* (money added) mixed with *debits* (money taken away). This trick involves taking every subtraction and turning it into the addition of a negative number. Remember that adding a negative is the same as subtracting a positive. That is, for any two integers a and b,

$$a - b = a + (-b)$$

Because the right-hand side of this equation is an addition problem, we can apply the commutative law and get

$$a + (-b) = -b + a$$

Now we can combine the above two equations into a three-way equation:

$$\begin{aligned} a - b &= a + (-b) \\ &= -b + a \end{aligned}$$

Then we can get rid of the middle term and write

$$a - b = -b + a$$

Let's call this the "inside-out commutative law for subtraction." That's not a formal name, but you might find it useful as a memory aid.

Are you confused?

Imagine that you have a checking account and your balance on January 1 was exactly \$700. Consider that as a "starting deposit." By the end of June, you've made 15 deposits and written 20 checks. You want to figure out your balance as of June 30. You convert all the checks to "negative deposits." For example, a check for \$25 becomes a "negative deposit" of –\$25. Now you can add all the "positive deposits" and "negative deposits" in any order, and you'll always end up with the same final balance—if you don't make any calculation errors!

Here's a challenge!

Suppose you have bought a new car and you want to go for a test drive. You live on a flat plain that seems to stretch forever in all directions. You start driving on a straight highway that runs north and south for hundreds of miles on either side of your home town. You drive 25 miles north, then turn around and drive 45 miles south. Then you turn around again, driving 50 miles north. Then you go 7 miles south, 12 miles north, 49 miles south, and finally 5 more miles south. How far from your home town, and it what direction, will you finish? Solve this problem in two different ways.

Solution

First, consider driving north as "positive mileage" and driving south as "negative mileage." Then you drive the following distances in miles, and in this order:

$$25, -45, 50, -7, 12, -49, -5$$

Add these all up:

$$25 + (-45) + 50 + (-7) + 12 + (-49) + (-5) = -19$$

That's 19 miles south of your home town. Now imagine that driving south is "positive mileage" and driving north is "negative mileage." Then you your trip is a sum of displacements like this:

$$-25 + 45 + (-50) + 7 + (-12) + 49 + 5 = 19$$

Again, that's 19 miles south.

The Associative Law for Addition

Another major rule that applies to addition involves how the addends are grouped when you have three or more numbers. You can lump the addends together any way you want, and you'll always end up with the same result. The simplest case of this rule, called the *associative law for addition,* involves sums of three integers.

It works when you add

Here's how a mathematician would state the associative law for three addends. For any three integers a, b, and c

$$(a + b) + c = a + (b + c)$$

For instance:

$$(3 + 5) + 7 = 8 + 7 = 15$$

and

$$3 + (5 + 7) = 3 + 12 = 15$$

This works whether the numbers are positive, negative, or 0. It also works if there are four or more numbers in a sum.

It fails when you subtract

In subtraction, the grouping *does* matter. It's easy to see why you can't apply the associative law to subtraction and get away with it. Consider this:

$$(3 - 5) - 7 = -2 - 7 = -9$$

but

$$\begin{aligned} 3 - (5 - 7) &= 3 - (-2) \\ &= 3 + 2 = 5 \end{aligned}$$

In formal terms, we would say that for any three integers a, b, and c, it is *not* always true that

$$(a - b) - c = a - (b - c)$$

The associative law *hardly ever* works with subtraction.

Mixing the signs

How about mixed addition and subtraction? Let's try it with some integers.

$$(3 + 5) - 7 = 8 - 7 = 1$$

and

$$3 + (5 - 7) = 3 + (-2) = 1$$

It works in this case. What happens if we switch the positions of the signs?

$$(3 - 5) + 7 = -2 + 7 = 5$$

but

$$3 - (5 + 7) = 3 - 12 = -9$$

This time, it fails! Now we know that with mixed signs, the associative law sometimes works, but not always. If something is to be called a *law* in mathematics, "sometimes" does not suffice. "Usually" won't do the job either. Even "almost always" is not good enough. In order to be a law, something has to work *all the time.*

It's easy to prove that a law does not hold in every possible case. You only have to find one case where it fails, called a *counterexample*, to show that something can't be called a law. Proving that a law always works is more difficult. You can't do it using specific integers, because you'd have to try an infinite number of cases one at a time. You have to use airtight logic. That's what proofs are all about.

Add, then subtract

The associative law can work indirectly when subtraction is involved, if you change every subtraction into addition. It's the same trick as with the commutative law. For any three integers a, b, and c

$$\begin{aligned}(a+b)-c &= (a+b)+(-c)\\ &= a+[b+(-c)]\\ &= a+(b-c)\end{aligned}$$

We use square parentheses, called *brackets*, to indicate a grouping with another grouping inside. This looks messy, but a little junk is easy to tolerate when you realize that we've just proved something significant. We've shown that you can *always* use the associative law with mixed signs when the first operation is addition and the second one is subtraction.

Subtract, then add

Now let's look the general situation where we have seen that direct application of the associative law does not always work. How can we modify it to make it work? We can rewrite an expression where the first operation is subtraction and the second one is addition, so it becomes all addition, like this:

$$(a-b)+c = [a+(-b)]+c$$

We can use the associative law for addition to get

$$a+[(-b)+c]$$

Now we can use the commutative law for addition inside the brackets to get

$$a+[c+(-b)]$$

We can simplify this to

$$a+(c-b)$$

Now we know that when the first operation is subtraction and the second is addition,

$$(a-b)+c = a+(c-b)$$

Subtract, then subtract again

Finally, let's explore the situation where we subtract twice. We can rearrange an expression like that as follows:

$$(a-b)-c = [a+(-b)]+(-c)$$

We can use the associative law for addition to get

$$a + [(-b) + (-c)]$$

We can simplify this to

$$a + (-b - c)$$

Now we know that when both operations are subtraction,

$$(a - b) - c = a + (-b - c)$$

Are you confused?

We haven't started to dissect the anatomy of multiplication yet. That will come in the next chapter. But you've had basic multiplication in your arithmetic classes, so let's "cheat" for a moment and take advantage of that. What are you actually doing when you change c to $-c$? Here's an alternative to the "number reflector" idea. When you want to find the negative (also called its *additive inverse*) of any integer, multiply by -1. It works like this:

$$c \times (-1) = -c$$

and

$$-c \times (-1) = c$$

Here's a challenge!

Based on the commutative law for the sum of two integers and the associative law for the sum of three integers, show that for any three integers a, b, and c

$$a + b + c = c + b + a$$

Solution

If you can manipulate the left-hand side of this equation to get the expression on the right-hand side, that's good enough. Because these statements are not very complicated and the proof is not too hard, you can write it as a table with statements on the left and reasons on the right. Table 4-1 shows how it's done. This is a simple *statements/reasons (S/R) proof.*

Table 4-1. Here is a proof that shows how you can reverse the order in which three integers *a*, *b*, and *c* are added, and get the same sum. As you read down the left-hand column, each statement is equal to all the statements above it.

Statements	Reasons
$a + b + c$	Begin here
$a + (b + c)$	Group the second two integers
$a + (c + b)$	Commutative law for the sum of b and c
$(c + b) + a$	Commutative law for the sum of a and $(c + b)$
$c + b + a$	Ungroup the first two integers
Q.E.D.	Latin *Quod erat demonstradum*, translated into English as "Which was to be proved"

Practice Exercises

This is an open-book quiz. You may (and should) refer to the text as you solve these problems. Don't hurry! You'll find worked-out answers in App. A. The solutions in the appendix may not represent the only way a problem can be figured out. If you think you can solve a particular problem in a quicker or better way than you see there, by all means try it!

1. Evaluate and compare these two sums:

$$a = |-3 + 4 + (-5) + 6|$$

and

$$b = |-3| + |4| + |-5| + |6|$$

What general fact can you deduce from the results?

2. To illustrate the importance of the placement of parentheses in a mixed sum and difference, evaluate the following two expressions. In long strings of sums and differences, you should first perform the operations inside the parentheses from left to right, and then perform the operations outside the parentheses from left to right. Here are the expressions:

$$(3 + 5) - (7 + 9) - (11 + 13) - 15$$

and

$$3 + (5 - 7) + (9 - 11) + (13 - 15)$$

3. Using the rules explained in the previous exercise, how should you evaluate the string of sums and differences if there are no parentheses at all? Here it is:

$$3 + 5 - 7 + 9 - 11 + 13 - 15$$

4. Suppose someone tells you that there was a significant trend in the mid-winter average temperatures in the town of Hoodopolis during the period 1998 through 2005. You want to find out if this is true. You come across some old heating bills from the utility company that show how much warmer or cooler a given month was, on the average,

compared to the same month in the previous year. You look at the records for January and find out that in Hoodopolis,

- January 2005 averaged 5 degrees cooler than January 2004.
- January 2004 averaged 2 degrees warmer than January 2003.
- January 2003 averaged 1 degree cooler than January 2002.
- January 2002 averaged 7 degrees warmer than January 2001.
- January 2001 averaged the same temperature as January 2000.
- January 2000 averaged 6 degrees cooler than January 1999.
- January 1999 averaged 3 degrees warmer than January 1998.

What was the difference in the average temperature between January 2005 and January 1998 in the town of Hoodopolis?

5. Show at least one situation where can you say that

$$a - b = b - a$$

Don't use the trivial case where a and b are both equal to 0.

6. Show at least one situation where you can say that

$$(a - b) + c = a - (b + c)$$

where a, b, and c are integers. Don't use the trivial case where a, b, and c are all 0.

7. Show at least one situation where you say that

$$(a - b) - c = a - (b - c)$$

where a, b, and c are integers. Don't use the trivial case where a, b, and c are all 0.

8. Based on the commutative law for the sum of two integers and the associative law for the sum of three integers, construct an S/R proof showing that for any four integers a, b, c, and d

$$a + b + c + d = d + c + b + a$$

Here's a hint: use the solution to the last "challenge" problem in this chapter as a shortcut. A previously proved fact, when used to prove something new, is called a *lemma.*

9. Based on the associative law for the sum of three integers, prove that for any four integers a, b, c, and d

$$(a + b + c) + d = a + (b + c + d)$$

Do this in narrative form. Don't use the S/R table method. Here's a hint: "zip up" the sum $b + c$, and call it by another name.

10. Simplify and compare these expressions:

$$(a + b - c) + (a - b + c)$$

and

$$a + (b - c) + (a - b) + c$$

CHAPTER

5

Multiplication and Division

In this chapter we'll explore the mechanics of *multiplication* and *division.* Multiplication is a shortcut for repeated addition. Division is something like "undoing" multiplication. These two operations, like addition and subtraction, obey certain rules as long as they aren't "stretched" too far!

Moving Out and In

Think of the number line again, where the integers are points above and below a "number reflector" plane that goes through 0. As you move upward, the numerical value increases; as you go downward, it decreases. As you get farther in either direction from the "number reflector," the absolute value increases.

To multiply, move out

When you add an integer to itself, you multiply it by 2. When you add an integer to itself twice, you multiply it by 3. When you add an integer to itself 3 times, you multiply it by 4. You can keep going with this. Here's an example of what happens when the integer is positive:

$$5 + 5 = 5 \times 2$$
$$5 + 5 + 5 = 5 \times 3$$
$$5 + 5 + 5 + 5 = 5 \times 4$$

Here's what happens when the integer is negative:

$$-5 + (-5) = -5 \times 2$$
$$-5 + (-5) + (-5) = -5 \times 3$$
$$-5 + (-5) + (-5) + (-5) = -5 \times 4$$

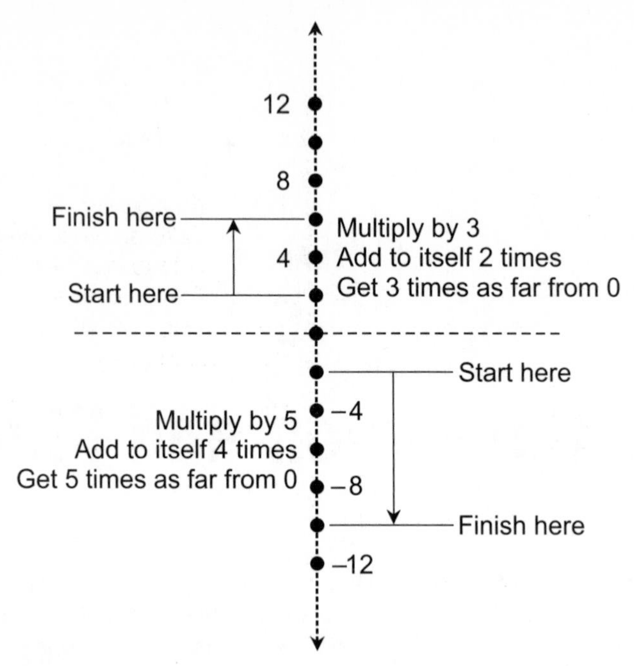

Figure 5-1 On top, 2 is multiplied by 3. On the bottom, −2 is multiplied by 5. To avoid clutter, only the even-integer points are shown on the number line.

Suppose p is an integer and n is a positive integer. From the above facts, you can see that whenever you multiply p by n, you add p to itself $(n - 1)$ times. On the number line, you move away from the "number reflector" $(n - 1)$ times by a distance equal to the absolute value of p. If your starting integer p is above the "number reflector," you move up; if your starting integer p is below the "number reflector," you move down. Figure 5-1 illustrates how this works for 2×3 and -2×5. The "number reflector" is shown as a horizontal, dashed line.

What if n is negative instead of positive? To multiply p by n in this situation, first take the additive inverse (negative) of your starting integer p, and then move away from the "number reflector" $|n| - 1$ times by a distance equal to the absolute value of p. Figure 5-2 shows how this works for $2 \times (-3)$ and $-2 \times (-5)$.

When you multiply two quantities, you get a *product*. In a multiplication problem, the first quantity (the one to be multiplied) is sometimes called the *multiplicand*, and the second quantity (the one you are multiplying by) is sometimes called the *multiplier*. More often, they are both called *factors*.

To divide, move in

Now imagine the multiplication process in reverse. For examples, look at Figs. 5-1 and 5-2 and think backward!

In Fig. 5-1, suppose you start with 6 and you want to divide it by 3. You reduce your distance from the "number reflector" by a factor of 3 but stay on the same side, so you end up

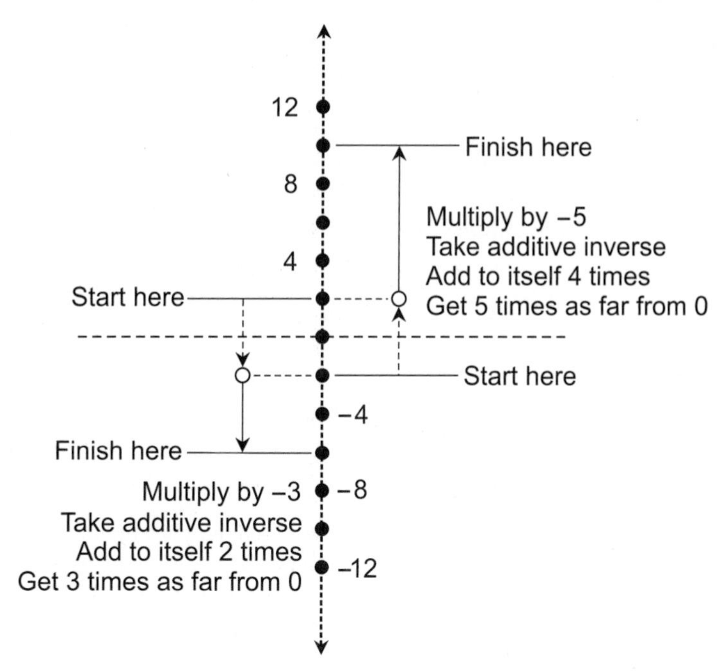

Figure 5-2 At lower left, 2 is multiplied by −3. At upper right, −2 is multiplied by −5. To avoid clutter, only the even-integer points are shown on the number line.

at the point representing 2. If you start with −10 and divide by 5, you cut your distance from the "number reflector" by a factor of 5 and stay on the same side, so you end up at −2.

Dividing by a negative number is trickier. In Fig. 5-2, suppose you start with −6 and divide it by −3. You jump to the other side of the "number reflector" and then reduce your distance from it by a factor of 3, finishing at 2. If you start with 10 and divide by −5, you jump to the other side of the "number reflector" and then cut your distance from it by a factor of 5, finishing at −2.

In these situations, the integers divide each other "cleanly" without remainders. Remainders occur when one nonzero integer divides another integer and the result is not an integer. "Messy division" produces fractions, which we'll study in Chap. 6. An example is the division of 5 by 3. Another example is the division of −7 by −12.

You can add, subtract, or multiply any two integers and always end up with another integer. In math jargon, the operations of addition, subtraction, and multiplication are *closed over the set of integers.* When you do a division problem with integers, you don't always get another integer. Therefore, the operation of division is not closed over the set of integers.

When you divide a quantity by another quantity, you get a *quotient.* Sometimes it is called a *ratio.* In a division problem, the first number is sometimes called the *dividend,* and the second number is sometimes called the *divisor.*

Division by 0

What happens if you try to divide an integer by 0? You don't get any integer, or any other known type of number. Look at this problem "inside-out." What must you multiply 0 by if you want to get, say, 3? How many times must you add 0 to itself to get anything but 0? No integer can do this trick. In fact, *no known number* solves this problem.

Most mathematicians will tell you that division by 0 is "not defined." That's what my 7th-grade math teacher kept saying, and I pestered her about it. I would ask, "Why not?" or retort, "Let's define it, then!" She would repeat herself, "Division by 0 is *not defined.*" I did not take her seriously, so I started trying to make division by 0 work. I never came up with a well-defined way to do it. But I came pretty close, and had a lot of fun trying.

Are you confused?

Have you heard that dividing a positive integer by 0 gives you "infinity"? If not, you probably will some day. Be skeptical! The first thing you must do to figure out if it's really true is to define "infinity." That's not easy. No meaningful, enduring definition of "infinity" produced by mathematicians has ever had anything to do with division by 0.

Again, look at the problem "inside-out." If you want to multiply 0 by any positive integer n, you must add 0 to itself $(n - 1)$ times. No matter how large you make n, you always get 0 when you add it to itself $(n - 1)$ times. Why should adding 0 to itself forever make any difference? It's tempting to suppose that it might, but that doesn't prove that it will. In mathematics, we need proof before we can claim something is true!

Manipulating equations

Whenever we add or subtract a certain quantity to or from both sides of the equation, we still have a valid equation. The same is true if we multiply both sides of an equation by a certain quantity, or divide either side by a certain quantity other than 0.

The quantity you add, subtract, or multiply by can be a number, a variable, or a complicated expression, as long as it is the same for the left-hand side of the equation as for the right-hand side. If you divide both sides of an equation by anything, it's best to stick to nonzero numbers. If you divide both sides of an equation by a variable or an expression containing a variable, you can get into trouble, as you'll see in Chap. 11.

Keep these rules in mind. That way, you won't get confused later on when we do something like divide both sides of an equation by 999, or multiply both sides by $(a + b)$. The fact that we can do these things makes solving equations and proving various facts far easier than they would be otherwise.

Here's a challenge!

In terms of the number line and displacements, show what happens when you multiply the integer -1 over and over, endlessly, by -2.

Solution

Figure 5-3 illustrates this process. Because the multiplier is negative, we jump to the opposite side of the "number reflector" each time we multiply. Then the result becomes a new multiplicand. Because the

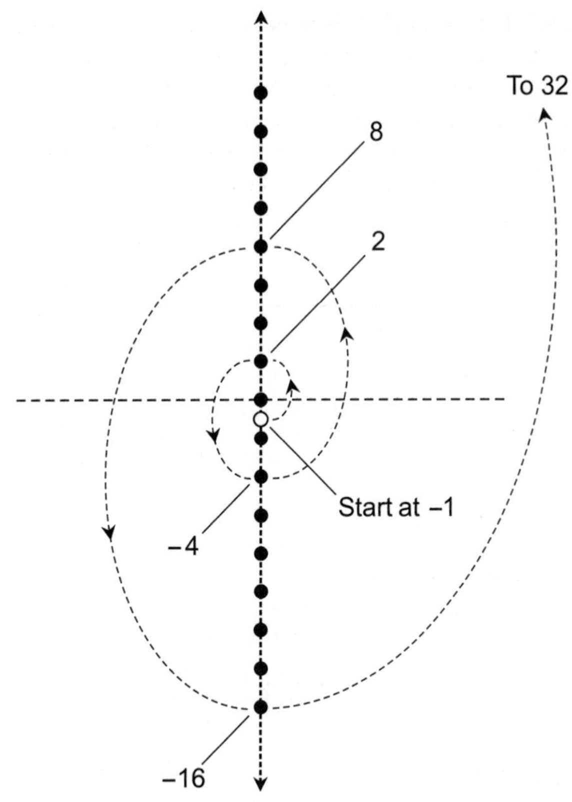

Figure 5-3 Here is what happens when we start at –1 and multiply by –2 over and over. We jump back and forth across the "number reflector," doubling our distance from it with each jump. To avoid clutter, only the even-integer points are shown on the number line.

absolute value of the multiplier is 2, we double our distance from the "number reflector" with each jump. Expressed as equations, we have

$$-1 \times (-2) = 2$$
$$2 \times (-2) = -4$$
$$-4 \times (-2) = 8$$
$$8 \times (-2) = -16$$
$$-16 \times (-2) = 32$$
$$\downarrow$$

and so on, forever

Identity, Grouping, and Signs

Let's review how signs work in multiplication and division. Then we'll proceed to the three major laws that govern the interplay between multiplication, division, addition, and subtraction.

Notation for multiplication

When you want to multiply two numbers, you can use the familiar "times sign" and put the numerals for the factors on either side. This symbol (×) looks like a tilted cross or a letter "x." Another symbol you'll often see is the small, elevated dot (·). When a number is multiplied by a variable, or when a variable is multiplied by another variable, you'll see their symbols run together without any space between. Parentheses are placed around complicated expressions when they are multiplied by each other.

- When you see 3×7, it means 3 times 7.
- When you see $-3 \cdot 7$, it means -3 times 7.
- When you see $-4a$, it means -4 times a.
- When you see ab, it means a times b.
- When you see abc, it means a times b times c.
- When you see $a(b - c)$, it means a times $(b - c)$.
- When you see $(a + b)(c + d)$, it means $(a + b)$ times $(c + d)$.

Notation for division

In this book, we'll use the forward slash (/) to indicate division. In arithmetic, you sometimes see the dash with two dots (÷), but that's rarely used in algebra. When expressions are complicated, the dividend (the number you want to divide) can be placed on top of a long horizontal line, and the divisor (the number you divide by) is placed underneath. As with multiplication, parentheses are placed around complicated expressions when they are divided by each other.

- When you see $8/2$, it means 8 divided by 2.
- When you see $-4/a$, it means -4 divided by a.
- When you see $a/(-4)$, it means a divided by -4.
- When you see a/b, it means a divided by b.
- When you see $a/(b - c)$, it means a divided by $(b - c)$.
- When you see $(a + b)/(c + d)$, it means $(a + b)$ divided by $(c + d)$.

The identity element

You can multiply or divide any integer by 1, and it won't change the value. Because multiplying or dividing by 1 always gives you the same number again, 1 is called the *multiplicative identity element.* (For some reason I've never heard it called the "divisive identity element," but technically this term is okay.) For any integer a

$$a1 = a$$

and

$$a/1 = a$$

The sign-changing element

When you multiply or divide any integer by −1, you reverse the sign but do not change the absolute value. A positive integer becomes negative, and a negative integer becomes positive, but the distance from 0 on the integer line stays the same. For any integer a

$$a(-1) = -a$$

and

$$a/(-1) = -a$$

Conversely,

$$-a(-1) = a$$

and

$$-a/(-1) = a$$

Note that $a(-1)$ here means a times −1, not a minus 1. Those parentheses are important! The integer −1 can be called the *multiplicative sign-changing element* or the *divisive sign-changing element.*

Parentheses in simple products and quotients

Look at these expressions:

$$3 \times (-5) = a$$

and

$$15/(-3) = b$$

If you don't write the parentheses, the above expressions are

$$3 \times -5 = a$$

and

$$15/-3 = b$$

Expressions like these might be clear enough to you, but they would confuse some people. Don't be afraid to add parentheses to an expression if you think they will prevent ambiguity. Just be sure that for every opening parenthesis you put in, you include a corresponding closing parenthesis later in the expression.

Signs in multiplication and division

Here is a set of rules for multiplication and division by any integer except −1, 0, or 1. Remember that "more" always means more positive, or moving upward on the integer line, and "less" always means more negative, or moving down. Study these rules carefully. If any of them confuse you, "plug in" some actual numbers and test them.

- When you multiply a positive integer by 2 or more, the result stays positive and the absolute value increases (it gets farther from 0).
- When you divide a positive integer by 2 or more, the result stays positive and the absolute value decreases (it gets closer to 0).
- When you multiply a negative integer by 2 or more, the result stays negative and the absolute value increases (it gets farther from 0).
- When you divide a negative integer by 2 or more, the result stays negative and the absolute value decreases (it gets closer to 0).
- When you multiply a positive integer by −2 or less, the result becomes negative and the absolute value increases (it gets farther from 0).
- When you divide a positive integer by −2 or less, the result becomes negative and the absolute value decreases (it gets closer to 0).
- When you multiply a negative integer by −2 or less, the result becomes positive and the absolute value increases (it gets farther from 0).
- When you divide a negative integer by −2 or less, the result becomes positive and the absolute value decreases (it gets closer to 0).

"Homogenize" these!

You would do well to memorize (or, as my dad would always say, "homogenize") two general facts. For any two integers a and b

$$a \times (-b) = -ab = -(ab)$$

and

$$a/(-b) = -a/b = -(a/b)$$

Are you confused?

Suppose you see an expression with addition, subtraction, multiplication, and division all mixed up, but with no parentheses. Here's an example.

$$2 + 48/4 \times 6 - 2 \times 5 + 12/2 \times 2 - 5$$

You'll get an answer that depends on which operations you do first. *Do not* approach a mixed-operation problem like this by simply grinding out the arithmetic from left to right. You must use certain *rules of precedence*. In this order:

- Group all the multiplications
- Do all the multiplications from left to right

Table 5-1. This is a step-by-step simplification of a complicated expression containing addition, subtraction, multiplication, and division without any parentheses to indicate the order in which the operations should be done. As you read down the left-hand column, each statement is equal to all the statements above it.

Statements	Reasons
$2 + 48 / 4 \times 6 - 2 \times 5 + 12 / 2 \times 2 - 5$	Begin here
$2 + 48 / (4 \times 6) - (2 \times 5) + 12 / (2 \times 2) - 5$	Group all the multiplications
$2 + 48 / 24 - 10 + 12 / 4 - 5$	Do all the multiplications
$2 + (48/24) - 10 + (12/4) - 5$	Group all the divisions
$2 + 2 - 10 + 3 - 5$	Do all the divisions
$2 + 2 + (-10) + 3 + (-5)$	Convert all the subtractions to negative additions
-8	Do the additions from left to right

- Group all the divisions
- Do all the divisions from left to right
- Convert all the subtractions to negative additions
- Do the additions from left to right

When you use these rules correctly, the above problem simplifies as shown in Table 5-1. As you read downward, each expression is equal to the one above it.

Here's a challenge!

Start with the integer 5, multiply by −4, then divide that result by −2, then multiply that result by 8, then divide that result by −20, and finally divide that result by −4. What do you end up with?

Solution

We can break this down step-by-step, paying careful attention to signs and using parentheses when we need them:

$$5 \times (-4) = -20$$

$$-20/(-2) = 10$$

$$10 \times 8 = 80$$

$$80/(-20) = -4$$

$$-4/(-4) = 1$$

The Commutative Law for Multiplication

In basic arithmetic, you learned that you can multiply numbers in any order and always get the same product. But you can't expect to do the same thing if there is division anywhere in the process.

It works when you multiply

The fact that you can multiply two integers in either order and get the same result is called the *commutative law for multiplication.* For any two integers a and b

$$ab = ba$$

If both factors have the same sign (positive or negative), the product is positive. If the factors have opposite signs, the product is negative. The commutative law also works if there are three or more factors. You can rearrange them in any order you want. If there are an even number of negative factors, the product is positive. If there are an odd number of negative factors, the product is negative.

It fails when you divide

When you divide an integer by another integer, the order is important. If you divide 20 by 4, you get 5. But if you divide 4 by 20, you don't get 5. You don't even get an integer. A more dramatic example is the division of 0 by any other integer. If you divide 0 by −3, you get 0. But if you divide −3 by 0, you get an undefined quantity!

Cross-multiplication

Here's a fact that you will find useful in algebra. It's called the *rule of cross-multiplication.* Suppose you have two ratios of integers, a/b and c/d, and you're told that they're equal. You're also assured that neither b nor d is equal to 0. You write

$$a/b = c/d$$

You can multiply the dividend on the left-hand side of this equation (here, that's a) by the divisor on the right-hand side (d), and get the same result as when you multiply the divisor on the left-hand side (b) by the dividend on the right-hand side (c). In formal terms,

$$\text{If } a/b = c/d, \text{ then } ad = bc$$

This rule works in the reverse sense, too. For any four integers a, b, c, and d, where b is not equal to 0 (written $b \neq 0$) and $d \neq 0$,

$$\text{If } ad = bc, \text{ then } a/b = c/d$$

When a rule works in both logical directions, mathematicians use the expression "if and only if" and abbreviate it as "iff." Now we know that for any four integers a, b, c, and d, where $b \neq 0$ and $d \neq 0$,

$$ad = bc \text{ iff } a/b = c/d$$

Are you confused?

Suppose you see an expression where you have to divide repeatedly, with no parentheses telling you which division to do first. Here is an example:

$$200/2/5/4$$

In a case like this, proceed from left to right. Here, that means you should take 200 and divide it by 2, getting 100. Then divide 100 by 5, getting 20. Finally, divide 20 by 4, getting 5.

Here's a challenge!

Under what circumstances can we divide an integer a by an integer b, and get the same quotient (or ratio) as when we divide b by a? Assume that $a \neq 0$ and $b \neq 0$.

Solution

We have two integers a and b, and we are told that

$$a/b = b/a$$

This equation is always true when a and b are the same. In that case, we can substitute a for b and get

$$a/a = a/a$$

which simplifies to $1 = 1$. That's trivial!

Now suppose that a and b are additive inverses. Therefore, $-a = b$. Here, we can substitute $-a$ for b in the original equation and get

$$a/(-a) = (-a)/a$$

which simplifies to $-1 = -1$. We know this because it is an application of one of those two facts we're supposed to have "homogenized" earlier in this chapter.

What's the verdict? If we have two nonzero integers a and b whose absolute values are the same, meaning that they're either identical or are additive inverses of each other, then $a/b = b/a$. In symbols, we can write

$$\text{If } |a| = |b|, \text{ then } a/b = b/a$$

The Associative Law for Multiplication

Another important rule that applies to multiplication involves how the factors are grouped when you have three or more of them. You can lump the factors together any way you want, and you'll always end up with the same product. The simplest case, called the *associative law for multiplication,* involves a product of three factors.

It works when you multiply

Here's how a mathematician would formally state the associative law in its most basic form. For any three integers *a, b,* and *c*

$$(ab)c = a(bc)$$

For instance:

$$(3 \times 5) \times 2 = 15 \times 2 = 30$$

and

$$3 \times (5 \times 2) = 3 \times 10 = 30$$

This works whether the numbers are positive, negative, or 0. It also works if there are more than three numbers in a product, as long as there aren't infinitely many.

It fails when you divide

In division, as in subtraction, the way in which you group the integers or variables is important. Consider this:

$$(16/4)/2 = 4/2 = 2$$

but

$$16/(4/2) = 16/2 = 8$$

For any three integers *a, b,* and *c*, it is *not* necessarily true that

$$(a/b)/c = a/(b/c)$$

The associative law *hardly ever* works with division.

Are you confused?

Look at another expression where we have to divide more than once, and see what happens when we insert parentheses in different places. Let's try this:

$$4{,}000/40/10/5$$

Going straightaway from left to right, we get 4,000/40 = 100, then 100/10 = 10, and finally 10/5 = 2. Now let's think of it this way:

$$(4{,}000/40)\ /\ (10/5)$$

In this case, it simplifies to 100/2 = 50. Now let's try this:

$$4{,}000 \,/\, (40/10) \,/\, 5$$

We do the division in parentheses first, getting

$$4{,}000/4/5$$

Starting at the left, we get 4,000/4 = 1,000. Then dividing 1,000 by 5 gives us 200.

Here's a challenge!

Here is a riddle that ought to get your brain running at full speed. Consider this infinite product:

$$1 \times (-1) \times 1 \times (-1) \times 1 \times (-1) \times 1 \times (-1) \times \cdots$$

If we start multiplying from left to right, we get a sequence of products that come out like this, in order:

$$1, -1, -1, 1, 1, -1, -1, 1, \ldots$$

The products switch back and forth between 1 and −1, endlessly. The final product therefore cannot be defined. No integer can have two values "at the same time." But suppose we try to apply the commutative law to the original expression infinitely many times! We can then rearrange the factors to get this:

$$(-1) \times (-1) \times 1 \times 1 \times (-1) \times (-1) \times 1 \times 1 \times \cdots$$

Now imagine that we try to group the integers in pairs, infinitely many times, like this:

$$[(-1) \times (-1)] \times [1 \times 1] \times [(-1) \times (-1)] \times [1 \times 1] \times \cdots$$

That gives us

$$1 \times 1 \times 1 \times 1 \times \cdots$$

The sequence of products in this case is

$$1, 1, 1, 1, \ldots$$

Now it seems as if the product of the whole thing is equal to 1! What's going on?

Solution

We derived a contradiction here because we improperly used the commutative and grouping laws. First, we tried to apply these rules to a product that is undefined in its most basic form. Second, we acted as if these rules can be used in a single expression *infinitely many times*, without really knowing if we can get away with such tricks. Evidently we can't!

Table 5-2. Here is a proof that shows how you can reverse the order in which three integers *a*, *b*, and *c* are multiplied, and get the same product. As you read down the left-hand column, each statement is equal to all the statements above it.

Statements	Reasons
abc	Begin here
$a(bc)$	Group the second two integers
$a(cb)$	Commutative law for the product of b and c
$(cb)a$	Commutative law for the product of a and (cb)
cba	Ungroup the first two integers
Q.E.D.	Mission accomplished

Here's another challenge!

Based on the commutative law for multiplication of two integers, and on the associative law for multiplication of three integers, show that for any three integers a, b, and c

$$abc = cba$$

Solution

This works out just like the "challenge" at the end of Chap. 4. Simply change all the instances of addition to multiplication. Table 5-2 is an S/R proof.

The Distributive Laws

When you come across a sum or difference that is multiplied by a single number or variable, you will sometimes want to expand it into a sum or difference of products. To do this, you can use the *distributive laws.*

Multiplication over addition

Suppose you have three integers a, b, and c arranged so that you must multiply a by the sum of b and c. The *left-hand distributive law of multiplication over addition* tells you that this is equal to the product ab plus the product bc. This comes out simpler if you write it down:

$$a(b + c) = ab + ac$$

Now imagine multiplying the sum of a and b by c. The *right-hand distributive law of multiplication over addition* says that

$$(a + b)c = ac + bc$$

Multiplication over subtraction

The distributive laws also work with subtraction. For any three integers *a, b,* and *c*

$$a(b - c) = ab - ac$$

and

$$(a - b)c = ac - bc$$

It's not hard to show how these follow from the laws for addition. If you want to do a rigorous job, the process is rather long. Table 5-3 breaks the derivation down into an S/R process, showing every logical step, for the *left-hand distributive law of multiplication over subtraction.* If you want to do a proof for the *right-hand distributive law of multiplication over subtraction,* consider it a bonus exercise!

The left-hand law fails with division

The left-hand distributive laws do not work for division over addition or for division over subtraction. To see why, all you have to do is produce examples of failure. That's easy! Consider this:

$$24/(4 + 2) = 24/6 = 4$$

but

$$24/4 + 24/2 = 6 + 12 = 18$$

And this:

$$24/(4 - 2) = 24/2 = 12$$

Table 5-3. Derivation of the left-hand distributive law for multiplication over subtraction. As you read down, each statement is equal to all the statements above it. Warning: Don't mistake the expression (–1) for the subtraction of 1! The parentheses emphasize that –1 is a factor in a product.

Statements	Reasons
$a(b - c)$	Begin here
$a[b + (-c)]$	Convert the subtraction to the addition of a negative
$ab + a(-c)$	Left-hand distributive law of multiplication over addition
$ab + ac(-1)$	Principle of the sign-changing element
$ab + a(-1)c$	Commutative law for multiplication
$ab + (-1)ac$	Commutative law for multiplication (again)
$ab + (-ac)$	Principle of sign-changing element (the other way around)
$ab - ac$	Convert the addition of a negative to a subtraction
Q.E.D.	Mission accomplished

but

$$24/4 - 24/2 = 6 - 12 = -6$$

The right-hand law works with division

If you have a sum or difference as the dividend and the single number as the divisor, you can use the distributive law for division over addition or division over subtraction. If a and b are any integers, and if c is any nonzero integer, then

$$(a + b)/c = a/c + b/c$$

and

$$(a - b)/c = a/c - b/c$$

Are you confused?

To help get rid of possible confusion about how the distributive laws operate when integers are negative, try an example where $a = -2$, $b = -3$, and $c = -4$. First, work out the expression where you multiply a times $(b + c)$:

$$-2 \times [-3 + (-4)] = -2 \times (-7) = 14$$

Now work out the expression where you add ab and ac:

$$[-2 \times (-3)] + [-2 \times (-4)] = 6 + 8 = 14$$

Here's a challenge!

With the aid of the commutative and distributive laws, prove that for any four integers a, b, c, and d

$$(a + b)(c + d) = ac + ad + bc + bd$$

Solution

Let's do this as a narrative. Even though S/R proofs look neat, narrative proofs are often preferred by mathematicians. We'll start with the left-hand side of the above equation:

$$(a + b)(c + d)$$

Let's think of the sum $a + b$ as a single unit, and call it e. Now we can rewrite this as

$$e(c + d)$$

The left-hand distributive law for multiplication over addition allows us to rewrite this as

$$ec + ed$$

Next, we expand e back into its original form and substitute it into the above expression twice, getting

$$(a + b)c + (a + b)d$$

We can apply the right-hand distributive law twice, and rewrite this as

$$ac + bc + ad + bd$$

Now we employ the commutative law in a generalized way, obtaining

$$ac + ad + bc + bd$$

We know this is equal to the expression we began with, because we just got done "morphing" it using known tactics, one step at a time. Therefore

$$(a + b)(c + d) = ac + ad + bc + bd$$

Q.E.D.

Practice Exercises

This is an open-book quiz. You may (and should) refer to the text as you solve these problems. Don't hurry! You'll find worked-out answers in App. A. The solutions in the appendix may not represent the only way a problem can be figured out. If you think you can solve a particular problem in a quicker or better way than you see there, by all means try it!

1. How does the absolute value change if you multiply an integer by −3?
2. How does the absolute value change if you start with an integer and multiply by −3 over and over without end?
3. Evaluate the following expression:
   $$4 + 32 / 8 \times (-2) + 20 / 5 / 2 - 8$$
4. Do an S/R proof showing that for any four integers a, b, c, and d
   $$abcd = dcba$$
   Here's a hint: you solved a problem like this for addition in Chap. 4. This proof proceeds in the same way.
5. Start with the integer −15, multiply by −45, then divide that result by −25, then multiply that result by −9, then divide that result by −81, and finally multiply that result by −5. What do you end up with?
6. Show at least one situation where you can say that
   $$a/(b/c) = (a/b)/c$$
   where a, b, and c are integers. Do not use the trivial case $a = 1$, $b = 1$, and $c = 1$. Find something more interesting!

7. Show at least one situation where you can say that

$$(ab)/c = a(b/c)$$

where a, b, and c are integers. Do not use the trivial case $a = 1$, $b = 1$, and $c = 1$. Find something more interesting!

8. Suppose you have learned the left-hand distributive law for multiplication over addition, but you have never heard about the right-hand version. Show that for any three integers n, m, and p, the right-hand distributive law for multiplication over addition will always work:

$$(m + n)p = mp + np$$

Use the narrative form, not an S/R table.

9. Construct an S/R table showing that for any two integers d and g

$$-(d + g) = -d - g$$

10. Prove that when you want to find the negative of the subtraction of one quantity from another, you can simply switch the order of the subtraction. To do this, put together an S/R table showing that for any two integers h and k

$$-(h - k) = k - h$$

CHAPTER

6

Fractions Built of Integers

You can always divide an integer by another integer, except when the divisor is 0. Then you get a *fraction.* A fraction might not be an integer, but it's still a number.

"Messy" Quotients

Figure 6-1 shows how you move around on the number line when you divide 3 by −2. First, you take the additive inverse of 3, because you're dividing by a negative quantity. That puts you at the point corresponding to −3. Then you reduce your distance from 0 by a factor equal to the absolute value of the divisor. That absolute value is 2, so you must move halfway from −3 to 0. That puts you at a point between −1 and −2.

Remainders

In the situation of Fig. 6-1, the finishing point is midway between the point for −1 and the point for −2. When you divide 3 by −2, you get −1 with a remainder 1. In this context, the word "remainder" means "portion left over."

The divisor in this situation is −2, so you have a remainder of 1/(−2), which has the same numerical value as −1/2. The remainder is always added to the whole-number part of the quotient to get the final version of the quotient. Here, the arithmetic works out like this:

$$\begin{aligned} 3/(-2) &= -1 + (-1/2) \\ &= -(1 + 1/2) \\ &= -1\text{-}1/2 \end{aligned}$$

The last quantity above is read "negative one and a half." The whole-number part is separated from the fractional part by a dash. This dash is *not* a minus sign! (The minus sign is much longer.) Some writers leave a space between the whole-number part and the fractional part of an expression like this. But that can also be confusing, because it could make the above result look like −11/2. That would mean −5-1/2, not −1-1/2.

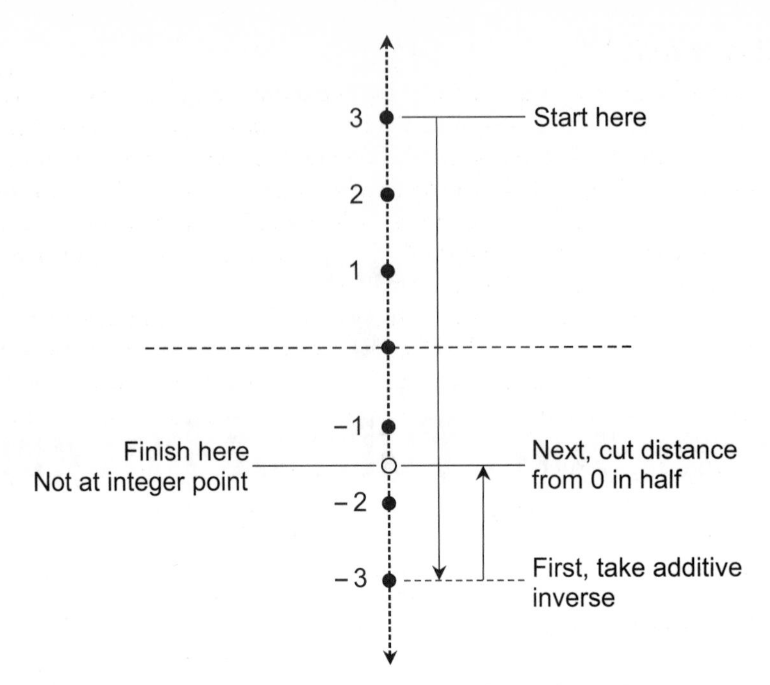

Figure 6-1 When you divide 3 by −2 and follow the process on the number line, you end up at a point that doesn't correspond to an integer.

Improper or proper?

In a fraction, the dividend is called the *numerator*, and the divisor is called the *denominator*. In the situation shown by Fig. 6-1, the numerator you start out with is 3, and the denominator is −2. You can change this so the numerator is −3 with a denominator of 2. That's easier to understand. You can always take the additive inverse of both the numerator and denominator in a fraction, and you'll still have the same numerical value.

When the absolute value of the numerator in a fraction is larger than, or equal to, the absolute value of the denominator, some people call it an *improper fraction*. There's nothing really inappropriate about this type of fraction, but in everyday usage, such a fraction can seem bizarre. No one ever says anything like, "It's 7/3 times as far to Happyville as it is to Bluesdale." Instead they would say, "It's 2-1/3 times as far to Happyville as it is to Bluesdale."

When the numerator in a fraction is an exact integer multiple of the denominator, the fraction divides out to a plain integer. Otherwise, an improper fraction can always be changed to an integer plus or minus a fraction. You divide the numerator by the denominator, getting the whole-integer part. Then you divide the remainder by the denominator to get the fractional part.

When the absolute value of the numerator in a fraction is less than the absolute value of the denominator, you have a *proper fraction*. In this context, the word "proper" does not imply anything more technically acceptable than "improper." It means that the fraction can't be changed into an integer plus or minus a fraction. A proper fraction can also be called a *simple fraction*.

Fraction or ratio?

Sometimes a fraction is called a *ratio*. These two terms are almost synonymous, but not quite. The term "fraction" implies that a particular quantity is a part of some other quantity. The term "ratio" expresses how two quantities are related in terms of their relative size or value.

Think of the Happyville-Bluesdale situation again. If you say, "It's *seven-thirds* times as far to Happyville as it is to Bluesdale," then you're using 7/3 as a fraction. If you say, "The ratio of the distance to Happyville compared with the distance to Bluesdale is *seven to three*," then you're saying the same thing, but in a different way.

You can write, "The ratio of the distance to Happyville compared with the distance to Bluesdale is 7/3," and read "7/3" as "seven to three." If you want to make clear that you're talking about a ratio, you can use a colon instead of a slash to separate the 7 and the 3. You would then write something like, "The Happyville-to-Bluesdale distance ratio is 7:3," reading "7:3" as "seven to three."

Ratios are always expressed in terms of two integers, one divided by the other. They're never expressed as a whole integer plus or minus a proper fraction.

Are you confused?

Proper fractions are always larger than −1 but smaller than 1. A mathematician would use the "strictly larger than" *inequality* (>) and the "strictly smaller than" symbol (<) to express this fact. If q is a proper fraction, then

$$q > -1 \qquad \text{and} \qquad q < 1$$

You can also write

$$-1 < q < 1$$

The proper-fraction interval

Figure 6-2 shows the "realm of proper fractions" on the number line. It contains the points for all possible fractions between, but not including, −1 and 1. The interval is shown as a shaded line. It's gray (not black) for a subtle reason. The set of all proper fractions doesn't account for all the geometric points above −1 and below 1. Within that range, there are numbers that can't be expressed as fractions. The same thing is true everywhere along the number line. Ratios of integers can't account for all the points on a true geometric line. You'll learn more about that in Chap. 9.

You should also be acquainted with two other symbols. The "larger than or equal to" symbol looks like a "strictly larger than" symbol with a line under it (≥). The "smaller than or equal to" symbol looks like a "strictly smaller than" symbol with a line under it (≤). If you wanted to include −1 and 1 in the interval above, you would write

$$q \geq -1 \qquad \text{and} \qquad q \leq 1$$

Alternatively, you could write

$$-1 \leq q \leq 1$$

Figure 6-2 Proper fractions correspond to points between −1 and 1 on the number line. The open circles at −1 and 1 indicate that they are not included in the interval.

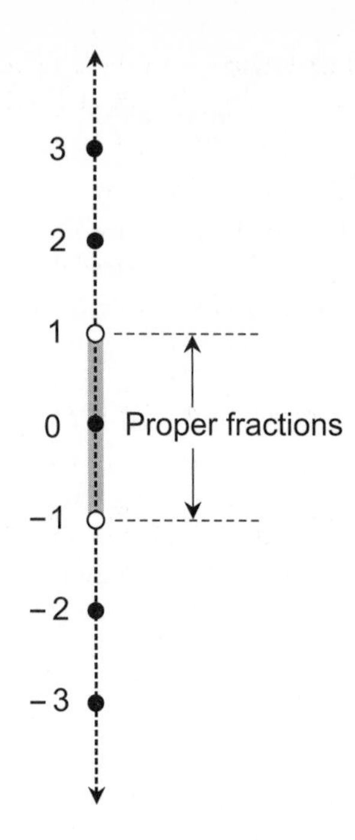

Here's a challenge!

Suppose a car dealer tells you that a certain sports car has a top speed that's "half again" as fast as the top speed of a certain pickup truck. What is the car's top speed compared to the truck's top speed, as a ratio of two integers? What is the truck's top speed to the car's top speed, as a ratio of two integers?

Solution

The term "half again" means "1-1/2 times as great." The ratio of the car's top speed to the truck's top speed is therefore 1-1/2 to 1, or 3/2 to 1. But that's not a ratio of two integers! The correct way to express the ratio is 3 to 2. You would write this as 3/2 or 3:2. When you want to express a ratio in the reverse sense, switch the numerator and the denominator. The ratio of the truck's top speed to the car's top speed is 2 to 3, which you can write as 2/3 or 2:3.

Technically, you can express both of these ratios in infinitely many other ways. In general, the ratio of the car's highest speed (call it c) to the truck's highest speed (call it t), expressed as a ratio between two integers, is

$$c/t = (3a)/(2a)$$

where a can be any integer except 0. The ratio of the truck's highest speed to the car's highest speed, in general, is

$$t/c = (2a)/(3a)$$

What if you want to specify the actual number of miles per hour that each vehicle can travel? Suppose the truck can go a maximum of 100 miles per hour on a straight, level road with no wind, and the car can go 150 miles per hour under the same conditions. You can then write the ratio of the car's top speed to the truck's top speed as 150/100 or 150:100. Conversely, the ratio of the truck's top speed to the car's top speed is 100/150 or 100:150. In these cases, $a = 50$ in the above equations:

$$c/t = (3 \times 50)/(2 \times 50)$$

and

$$t/c = (2 \times 50)/(3 \times 50)$$

"Reducing" a Fraction or Ratio

Any fraction or ratio can be expressed in countless ways, but one form is considered the most "elegant." That's the form in which the absolute values of the numerator and denominator are both as small as possible, and the denominator is positive. A fraction or ratio in this form is said to be in *lowest terms* or *lowest form.*

Negative denominators

A fraction with a negative denominator is okay in theory, but it's hard to think about. You can probably imagine "negative three-fifths" without much trouble, but how about "three negative fifths"? That's tough for almost everybody. Why bother with such ugly fractions? Both of the fractions or ratios $-3/5$ and $3/(-5)$ have the same numerical value. Why not use the one that makes more sense?

When you see a fraction or ratio with a negative denominator, you can multiply both the numerator and the denominator by -1. That will turn the denominator positive while multiplying the value of the entire fraction by $-1/(-1)$. Of course, $-1/(-1)$ is equal to 1, which is the multiplicative identity element. So you end up with the same number in a form that is easier to comprehend.

Finding common factors

When a fraction is not in lowest terms, it means that the numerator and denominator can both be divided by at least one integer, called a *common factor*, and no remainder will be left in either case. For example, 6/10 is not in lowest terms. We can divide both the numerator and denominator by 2 and get 3/5:

$$(6/2)/(10/2) = 3/5$$

If we start with $-6/(-10)$, we can divide both the numerator and the denominator by -2 and get 3/5 again:

$$[-6/(-2)]/[-10/(-2)] = 3/5$$

If we divide both the numerator and the denominator of a fraction by the same integer and get integers in both places, we have the same fraction in a *lower form.*

Getting the lowest form

Even after we have reduced a fraction to a lower form, we might not have it in lowest terms. We can always get the lowest form if we are willing to go through a rather tedious process. A mathematician might call this the "brute-force approach." It isn't elegant but it always works, and we don't need any intuition to grind it out.

We start by factoring both the numerator and denominator into products of primes. If the original numerator is negative, we attach an extra "factor" of −1 to its product of primes, making sure all the prime factors are positive. We do the same thing with the denominator if it is negative. Once we have factored both the numerator and the denominator into products of primes, we look at those products closely. If the same prime appears in both the numerator and the denominator, then that prime is a *common prime factor*. We remove all the common prime factors from both the numerator and denominator. That leaves us with a smaller product of primes in the numerator, and a completely different product of primes in the denominator. We multiply all the factors in the numerator together, and do the same thing with the factors in the denominator. If we end up with a negative denominator, we multiply both the numerator and the denominator by −1.

Let's reduce the fraction 210/(−390) to lowest terms according to this set of rules. First, we convert both the numerator and the denominator into products of primes, and attach an extra "factor" of −1 to the denominator. The numerator then becomes

$$210 = 2 \times 3 \times 5 \times 7$$

and the denominator becomes

$$390 = -1 \times 2 \times 3 \times 5 \times 13$$

Next, we use these products to build a fraction in which both the numerator and the denominator consist of prime factors, and the denominator has the extra "factor" −1:

$$(2 \times 3 \times 5 \times 7) \,/\, (-1 \times 2 \times 3 \times 5 \times 13)$$

The common prime factors are 2, 3, and 5. We remove these from both the numerator and the denominator, getting

$$7/(-1 \times 13)$$

That's 7/(−13). We finish up by multiplying both the numerator and the denominator by −1 to obtain −7/13. This is the lowest form.

You can check to see that −7/13 = 210/(−390) by dividing 210 by −7 and −390 by 13. You'll get the same integer, −30, in either case.

Are you confused?

When you try to factor the numerator and denominator of a fraction into products of primes, you might find that one or the other is already prime or is the negative of a prime. Maybe both are like that! This means the original fraction is in lowest form, except when the denominator is negative. If the denominator is negative, you simply change both the numerator and the denominator to their additive inverses.

Here's a challenge!

Suppose that *a, b, c, d, e,* and *f* are all primes, and no two of them are the same. Reduce the following ratio to its lowest form:

$$-ab^3cde/(-ab^2ce^3f)$$

An exponent of 2 after any factor means that the factor appears twice. An exponent of 3 after any factor means that the factor appears three times. So, for example, b^2 means *bb,* and b^3 means *bbb.*

Solution

This problem is "messy" but not difficult. Let's expand the numerator and the denominator out so we don't have any exponents, and let's put small elevated dots (they can represent multiplication, remember!) between the factors to make them easy to tell apart. Let's write the numerator out in full, scratch a long dashed line underneath it, and then below that, write the denominator out in full. We get

$$-1 \cdot a \cdot b \cdot b \cdot b \cdot c \cdot d \cdot e$$

$$-1 \cdot a \cdot b \cdot b \cdot c \cdot e \cdot e \cdot e \cdot f$$

Now it's easy to see which factors are duplicated in the numerator and the denominator. They are -1, *a*, *b* (twice), *c*, and *e*. When we remove these factors from the expressions above and below the dashed line, we get

$$b \cdot d$$

$$e \cdot e \cdot f$$

This can be written as bd/e^2f. The parentheses in the denominator are no longer needed because the minus sign is gone. This ratio is in lowest form because both the numerator and the denominator are products of primes, and none of the primes in the numerator and denominator are duplicates.

Multiplying and Dividing Fractions

When you want to multiply two fractions, the process is simple. Dividing one fraction by another is a little more complicated, but not much. Let's look at how the processes work, using variables rather than numerical examples.

A fraction times a fraction

To multiply a fraction by another fraction, first be sure they are both "pure fractions." That means they should both consist of an integer divided by a positive integer. The absolute value of the numerator doesn't have to be smaller than the absolute value of the denominator, but one fraction should be of the form a/b and the other should be of the form c/d, where *a* and *c* are integers (positive, negative, or 0), and *b* and *d* are positive integers. If either fraction has a negative denominator, you know what to do!

Once you've "prepped" the fractions and you want to multiply them together, multiply the individual numerators to get the numerator of the product, and multiply the individual denominators to get the denominator of the product:

$$(a/b)(c/d) = ac/bd$$

If either of the original fractions is not in lowest terms, the product won't be either. You can reduce the product to lowest terms after you've multiplied two fractions, but you don't necessarily have to. If the multiplication is part of a multiple-step calculation process, you might as well wait until the entire process is complete before you worry about lowest terms.

The reciprocal of a fraction

The *reciprocal* of an integer, also called the *multiplicative inverse,* is the quantity by which you must multiply the original integer to get 1. The reciprocal of any integer is equal to 1 divided by that integer. That means 0 has no reciprocal, 1 is its own reciprocal, and −1 is also its own reciprocal. The reciprocal of every other integer lies somewhere between (but not including) 0 and 1, or else somewhere between (but not including) −1 and 0.

Suppose you have a fraction or ratio a/b, where a and b are integers and neither of them is equal to 0. Then the reciprocal of a/b is equal to b/a. It's easy to see why this is true when you multiply a/b times b/a. To do that, multiply the numerators and the denominators:

$$(a/b)(b/a) = ab/ba$$

The commutative law for multiplication can be used to switch around the factors in the denominator in the right-hand side of this equation, so you get

$$(a/b)(b/a) = ab/ab$$

Any nonzero quantity divided by itself is equal to 1. You have already been assured that $a \neq 0$ and $b \neq 0$, so you know that $ab \neq 0$. Therefore

$$(a/b)(b/a) = ab/ab = 1$$

This shows that the reciprocal of a/b is equal to b/a.

A fraction divided by a fraction

When you want to divide a fraction by another fraction, you can find the reciprocal of the second fraction (the divisor) and then multiply the first fraction (the dividend) by it. Suppose a is an integer, c is a nonzero integer, and b and d are positive integers. If you want to divide the quantity a/b by the quantity c/d, you can do it like this:

$$\begin{aligned}(a/b)/(c/d) &= (a/b)(d/c)\\ &= ad/bc\end{aligned}$$

The original expression in this equation, $(a/b)/(c/d)$, is called a *compound fraction.* It gets that name from the fact that it's a fraction made of other fractions! You can also think of it as a ratio of ratios.

Table 6-1. Derivation of a formula for repeated division of fractions. As you read down the left-hand column, each statement is equal to all the statements above it.

Statements	Reasons
$[(a/b)/(c/d)]/(e/f)$	Begin here
$(ad/bc)/(e/f)$	Apply the formula for division of a/b by c/d
$(g/h)/(e/f)$	Temporarily let $ad = g$ and $bc = h$, and substitute the new names in the previous expression
gf/he	Apply the formula for division of g/h by e/f
adf/bce	Substitute ad for g and bc for h in the previous expression

Are you confused?

Any ratio of two integers, where the denominator is nonzero, is known as a *rational number*. The set of rational numbers, symbolized R, contains all such integer ratios that can exist.

It's always possible to get any ratio of this kind into lowest terms. If you have some rational number r, you can always convert it to the form m/p, where m is an integer and p is a positive integer, with the resulting fraction is in lowest terms. The term *rational* in this context comes from the word "ratio."

If you stumble across the ratio $(-6)/(-7)$, for example, you can write it as $6/7$ and it represents the same number. If you see $6/(-7)$, you can rewrite it as $-6/7$.

Here's a challenge!

Let a, b, c, d, e, and f all be nonzero integers. Suppose you start with a/b and divide it by c/d, and then divide that result by e/f. Write an expression for the final quotient.

Solution

Table 6-1 shows how this can be done, in the form of an S/R derivation.

Adding and Subtracting Fractions

When you want to add or subtract two integers, the process is straightforward. Fractions are more involved. You've probably had plenty of practice adding and subtracting fractions in arithmetic courses. Let's look at these problems from a point of view a little closer to algebra, using variables instead of specific numerical examples.

Getting a common denominator

When you have two fractions that you want to add or subtract, you should be sure that neither fraction has a negative denominator. If one of them does, convert it into the equivalent form that has a positive denominator. Then you can modify the fractions so they have the same denominator, called the *common denominator*.

Suppose you have two fractions a/b and c/d in which a and c are integers, and b and d are positive integers. You want to add them, so you write

$$a/b + c/d$$

Now multiply a/b by d/d, which is equal to 1. Then multiply c/d by b/b, which is also equal to 1. (Remember, anything multiplied by 1 is equal to itself.) The above expression turns into this:

$$(a/b)(d/d) + (c/d)(b/b)$$

Next, multiply the products of the fractions on each side of the plus sign, getting

$$ad/bd + cb/db$$

Now apply the commutative law for multiplication to the numerator cb and the denominator db, morphing the above expression into

$$ad/bd + bc/bd$$

This produces a sum of two fractions with the common denominator bd. These two fractions have the same numerical values as the original ones, but they are in "higher terms."

A fraction plus another fraction

Once you've found a common denominator for a sum of two fractions, adding the fractions is easy. Simply add the numerators, and put them over the common denominator. In the above situation, then

$$ad/bd + bc/bd = (ad + bc)/bd$$

An equation such as this, which describes a general solution to a math problem, is called a *formula*. This particular formula is worth "homogenizing" into your brain! If a and c are integers, and b and d are positive integers, then

$$a/b + c/d = (ad + bc)/(bd)$$

If you'd rather see it in words, the process goes like this:

- Multiply the numerator of the first fraction by the denominator of the second.
- Multiply the denominator of the first fraction by the numerator of the second.
- Add these two products together.
- Divide this sum by the product of the denominators.

Are you confused?

When you add two fractions using the above method, the result might not be in lowest terms. As a final step in the addition of two fractions, you can reduce the result to lowest terms. But you don't always have to.

By now you must wonder, "Does it matter whether or not a fraction is in lowest terms?" The answer is, "It depends." If you want to express a fraction in the simplest or most "elegant" possible way, you should reduce it. But you will sometimes come across situations where it's better to put a fraction, ratio, or proportion in a form other than lowest terms.

Suppose you want to describe how many people in Country X have green eyes, *per 100,000 population.* You might discover that one out of every 10 people has green eyes. You could say that 1/10 of the people have green eyes, or the ratio of people with green eyes to all the people in Country X is 1:10. But if you use 100,000 population as a basis, you'll have to say that the proportion of people with green eyes is 10,000 per 100,000 population in Country X. That's not even close to the lowest form, but it conveys the intended meaning better.

In pure theory, it doesn't matter if a fraction, ratio, or proportion is in lowest terms or not. These days, nearly everyone uses computers in complicated calculations, and the machines don't care about lowest terms. You can input the numbers as they are, and the computer will output the data in any form you want.

Here's a challenge!

Start with the general equation for adding two fractions:

$$a/b + c/d = (ad + bc)/(bd)$$

Based on this, and on the rules you already know, prove that

$$a/b - c/d = (ad - bc)/(bd)$$

as long as $b \neq 0$ and $d \neq 0$.

Solution

Table 6-2 shows how the subtraction formula is derived from the addition formula. Near the end of this proof parentheses and brackets aren't enough for grouping of an expression, so braces must be used! They look exactly the same as they braces you use to enclose lists of symbols representing set elements, but the purpose is different.

Table 6-2. Derivation of a general formula for the subtraction of one fraction from another, based on the formula for addition of fractions. As you read down the left-hand column, each statement is equal to all the statements above it.

Statements	Reasons
$a/b - c/d$	Begin here
$a/b + [-(c/d)]$	Convert subtraction to addition of a negative
$a/b + [-1(c/d)]$	Principle of the sign-changing element
$a/b + (-1/1)(c/d)$	"Divisive identity element": substitute $-1/1$ for -1
$a/b + (-1)c/1d$	Multiplication of fractions to right of plus sign
$a/b + (-1)c/d$	Multiplicative identity element: substitite d for $1d$
$[ad + b(-1)c]/bd$	Formula for addition of two fractions, considering $(-1)c$ as a single quantity
$[ad + (-1)bc]/bd$	Commutative law for multiplication
$[ad + (-1)(bc)]/bd$	Group elements b and c with parentheses
$\{ad + [-(bc)]\}/bd$	Principle of sign-changing element (the other way around)
$(ad - bc)/bd$	Convert addition of a negative to subtraction
Q.E.D.	Mission accomplished

Practice Exercises

This is an open-book quiz. You may (and should) refer to the text as you solve these problems. Don't hurry! You'll find worked-out answers in App. A. The solutions in the appendix may not represent the only way a problem can be figured out. If you think you can solve a particular problem in a quicker or better way than you see there, by all means try it!

1. The highest wind speed in a "category 1" hurricane is 95 miles per hour. The highest wind speed in a "category 4" hurricane is 155 miles per hour. What is the ratio of these wind speeds, expressed in lowest terms between two integers?
2. On the absolute temperature scale, the coldest possible temperature is 0, defined in units called *kelvins*. On this scale, pure water at sea level freezes at 273 kelvins and boils at 373 kelvins (to the nearest kelvin). Based on this information, what is the ratio of the absolute temperature of the boiling point to the absolute temperature of the freezing point, expressed in lowest terms between two integers?
3. Is the fraction 231/230 in lowest terms? If not, reduce it to lowest terms.
4. Is the fraction −154/165 in lowest terms? If not, reduce it to lowest terms.
5. If two fractions are in lowest terms and they are multiplied by each other, is the product always in lowest terms? If so, prove it. If not, provide an example. If the product is sometimes in lowest terms but not always, provide examples of both situations.
6. If two fractions are in lowest terms and one is divided by the other, is the quotient always in lowest terms? If so, prove it. If not, provide an example. If the quotient is sometimes in lowest terms but not always, provide examples of both situations.
7. In one of the "challenge" problems, you found a general expression for

   $$[(a/b)/(c/d)]/(e/f)$$

   when a, b, c, d, e, and f are all nonzero integers. You divided a/b by c/d, and then divided the result by e/f. Now find a general expression for

   $$(a/b)/[(c/d)/(e/f)]$$

   Once you've done this, you'll know what happens when you divide c/d by e/f first, and then divide a/b by the result. Compare this with the formula you got when you solved the "challenge" problem.
8. Imagine that you have a fraction of the form a/b and another of the form c/d, where a and c are integers, and b and d are positive integers. Show that the commutative law works for multiplication of these fractions, based on your knowledge of the commutative law for integers.
9. Imagine that you have fraction of the form a/b, another of the form c/d, and a third of the form e/f, where a, c, and e are integers, and b, d, and f are positive integers. Show that the associative law works for multiplication of these fractions, based on your knowledge of the associative law for integers.
10. Give at least one example of a situation in which the following equation is true:

    $$(a/b)/(c/d) = (c/d)/(a/b)$$

    where a, b, c, and d are integers. Don't use any of the trivial cases where all four of the integers have absolute values of 1.

CHAPTER

7

Decimal Fractions

Now that you know how a rational number can be expressed as a ratio of two integers, let's look at the other common way these numbers are symbolized. Since the middle of the twentieth century, calculators and computers have replaced "manual" methods of calculation. These machines use *decimal fractions*.

Powers of 10

A *positive-integer power* is a quantity multiplied by itself a certain number of times. If a nonzero quantity is divided by itself once, it is said to be "raised" to the *zeroth power*, and the result is always 1. A *negative-integer power* is a nonzero quantity divided by itself more than once. Powers are denoted by exponents. Decimal notation is based on integer powers of 10. The number 10 is called the *exponential base*. It can also be called simply the *base* or the *radix*.

Orders of magnitude

Figure 7-1 is a number line showing the powers of 10, from 10^5 (the largest number in the illustration) down to 10^{-5} (the smallest). Each multiple of 10 is called an *order of magnitude*. For example, 10^3 is one order of magnitude larger than 10^2, and 10^{-2} is three orders of magnitude larger than 10^{-5}.

This number line differs from the ones you've seen so far. All the values here are positive. As you go upward on the line, the numerical value increases faster and faster, so you race off toward "infinity" more rapidly than you do on a conventional number line. As you go downward, the value decreases at a slower and slower rate, "closing in" on 0 but never quite getting there.

If you expand on this idea, you can "build" any positive rational number by taking single-digit multiples of powers of 10, and adding them up. Every positive number in this form has its negative "twin." To show the negative rational numbers, you can make up a separate number line for them. In order to account for 0, you can give it a special point that isn't on either line. Figure 7-2 portrays all the rational numbers using this system. "Infinity" and "negative infinity" are not entitled to points here, because neither of them is a rational number!

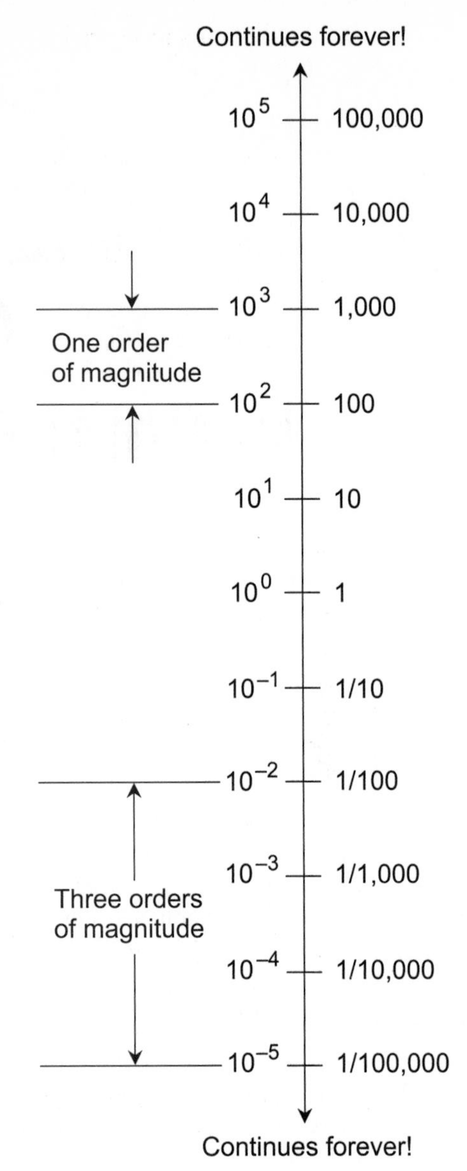

Figure 7-1 This is part of the positive rational number line, showing values from 100,000 down to 1/100,000. The powers of 10 range from 5 down to -5.

Start at the decimal point

The "cornerstone" on which any decimal numeral is "built" is the *decimal point.* It looks like an ordinary period (.). Digits for positive powers of 10 are written to the left of the decimal point. Digits for negative powers of 10 are written to the right of the point.

Consider the decimal numeral 362.7735. Let's break it down. Starting at the decimal point and working toward the left:

$$(2 \times 10^0) + (6 \times 10^1) + (3 \times 10^2) = 2 + 60 + 300 = 362$$

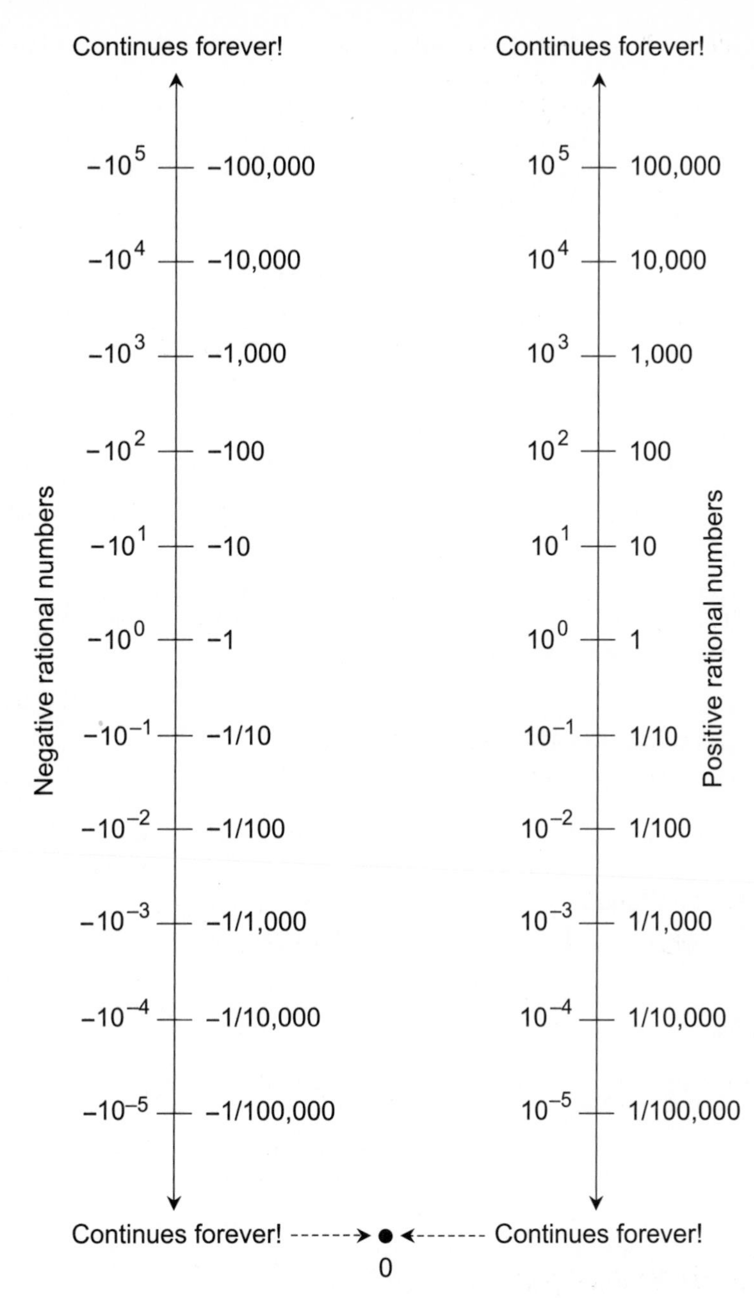

Figure 7-2 The entire set of rational numbers can be portrayed in power-of-10 form as two lines, one for the positive values and the other for the negative values, with a separate point for 0.

Starting at the decimal point and working toward the right:

$$(7 \times 10^{-1}) + (7 \times 10^{-2}) + (3 \times 10^{-3}) + (5 \times 10^{-4})$$
$$= 7/10 + 7/100 + 3/1{,}000 + 5/10{,}000 = 0.7735$$

When we add the whole number to the decimal fraction, we get

$$362 + 0.7735 = 362.7735$$

Are you confused?

You will sometimes hear scientists talk rather loosely about orders of magnitude. They mean to say that the *absolute value* of one quantity is some power of 10 times bigger or smaller than the *absolute value* of the other quantity. Here are a few examples:

- 45,300 is one order of magnitude larger than 4,530
- 0.56 is two orders of magnitude smaller, in absolute terms, than −56
- −0.565 is three orders of magnitude smaller, in absolute terms, than 565
- −88,888 is four orders of magnitude larger, in absolute terms, than −8.8888

If one or both quantities is negative, you should be sure to include the phrase "in absolute terms" so there's no confusion about the meanings of "smaller" or "larger."

When you want to portray a decimal number greater than or equal to 0 but smaller than 1, it is customary to write a single numeral 0 to the left of the decimal point. If the number is greater than −1 but smaller than 0, you should write the minus sign first, then a single 0, and then the decimal point.

Here's a challenge!

Express the scheme for "building" a decimal numeral in general terms, rather than merely providing examples.

Solution

Imagine two sets of single-digit numerals, called set A and set B. Suppose A has m elements and B has n elements, named as follows:

$$A = \{a_1, a_2, a_3, \ldots, a_m\}$$

and

$$B = \{b_1, b_2, b_3, \ldots, b_n\}$$

Now imagine these single-digit numerals arranged around a decimal point like this:

$$a_m \ldots a_3\ a_2\ a_1 \,.\, b_1\ b_2\ b_3 \ldots b_n$$

The string of numerals to the left of the point represents the sum

$$S_A = (a_m \times 10^{m-1}) + \cdots + (a_3 \times 10^2) + (a_2 \times 10^1) + (a_1 \times 10^0)$$

The string of numerals to the right of the point represents the sum

$$S_B = (b_1 \times 10^{-1}) + (b_2 \times 10^{-2}) + (b_3 \times 10^{-3}) + \cdots + (b_n \times 10^{-n})$$

The entire number is represented by the sum $S = S_A + S_B$. Mathematicians like to use the nonitalic, uppercase Greek letter *sigma* (Σ) to represent sums of many numbers. We can get fancy and write the above expressions like this:

$$\Sigma_A = (a_m \times 10^{m-1}) + \cdots + (a_3 \times 10^2) + (a_2 \times 10^1) + (a_1 \times 10^0)$$

and

$$\Sigma_B = (b_1 \times 10^{-1}) + (b_2 \times 10^{-2}) + (b_3 \times 10^{-3}) + \cdots + (b_n \times 10^{-n})$$

The entire number is therefore the sum $\Sigma = \Sigma_A + \Sigma_B$. It's a good idea to remember this *summation symbol.* You'll come across it when you study higher algebra, precalculus, and calculus.

Terminating Decimals

All of the decimal expressions you'll see in everyday situations have a finite number of digits to the right of the decimal point. The most common example is the notation for dollars and cents, where there are always two digits to the right of the decimal point. Any further digits, if you want to write them, are all ciphers. A numeral of this sort is called a *terminating decimal.*

Tenths, hundredths, or whatever

In a terminating decimal, the digits to the right of the decimal point always represent a fraction having a denominator that is some power of 10. If there's one digit, it represents 10ths; if there are two digits, they represent 100ths; if there are three digits, they represent 1,000ths; and so on, as far as you want to go. For example,

- 0.7 represents 7/10
- 0.72 represents 72/100
- 0.729 represents 729/1,000
- 0.7294 represents 7,294/10,000
- 0.72941 represents 72,941/100,000

When you have a denominator of 10^n, where n is a positive integer, it is the equivalent of multiplying by 10^{-n}. You can also write the above examples as

- 0.7 represents 7×10^{-1}
- 0.72 represents 72×10^{-2}
- 0.729 represents 729×10^{-3}
- 0.7294 represents $7{,}294 \times 10^{-4}$
- 0.72941 represents $72{,}941 \times 10^{-5}$

Don't be fooled here! These examples don't represent changes in the order of magnitude. Instead, they all represent numbers that are very close to each other. The values approach the last number in either list, which is 0.72941. If that's as far as you want to go, then adding any more digits to the right of the 1 will only clutter the page with *ciphers* (zeros). The following numerals all represent exactly the same number to a pure mathematician:

0.72941
0.729410
0.7294100
0.72941000
0.729410000
0.7294100000
↓
and so on, forever

Physicists or engineers see the above numbers differently. To them, those extra ciphers are important, because they represent increasing precision or accuracy. They're extra *significant figures*. Let's not worry about that right now.

Commas and extra ciphers

In any decimal expression, the number of digits to the left of the decimal point is always finite. If you "chop off" the digits to the right of the point, the digits to the left represent an integer. You might add ciphers to the left-hand end of the digit string without changing the value, but there's rarely any reason to do that. You won't often see a numeral like this:

00,000,004,580,103.7864892022

Instead, it would be written as

4,580,103.7864892022

In a decimal numeral, commas are not customarily inserted to the right of the point, no matter how many digits there are.

What about lowest terms?

When you see a decimal expression that ends after a certain number of digits to the right of the point, those digits always express a fraction with a denominator that is some power of 10. This fraction might be in lowest terms, but often it is not. For example, in the decimal 66.31, the fractional part is 31/100, which is in lowest terms because 31 is prime. However, in the decimal 66.35, the fractional part is 35/100. If reduced to lowest terms, that would be 7/20.

When you write or see a decimal expression, you shouldn't worry about reducing the fractional part to lowest terms unless the nature of the problem demands it. Those digits to the right of the point are always supposed to represent 10ths, 100ths, 1,000ths, and so on.

Are you confused?

Numbers such as 4,580,103.7864892022 can be hard to read, especially when you see them on a calculator display that does not insert the commas. You can insert spaces on either side of the point, and also after every third digit to the right of the decimal point. Those spaces will make the whole thing easier to read. The above numeral would then look like this:

4,580,103 **.** 786 489 202 2

Be careful when you insert spaces into a numeral! In this example, the lonely digit 2 at the end might confuse some people. Also, note that the spaces between these digits don't correspond to the places you'd put the commas if you were to express them as a fraction. (You'll see this in the "challenge" example below.)

Here's a challenge!

Break down 4,580,103.7864892022 into a sum of a single integer and a single fraction.

Solution

Let's look to the left of the point first. The string of numbers is one big integer:

4,580,103

Now let's look to the right of the point. There are 10 digits here. That means the denominator of the fraction should be denoted as 1 with 10 ciphers after it, producing the fraction

7,864,892,022 / 10,000,000,000

The entire number is the sum of these:

4,580,103 + 7,864,892,022 / 10,000,000,000

Endless Decimals

Whenever you write out a decimal expression in the "real world," it's always a terminating decimal. But in theory, the digits to the right of the decimal point can continue forever, so you can never reach a spot where every digit further to the right is a cipher. This always happens when you divide a prime number larger than 5 by any other prime larger than 5. It can happen in other cases, as well.

Endless repeating decimals

Your calculator can give you a glimpse of what an *endless repeating decimal,* also called a *nonterminating repeating decimal,* looks like. The calculator program in a personal computer is excellent for this purpose, because it displays a lot of digits.

A good way to see the difference between a terminating decimal and an endless repeating decimal is to use the "$1/x$" key on your calculator and start with 2 for x. Then try it with 3, 4, 5, and so on, watching the results:

$$1/2 = 0.5$$
$$1/3 = 0.333333333333...$$
$$1/4 = 0.25$$
$$1/5 = 0.2$$
$$1/6 = 0.166666666666...$$
$$1/7 = 0.142857142857...$$
$$1/8 = 0.125$$
$$1/9 = 0.111111111111...$$

The fractions 1/2, 1/4, 1/5, and 1/8 all work out as terminating decimals. The fractions 1/3, 1/6, 1/7, and 1/9 divide out as endless repeating decimals. The presence of an ellipsis (three periods) indicates that the pattern continues forever. Note the uniqueness of 1/7, which goes through a repeating cycle of the six digits 142857.

Are you confused?

When you divide an integer by another integer, and if the two integers are large enough, your calculator display might not show enough digits to let you see the pattern of repetition. Take a calculator that can show 10 digits, and divide out this fraction:

$$138{,}297{,}004{,}792/999{,}999{,}999{,}999$$

Now suppose you show your calculator display to a friend, tell her it's the quotient of two integers you entered, and then ask her what fraction you put in. Even if she has a Ph.D. in math, she will not be able to figure it out. The pattern here is too big for the display.

Once in a while you'll come across a situation where an integer is divided by another integer and you can't see the pattern in the decimal expression because the repeating sequence has too many digits. But there is always a pattern whether you can see it or not. That's because *any* rational number can be expressed as either a terminating decimal or an endless repeating decimal.

Endless nonrepeating decimals

You might wonder whether there are any decimal numbers that go on forever with digits to the right of the decimal point, but that don't produce a repeating sequence. The answer is "Yes." Examples are easy to find.

Consider the circumference of a perfect circle divided by its diameter. This value is always the same, no matter how large or small the circle happens to be. In ancient times, people knew that the circumference of a circle is slightly more than 3 times its diameter. They tried to define it as a fractional ratio—that is, as a quotient of two integers—but the best they could do was to come close. For a long time they thought it was 22/7. Eventually, mathematicians were able to prove that this number, which can be so easily defined in terms of geometry, *cannot* be defined as a ratio between two integers!

If you've taken any geometry, you know that the circumference of a circle divided by its diameter is symbolized by the small Greek letter *pi* (π). Many calculators have a key you can punch to get π straightaway. You'll never find any repeating pattern of digits in the decimal expansion of π. Even if you spend the rest of your life trying, you will fail. That's because π is not a rational number. Mathematicians call quantities such as π *irrational numbers.* You'll learn more about them in Chap. 9.

Here's a challenge!

Imagine a decimal expression that has an endlessly repeating triplet of digits. We can write it down in this form:

0 . ### ### ### ...

where ### represents the sequence of three digits that repeats. The spaces on either side of the decimal point, and after each triplet of pound signs, are inserted to make the expression clear. Our mission is to show that this decimal numeral represents the fraction

###/999

Solution

Let's call the "mystery fraction" *m.* Our task is to find a fractional expression for *m.* This process is straightforward, but it takes several steps. Follow along closely, and you shouldn't have any trouble understanding how it works. We've been told that

m = 0 . ### ### ### ...

We can break this into a sum of two decimal expressions, one terminating and the other endless, like this:

m = (0 . ###) + (0 . 000 ### ### ### ...)

Note that the first addend here is ###/1,000. The second addend happens to be the original mystery number, *m,* divided by 1,000. Therefore,

m = (###/1,000) + (*m*/1,000)

The two fractions on the right-hand side of the equals sign have a common denominator, so they're easy to add. We get

m = (### + *m*)/1,000

Now we can multiply each side of this equation by 1,000 and then manipulate the right-hand side, getting

1,000 *m* = 1,000 (### + *m*)/1,000
= (### + *m*) (1,000/1,000)
= ### + *m*

Therefore,

$$1{,}000\ m = \#\#\# + m$$

Subtract m from the expressions on both sides of the equals sign, obtaining

$$1{,}000m - m = \#\#\# + m - m$$

This simplifies to

$$999\ m = \#\#\#$$

Finally, we divide each side by 999, getting

$$m = \#\#\#/999$$

Mission accomplished! *Q.E.D.*

Conversions

Every rational number can be expressed in two ways: the *ratio form* as an integer divided by another integer, and the *decimal form* as a string of digits with a decimal point somewhere. If you have a rational number in one form, you can always convert it to the other.

Ratio to decimal

When you see a ratio of integers, you can convert it to decimal form if you have a calculator that can display enough digits. But that's the catch! Even a good calculator can fall short in this respect. If you have a calculator with a 10-digit display and you divide 1 by 7, you will not even see two full repetitions of the pattern. If you didn't know better from having seen the decimal expansion of 1/7 earlier in this chapter, you might not be able to deduce it from a 10-digit calculator alone. If you have a good computer calculator program, you're better off. But even the best calculators can be overwhelmed if you give them a "bad" enough ratio. Try 51/29, for example!

Fortunately, you won't have to perform ratio-to-decimal conversions very often. When you come across a problem where you have to do it, the calculator program in any good personal computer will usually work. In the extreme, you can always resort to old-fashioned, manual long division. You can also write, or find, a computer program to grind out thousands of digits and look for patterns.

Terminating decimal to ratio

When you see a terminating decimal expression and you want to convert it to a ratio of integers, you can do it in steps. Here's an example. Imagine that you are given this decimal numeral and are told to put it into ratio form as a quotient of two integers:

$$3{,}588\,.\,7601811$$

The extra spaces on either side of the decimal point are there to make it easy to distinguish the digit string to its left from the digit string to its right.

First, take the part of the decimal expression to the right of the point. Put those digits into the numerator of a fraction. Then count the number of digits in the string. Suppose that number is n. In this case, you have 7601811. That's a string of seven digits, so $n = 7$. In the denominator of the fraction, write a 1 and then n ciphers. The fractional part is therefore

$$7{,}601{,}811/10{,}000{,}000$$

Second, take the part of the decimal expression to the left of the point. Put those digits into the numerator of a new fraction. In the denominator, put 1. In this case, the result is

$$3{,}588/1$$

The third part of the process is a little tricky! Add n ciphers after the 1 in the denominator you just put down, so that denominator is identical to the denominator you "built" for the decimal part of the expression. Then also add n ciphers in the numerator for the whole-number part. When you do this, you multiply the whole-number part of the expression by a certain number and then divide that number by itself. That's just a fancy (or maybe you'd rather say messy) way of multiplying by 1. In this case you get

$$35{,}880{,}000{,}000/10{,}000{,}000$$

Fourth, add the fraction you "built" for the whole-number part of the decimal expression to the fraction you "built" for the decimal part. This should be easy, because you have engineered things to get a common denominator! In this case, it's 10,000,000. Adding the numerators produces

$$35{,}880{,}000{,}000 + 7{,}601{,}811 = 35{,}887{,}601{,}811$$

That's the numerator of the ratio you want. The denominator is 10,000,000, so the complete ratio is

$$35{,}887{,}601{,}811/10{,}000{,}000$$

If you'd like to check this, divide the ratio out on a calculator that can display at least 11 digits. You should get the original decimal expression.

Endless repeating decimal to ratio

The solution to the "challenge" problem in the last section should give you an idea of how to convert any endlessly repeating decimal to a ratio of two integers. You can generalize on the number of digits in the repeating pattern, from one up to as many as you want.

When you encounter a decimal expression that has a sequence of digits that repeats without end, first split the whole-number part from the decimal part. Call the whole-number part a.

Then write down the part of the expression to the right of the decimal point in this form, with the point on the extreme left:

$$. b_1 \ b_2 \ b_3 \ ... \ b_k \ b_1 \ b_2 \ b_3 \ ... \ b_k \ b_1 \ b_2 \ b_3 \ ... \ b_k$$

where $b_1 \ b_2 \ b_3 \ ... \ b_k$ represents the sequence of k digits that repeats. (Each b with a subscript represents a single digit.) The extra spaces after the decimal point, and between each digit, are there to make the expression easy to read. The fractional part of the expression is

$$b_1 \ b_2 \ b_3 \ ... \ b_k/999 \ ... \ 999$$

where the denominator has k digits, all 9s. Now you can put back the whole-number part, getting the number in this form:

$$a\text{-}b_1 \ b_2 \ b_3 \ ... \ b_k/999 \ ... \ 999$$

Here, the dash after the a is there only to separate the whole-number part of the expression from the fractional part. It is *not* a minus sign!

Now convert a to a fraction with a denominator consisting of k digits, all 9s. All you have to do is multiply a by the number 999 ... 999, put the result into the numerator of a fraction, and then put 999 ... 999 in the denominator, getting

$$999 \ ... \ 999 \times a/999 \ ... \ 999$$

Add this to the fraction you got by converting the decimal part of the original expression. That gives you

$$(999 \ ... \ 999 \times a/999 \ ... \ 999) + (b_1 \ b_2 \ b_3 \ ... \ b_k/999 \ ... \ 999)$$

You have a common denominator now, so you can easily add to get

$$[(999 \ ... \ 999 \times a) + (b_1 \ b_2 \ b_3 \ ... \ b_k)]/999 \ ... \ 999$$

Remember that the expression 999 ... 999 always stands for a sequence of k digits, all 9s.

Are you confused?

The notation shown above is messy, and it's easy to "get lost." Try reading it over a few times and it should become clearer to you. A specific example, showing the process in "real life action," can help. Let's convert the following expression to a ratio of integers:

$$23 \ . \ 860486048604 \ ...$$

Again, the extra spaces on either side of the decimal point are there only to make it easy to distinguish between the whole-number part of the expression and the decimal part.

The decimal portion is a sequence of the digits 8, 6, 0, and 4 that endlessly repeats. We can tell right away that this is 8,604/9,999.

The whole-number portion, 23, can be multiplied by 9,999, and the result put into the numerator of a fraction. The denominator should then be 9,999, so we get

$$(23 \times 9{,}999)/9{,}999 = 229{,}977/9{,}999$$

That's just the whole number 23 expanded into 9,999ths. Now we add the decimal part back in, so the entire number becomes

$$\begin{aligned} 229{,}977/9{,}999 + 8{,}604/9{,}999 &= (229{,}977 + 8{,}604)/9{,}999 \\ &= 238{,}581/9{,}999 \end{aligned}$$

If you use a calculator that can display a lot of digits to divide out this fraction, you should get the original expression: 23 followed by a decimal point, and then the sequence of digits 8, 6, 0, and 4 repeating.

Here's a challenge!

There's a less formal, but much quicker, way to do the decimal-to-ratio conversion described in the section "Terminating decimal to ratio" earlier in this chapter. How does it work?

Solution

For reference, here is the original decimal expression again, with extra spaces on either side of the point for easy reading:

$$3{,}588\ .\ 7601811$$

Move the point to the right until it's at the end of the string of digits, leaving nothing beyond. Then delete the point. You'll get the whole number

$$35{,}887{,}601{,}811$$

Now make this the numerator of a fraction. Count the number of places you moved to the right to get the point to the end of the string of digits. (In this case, it's seven places.) Then in the denominator of the fraction, write down a 1 followed by that number of ciphers. The result:

$$35{,}887{,}601{,}811/10{,}000{,}000$$

You can apply this method to any decimal expression you'll ever see. The end result might not be in lowest terms, but you can reduce it to lowest terms if you want.

Practice Exercises

This is an open-book quiz. You may (and should) refer to the text as you solve these problems. Don't hurry! You'll find worked-out answers in App. A. The solutions in the appendix may not represent the only way a problem can be figured out. If you think you can solve a particular problem in a quicker or better way than you see there, by all means try it!

1. Draw a number line in power-of-10 style that shows the rational numbers from 10 to 100,000. How many orders of magnitude is this?
2. Draw a number line in power-of-10 style that shows the rational numbers from 30 to 300,000. How many orders of magnitude is this?
3. How many orders of magnitude larger than 330 is 75,000,000? Here's a hint: Express the answer by saying "75,000,000 is between n and $n + 1$ orders of magnitude larger than 330," where n is a whole number.
4. Write the following decimal expressions as combinations of integers and fractions. Reduce the fractional part to lowest terms.
   (a) 4.7
   (b) −8.35
   (c) 0.02
   (d) −0.29
5. Express the numbers from the solutions to Prob. 4 as ratios of integers, with the denominator always positive.
6. Write the following ratios as decimal expressions.
   (a) 44/16
   (b) −81/27
   (c) 51/13
   (d) −45/800
7. Convert the fraction 1/17 to another fraction whose denominator is a string of 9s.
8. Convert the expression 2.892892892 ... to a ratio of integers.
9. Suppose that somebody tells us there are two integers so large that it would take a person billions of years to write either of them out by hand. We are also told that both of these integers are prime numbers, and that the decimal expansion of their ratio (that is, one of these primes divided by the other) is an endless sequence of digits. Is there a repeating pattern to the digits in the decimal expansion?
10. Imagine that we come across a gigantic string of digits—miles long if we try to write it out—and we can't see any pattern. We use a computer to examine this number to a thousand decimal places, then a million, then a billion, and we still can't find a pattern. Can we ever know for sure whether or not a pattern actually exists, so we can decide whether or not the number is rational? Here's a hint: This exercise is meant to force your imagination into overdrive!

CHAPTER

8

Powers and Roots

Now it's time to review and expand your knowledge of *powers* and *roots*. When you take a number to an integer power, it's like repeated multiplication. When you take a number to an integer root, it's like repeated division. But powers and roots go deeper than that! With a few exceptions, you can raise anything to a rational-number power and get a meaningful result.

Integer Powers

The simplest *powers*, also called *exponential operations*, involve multiplying a number or quantity by itself a certain number of times. The power is written as a superscript after the quantity to be "operated on." This operation is sometimes called *raising to a power*.

Positive integer powers

If a is any number and p is a positive integer, the expression a^p means a to the pth power, which is a multiplied by itself p times. More generally, a doesn't have to be a number. It can be a variable or a complicated expression containing numbers and variables. Here are some examples of quantities raised to positive integer powers:

$$4^2$$

$$x^4$$

$$(k+4)^7$$

$$(abc)^4$$

$$(m/n)^{12} \qquad \text{where } n \neq 0$$

$$(x^2 - 2x + 1)^5$$

Note that in the last expression, the quantity raised to the 5th power actually contains a variable raised to a different power.

The 0th power

By convention, anything raised to the 0th power is equal to 1. Anything except 0 itself, that is! The quantity 0^0 is not defined. You'll see why any nonzero quantity raised to the 0th power is equal to 1 later in this chapter. You'll also see why 0^0 is not defined. Here are some expressions that are all equal to 1:

$$4^0$$
$$x^0 \quad \text{where } x \neq 0$$
$$(k+4)^0 \quad \text{where } k \neq -4$$
$$(abc)^0 \quad \text{where } a \neq 0,\ b \neq 0, \text{ and } c \neq 0$$
$$(m/n)^0 \quad \text{where } m \neq 0 \text{ and } n \neq 0$$
$$(x^2 - 2x + 1)^0 \quad \text{where } x \neq 1$$

In every expression except the topmost one, there are constraints on the variables in the quantities being raised to the 0th power. These keep the values of the quantities from being equal to 0. If you're not sure about the reason for the constraint on x in the last expression, hold on a minute and you'll see.

Whenever you find any expression containing variables, and that expression is to be raised to the 0th power, be sure you never let that expression attain a value of 0. This is especially important if such an oversight is a step in solving a problem! You would be throwing an undefined quantity into a sensitive process. You've heard what computer programmers say about putting nonsense into a machine! The same thing happens in mathematics.

Negative integer powers

If a is any number and n is a negative integer, the expression a^n means the reciprocal of the quantity a raised to the power of $|n|$. For example,

$$a^{-1} = 1/(a^1) = 1/a$$
$$a^{-3} = 1/(a^3)$$
$$a^{-20} = 1/(a^{20})$$

As before, a can be almost any expression you can imagine. Note that the -1st power of any quantity is the same thing as its reciprocal (multiplicative inverse). You'll often see this notation used because it can be a lot less "messy" than writing 1, then a slash, and then a complicated expression. Here are some examples of numbers or quantities raised to negative integer powers:

$$4^{-2}$$
$$x^{-4} \quad \text{where } x \neq 0$$
$$(k+4)^{-7} \quad \text{where } k \neq -4$$
$$(abc)^{-4} \quad \text{where } a \neq 0,\ b \neq 0, \text{ and } c \neq 0$$
$$(m/n)^{-12} \quad \text{where } m \neq 0 \text{ and } n \neq 0$$
$$(x^2 - 2x + 1)^{-5} \quad \text{where } x \neq 1$$

The constraints are imposed, as with the 0th power, to keep the "powerized" quantities from being equal to 0. If you take a negative integer power of 0, you end up with 1/0, and that's not defined. For example, $0^{-7} = 1/(0^7) = 1/0$.

Are you confused?

You should be able to see, without much trouble, how the second, third, fourth, and fifth quantities above can be equal to 0. Here they are:

$$x = 0 \qquad \text{when } x = 0, \text{ of course!}$$
$$(k + 4) = 0 \qquad \text{when } k = -4$$
$$(abc) = 0 \qquad \text{when } a = 0,\ b = 0, \text{ or } c = 0$$
$$(m/n) = 0 \qquad \text{when } m = 0$$

But what about the sixth and last expression? It's not immediately clear that

$$(x^2 - 2x + 1) = 0 \qquad \text{when } x = 1$$

You can "plug in" the value 1 for x and see that you get 0 when you add everything up. But would you have "plugged in" 1 at random, thinking it might cause the whole expression to equal 0? Probably not!

When you deliberately allow this whole expression to be equal to 0, you get something called a quadratic equation. Such equations can be solved in various ways. It turns out that there is one solution to the equation

$$(x^2 - 2x + 1) = 0$$

That happens to be $x = 1$. Don't worry about how this type of equation can be solved right now. You'll learn how to do it later in this book. For the moment, pay attention to the important message: Beware of taking 0 to the 0th power, and beware of dividing by 0! Don't let these things happen, even accidentally.

Here's a challenge!

What happens if you start with 2 and raise it to successively higher positive integer powers? What happens if you raise 2 to integer powers that get larger and larger negatively?

Solution

If you start with 2 and raise it to positive integer powers, you get

$$2^1 = 2$$
$$2^2 = 2 \times 2 = 4$$
$$2^3 = 2 \times 2 \times 2 = 8$$
$$2^4 = 2 \times 2 \times 2 \times 2 = 16$$
$$\downarrow$$
and so on, forever

The value keeps getting larger without limit, doubling every time. The sequence of values is said to *diverge.* In this particular example, it "approaches infinity." But if you start with 2 and raise it to integer powers that get larger and larger negatively, this is what happens:

$$2^{-1} = 1/2$$
$$2^{-2} = 1 / (2 \times 2) = 1/4$$
$$2^{-3} = 1 / (2 \times 2 \times 2) = 1/8$$
$$2^{-4} = 1 / (2 \times 2 \times 2 \times 2) = 1/16$$
$$\downarrow$$

and so on, forever

The value keeps getting smaller and smaller, becoming half its former size every time you decrease the integer power by 1, but always remaining positive. This sequence of values is said to *converge.* In this case it "approaches 0."

Reciprocal-of-Integer Powers

Now that we've seen what happens when a number is raised to an integer power, let's find out what goes on when a number is raised to a power that is the reciprocal of an integer.

Integer roots are reciprocal-of-integer powers

Suppose we take some number or quantity a, and raise it to the power $1/p$ where p is a positive integer. We write this as

$$a^{1/p}$$

We can surround the exponent with parentheses for clarity. If we do that to the above expression, we get

$$a^{(1/p)}$$

In this case, the parentheses are not technically necessary because the whole ratio is written as a superscript anyway.

When we take a reciprocal-of-integer power of a quantity, the result is often called a *root.* If you have a number and raise it to the power $1/p$, it is the same thing as taking the pth root of that number. If p is a positive integer, then the pth root of a quantity is something we must multiply by itself p times in order to get that quantity.

The square root

If the general formulas above confuse you, it can help if we look at an example. We know that

$$5^2 = 5 \times 5 = 25$$

The second power is often called the *square*, so we can say, "5 *squared* equals 25." By definition then

$$25^{1/2} = 5$$

We would say, "The *square root* of 25 is equal to 5." In general, for any two numbers a and b, and for any positive integer p, we can say this:

$$\text{If } a^p = b\text{, then } b^{1/p} = a$$

The reason the 2nd power is called the square and the 1/2 power is called the square root can be explained in terms of the dimensions and area of a perfect *geometric square*. For any geometric square, the interior area is equal to the 2nd power of the length of any one of the edges, as shown in Fig. 8-1. That's why the 2nd power is called the square. Looking at it the other way, the length of any one of the edges is equal to the 1/2 power of the interior area. That's why the 1/2 power is called the square root. In the figure, the *radical notation* for square root is shown, in addition to the 1/2 power notation. The radical consists of a *surd* symbol (√) with a line extending over the top of the quantity of which the square root is taken.

The cube root

Now let's see what happens when $p = 3$, so $1/p = 1/3$. We can easily figure out what happens when we raise a number, say 4, to the 3rd power:

$$4^3 = 4 \times 4 \times 4 = 64$$

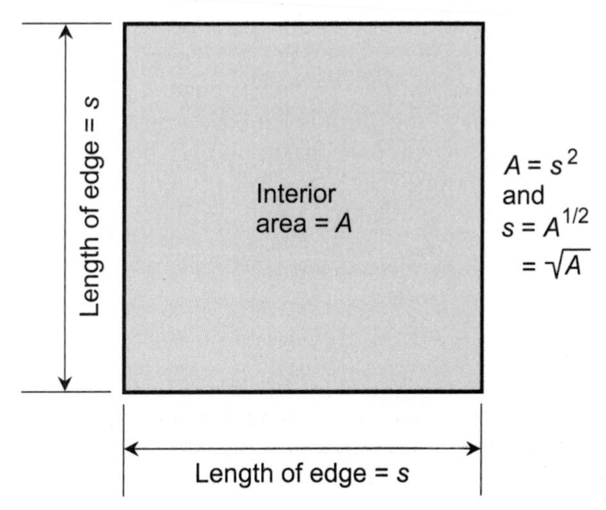

Figure 8-1 The area of a geometric square is equal to the 2nd power, or square, of the length of any edge. Therefore, the length of any edge is equal to the 1/2 power, or square root, of the area.

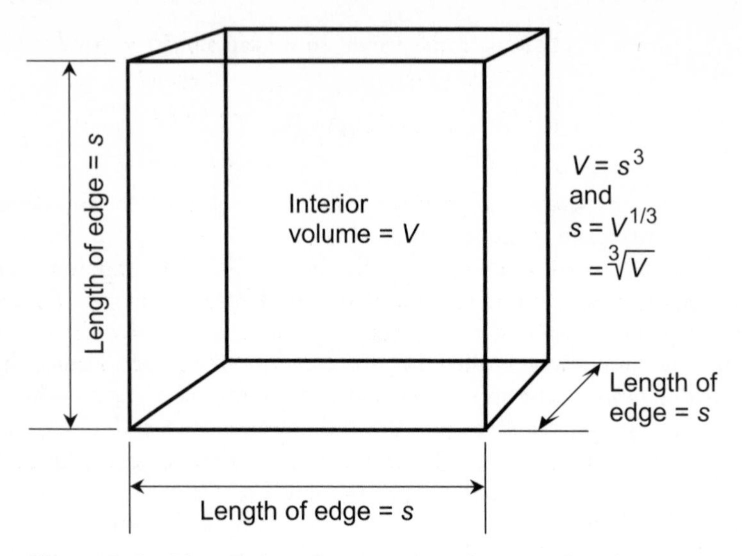

Figure 8-2 The volume of a geometric cube is equal to the 3rd power, or cube, of the length of any edge. Therefore, the length of any edge is equal to the 1/3 power, or cube root, of the volume.

The third power is often called the *cube.* We can say, "4 *cubed* equals 64." Now if we go with the reciprocal power and work backwards, we get

$$64^{1/3} = 4$$

This can be read as, "The *cube root* of 64 equals 4."

The 3rd power is called the cube and the 1/3 power is called the cube root because of the relationship between the edges and the interior volume of a *geometric cube.* For any perfect cube, the volume is equal to the 3rd power of the length of any edge (Fig. 8-2). Going the other way, the length of any edge is equal to the 1/3 power of the volume. The figure also shows the radical notation for the cube root. The fact that the radical refers to the cube root, rather than the square root, is indicated by the small numeral 3 in the upper-left part of the radical symbol.

Higher roots

When p is a positive integer equal to 4 or more, people write or talk about the numerical powers and roots directly. That's because *geometric hypercubes* having 4 dimensions or more are not commonly named. A 4-dimensional hypercube is technically called a *tesseract,* but you should expect incredulous stares from your listeners if you say "2 *tesseracted* is 16" or "The *tesseract root* of 81 is 3."

Here are some examples of higher powers and roots. You can check the larger ones on your calculator if you like.

$$2^4 = 16 \qquad \text{so } 16^{1/4} = 2$$

$$3^4 = 81 \qquad \text{so } 81^{1/4} = 4$$

$$5^6 = 15{,}625 \qquad \text{so } 15{,}625^{1/6} = 5$$
$$-3^7 = -2{,}187 \qquad \text{so } (-2{,}187)^{1/7} = -3$$
$$(-5)^9 = -1{,}953{,}125, \text{ so } (-1{,}953{,}125)^{1/9} = -5$$
$$6^4 = 1{,}296 \qquad \text{so } 1{,}296^{1/4} = 6$$
$$(-6)^4 = 1{,}296 \qquad \text{so } 1{,}296^{1/4} = -6 \ldots \textit{What?}$$

The radical notation can be used for any integer root. For the $1/n$ power, a small numeral n is placed in the upper left part of the radical symbol. If you use this notation, you must be sure that the radical symbol extends completely over the quantity of which you want to take the root. If you use the fractional notation, parentheses, brackets, and braces should be used to define the quantity of which you want to take the root.

Are you confused?

Now you will ask, "Can 6 and –6 both be valid 4th roots of 1,296?" The answer is "Yes." Both 6 and –6 will work here:

$$6 \times 6 \times 6 \times 6 = 1{,}296$$

and

$$(-6) \times (-6) \times (-6) \times (-6) = 1{,}296$$

If you multiply any negative number by itself an even number of times, you'll get a positive number. Therefore, if you have some number a and its additive inverse $-a$, and then you raise both of those numbers to an even positive integer power p, you will get

$$(-a)^p = a^p$$

every time! If we call $(-a)^p$ or a^p by some other name such as b, then the pth root of b is ambiguous. That would mean, for example,

$$16^{1/4} = 2 \text{ and } -2$$
$$81^{1/4} = 3 \text{ and } -3$$
$$15{,}625^{1/6} = 5 \text{ and } -5$$

It could even mean something as simple, and yet as troubling, as

$$1^{1/2} = 1 \text{ and } -1$$

Mathematicians get around this problem by saying that whenever "two numbers at once" are the result of a reciprocal power, the positive value is the correct one, unless otherwise specified. That means

$$16^{1/4} = 2$$
$$81^{1/4} = 3$$

$$15{,}625^{1/6} = 5$$
$$1^{1/2} = 1$$

You can indicate that you want to use the negative value by placing minus signs like this:

$$-(16^{1/4}) = -2$$
$$-(81^{1/4}) = -3$$
$$-(15{,}625^{1/6}) = -5$$
$$-(1^{1/2}) = -1$$

Sometimes you will actually want to let either the positive or the negative value be used. In cases of that sort, you should throw a *plus-or-minus sign* (±) into the mix, like this:

$$\pm(16^{1/4}) = \pm 2$$
$$\pm(81^{1/4}) = \pm 3$$
$$\pm(15{,}625^{1/6}) = \pm 5$$
$$\pm(1^{1/2}) = \pm 1$$

Here's another possible confusion-maker. Always pay special attention to where the parentheses are placed if you see a negative number raised to a power. Also, be careful if there are no parentheses at all. If there's any doubt, it's best to place extra parentheses in an expression so everyone knows exactly what it means. For example,

$$(-2)^3 = (-2) \times (-2) \times (-2) = -8$$

and

$$\begin{aligned} -2^3 &= -(2^3) \\ &= -(2 \times 2 \times 2) \\ &= -8 \end{aligned}$$

In contrast to this,

$$(-2)^4 = (-2) \times (-2) \times (-2) \times (-2) = 16$$

but

$$\begin{aligned} -2^4 &= -(2^4) \\ &= -(2 \times 2 \times 2 \times 2) \\ &= -16 \end{aligned}$$

Negative reciprocal powers

We still have not explored what happens when we raise a number to a negative reciprocal-of-integer power. You can probably figure out the meaning of an expression such as $125^{-1/3}$, or

125 to the $-1/3$ power. You take the $1/3$ power of 125, which is 5, and then take the reciprocal of that, which is $1/5$. Mathematically, it goes like this:

$$\begin{aligned} 125^{-1/3} &= 125^{-(1/3)} \\ &= 1/(125^{1/3}) \\ &= 1/5 \end{aligned}$$

Even roots of negative numbers

What happens when you take an even root of a negative number? The simplest example of this sort of problem is the square root of -1, but there are plenty of others. What can you multiply by itself to get -1? Nothing that we've defined yet! What is the $1/4$ power of 16? Again, nothing we know of so far.

Mathematicians have defined quantities like this. We will explore them in Chap. 21. They're called *imaginary numbers.* They have some fascinating properties. Unlike division by 0 or the 0th root of 0, even roots of negative numbers can be "tamed." They are commonly used in science and engineering.

Here's a challenge!

State the rule for negative reciprocal powers in general terms, where a is the base (the number to be raised to the power) and p is a positive integer.

Solution

The power to which we want to raise the base is $-1/p$, where p is some positive integer. (We know that $-1/p$ will be negative, because a negative divided by a positive always gives us a negative.) If we use the method from the above example where we evaluated $125^{-1/3}$, then we have

$$a^{-1/p} = a^{-(1/p)} = 1/(a^{1/p})$$

Multiplying and Dividing with Exponents

When we have a certain base number raised to two different powers, we get two different quantities. But the fact that those quantities have the same base lets us take shortcuts in multiplication and division.

Multiply by adding

Let's state the general case first, and then check out an example. Imagine a number and call it a. This can be any number except 0. It doesn't have to be an integer or even a rational number. Now imagine the quantities a^m and a^n, where m and n are integers. If you multiply these two quantities, you get the same result as if you add m to n, and then raise the base a to that power. We can write this as an equation:

$$a^m a^n = a^{(m+n)}$$

Let's call this the *addition-of-exponents (AOE) rule.* We can see how it works by trying out an example with specific numbers. Let $a = 3$, $m = 2$, and $n = -4$. Then

$$
\begin{aligned}
3^2 \times 3^{-4} &= 9 \times 1/81 \\
&= 9/81 \\
&= 1/9
\end{aligned}
$$

and

$$
\begin{aligned}
3^{[2+(-4)]} &= 3^{(2-4)} \\
&= 3^{-2} \\
&= 1/(3^2) \\
&= 1/9
\end{aligned}
$$

Divide by subtracting

Think of a nonzero number, b, along with two quantities b^p and b^q, where p and q are integers. If you divide the first of these quantities by the second, you get the same result as if you subtract q from p, and then raise b to that power. Mathematically:

$$b^p / b^q = b^{(p-q)}$$

Let's call this the *subtraction-of-exponents (SOE) rule.* Now we'll work out an example. Suppose we have $b = 10$, $p = 5$, and $q = 3$. Then

$$
\begin{aligned}
10^5/10^3 &= 100{,}000/1{,}000 \\
&= 100
\end{aligned}
$$

and

$$
\begin{aligned}
10^{(5-3)} &= 10^2 \\
&= 100
\end{aligned}
$$

Are you confused (yet)?

Was that too easy for you? Let's try a slightly tougher example. Let $b = -2$, $p = 3$, and $q = 4$. Then

$$
\begin{aligned}
(-2)^3/(-2)^4 &= -8/16 \\
&= -1/2
\end{aligned}
$$

and

$$
\begin{aligned}
(-2)^{(3-4)} &= (-2)^{-1} \\
&= 1/(-2) \\
&= -1/2
\end{aligned}
$$

The AOE and SOE rules work not only when the exponents are integers, but for any rational numbers. You might call these facts the *generalized addition-of-exponents (GAOE) rule* and the *generalized subtraction-of-exponents (GSOE) rule.*

Here's a challenge!

Using the SOE rule, provide a demonstration of why any nonzero quantity to the 0th power is equal to 1. Also show why 0^0 is not defined.

Solution

Look again at the formula that "translates" division of quantities into subtraction of exponents. That formula is

$$b^p / b^q = b^{(p-q)}$$

where b is the base and p and q are integers. Now let's think of the formula in reverse. We can transpose the left-hand and right-hand sides of the equation to get

$$b^{(p-q)} = b^p / b^q$$

There's nothing in the "rule book" that says we can't have p and q be the same. Let's do that, and call them both p. Then we have

$$b^{(p-p)} = b^p / b^p$$

The left-hand side of this equation is b raised to the $(p - p)$th power, which must be b raised to the 0th power because $p - p$ is always 0. The right-hand side is b^p divided by itself, which has to equal 1 as long as $b \neq 0$.

Now we get to the 0^0 situation. Let's violate the "rule book" and let $b = 0$ in the above equation. Then we get

$$0^{(p-p)} = 0^p / 0^p$$

No matter what nonzero value we choose for p, we get 0^0 on the left-hand side of this equation, and 0/0 on the right. So

$$0^0 = 0/0$$

The quantity 0/0 is not defined, so 0^0 can't be, either.

Multiple Powers

Numbers can be raised to powers more than once. In this section we'll see what happens when you raise a quantity to a power, and then raise the result to another power.

When exponents multiply

Imagine that we have a number a that is not equal to 0. Suppose p and q are integers. What happens if we raise a to the pth power, and then raise the result to the qth power? Mathematically, we get this expression:

$$(a^p)^q$$

It is tempting to suppose that the result of this operation will always produce a huge number. That can happen if the absolute value of a is larger than 1, and if p and q are both positive and more than 1. For example:

$$\begin{aligned}(4^5)^6 &= 1{,}024^6 \\ &= 1{,}152{,}921{,}504{,}606{,}846{,}976\end{aligned}$$

It doesn't always work out that way, however. If the absolute value of a is between 1 and 0, and if p and q are both positive and more than 1, the number may be quite close to 0. For example,

$$\begin{aligned}[(-0.1)^3]^5 &= (-0.001)^5 \\ &= -0.000000000000001\end{aligned}$$

When you have any expression of this sort, you can get the same result if you take the base a to the power of the product of the exponents pq. That is,

$$(a^p)^q = a^{pq}$$

Let's call this the *multiplication-of-exponents (MOE) rule*. Looking at the numerical examples we just saw, and putting them in this form, illustrates this:

$$\begin{aligned}(4^5)^6 &= 4^{5\times 6} \\ &= 4^{30} \\ &= 1{,}152{,}921{,}504{,}606{,}846{,}976\end{aligned}$$

and

$$\begin{aligned}[(-0.1)^3]^5 &= (-0.1)^{3\times 5} \\ &= (-0.1)^{15} \\ &= -0.000000000000001\end{aligned}$$

You can evaluate expressions with large exponents quickly by using a calculator with an "x to the yth power" key.

Rational-number powers

When you raise an *exponentiated quantity* (i.e., anything to a power) to another power, either or both of the exponents can be negative. The MOE rule still applies. In fact, we can let the exponents p and q be any rational numbers we want. That gives us the powerful, far-reaching *generalized multiplication-of-exponents (GMOE) rule*! If a, p, and q are rational numbers and $a \neq 0$, then

$$(a^p)^q = a^{pq}$$

We now have a way to evaluate an expression where we raise a number to a certain power, and then take a root of the result. Remember that a root is a reciprocal power. So, if we encounter an exponent that takes the form r/s, we can call this the product of r and $1/s$, and then use the GMOE rule:

$$(a^r)^{1/s} = a^{r(1/s)} = a^{r/s}$$

That's how we'd evaluate the sth root of a^r. But it also tells us something more: when we take a base number to a rational-number, noninteger power, it's the same thing as taking the base to an integer power and then taking an integer root of the result. Remember, a rational number is a quotient of two integers! If we reverse the order of the terms in the above three-way equation, we get

$$a^{r/s} = a^{r(1/s)} = (a^r)^{1/s}$$

This is a heavy dose of abstract math! Let's look at a couple of specific cases where integers are raised to rational-number powers. First, this:

$$\begin{aligned} 10^{6/3} &= 10^{6\times(1/3)} \\ &= (10^6)^{1/3} \\ &= 1{,}000{,}000^{1/3} \\ &= 100 \end{aligned}$$

If you're astute, you can solve this a lot quicker by noting that $6/3 = 2$, so

$$\begin{aligned} 10^{6/3} &= 10^2 \\ &= 100 \end{aligned}$$

Usually, rational-number powers aren't this easy to evaluate. The results often produce numbers that aren't even rational. Consider this example:

$$\begin{aligned} 2^{3/2} &= 2^{3\times(1/2)} \\ &= (2^3)^{1/2} \\ &= 8^{1/2} \\ &= 2.8284\ \ldots \end{aligned}$$

This is an *endless nonrepeating decimal.* It cannot be expressed as a ratio of integers, and is not a rational number. You'll learn more about these types of numbers in the next chapter.

Are you confused?

Let's review the most important points in this chapter. They can be condensed into six statements.

- If a is any nonzero number and $-n$ is a negative integer, the expression a^{-n} means you should raise a to the power of $|n|$, and then take the reciprocal of the result.
- If a is any nonzero number and m and n are rational numbers, then a^m times a^n is the same as a raised to the power of $(m + n)$.
- If a is any nonzero number and m and n are rational numbers, then a^m divided by a^n is the same as a raised to the power of $(m - n)$.
- If a is any nonzero number and p is any nonzero integer, then the pth root of a is the same as raising a to the power of $1/p$.
- If a, p, and q are rational numbers and a is nonzero, then if you raise a to the pth power and take the result to the qth power, it's the same as raising a to the power of pq.
- If a, p, and q are rational numbers with a and q nonzero, then if you raise a to the pth power and take the qth root of the result, it's the same as raising a to the power of p/q.

Here's a challenge!

What do you get if you take the −5/2 power of 6? Mathematically, evaluate this expression and use a calculator to figure out the result to several decimal places:

$$6^{(-5/2)}$$

Solution

Let's apply the GMOE rule to this problem. It can be tricky because of the minus sign, and we have to be sure we remember the difference between negative powers and reciprocal powers. Let's go:

$$\begin{aligned} 6^{(-5/2)} &= 6^{-5\times(1/2)} \\ &= (6^{-5})^{1/2} \\ &= [1/(6^5)]^{1/2} \\ &= (1/7{,}776)^{1/2} \end{aligned}$$

Now it's time to use a calculator! Remember that the 1/2 power is the same as the square root. First we take the reciprocal of 7,776, getting a decimal point, three ciphers, and a long string of digits. Then we hit the square root key with the string of digits still in the display, getting

$$0.01134023\ ...$$

The digits go on without end, and there's no apparent pattern. As things turn out, this is not a rational number.

Practice Exercises

This is an open-book quiz. You may (and should) refer to the text as you solve these problems. Don't hurry! You'll find worked-out answers in App. A. The solutions in the appendix may not represent the only way a problem can be figured out. If you think you can solve a particular problem in a quicker or better way than you see there, by all means try it!

1. What happens when a negative number is raised to an even positive integer power? An odd positive integer power?

2. What happens if we

   (a) Raise −2 to increasing integer powers, starting with 1?

   (b) Do the same thing with a base of −1?

   (c) Do the same thing with a base of −1/2?

3. What happens if we

   (a) Raise −2 to decreasing negative integer powers, starting with −1?

   (b) Do the same thing with a base of −1?

   (c) Do the same thing with a base of −1/2?

4. Suppose we come across the following expression. We want to simplify it to a sum of individual terms. How can we do this? Here's a hint: Use the results of the final "challenge" in Chap. 5.

$$(y + 1)^2$$

5. If we use the same technique as in the previous problem, we can also simplify the following expression. How?

$$(y - 1)^2$$

6. Use the GAOE rule to prove, in narrative form, that for any number a except 0, and for any rational numbers p, q, and r

$$a^p a^q a^r = a^{(p+q+r)}$$

7. Use the GMOE rule to prove, in the form of an S/R table, that for any number a except 0, and for any rational numbers p, q, and r

$$[(a^p)^q]^r = a^{pqr}$$

8. Prove, in narrative form, that if x is any number except 0, and if r and s are rational numbers, then

$$(x^r)^s = (x^s)^r$$

9. Show that if you take the 4th root of any positive number and then square the result, it is the same as taking the square root of the original number.

10. Show that if you take the 6th power of any positive number and then take the cube root of the result, it is the same as squaring the original number.

CHAPTER

9

Irrational and Real Numbers

We started out with the natural numbers (or *naturals*), and took their negatives to get the integers. Then we divided integers by each other to come up with rational numbers (or *rationals*). In this chapter, we'll study the irrational numbers (or *irrationals*) and real numbers (or *reals*), and compile a full set of rules for working with *real variables.*

The Number Hierarchy

Mathematicians have known for centuries that there are plenty of irrational numbers that can't be expressed as ratios of integers. Let's see how they behave compared with the rational numbers.

Rational-number "density"

Suppose we assign rational numbers to points on a horizontal line so the distance of any point from the origin is directly proportional to its absolute value. If a point is on the left-hand side of the point representing 0, then that point corresponds to a negative number; if it's on the right-hand side, it corresponds to a positive number.

If we take any two points a and b on the line that correspond to rational numbers, then the point midway between them corresponds to the rational number $(a + b)/2$. (Do you recognize this as the formula for the *average,* or *arithmetic mean,* of two numbers?) We can keep cutting an interval in half, and if the end points are both rational numbers, then the midpoint is another rational number. Figure 9-1 shows an example of this, starting with the interval between 1 and 2.

It is tempting to suppose that the points on a *rational-number line* are "infinitely dense." The point midway between any two rational-number points always corresponds to another rational number. But do the rational numbers account for *all* of the points along a true geometric line? The answer is "No. They don't even come close!"

Irrational numbers

As we have seen, an irrational number can't be expressed as the ratio of two integers. Examples of irrational numbers include:

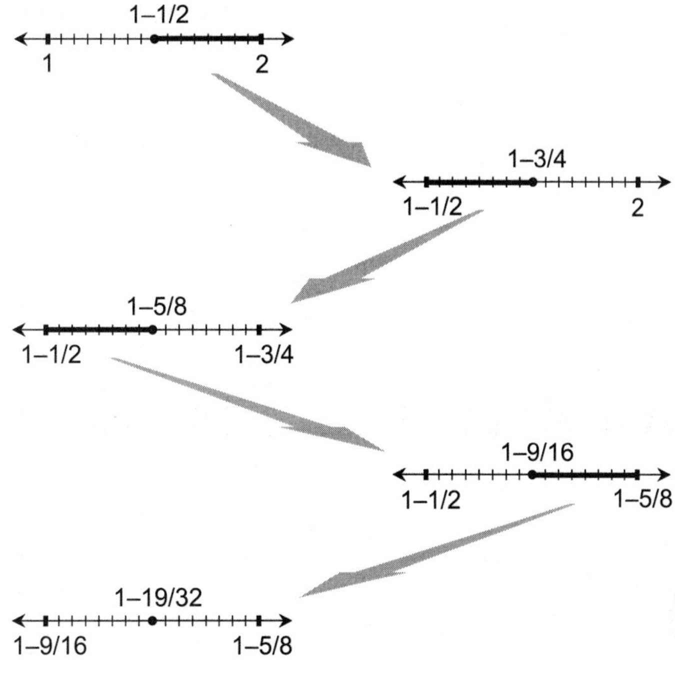

Figure 9-1 An interval on the rational-number line can be cut in half over and over, and you can always find infinitely many numbers in it.

- The length of the diagonal of a square that measures 1 unit on each edge.
- The length of the diagonal of a cube that measures 1 unit on each edge.
- The ratio of a circle's circumference to its diameter.
- The decimal number 0.010010001000010000001...

Whenever we try to express an irrational a number in decimal form, the result is an endless nonrepeating decimal. (The last item in the above list has a pattern of sorts, but it is not a repeating pattern like the decimal expansion of a rational.) No matter how many digits we write down to the right of the decimal point, the expression is an approximation of the actual value. A pattern can never be found that allows us to convert the expression to a ratio of integers.

The set of irrationals can be denoted S. This set is disjoint from the set Q of rationals. No irrational number is rational, and no rational number is irrational. In set notation,

$$S \cap Q = \varnothing$$

Real numbers

The set of *real numbers,* denoted by R, is the union of the set Q of all rationals and the set S of all irrationals:

$$R = Q \cup S$$

We can envision the reals as corresponding to points on a continuous, straight, infinitely long line, in the same way as we can imagine the rationals. But there are more points on a *real-number line* than there are on a rational-number line. (Whether or not the real numbers can be paired off one-to-one with the points on a true geometric line is a question that goes far beyond the scope of this book!) The set of real numbers is related to the sets of rational numbers Q, integers Z, and natural numbers N like this:

$$N \subset Z \subset Q \subset R$$

The operations of addition, subtraction, and multiplication can be defined over R. If # represents any of these operations and x and y are elements of R, then:

$$x \# y \in R$$

This is a fancy way of saying that whenever you add, subtract, or multiply a real number by another real number, you always get a real number. This is not generally true of division, *exponentiation* (raising to a power), or taking a root. You can't divide by 0, take 0 to the 0th power, or take the 0th root of anything and get a real number. Also, you can't take an even-integer root of a negative number and get a real number.

Russian dolls

Now we can see the full hierarchy of number types. We started with the set of naturals, N. Then we built the set of integers, Z, by introducing the notion of negative values. From there, we generated the set of rationals, Q, by dividing integers by each other. Now, we have found out about the set of irrationals, S, and the set of reals, R. The sets N, Z, Q, and R fit inside each other like Russian dolls:

$$N \subset Z \subset Q \subset R$$

The set S is a proper subset of R, but it's "standoffish" in the sense that it does not allow any of the rationals, integers, or naturals into its "realm." The Venn diagram of Fig. 9-2 shows how all these sets are related.

Later on, we'll learn about a set of numbers that's even larger than the reals. Those are the *imaginary numbers* and *complex numbers*. They result from taking the square roots of negative reals and adding those quantities to other real numbers.

Are you confused?

Take the interval with end points on the number line corresponding to 1 and 2, as shown in Fig. 9-1. It seems reasonable to think that there must be enough rational numbers (there are infinitely many, after all) to account for every possible point in this interval. If that were true, then you could take any point P in the interval and slice finer and finer intervals around it, with a rational number at the middle of each interval, until finally you got an interval with P right in the middle. But you *can't* always do this!

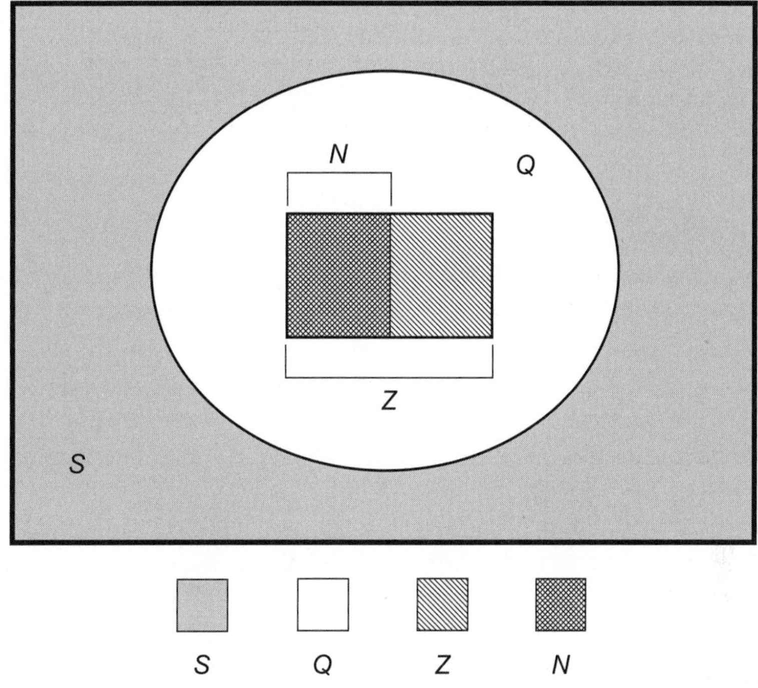

Figure 9-2 The relationship among the sets of real numbers (*R*), irrationals (*S*), rationals (*Q*), integers (*Z*), and natural numbers (*N*).

You can define a point between 1 and 2 that doesn't correspond to any rational number. Take a square measuring exactly 1 unit on each edge, known as a *unit square,* and place it on the number line so one corner is at the point for 0 and the other corner is on the line between the points for 1 and 2, as shown in Fig. 9-3. The opposite corner of the square falls exactly on the point for the square root of 2, because the diagonal of a unit square is precisely $2^{1/2}$ units long. (That fact comes from basic geometry.) The number $2^{1/2}$ is irrational.

If you build a line using only the points corresponding to rational numbers, that line will be full of "holes."

Are you still confused?

The rational numbers, when depicted as points along a line, are "dense," but not as dense as the points on a line can get. No matter how close together two rational-number points on a line might be, there is always another rational-number point between them. But there are points on a continuous geometric number line that don't correspond to any rational quantity. The set of reals is more "dense" than the set of rational numbers.

"All right," you say. "This game is strange, but I'll play along. How many times more dense is the real-number line than the rational line? Twice? A dozen times? A hundred times?" The answer might astonish you. The set of real numbers, when assigned to points on a line, is *infinitely* more "dense" than the set of rational numbers.

Here's a crude analogy. Suppose you're exploring the "planet Maths" and you stumble across a rational-number line and a real-number line lying in a field. Both lines resemble straight, infinitely long, thin, rigid

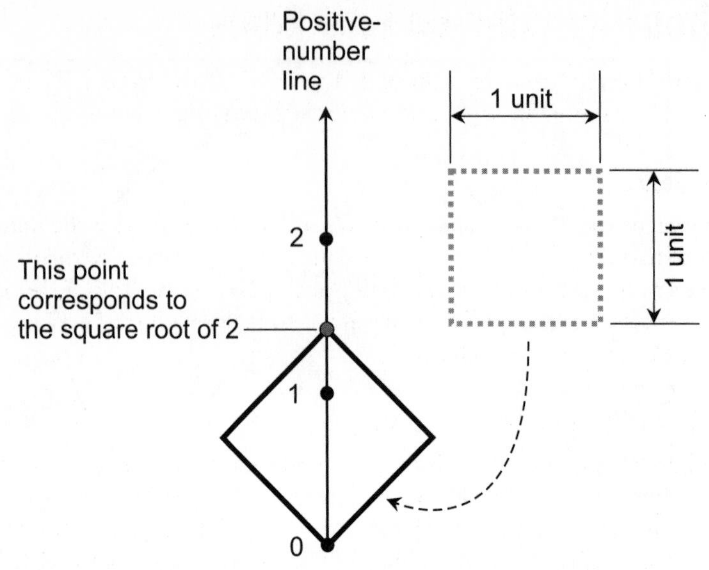

Figure 9-3 If you take a unit square and place it diagonally along the positive-number line with one corner at the point for 0, the opposite corner is at the point for $2^{1/2}$, which is irrational.

wires. The rational-number line is gray. When you pick it up, you find that it's weightless. The real-number line is black. When you lift it, you discover that it's heavy. The real-number line contains more "stuff."

Here's a challenge!

Let's try a little exercise that involves some plain-language logic, some set theory, and some Venn diagram reading skill. Which of the following statements are true, based on our knowledge of the number hierarchy?

- All rational numbers are real.
- All integers are real.
- Some integers are irrational.
- Some irrational numbers are real.
- All irrational numbers are real.

Solution

We can figure all of these out by looking at Fig. 9-2. The answers, bullet-by-bullet, are as follows.

- Set Q is entirely contained in set R. Therefore, all rationals are real.
- Set Z is entirely contained in set R. Therefore, all integers are real.
- Set Z is completely separate from set S. Therefore, no integers are irrational.
- Set S has elements in common with set R. Therefore, some irrationals are real.
- Set S is entirely contained in set R. Therefore, all irrationals are real.

More About Irrationals and Reals

Let's explore "infinity" for a few minutes. Then we'll prove that the square root of 2 is irrational. Get into the mood for some serious abstract thinking!

Number lists

Mathematicians use the symbol $\aleph_0$ (called *aleph-null*) to describe the number of elements in the set N of natural numbers. This is the same as the number of elements in the set Z of integers, as we saw in the solution to the final practice exercise in Chap. 3. We can create "implied lists" of both sets, and be confident that if we go far enough out, we'll always hit any natural number or integer we care to choose.

Cardinality of a set

The number of elements in a set is called the *cardinality* of the set. The cardinality of N is $\aleph_0$, and the cardinality of Z is also $\aleph_0$. Even the elements of Q, the set of rationals, can be defined in terms of an "implied list." Figure 9-4 is an example. So, as counterintuitive as it may seem, the cardinality of Q is the same as the cardinality of N or Z, that is, $\aleph_0$. If we can make an "implied list" of the elements in an infinite set, that set is said to be *denumerably infinite* (or simply *denumerable*), and by definition it has cardinality $\aleph_0$.

The irrationals and reals can't be "listed"

The elements of S or R cannot be denoted in any type of list. We can't even make an "implied list" of all the nonnegative irrational numbers smaller than 1 in decimal-expansion form. Table 9-1, and the following explanation, should give you some idea of why this is so.

Suppose we try to list the irrational numbers between 0 and 1 (including 0, but not 1) by writing down the numerals in their expanded-decimal forms. To the left of the decimal point, every numeral will have a single 0 and nothing else. The first numeral will have an endless string of digits from the set {0, 1, 2, 3, 4, 5, 6, 7, 8, 9} to the right of the decimal point. Let's call those digits a_{11}, a_{12}, a_{13}, and so on. The second number will have an endless string of digits that we can call a_{21}, a_{22}, a_{23}, and so on, different from the first string. We can keep on listing irrational numbers like this forever. The nth number in our list (corresponding to the nth row in Table 9-1) will be of the form

$$0.a_{n1}a_{n2}a_{n3}\ \ldots$$

Now imagine that we have listed one irrational number for every possible value of n. (This could not actually be done by any mortal human, because it would take forever. But in the world of mathematics, our imaginations let us do infinitely many tasks in a finite amount of time!) Suppose that no two irrational numbers in this list are the same. It is tempting to believe that this list of irrational numbers, taking the form of a matrix that extends forever to the right and downward from what we see in Table 9-1, must contain all possible irrationals. After all, there are infinitely many of them, and we haven't listed any of them twice.

But no! Even this infinite list is not complete. There are still more irrationals. Here is one of them. Imagine building an irrational number of this form:

$$0.b_1b_2b_3\ \ldots$$

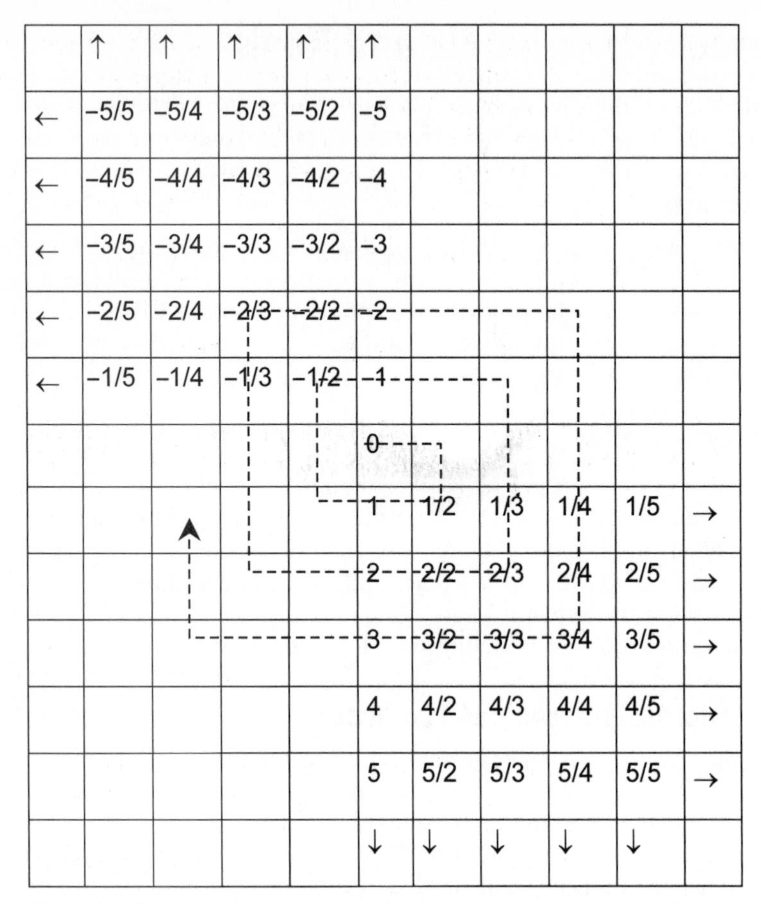

Figure 9-4 All the elements in the set of rational numbers can be arranged in an "implied list." Follow the dashed "square spiral" in this two-dimensional list, starting at the center and going outward as shown. Eventually, you'll hit the box for any ratio of integers that can exist.

where the digit b_1 is different from a_{11} in the first list entry, the digit b_2 is different from a_{22} in the second list entry, the digit b_3 is different from a_{33} in the third list entry, and so on forever. This new irrational number cannot be in the list denoted by Table 9-1. No matter which of the entries in that list (with the *a*'s and subscripts) we choose, our new number (with the *b*'s and subscripts) has at least one digit that doesn't match.

When we can't make a list—or even concoct an infinitely long "implied list" in the mathematical cosmos—of all the elements of a set, the set is said to be *non-denumerable.*

Are you confused?

The notion of non-denumerability is difficult to grasp. Don't feel bad if you don't fully understand it. It's a little like trying to envision space with more than three dimensions. Some things that can be defined in

Table 9-1. If we try to create an "implied list" of all the irrational numbers between 0 and 1 as endless, nonrepeating decimal expansions, we are destined to fail. Each digit in the bottom row (b with subscript) is chosen so that it's different from the boldface digit above it in the same column (a with subscript). Mathematically, for every positive integer subscript n in the table, $b_n \neq a_{nn}$, the endless decimal $0.b_1b_2b_3$ ... can't be in the "a" list, even though that list is infinitely long.

0	.	$\mathbf{a_{11}}$	a_{12}	a_{13}	a_{14}	a_{15}	a_{16}	a_{17}	$\rightarrow$
0	.	a_{21}	$\mathbf{a_{22}}$	a_{23}	a_{24}	a_{25}	a_{26}	a_{27}	$\rightarrow$
0	.	a_{31}	a_{32}	$\mathbf{a_{33}}$	a_{34}	a_{35}	a_{36}	a_{37}	$\rightarrow$
0	.	a_{41}	a_{42}	a_{43}	$\mathbf{a_{44}}$	a_{45}	a_{46}	a_{47}	$\rightarrow$
0	.	a_{51}	a_{52}	a_{53}	a_{54}	$\mathbf{a_{55}}$	a_{56}	a_{57}	$\rightarrow$
0	.	a_{61}	a_{62}	a_{63}	a_{64}	a_{65}	$\mathbf{a_{66}}$	a_{67}	$\rightarrow$
0	.	a_{71}	a_{72}	a_{73}	a_{74}	a_{75}	a_{76}	$\mathbf{a_{77}}$	$\rightarrow$
$\downarrow$	$\downarrow$	$\downarrow$	$\downarrow$	$\downarrow$	$\downarrow$	$\downarrow$	$\downarrow$	$\downarrow$	$\downarrow \rightarrow$
0	.	b_1	b_2	b_3	b_4	b_5	b_6	b_7	$\rightarrow$

mathematics, and in which problems can be neatly worked out, simply defy any attempt at "seeing them in the mind's eye." It's good enough to remember that there are more real numbers than rational numbers—a lot more.

It can help if you stop thinking of "infinity" as something you can count toward. Instead, think of "infinity" as an expression of the size of a set. The cardinality of the set R of real numbers is greater than $\aleph_0$, the cardinality of N, Z, or Q.

You might wonder if there are any "infinities" larger than the cardinality of the set of reals. The mathematician Georg Cantor's answer to this question was "Yes, infinitely many!" He called these "infinities" *transfinite cardinals*. During his lifetime, Cantor was scorned by some of his fellow mathematicians for his theory of transfinite cardinals. Now they are commonly accepted in advanced mathematics.

Here's a big challenge!

You are now invited to follow along with an "extra-credit" proof. It will take some time and effort to understand it. But why not try it? It doesn't involve anything more complicated or sophisticated than facts of arithmetic you already know. Let's *prove* that the value of $2^{1/2}$ cannot be represented as a ratio of integers in lowest terms, and therefore that it's an irrational number. We'll need two lemmas. Let's accept them on faith. Both of them can be proved, but that would be a distraction right now.

- The square of an integer is always an integer. Let's call this the *integer-squared rule.*
- The square of an odd integer is always an odd integer. Let's call this the *odd-integer-squared rule.*

Solution

Whenever we suspect that a proof might be involved, it's tempting to try *reductio ad absurdum.* Let's use it now. Table 9-2 does the job. If you can't follow this proof as a whole, don't worry. Try to understand each step, one at a time.

Table 9-2. An S/R proof that $2^{1/2}$ is not a rational number.

Statements	Reasons
Assume $2^{1/2}$ is rational	We begin with this statement and will prove it false
The quantity $2^{1/2}$ can be represented as a ratio of two integers, p and q, in lowest terms	Definition of rational number
$2^{1/2} = p/q$	This is a mathematical statement of the claim made above
$(2^{1/2})^2 = (p/q)^2$	Square both sides of the equation in the previous line
$2 = p^2/q^2$	Use arithmetic to manipulate the equation in the previous line
$2q^2 = p^2$	Multiply each side of the equation in the previous line by q^2
$q^2 = p^2/2$	Divide each side of the equation in the previous line by 2
We know q is an integer, so q^2 is an integer	Integer-squared rule
We know p is an integer, so p^2 is an integer	Integer-squared rule
$p^2/2$ is an integer	We know this because $q^2 = p^2/2$, and q^2 is an integer
p^2 is an even integer	This follows from the definition of even integer
p is an even integer	According to the odd-integer-squared rule, if p were odd, then p^2 would be odd; but p^2 is even
$p/2$ is an integer	This follows from the definition of even integer
Call $p/2 = t$	This will make things a little simpler
$p = 2t$	Multiply each side of the equation in the previous line by 2
$2q^2 = (2t)^2$	Substitute $2t$ for p in the equation $2q^2 = p^2$ from earlier in this proof
$2q^2 = 4t^2$	Simplify the right-hand side of the equation in the previous line
$q^2/2 = t^2$	Divide each side of the equation in the previous line by 4
We know t is an integer, so t^2 is an integer	Integer-squared rule
$q^2/2$ is an integer	This follows from the previous two lines
q^2 is an even integer	This follows from the definition of even integer
q is an even integer	According to the odd-integer-squared rule, if q were odd, then q^2 would be odd; but q^2 is even
$q/2$ is an integer	This follows from the definition of even integer
The quotient p/q is a ratio of integers in lowest terms	Part of the assumption we made at the beginning of this proof
$p/2$ is an integer, and $q/2$ is an integer	We have proven both of these facts
$(p/2)/(q/2)$ is a ratio of integers	This follows from the statement immediately above this line
The ratio p/q is not given in lowest terms	$(p/2)/(q/2)$ is in lower terms than p/q, but the two expressions represent the same quantity
We have produced a logical absurdity	The preceding line contradicts our original assumption about p/q
$2^{1/2}$ is not rational	*Reductio ad absurdum*
Q.E.D.	Mission accomplished

How Real Variables Behave

Most of the properties of the natural numbers, integers, and rationals also apply to the reals. To get ready for the algebra to come, let's put together a collection of these properties. All the old rules are here, along with a few new or expanded ones. Don't try to memorize these rules. Just look them over for a little while. You can always come back to this section for reference if you get stuck later on.

Naming variables

You'll notice that some of the variables here have different letter names than we used before. In algebra, real variables are usually given letters from near the end of the alphabet, especially t and after. Other letters more often represent natural-number variables (n and m are especially common), integer variables, and rational-number variables. But there's no absolute law about this.

Additive identity

For every real number x,

$$x + 0 = x$$

and

$$0 + x = x$$

Multiplicative identity

For every real number x,

$$x1 = x$$

and

$$1x = x$$

Additive inverse

For every real number x,

$$x + (-x) = 0$$

and

$$(-x) + x = 0$$

Multiplicative inverse (reciprocal)

For every nonzero real number x,

$$x(1/x) = 1$$

and

$$(1/x)x = 1$$

Commutative law for addition

For all real numbers x and y,

$$x + y = y + x$$

Commutative law for multiplication

For all real numbers x and y,

$$xy = yx$$

Associative law for addition

For all real numbers x, y, and z,

$$(x + y) + z = x + (y + z)$$

Associative law for multiplication

For all real numbers x, y, and z,

$$(xy)z = x(yz)$$

Distributive laws

For all real numbers x, y, and z,

$$x(y + z) = xy + xz$$

and

$$(x + y)z = xz + yz$$

These rules also work for subtraction:

$$x(y - z) = xy - xz$$

and

$$(x - y)z = xz - yz$$

Variants of these rules work for division as long as the divisor (or denominator) consists of a single nonzero variable, and never a sum or difference. Therefore

$$(x + y)/z = x/z + y/z$$

and

$$(x - y)/z = x/z - y/z$$

Zero numerator

For all nonzero real numbers x, if 0 is divided by x, then the quotient is equal to 0:

$$0/x = 0$$

Zero denominator

For all real numbers x, if x is divided by 0, then the quotient is undefined:

$$x/0 = \text{(undefined)}$$

Multiplication by zero

Whenever a real number x is multiplied by 0, the product is equal to 0:

$$x0 = 0$$

and

$$0x = 0$$

0th power

Whenever a nonzero real number x is taken to the 0th power, the result is equal to 1:

$$x^0 = 1 \qquad \text{when } x \neq 0$$

Product of signs

When two real numbers with plus signs (meaning positive, or larger than 0) or minus signs (meaning negative, or less than 0) are multiplied by each other, the following rules apply:

$$(+)(+) = (+)$$
$$(+)(-) = (-)$$
$$(-)(+) = (-)$$
$$(-)(-) = (+)$$

Quotient of signs

When two real numbers with plus or minus signs are divided by each other, the following rules apply:

$$(+) / (+) = (+)$$
$$(+) / (-) = (-)$$
$$(-) / (+) = (-)$$
$$(-) / (-) = (+)$$

Power of signs

When a real number with a plus sign or a minus sign is raised to a positive integer power, n, the following rules apply:

$$(+)^n = (+)$$
$$(-)^n = (-) \text{ if } n \text{ is odd}$$
$$(-)^n = (+) \text{ if } n \text{ is even}$$

Reciprocal of reciprocal

For all nonzero real numbers, x, the reciprocal of the reciprocal is equal to the original number:

$$1/(1/x) = x \qquad \text{when } x \neq 0$$

More Rules for Real Variables

Here are some more sophisticated rules for expression morphing. As with the facts in the previous section, you don't have to try to memorize these. It can help if you work out a few examples using numbers in place of the variables. Some of the restrictions here, in which variables are not allowed to equal 0, are a little stronger than necessary to keep things straightforward and to be sure we stay safe!

Product of sums

For all real numbers w, x, y, and z,

$$(w+x)(y+z) = wy + wz + xy + xz$$

Cross multiplication

For all real numbers w, x, y, and z where neither x nor z is equal to 0,

$$w/x = y/z \text{ if and only if } wz = xy$$

Reciprocal of product

For all nonzero real numbers x and y,

$$1/(xy) = (1/x)(1/y)$$

This can also be written as

$$(xy)^{-1} = x^{-1}y^{-1}$$

Product of quotients

For all real numbers w, x, y, and z where $x \neq 0$ and $z \neq 0$,

$$(w/x)(y/z) = (wy)/(xz)$$

Reciprocal of quotient

For all nonzero real numbers x and y,

$$1/(x/y) = y/x$$

This can also be written as

$$(x/y)^{-1} = y/x$$

Quotient of products

For all real numbers w, x, y, and z where $x \neq 0$, $y \neq 0$, and $z \neq 0$,

$$\begin{aligned}(wx)/(yz) &= (w/y)(x/z) \\ &= (w/z)(x/y)\end{aligned}$$

Quotient of quotients

For all real numbers w, x, y, and z where $x \neq 0$, $y \neq 0$, and $z \neq 0$,

$$\begin{aligned}(w/x)/(y/z) &= (w/x)(z/y) \\ &= (w/y)(z/x) \\ &= (wz)/(xy)\end{aligned}$$

Sum of quotients

For all real numbers w, x, y, and z where $x \neq 0$ and $z \neq 0$,

$$w/x + y/z = (wz + xy)/(xz)$$

Integer roots

Suppose that x is a positive real number. Also suppose that n is a positive integer. Then the nth root of x can also be expressed as the $1/n$ power of x. The second root (or square root) is the same thing as the 1/2 power, the third root (or cube root) is the same thing as the 1/3 power, the fourth root is the same thing as the 1/4 power, and so on.

Rational-number powers

Suppose that x is a real number. Also suppose that m and n are integers, and $n \neq 0$. Then

$$x^{m/n} = (x^m)^{1/n} = (x^{1/n})^m$$

Negative powers

Let x be a nonzero real number. Let y be any real number. Then

$$x^{-y} = (1/x)^y = 1/(x^y)$$

Sum of powers

For all nonzero real numbers x, y, and z,

$$x^{(y+z)} = x^y x^z$$

Difference of powers

For all nonzero real numbers x, y, and z,

$$x^{(y-z)} = x^y / x^z$$

Product of powers

For all nonzero real numbers x, y, and z,

$$x^{yz} = (x^y)^z = (x^z)^y$$

Quotient of powers

For all nonzero real numbers x, y, and z,

$$x^{y/z} = (x^y)^{1/z} = (x^{1/z})^y$$

Power of product

For all nonzero real numbers x, y, and z,

$$(xy)^z = x^z y^z$$

Power of quotient

For all nonzero real numbers x, y, and z,

$$(x/y)^z = x^z/y^z$$

Power of reciprocal

Let x be a nonzero real number. Let y be any real number. Then

$$(1/x)^y = 1/(x^y)$$

Square of sum

For all real numbers x and y,

$$(x + y)^2 = x^2 + 2xy + y^2$$

Square of difference

For all real numbers x and y,

$$(x - y)^2 = x^2 - 2xy + y^2$$

Are you confused?

Some of the rules above involve real-number exponents. That can include irrationals! You may wonder how can anybody define such a thing as a number raised to the power of pi (π), for example. You'll understand irrational exponents better when you learn about logarithmic and exponential functions in Part 3. For now, here's a little glimpse into the mystery.

Imagine that you stumble across the expression 2^π in some mathematical adventure. The value of π, to five decimal places, is 3.14159, but you know it can't be expressed exactly as a ratio of integers, because it is irrational. But you can "zoom in" on it. Divide out the fractions 59/19 and 60/19 with a calculator to obtain their decimal expansions. These are both rational numbers, and you've learned how you can take any number to a rational-number power. If you expand them to five decimal places, and then display π to the same number of places (using a calculator with a π key), you will see that π is between these two rationals:

$$59/19 = 3.10526 \ldots$$
$$\pi = 3.14159 \ldots$$
$$60/19 = 3.15789 \ldots$$

That means

$$59/19 < \pi < 60/19$$

You should now be able to imagine, without too much trouble, that

$$2^{59/19} < 2^{\pi} < 2^{60/19}$$

The "mystery number" is in that interval somewhere. It's a real number, and it corresponds to its own special point on a real-number line. You can indeed raise a number, variable, or expression to an irrational power.

Here's a challenge!

Suppose x, y, and z are real numbers, with $x \neq 0$ and $y \neq 0$. Show that if you take x to the yth power and then take the result to the zth power, you do not necessarily get the same result as if you take x to the power of y^z.

Solution

All we have to do is find a numerical example where the two expressions don't agree. There are plenty! Let $x = 2$, $y = 9$, and $z = 1/2$. Then $x^y = 2^9 = 512$. If we raise 512 to the 1/2 power, we get 22.627 ..., an endless, nonrepeating decimal. Now consider y^z. That's $9^{1/2}$, or 3. If we raise x, which is 2, to the power of 3, we get 8.

Practice Exercises

This is an open-book quiz. You may (and should) refer to the text as you solve these problems. Don't hurry! You'll find worked-out answers in App. A. The solutions in the appendix may not represent the only way a problem can be figured out. If you think you can solve a particular problem in a quicker or better way than you see there, by all means try it!

1. Which of the following quantities can we reasonably suspect is irrational?
   (a) $16^{3/4}$
   (b) $(1/4)^{1/2}$
   (c) $(-27)^{-1/3}$
   (d) $27^{1/2}$

2. Suppose we have an irrational number and we display the first few digits of its endless nonrepeating decimal expansion. If we multiply this number by 10, is the result still irrational? What if we multiply it by 100, or 1,000, or any natural-number power of 10? Will the results always be irrational?

3. What is the cardinality of

   (a) The set of even natural numbers?

   (b) The set of naturals divisible by 10 without a remainder?

   (c) The set of naturals divisible by 100 without a remainder?

   (d) The set of naturals divisible by any natural power of 10 without a remainder?

4. Consider this equation in real variables x and y:

$$36x + 48y = 216$$

   How can we simplify this so it has the minimum possible number of symbols (variables and digits)?

5. The nonnegative square root of 18 can be simplified, or *resolved*, into a product of a natural number and an irrational number. What are these numbers?

6. The nonnegative square root of 83 cannot be resolved into a product of a natural number and an irrational number, other than the trivial case $1 \times 83^{1/2}$. How can we tell?

7. The numbers $50^{1/2}$ and $2^{1/2}$ are both irrational. But the ratio of $50^{1/2}$ to $2^{1/2}$ is a natural number. What natural number?

8. Using the sum of quotients rule, add the fractions 7/11 and −5/17, and then reduce this sum to lowest terms.

9. Using the product of sums rule, multiply out the product $(x + y)(x - y)$.

10. Using the rules we have learned so far, derive a formula for multiplying out the following expression, where u, v, w, x, y, and z are real numbers:

$$(u + v + w)(x + y + z)$$

CHAPTER 10

Review Questions and Answers

Part One

This is not a test! It's a review of important general concepts you learned in the previous nine chapters. Read it though slowly and let it "sink in." If you're confused about anything here, or about anything in the section you've just finished, go back and study that material some more.

Chapter 1

Question 1-1

What's the difference between a number and a numeral?

Answer 1-1

A number is an abstraction. Numerals are tangible or physical symbols that represent numbers, such as 30 or 7 as they appear on this page. A numeral is not a number, just as your name is not you.

Question 1-2

In the Roman numeration system, how would we write the equivalents of the following decimal numerals?

(a) 10 (b) 30 (c) 40 (d) 50
(e) 100 (f) 300 (g) 400 (h) 600

Answer 1-2

The Roman equivalents are:

(a) X (b) XXX (c) XL (d) L
(e) C (f) CCC (g) CD (h) DC

Question 1-3

The Hindu-Arabic system uses the numeral 0 to represent a quantity of nothing. Why is this significant?

Answer 1-3

The numeral 0 serves as a placeholder when writing numerals to represent large numbers. It makes arithmetic easier than it was with the Roman system, or with systems that used simple counting. Eventually, it made higher mathematics, such as algebra, possible.

Question 1-4

What's the difference between an "English billion" and a "U.S. billion"?

Answer 1-4

In England, the term "billion" usually refers to the quantity represented by the numeral 1,000,000,000,000. In the United States, it means the quantity represented by 1,000,000,000. The "English billion" is called a "trillion" in the United States.

Question 1-5

What's the difference between the decimal, binary, octal, and hexadecimal numeration systems?

Answer 1-5

The decimal system works in base ten, where the numeral 10 represents the number of dots in the following group:

• • • • • • • • • •

If we use decimal numerals to represent quantities, then the binary system works in base 2, the octal system works in base 8, and the hexadecimal system works in base 16.

Question 1-6

The hexadecimal system needs some extra digits besides the usual 0 through 9 to denote numbers. What are these digits, and what are their decimal equivalents?

Answer 1-6

Here are the extra digits and their decimal equivalents:

- Hexadecimal A stands for decimal 10.
- Hexadecimal B stands for decimal 11.
- Hexadecimal C stands for decimal 12.
- Hexadecimal D stands for decimal 13.
- Hexadecimal E stands for decimal 14.
- Hexadecimal F stands for decimal 15.

Question 1-7

In the octal numeration system, what follows 7? What follows 47? What follows 77? What follows 577?

Answer 1-7

In the octal system, 7 is followed by 10, 47 is followed by 50, 77 is followed by 100, and 577 is followed by 600.

Question 1-8

What is the main advantage of the binary numeration system?

Answer 1-8

It has only two states, which can be represented by the on/off states of simple switches. This makes the binary system ideal for use in electronic systems such as computers.

Question 1-9

In the binary numeration system, what follows 10? What follows 100? What follows 111? What follows 1101?

Answer 1-9

In the binary system, 10 is followed by 11, 100 is followed by 101, 111 is followed by 1000, and 1101 is followed by 1110. Note that there are no commas in large binary numerals, as there usually are in large decimal numerals.

Question 1-10

In the hexadecimal numeration system, what follows 10? What follows FF? What follows 999? What follows FFF?

Answer 1-10

In the hexadecimal system, 10 is followed by 11, FF is followed by 100, 999 is followed by 99A, and FFF is followed by 1,000.

Chapter 2

Question 2-1

What's the difference between a set and an element?

Answer 2-1

A set is a collection of elements. The elements of a set are also called its members. If we want to call a number or variable x an element of a set X, then we write $x \in X$. We can also say that x belongs to X, or that x is in X. If some other number or variable y is *not* an element of set X, then we write $y \notin X$. A set can be an element of another set.

Question 2-2

What's the difference, if any, between the following sets?

$$A = \{x, y, z\}$$

$$B = \{z, y, x\}$$

$$C = \{y, z, x\}$$

$$D = \{x, y, z, y, x\}$$

$$E = \{y, x, z, z, z, z, z, \ldots\}$$

Answer 2-2

These sets are all the same. When the elements of a set are listed, the order of the list isn't important. Once we've listed something as an element of a set, we can list it again, or 10 times more, or even infinitely many times, and it doesn't matter.

Question 2-3

Imagine two sets, one called R and the other called S. They share three elements, called x, y, and z. However, they both contain elements other than x, y, or z. What name can we give to the set $\{x, y, z\}$?

Answer 2-3

The set $\{x, y, z\}$ contains all the elements that belong to both R and S. By definition, the set $\{x, y, z\}$ is the intersection of R and S, written $R \cap S$.

Question 2-4

What does the entire gray shaded region in Fig. 10-1 represent?

Answer 2-4

This region represents the set of all elements in set P, set Q, or both. That's the union of sets P and Q, written $P \cup Q$.

Figure 10-1 Illustration for Questions and Answers 2-4 and 2-5.

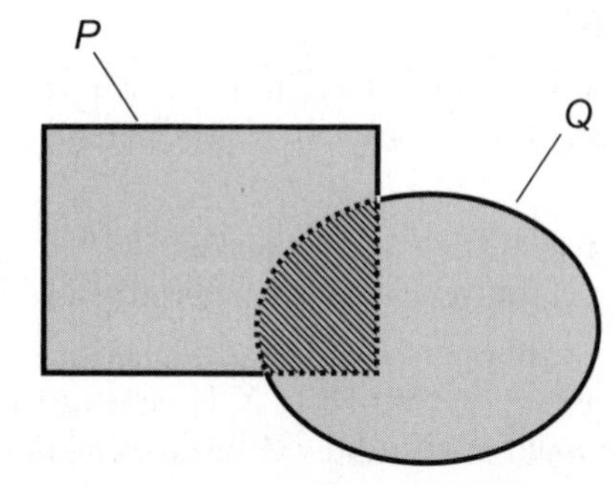

Question 2-5

What does the hatched region in Fig. 10-1 represent?

Answer 2-5

It portrays the set of elements belonging to both P and Q. That's the intersection of the two sets, written $P \cap Q$.

Question 2-6

What's the union of two disjoint sets X and Y? What's the intersection of two disjoint sets X and Y?

Answer 2-6

The union of two sets X and Y is the set of all elements in either X or Y. Their intersection is the set of elements in both X and Y. If X and Y are disjoint, then there are no elements belonging to them both, so the intersection is the null set.

Question 2-7

What's the union of two congruent sets? The intersection of two congruent sets?

Answer 2-7

By definition, if two sets X and Y are congruent, then they have exactly the same elements. Every element in X is also in Y, and every element in Y is also in X. The two sets overlap totally. Their union is the set of all elements in either X or Y. That's the same as either X or Y by itself. Their intersection is the set of elements in both X and Y. That's also the same as either X or Y by itself. In this case, all the following sets are identical:

$$X$$

$$Y$$

$$X \cup Y$$

$$X \cap Y$$

Question 2-8

Can a set be a subset of itself? A proper subset of itself?

Answer 2-8

Any set is a subset of itself. But no set is a proper subset of itself. By definition, a proper subset of any set S can't contain all of the elements in S.

Question 2-9

Is there any set that is a subset of every possible set?

Answer 2-9

Yes. The null set is a subset of any set we can imagine, even itself.

Question 2-10

How can we describe the relationship between the following sets? How would we write it in symbolic form?

$$A = \{4, 5, 6, 7, 8, 9, 10, ...\}$$
$$B = \{7, 8, 9, 10, 11, 12, ...\}$$

Answer 2-10

In this case, B is a proper subset of A. That's because every element in B is also in A, but there are some elements in A that are not in B. We write this fact as $B \subset A$.

Chapter 3

Question 3-1

The natural numbers are sometimes defined in terms of sets. How can we do this?

Answer 3-1

We can define the number 0 as the set containing nothing. That's the null set:

$$0 = \{\ \} = \varnothing$$

Once we've defined the number 0, then we can define the number 1 as the set containing the number 0, like this:

$$1 = \{\{\ \}\} = \{\varnothing\} = \{0\}$$

After that, we can build the rest of the natural numbers upon each other:

$$2 = \{0, 1\}$$
$$3 = \{0, 1, 2\}$$
$$4 = \{0, 1, 2, 3\}$$
$$\downarrow$$
$$n + 1 = \{0, 1, 2, ..., n\}$$
$$\downarrow$$

and so on, forever

Question 3-2

How can we write the natural numbers 0 through 4 purely in terms of set braces and the null set symbol?

Answer 3-2

We start with 0, which is equal to $\varnothing$ by definition. The numbers are built upon each other as sets within sets, like this:

$$0 = \varnothing$$
$$1 = \{0\} = \{\varnothing\}$$

$$2 = \{0, 1\} = \{\varnothing, \{\varnothing\}\}$$
$$3 = \{0, 1, 2\} = \{\varnothing, \{\varnothing\}, \{\varnothing, \{\varnothing\}\}\}$$
$$4 = \{0, 1, 2, 3\} = \{\varnothing, \{\varnothing\}, \{\varnothing, \{\varnothing\}\}, \{\varnothing, \{\varnothing\}, \{\varnothing, \{\varnothing\}\}\}\}$$

Question 3-3

How is the set of natural numbers symbolized? What elements does it contain?

Answer 3-3

In this book, we symbolize the set of natural numbers as N, and define it as

$$N = \{0, 1, 2, 3, 4, \ldots\}$$

The three dots, called an ellipsis, tell us that the sequence goes on without end. The set N is also known as the set of whole numbers. In some texts, N is defined without including 0:

$$N = \{1, 2, 3, 4, 5, \ldots\}$$

This is also called the set of counting numbers.

Question 3-4

According to the set-based definition of the natural numbers 0, 1, 2, 3, and so on, what *number* is represented by the entire set N?

Answer 3-4

Mathematicians call this an infinite ordinal or transfinite ordinal, and denote it using the lowercase Greek letter omega (ω). We can imagine it as a form of "infinity."

Question 3-5

How can we generate the set of even natural numbers (call it N_{even}) from the set N of natural numbers? How can we generate the set of odd natural numbers (call it N_{odd}) from N_{even}?

Answer 3-5

We can generate N_{even} by taking each element of N and multiplying it by 2. We can generate N_{odd} from N_{even} by taking every element of N_{even} and adding 1.

Question 3-6

In the set N, what is a prime number? What's a composite number? Are there any natural numbers that are neither prime nor composite? Are there any natural numbers that are both prime and composite?

Answer 3-6

A prime number is a natural number larger than 1 (in other words, 2 or larger) that can only be factored into a product of itself and 1. A composite number is a natural number that's a product of two or more primes. All the nonprime numbers larger than 1 are composite. According to these definitions, the numbers 0 and 1 are neither prime nor composite. No natural number can be both prime and composite.

Question 3-7

In the set N, what is a perfect square? Can any perfect square be prime?

Answer 3-7

A perfect square is the result of multiplying a natural number by itself. The first few perfect squares are 0, 1, 4, 9, 16, 25, 36, 49, and 64. No perfect square can be prime. By definition, 0 and 1 are not prime. Any perfect square larger than 1 can be broken down into a product of two or more primes, so it's composite.

Question 3-8

How can we write down the set Z of integers as an "implied, two-ended list"? As an "implied, one-ended list"?

Answer 3-8

Here's an "implied, two-ended list" that can give any reader the basic idea concerning the elements of Z:

$$Z = \{..., -4, -3, -2, -1, 0, 1, 2, 3, 4, ...\}$$

To create the "implied, one-ended list," we start with 0 and then go through the positive and negative integers alternately, like this:

$$Z = \{0, 1, -1, 2, -2, 3, -3, 4, -4, ...\}$$

Question 3-9

How can we quickly and easily tell if a large natural number is divisible by 2, 3, 5, 9, or 10 without a remainder?

Answer 3-9

A natural number is divisible by 2 if its last digit is 0, 2, 4, 6, or 8. A natural number is divisible by 3 if the sum of its digits is a natural-number multiple of 3, by 5 if its numeral ends in 0 or 5, by 9 if the sum of its numeric digits is a natural-number multiple of 9, and by 10 if its numeral ends in 0.

Question 3-10

How can we find the prime factors of a large natural number n?

Answer 3-10

We start by finding the square root of n. We ignore the digits after the decimal point, so we have a whole number. We add 1 to that whole number and call the result s. Then we divide the original number n by all the primes less than or equal to s, one by one, starting with the largest prime and working our way down. If we ever get a whole-number quotient, then we know that the divisor and the quotient are both factors of n. Sometimes the quotient is prime, and sometimes it is not. If it isn't prime, then it can be factored down further. We keep dividing n by smaller and smaller primes until we get down to 2. Once we've found all the primes smaller than s that divide n without leaving a remainder, we have found the prime factors of n. Some of these prime factors may occur more than once.

Chapter 4

Question 4-1

What is the meaning of the term *absolute value* with respect to integers?

Answer 4-1

The absolute value is the extent to which a number differs from 0. If we imagine the integers as points on a straight line with a "number reflector" passing through the point for 0, then the absolute value of any integer is its distance from the "number reflector." Figure 10-2 shows a couple of examples. The absolute value of an integer is always positive or 0, because it is an expression of distance without taking direction into account. The absolute value of any natural number is equal to that natural number. The absolute value of any negative integer can be obtained by removing the minus sign.

Question 4-2

Suppose we have two integers a and b. How can we define $a + b$ on a vertical number line where values increase as we move upward?

Answer 4-2

We start by finding the point on the line that corresponds to a. Then we move upward along the line for a distance of b units. We end up at the point for $a + b$.

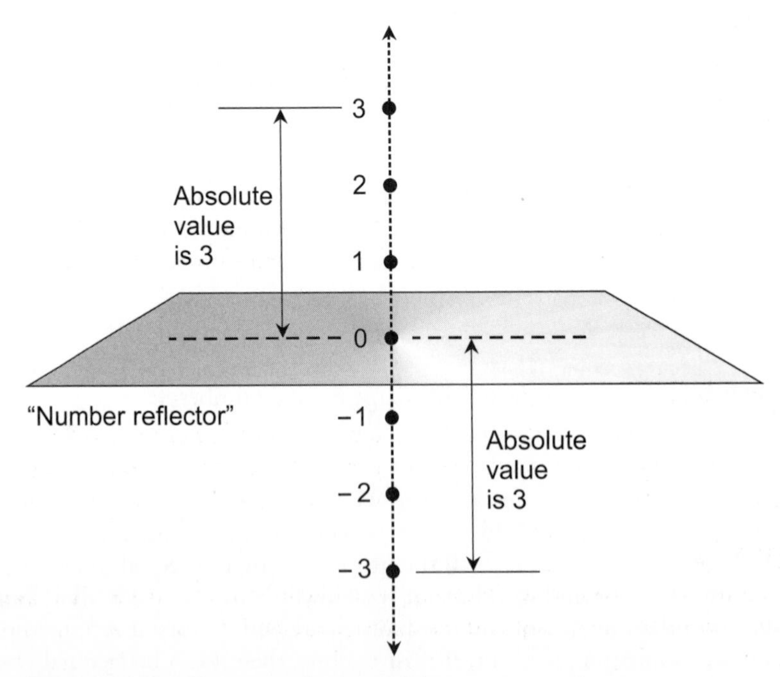

Figure 10-2 Illustration for Answer 4-1.

Question 4-3

Suppose we have two integers c and d. How can we define $c - d$ on a vertical number line where values increase as we move upward?

Answer 4-3

We start by finding the point on the line that corresponds to c. Then we move downward along the line for a distance of d units. We end up at the point for $c - d$.

Question 4-4

How do signs work when adding and subtracting positive and negative integers?

Answer 4-4

When we add a positive or subtract a negative, the result grows larger. When we subtract a positive or add a negative, the result grows smaller. For any two integers p and q,

$$p + (-q) = p - q$$

and

$$p - (-q) = p + q$$

Question 4-5

Suppose that we start with −6, add −8 to it, then subtract 12 from that, then subtract −5 from that, then add −2 to that, and finally subtract −23 from that. What's the result?

Answer 4-5

Let's work this through in steps, paying careful attention to signs and using parentheses when we need them:

$$-6 + (-8) = -6 - 8 = -14$$
$$-14 - 12 = -26$$
$$-26 - (-5) = -26 + 5 = -21$$
$$-21 + (-2) = -21 - 2 = -23$$
$$-23 - (-23) = -23 + 23 = 0$$

Question 4-6

What does the commutative law tell us about the sum of two integers? What does the associative law tell us about the sum of three integers? What do these two laws, taken together, allow us to do?

Answer 4-6

The commutative law tells us that when we add two integers, we can do it in either order and the sum will be the same. If a and b are integers, then

$$a + b = b + a$$

The associative law says that we can group a sum of three integers in a certain order by twos either way, and the result will always be the same. If a, b, and c are integers, then

$$(a + b) + c = a + (b + c)$$

In combination, the commutative and associative laws allow us to arrange and group a sum of integers in any possible way, and the result will always be the same.

Question 4-7

Does the commutative law for addition apply to sums of more than two integers? Does the associative law apply for sums of more than three integers?

Answer 4-7

Both of these laws will work for sums having as many addends as we want, as long as the number of addends is finite.

Question 4-8

Does the commutative law work for subtraction?

Answer 4-8

We cannot apply the commutative law to subtraction problems and expect valid results. Let's look at these two subtractions:

$$5 - 10 = -5$$

but

$$10 - 5 = 5$$

Question 4-9

Does the associative law work for subtraction done twice?

Answer 4-9

We can't apply the associative law to subtraction done twice and expect valid results. Here's an example showing its failure:

$$(5 - 10) - 15 = -5 - 15 = -20$$

but

$$\begin{aligned} 5 - (10 - 15) &= 5 - (-5) \\ &= 5 + 5 \\ &= 10 \end{aligned}$$

Question 4-10

Suppose a, b, and c are integers and we see the expression $a - b + c$ without any parentheses in it. Can we use parentheses for grouping, apply the associative law, and expect valid results in this situation?

Answer 4-10

No! Here's an example:

$$(5 - 10) + 15 = -5 + 15 = 10$$

but

$$5 - (10 + 15) = 5 - 25 = -20$$

Chapter 5

Question 5-1

When we multiply a positive integer c by another positive integer d, it's the equivalent of starting with c and then adding c repeatedly a certain number of times. How many times? Give an example.

Answer 5-1

The product cd is equivalent to starting with c and then adding c a total of $(d - 1)$ times. For example, we get 5×12 when we start with 5 and then add 5 over and over, a total of 11 times. Another way to look at this is to imagine that 5×12 is what we get when we start with 0 and then add 5 repeatedly, a total of 12 times.

Question 5-2

What happens if we multiply a positive integer p by a negative integer n? How can we describe that in terms of repeated addition?

Answer 5-2

It works the same way as it does when adding a positive integer. The only difference is that, as we keep adding the negative integer repeatedly, the sum gets smaller instead of larger.

Question 5-3

In which of the following quotients is there a remainder?

(a) 20/10 (b) 33/11 (c) 51/17 (d) 95/19
(e) 105/21 (f) 116/29 (g) 218/31 (h) 301/43

Answer 5-3

There is a remainder only in case (g). It's easy to see this by using a calculator to divide the numerator by the denominator in each of the expressions.

Question 5-4

Suppose a mathematician says, "The operations of addition, subtraction, and multiplication are closed over the set of integers, but the operation of division is not." What does he/she mean by this?

Answer 5-4

He/She means that we always get an integer if we add, subtract, or multiply one integer by another. But he/she also warns us that we don't always get an integer if we divide one integer by another.

Question 5-5

Which of the following "rules" is actually false? How can its wording be changed to make it right?

(a) When we multiply a positive integer by 2 or more, the result stays positive and the absolute value increases.

(b) When we divide a positive integer by 2 or more, the result stays positive and the absolute value decreases.

(c) When we multiply a negative integer by −2 or less, the result stays negative and the absolute value increases.

(d) When we divide a negative integer by −2 or less, the result becomes positive and the absolute value decreases.

Answer 5-5

All of the above "rules" are true except (c). Remember that if we multiply a negative by a negative, we get a positive! The correct way to state this rule would be, "When we multiply a negative integer by −2 or less, the result becomes positive and the absolute value increases."

Question 5-6

Sometimes we'll come across an expression that doesn't contain parentheses, brackets, or braces. This can be confusing if we don't know the order in which the operations should be done. Suppose we see this:

$$6 \times 8 - 14/2 + 3$$

What number does this represent?

Answer 5-6

Remember the rules for precedence of operations when we see expressions without parentheses. The steps go in this order:

- Do all the multiplications.
- Do all the divisions.
- Convert all the subtractions to negative additions.
- Do all the additions.

The product should be found first, and then the ratio. Then the subtraction should be changed to negative addition, and finally the additions should be done. This gives us

$$\begin{aligned} 6 \times 8 - 14/2 + 3 &= 48 - 14/2 + 3 \\ &= 48 - 7 + 3 \\ &= 48 + (-7) + 3 \\ &= 41 + 3 \\ &= 44 \end{aligned}$$

Question 5-7

How can we change the expression in Question 5-6 to indicate that the subtraction should be done first, then the multiplication, then the division, and finally the last addition? What will the result be then?

Answer 5-7

We should place an opening parenthesis to the left of the 8 and a closing parenthesis to the right of the 14, like this:

$$6 \times (8 - 14)/2 + 3$$

Now we've isolated the subtraction problem so it must be done first. We don't need to change it to negative addition in this case, because there's no risk of ambiguity with the subtraction part alone inside the parentheses. We proceed like this:

$$\begin{aligned} 6 \times (8 - 14)/2 + 3 &= 6 \times (-6)/2 + 3 \\ &= -36/2 + 3 \\ &= -18 + 3 \\ &= -15 \end{aligned}$$

Question 5-8

What does the commutative law tell us about the product of two integers? What does the associative law tell us about the product of three integers? What do these laws, taken together, allow us to do?

Answer 5-8

The commutative law tells us that when we multiply two integers, we can do it in either order and the product will be the same. If a and b are integers, then

$$ab = ba$$

The associative law says that we can group a product of three integers in a certain order by twos either way, and the result will always be the same. If a, b, and c are integers, then

$$(ab)c = a(bc)$$

Taken together, the commutative and associative laws allow us to arrange and group a product of integers in any possible way, and the result will always be the same.

Question 5-9

Does the commutative law for multiplication apply to products of more than two integers? Does the associative law apply for products of more than three integers?

Answer 5-9

Both of these laws will work for products having as many factors as we want, provided the number of factors is finite.

Question 5-10

What are the left-hand and right-hand distributive laws of multiplication over addition and subtraction? How do they work?

Answer 5-10

The distributive laws tell us how to work out products when one of the factors is a sum or difference. Imagine three integers *a, b,* and *c.* The left-hand distributive law of multiplication over addition says that

$$a(b + c) = ab + ac$$

The right-hand distributive law of multiplication over addition says that

$$(a + b)c = ac + bc$$

The left-hand distributive law of multiplication over subtraction says that

$$a(b - c) = ab - ac$$

The right-hand distributive law of multiplication over subtraction says that

$$(a - b)c = ac - bc$$

Chapter 6

Question 6-1

How can we express the following quotients in terms of an integer along with a proper fraction?

(a) 5/2 (b) −7/3 (c) 19/2
(d) −25/7 (e) 31/9 (f) 100/(−11)

Answer 6-1

The problems can be worked out as follows.

(a) The quotient is 2 with a remainder of 1 (out of 2). This is 2-1/2.
(b) The quotient is −2 with a remainder of −1 (out of 3). This is −2-1/3.

(c) The quotient is 9 with a remainder of 1 (out of 2). This is 9-1/2.

(d) The quotient is −3 with a remainder of −4 (out of 7). This is −3-4/7.

(e) The quotient is 3 with a remainder of 4 (out of 9). This is 3-4/9.

(f) The quotient is −9 with a remainder of 1 (out of −11). This is −9-1/11.

Remember that the short dashes, separating the integers from the fractions, are not minus signs! The minus signs are the longer dashes.

Question 6-2

Do the commutative and associative laws work for quotients or ratios?

Answer 6-2

Not in general. In most cases, we can't apply the commutative or associative laws to quotients or ratios and get valid results.

Question 6-3

What is the "brute force" method of reducing a ratio or fraction to its lowest terms?

Answer 6-3

We begin by factoring both the numerator and denominator into products of primes. If the original numerator is negative, we attach an extra "factor" of −1, making sure all the prime factors are positive. If the original denominator is negative, we do the same thing with it. Next, we remove all the common prime factors from both the numerator and denominator. Then we multiply all the factors in the numerator together, and do the same thing with the factors in the denominator. If we end up with a negative denominator, we multiply both the numerator and the denominator by −1.

Question 6-4

How can we use the "brute force" method to reduce 462/561 to lowest terms?

Answer 6-4

First, we factor the numerator and denominator into products of primes. This can take a little while, but we eventually get

$$(2 \times 3 \times 7 \times 11)/(3 \times 11 \times 17)$$

The common prime factors in the numerator and denominator are 3 and 11. When we remove these factors, we get

$$(2 \times 7)/17$$

Multiplying out the factors in the numerator gives us 14/17. That's the lowest form of 462/561.

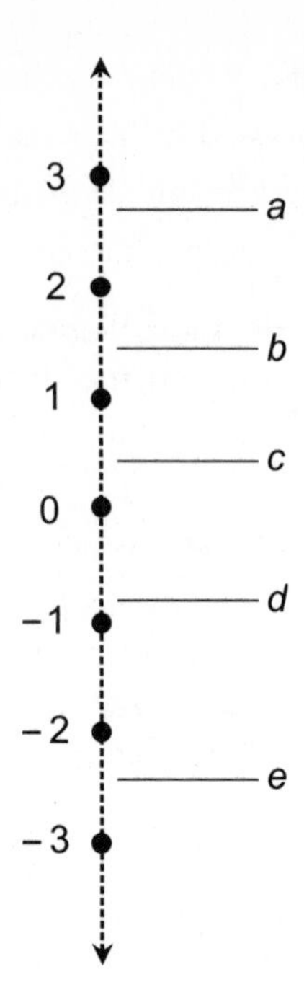

Figure 10-3 Illustration for Question and Answer 6-5.

Question 6-5

On the number line of Fig. 10-3, suppose a through e all represent ratios of integers. Which of these ratios, if any, would be proper fractions if written out with numerators and denominators?

Answer 6-5

The ratios represented by c and d would be proper fractions if written out. That's because the absolute values of the numerators would be less than the absolute values of the denominators. In the ratios represented by a, b, and e, the absolute values of the numerators would be larger than the absolute values of the denominators. That means they'd be improper fractions if written out.

Question 6-6

How can we quickly add two fractions m/n and p/q, where m and p are integers, and n and q are positive integers?

Answer 6-6

We can add these two fractions like this:

$$m/n + p/q = (mq + np)/nq$$

We can also express this in words:

- Multiply the numerator of the first fraction by the denominator of the second.
- Multiply the denominator of the first fraction by the numerator of the second.
- Add these two products together.
- Divide this sum by the product of the denominators.

Question 6-7

How can we quickly subtract a fraction p/q from a fraction m/n, where m and p are integers, and n and q are positive integers?

Answer 6-7

The difference can be found this way:

$$m/n - p/q = (mq - np)/nq$$

Stated in words:

- Multiply the numerator of the first fraction by the denominator of the second.
- Multiply the denominator of the first fraction by the numerator of the second.
- Subtract the second product from the first.
- Divide this difference by the product of the denominators.

Question 6-8

How do we multiply a fraction m/n by a fraction p/q, where m and p are integers, and n and q are positive integers?

Answer 6-8

We multiply the numerators to get the numerator of the product, and multiply the denominators to get the denominator of the product. As a formula:

$$(m/n)(p/q) = mp/nq$$

Question 6-9

How do we divide a fraction m/n by a fraction p/q, where m and p are integers, $p \neq 0$, and n and q are positive integers?

Answer 6-9

First, we invert the second fraction, making it q/p. Then we multiply the numerators to get the numerator of the product, and multiply the denominators to get the denominator of the product. As a formula:

$$(m/n)/(p/q) = mq/np$$

Question 6-10

What sort of fraction do we have on the left-hand side of the equation in Answer 6-9?

Answer 6-10

This is a ratio of fractions, also known as a compound fraction. It's a good idea to simplify expressions like this whenever we can (as the formula above shows), because compound fractions can be awkward and confusing.

Chapter 7

Question 7-1

What does it mean to raise a number to the power of 2? To the power of 3? To the power of *n*, where *n* is any positive integer?

Answer 7-1

When we raise a number to the power of 2, we multiply it by itself. This is also called squaring the number. When we raise a number to the power of 3, we multiply it by itself, and then multiply the result by the original number again. This is also called cubing the number. In general, if *b* is any number and we raise it to the power of *n*, we have

$$b^n = b \times b \times b \times \ldots \times b \; (n \text{ times})$$

Question 7-2

What is an order of magnitude in the decimal system?

Answer 7-2

An order of magnitude is a way to express how many times larger or smaller a certain quantity is, in terms of absolute value, compared to another quantity. In the base-10 system:

- When quantities differ by one order of magnitude, then the absolute value of one quantity is 10 times larger than the absolute value of the other.
- When quantities differ by two orders of magnitude, then the absolute value of one quantity is 10^2, or 100, times larger than the absolute value of the other.
- When quantities differ by *n* orders of magnitude, then the absolute value of one quantity is 10^n times larger than the absolute value of the other.

Question 7-3

What is the 0th (or zeroth) power of a nonzero number?

Answer 7-3

When we raise any number except 0 to the power of 0, we get 1. The 0th power of 0 is not defined. So, for example, $369^0 = 1$ and $(-87/16)^0 = 1$, but 0^0 is undefined.

Question 7-4

What's the difference between a terminating decimal and an endless decimal? What's the difference between an endless repeating decimal and an endless nonrepeating decimal?

Answer 7-4

A terminating decimal has a finite number of digits to the right of the decimal point. After that, they are no more nonzero digits. (If we want to add more digits, e.g., to indicate a certain level of accuracy in a physics experiment, then the digits will all be ciphers.) Here's an example:

$$25\text{-}4/100 = 25.04$$

Here's an example of an endless repeating decimal:

$$43/99 = 0.43434343...$$

Here, the digit sequence 43 repeats forever. We can repeatedly write down the digit pair 43 to the right of the decimal point, keeping at it for hours, days, or years; but the resulting decimal expression never reaches the precise value of 43/99. Now let's look at an example of an endless nonrepeating decimal:

$$\pi = 3.14159265...$$

The digits go on forever, but there is no pattern to them. We can let a computer grind out more and more digits, and the resulting decimal expression approaches (but never quite reaches) the exact value of π.

Question 7-5

How can we convert 356.0056034 into the sum of a whole number and a fraction?

Answer 7-5

We look to the left of the decimal point first. The entire string of numbers here is the integer 356. Now we look to the right of the point. There are seven digits. That means the denominator of the fraction should be 10^7 or 10,000,000, and the numerator should be the entire string of digits after the decimal point. We get

$$0056034/10{,}000{,}000$$

for the fractional part. The ciphers at the left in the numerator are useless in the fractional notation, so we can take them out and add a comma to the digits that remain, getting

$$56{,}034/10{,}000{,}000$$

The entire number is the sum of the integer part and the fractional part:

$$356 + 56{,}034/10{,}000{,}000$$

Question 7-6

When a number is written in decimal form, is the fractional equivalent (in 10ths, 100ths, 1,000ths, or whatever) in lowest terms?

Answer 7-6

Sometimes, but usually not. Consider 0.55, which converts to 55/100. This is not a fraction in lowest terms, because it can be reduced to 11/20. But 0.23 converts to 23/100. This is in lowest terms.

Question 7-7

How can we write down the fractions 1/2, 1/3, 1/4, ..., 1/10 as decimal expressions?

Answer 7-7

A calculator can be used for this purpose, but it must be able to display a lot of digits! Otherwise, old-fashioned long division is the best way to find these:

$$1/2 = 0.5$$

$$1/3 = 0.333333333333 \ldots$$

$$1/4 = 0.25$$

$$1/5 = 0.2$$

$$1/6 = 0.166666666666 \ldots$$

$$1/7 = 0.142857142857 \ldots$$

$$1/8 = 0.125$$

$$1/9 = 0.111111111111 \ldots$$

$$1/10 = 0.1$$

Question 7-8

How can we write the endless repeating decimal 0.458745874587... as a fraction?

Answer 7-8

First, we must identify the sequence of digits that repeats. Here, it's 4, 5, 8, and 7. There are four digits in the sequence, so we create a fraction with a denominator having four digits, all 9s. Then we make the repeating sequence into a four-digit numeral and use it as the numerator of the fraction. That gives us

$$0.458745874587\ldots = 4{,}587/9{,}999$$

Question 7-9

How can we make a decimal expression an order of magnitude larger or smaller?

Answer 7-9

To make it an order of magnitude larger, move the decimal point to the right by one place. To make it an order of magnitude smaller, move the decimal point to the left by one place. Here's an example. Start with

$$35{,}468.0337$$

To make this number an order of magnitude (or factor of 10) larger, we must multiply by 10. That's done by moving the decimal point to the right by one place and then repositioning the comma, getting

$$354{,}680.337$$

To make the original number an order of magnitude smaller, we divide by 10. That can be done by moving the decimal point to the left by one place and then repositioning the comma. We end up with

$$3{,}546.80337$$

Question 7-10

A negative integer power is a nonzero quantity divided by itself a certain number of times. How can we show the meaning of this by examples in the decimal system?

Answer 7-10

Let's start with the number 10. This is 10^1. When we divide 10 by itself, we get 10^0, which is 1. When we divide 10 by itself twice, we get 10^{-1}, which is 0.1. It goes on like this for negative integer powers:

10 divided by itself three times is 10^{-2}, which is $1/10^2$ or 0.01

10 divided by itself four times is 10^{-3}, which is $1/10^3$ or 0.001

10 divided by itself five times is 10^{-4}, which is $1/10^4$ or 0.0001

↓

and so on, as far as we want!

Chapter 8

Question 8-1

Suppose we start with 7 and raise it to the −1st power, then the −2nd power, then the −3rd power, then the −4th power, then the −5th power, and so on, endlessly. What happens to the result?

Answer 8-1

We get a sequence of fractions that converges toward 0. It starts at 1/7, and then keeps getting 1/7 as large with each succeeding power. Here's what happens:

$$7^{-1} = 1/7$$

$$7^{-2} = 1/7^2 = 1/49$$

$$7^{-3} = 1/7^3 = 1/343$$

$$7^{-4} = 1/7^4 = 1/2{,}401$$

$$7^{-5} = 1/7^5 = 1/16{,}807$$

↓

and so on, forever

Question 8-2

When we see a complicated expression raised to negative integer power, what must we watch out for?

Answer 8-2

We must be sure that the expression is not equal to 0, and can never attain a value of 0. Otherwise, we'll end up dividing by 0. That's forbidden!

Question 8-3

Suppose that we encounter these expressions. What restrictions must we place on x and y in each case?

(a) $(x - 5)^{-3}$ (b) $(x + y)^{-2}$
(c) $(3xy)^{-2}$ (d) $(x - y)^{-5}$

Answer 8-3

We must be sure the expressions inside the parentheses can never equal 0. Here's what we must do to stay safe:

(a) We can't let x be equal to 5.

(b) We can't let x be equal to $-y$.

(c) We can't let either x or y be equal to 0.

(d) We can't let x be equal to y.

Question 8-4

What is meant by the square root of a number? The cube root? The 4th root? The nth root, where n is a positive integer?

Answer 8-4

The square root of a number is a quantity that gives us the original number when squared (multiplied by itself or raised to the 2nd power). The cube root is a quantity that gives us the original number when cubed (raised to the 3rd power). The 4th root is a quantity that gives us the original number when raised to the 4th power. The nth root is a quantity that gives us the original number when raised to the nth power.

Question 8-5

Even-numbered roots can be ambiguous. For example, if we want to find the square root of 16, both 4 and –4 will work, because

$$4^2 = 4 \times 4 = 16$$

and

$$(-4)^2 = (-4) \times (-4) = 16$$

How can we prevent this sort of confusion?

Answer 8-5

For any positive integer n, we call the nth root of a number the $(1/n)$th power of that number. Then we insist that if n is even, the $(1/n)$th power is positive by default. We can specify that we want to use the negative value instead of the positive one (or along with it, as we'll do later in this book when we solve quadratic equations), but we must be clear about it.

Question 8-6

What happens when we take a positive odd-integer root of a negative number? A positive even-integer root of a negative number?

Answer 8-6

A positive odd-integer root of a negative number is another negative number. A positive even-integer root of a negative number is an imaginary number. We haven't worked with imaginary numbers yet.

Question 8-7

Suppose we have a nonzero number x, and two integers p and q. We want to multiply x^p by x^q. How can we express $x^p x^q$ as a single power of x?

Answer 8-7

We add the exponents p and q. Then we raise x to that power, getting

$$x^p x^q = x^{(p+q)}$$

Question 8-8

Suppose we have a nonzero number x, and two integers p and q. We want to divide x^p by x^q. How can we express x^p/x^q as a single power of x?

Answer 8-8

We find the difference $(p - q)$. Then we raise x to that power, getting

$$x^p/x^q = x^{(p-q)}$$

Question 8-9

Suppose we have a nonzero number x, and two integers p and q. We want to raise the quantity x^p to the qth power. How can we express $(x^p)^q$ as a single power of x?

Answer 8-9

We multiply p by q. Then we raise x to that power, getting

$$(x^p)^q = x^{pq}$$

Question 8-10

Suppose we have a nonzero number x, an integer p, and a nonzero integer q. We want to take the qth root of the quantity x^p. How can we express this as a single power of x?

Answer 8-10

Let's remember that the qth root of any quantity is the same as the $(1/q)$th power. This lets us apply the rule from Answer 8-9, like this:

$$(x^p)^{1/q} = x^{p(1/q)}$$
$$= x^{p/q}$$

We divide p by q and then raise x to that power. It's important that q never be 0! If it is, we'll get an exponent of $p/0$, and the entire expression will become meaningless.

Chapter 9

Question 9-1

No matter how close together two rational numbers happen to be, we can always find another rational number between them. How?

Answer 9-1

We can take the average of the original two numbers. If those two numbers are p and q, then the rational number $(p + q)/2$ is always greater than p but less than q.

Question 9-2

Imagine that we take a continuous, infinitely long geometric line and make it into a number line. Every rational number will then correspond to a unique point on this line. Does that mean every point on the line will represent a rational number?

Answer 9-2

No! Even though every rational number can be represented on the line, there are "extra" points on the line that don't correspond to any rational number.

Question 9-3

How many points on a true geometric number line do not correspond to any rational number? Give two examples.

Answer 9-3

There are infinitely many such points. A good example is the point corresponding to the positive square root of 2. We saw how that point can be located in the text of Chap. 9. Another example is π, the ratio of a circle's circumference to its diameter. We can locate this point by taking a circle with a diameter of exactly 1 unit, and then "rolling" it for one complete revolution, without slipping or sliding, upward along the number line as shown in Fig. 10-4.

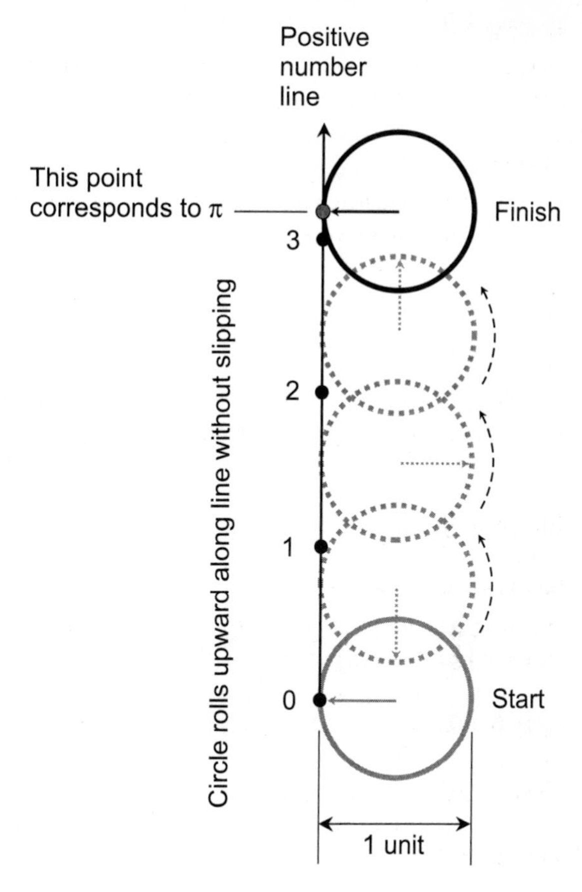

Figure 10-4 Illustration for Answer 9-3.

Question 9-4

In this book, here's how we symbolize various sets of numbers.

- The set of natural numbers is N.
- The set of integers is Z.
- The set of rational numbers is Q.
- The set of irrational numbers is S.
- The set of real numbers is R.

Which of these sets are proper subsets of which? Which pairs of sets are disjoint?

Answer 9-4

Here's how the sets are related:

- $N \subset Z$, $N \subset Q$, and $N \subset R$
- $Z \subset Q$ and $Z \subset R$
- $Q \subset R$
- $S \subset R$
- $N \cap S = \varnothing$, $Z \cap S = \varnothing$, and $Q \cap S = \varnothing$

Question 9-5

One of the following statements is false. Which one?

- Some real numbers are rational.
- Some real numbers are irrational.
- All integers are rational.
- Some integers are irrational.
- All irrationals are real.
- All integers are real.

Answer 9-5

The fourth statement is false. No integer is irrational. Every integer is rational because it can be expressed as a ratio of integers. (Any integer is equal to itself divided by the integer 1!)

Question 9-6

Which of the sets *N*, *Z*, *Q*, *S*, and *R* are denumerable? Which are not?

Answer 9-6

The sets *N*, *Z*, and *Q* are denumerable. That means the elements of each of these sets can be arranged in an "implied list," even though the "list" can't be written out in full because it's infinitely long. The sets *S* and *R* are not denumerable. Their elements can't be arranged in any sort of "implied list." Even an infinitely long "list" can't capture them all!

Question 9-7

What does it mean for an operation to be closed over the set of real numbers?

Answer 9-7

Imagine that we have an operation between two quantities. Let's call that operation "pound" and use the symbol #. The "pound" operation is closed over the set of real numbers if and only if, for any two real numbers x and y, the quantity $x \# y$ is also a real number.

Question 9-8

Which of the common arithmetic operations are closed over the set of reals? Which are not?

Answer 9-8

Addition, multiplication, and subtraction are closed over the set of real numbers. Division is not, because if we divide a real number by 0, we get an undefined quantity. Exponentiation (the raising of a real number to a real-number power) is not, because 0^0 is undefined.

Question 9-9

What about taking a real root of a real number? Is this operation closed over the set of reals?

Answer 9-9

No. Any even-integer root of a negative real number produces an imaginary number, and no imaginary number is real. The 0th root of any real number is undefined, because that's the equivalent of taking the number to the power of 1/0.

Question 9-10

What is meant by the cardinality of a set? What is the cardinality of the set N? The set Z? The set Q? The set R?

Answer 9-10

The cardinality of a set is the number of elements it contains. The sets N, Z, and Q have cardinality of $\aleph_0$ (aleph-null). The set R has cardinality larger than $\aleph_0$.

PART 2

Linear Equations and Relations

CHAPTER

11

Equations and Inequalities

You've learned how numbers and variables can be combined to form equations. Now it's time to get more adept with simple equations. You'll also learn the basic principles of *inequalities.*

Equation Morphing Revisited

In Chap. 5, we mentioned some rules for manipulating equations. Let's examine these rules in more detail. Keep in mind that in any equation, there are two or more parts separated by one or more equality symbols, also called equals signs. The most basic form of equation has a left side and a right side, with a single equality symbol between them.

Changing the order

An equation can be stated in any order. If there are only two expressions with an equals sign between them, we can reverse the right and left sides. If we see $a = b$, we can change it to $b = a$. If there are more than two expressions, we can rearrange them however we want.

Adding or subtracting a quantity

All the parts of an equation have the same value, so we can add the same quantity to each part and have those parts remain equal. Here's a simple example with numbers:

$$5/2 = 20/8$$

Both sides of this equation are equal to 2-1/2. If we add 7 to each side, we get

$$5/2 + 7 = 20/8 + 7$$

Simplifying each side of this gives us 9-1/2 = 9-1/2.

Now suppose we have some variable or quantity. We don't know its numerical value, but we can find it if we want. Let's call that quantity x. We can add x to both sides of any equation we might come across. For example,

$$5/2 + x = 20/8 + x$$

We can subtract the same quantity from both sides, or from all parts, of an equation and get another valid equation. Here's an example:

$$5/2 = 10/4 = 20/8$$

If we subtract 1/2 from each part of this, we get

$$5/2 - 1/2 = 10/4 - 1/2 = 20/8 - 1/2$$

Simplifying this gives us the three-way equation $2 = 2 = 2$.

Adding or subtracting equations

Just as we can add a quantity to, or subtract it from, all parts of an equation and get another valid equation, we can add or subtract entire equations to or from each other. Here's an example with numbers. Suppose we see these two equations:

$$5/2 = 20/8$$

and

$$3/2 = 12/8$$

If we add these two equations to each other by adding their left and right sides individually, we get

$$5/2 + 3/2 = 20/8 + 12/8$$

which simplifies to $8/2 = 32/8$, and further to $4 = 4$. We can subtract the second equation from the first, and get

$$5/2 - 3/2 = 20/8 - 12/8$$

which simplifies to $2/2 = 8/8$, and further to $1 = 1$.

Multiplying through by a quantity

We can multiply each side, or all parts, of an equation by the same quantity and get another valid equation. This is called *multiplying through*. Let's look at our original numeric example first:

$$5/2 = 20/8$$

If we multiply through by 8, we get

$$(5/2) \times 8 = (20/8) \times 8$$

which simplifies to 40/2 = 160/8, and further to 20 = 20.

Dividing through by a quantity

We can divide any equation through by a nonzero number, and we'll always get another valid equation. Dividing by any nonzero number is the same thing as multiplying by its reciprocal.

As we've already learned, when we divide an equation through by some complicated expression that contains one or more variables, we must be careful! If that expression can become equal to 0, we're in trouble. The danger is worsened by the fact that it can be difficult to tell when such trouble is taking place—until it's too late.

As an example of what can happen when we get careless about this, suppose we come across this equation and are told to find the value of x:

$$x^2 + x + 3 = 3$$

We might start by subtracting 3 from each side. That is perfectly legitimate. We then get

$$x^2 + x = 0$$

Remembering that any quantity squared is equal to that quantity multiplied by itself, we divide the equation through by x, getting

$$x^2/x + x/x = 0/x$$

Because $x/x = 1$ and $0/x = 0$ no matter what x might happen to be (or so we think in the excitement of the moment), we can simplify this to

$$x = 0$$

Confident that we have solved the equation, we "plug in" 0 for x in the original and check it out:

$$0^2 + 0 + 3 = 3$$

This simplifies to 3 = 3. "The problem has been solved," we say.

Not so fast! We've missed the other solution, which is $x = -1$. Check it out. Try "plugging in" −1 for x in the original equation, and see what happens. This oversight occurred because we made the mistake of dividing through by x when one of the two solutions to the original equation is $x = 0$. This blinded us to the existence of the other solution.

Are you confused?

To avoid mistakes like the one just described, we must never divide an equation through by any variable or expression that might happen to equal 0. The safest approach is to divide through only by nonzero numbers, and never by variables or expressions containing variables.

Here's a challenge!

Show that we cannot, in general, square both sides of an equation and end up with another equation that has the same solution.

Solution

To see what can go wrong when we square both sides of an equation, consider this:

$$x = -4$$

This is as simple as an equation can be. It states its own solution, which is −4. If we square both sides, we get

$$x^2 = 16$$

This new equation has two solutions, $x = -4$ and $x = 4$. It's clearly not equivalent to the original equation. We've gotten ourselves into trouble, just as we did when we inadvertently divided by 0. But this time, instead of missing a legitimate solution, we've generated a false one!

Inequalities

Most of the time in algebra, you'll work with equations. These are statements that involve quantities that are the same, or are supposed to be the same. But sometimes you'll need to express the fact that quantities differ, or at least the fact that they don't have to be the same. Such statements are called *inequalities.*

Not equal

When you want to indicate that two quantities are never equal, but you don't want to specify relative size or the extent to which they're different, you can use the "not equal to" symbol. You've already seen this in action. It's an equals sign with a slash through it (≠). Here are some examples of its use:

- To state that 3 is not equal to 7/2, write $3 \neq 7/2$.
- To state or require that x is never equal to 0, write $x \neq 0$.
- To state or require that x is never equal to y, write $x \neq y$.
- To state or require that $2x$ is never equal to x, write $2x \neq x$.

Strictly larger

When a certain quantity is always larger than (or greater than) some other quantity, the "strictly larger than" symbol is used. It looks like a letter V rotated a quarter-turn counterclockwise, or an arrowhead pointing to the right (>). When you use this symbol, remember that "larger" means "more positive" or "less negative."

- To state that 3 is strictly larger than $-7/2$, write $3 > -7/2$.
- To state or require that x is strictly larger than 0, write $x > 0$.
- To state or require that x is strictly larger than y, write $x > y$.
- To state or require that $2x$ is strictly larger than x, write $2x > x$.

Strictly smaller

When a certain quantity is always smaller than (or less than) some other quantity, the "strictly smaller than" symbol is used. It looks like a letter V rotated a quarter-turn clockwise, or an arrowhead pointing to the left (<). When you use this symbol, remember that "smaller" means "less positive" or "more negative."

- To state that -1 is strictly smaller than $7/2$, write $-1 < 7/2$.
- To state or require that x is strictly smaller than 0, write $x < 0$.
- To state or require that x is strictly smaller than y, write $x < y$.
- To state or require that $2x$ is strictly smaller than x, write $2x < x$.

Larger than or equal

When a certain quantity is always larger than or equal to some other quantity, the "larger than or equal" symbol is used. It looks like a Roman numeral IV rotated a quarter-turn counter-clockwise, or an arrowhead pointing to the right with a line underneath (≥).

- To state that 3 is larger than or equal to $-7/2$, write $3 \geq -7/2$.
- To state or require that x is larger than or equal to 0, write $x \geq 0$.
- To state or require that x is larger than or equal to y, write $x \geq y$.
- To state or require that $2x$ is larger than or equal to x, write $2x \geq x$.

Smaller than or equal

When a certain quantity is always smaller than or equal to some other quantity, the "smaller than or equal" symbol is used. It looks like a Roman numeral VI rotated a quarter-turn clockwise, or an arrowhead pointing to the left with a line underneath (≤).

- To state that -1 is smaller than or equal to $7/2$, write $-1 \leq -7/2$.
- To state or require that x is smaller than or equal to 0, write $x \leq 0$.
- To state or require that x is smaller than or equal to y, write $x \leq y$.
- To state or require that $2x$ is smaller than or equal to x, write $2x \leq x$.

Are you confused?

How can a quantity $2x$ can be strictly smaller than x, or smaller than or equal to x, as is mentioned twice in the above examples? Think for a moment about the meaning of "smaller" with respect to positive and negative numbers. Then remember what happens when you multiply a negative number by a positive number such as 2. Once you remember this, it's easy to see that if x is any negative number, then $2x$ is smaller than x. Now you can write

$$\text{If } x < 0, \text{ then } 2x < x$$

and

$$\text{If } x \leq 0, \text{ then } 2x \leq x$$

Check these facts out with some actual numbers and you'll see how they work. When any number is negative to begin with, doubling it makes it more negative, and therefore smaller.

Logical implication

Here is a new mathematical symbol. An "if/then" statement, such as those above, can be abbreviated using a double-shafted arrow pointing to the right, often with a little extra space on either side ($\Rightarrow$), between the "if" part of the statement and the "then" part. This arrow stands for the term *logically implies,* which in plain English translates to "means it is always true that." (It does *not* mean "causes"!) With the help of this symbol, the above facts can be shortened to

$$(x < 0) \Rightarrow (2x < x)$$

and

$$(x \leq 0) \Rightarrow (2x \leq x)$$

Try reading these statements by saying "logically implies" or "means it is always true that" when you see the arrow.

In any *logical implication* of this kind, the part of the statement to the left of the arrow is called the *antecedent.* The part of the statement to the right of the arrow is called the *consequent.*

Here's a challenge!

Write a pair of "if/then" statements that precisely define all the real numbers that, when divided by 10, become smaller than the original number.

Solution

Let's begin by seeking out all the real numbers that become strictly smaller when we divide them by 10. It's not difficult to see that any positive real will work. We can say that

$$(x > 0) \Rightarrow (x/10 < x)$$

If we start with a negative real and then divide it by 10, the result gets less negative, meaning that it becomes larger. All the negative reals therefore fail to "qualify." We want the number to get smaller, not larger! What about 0? If we divide 0 by 10, we end up with 0 again, so 0 does not "qualify" either. We want the number to get strictly smaller! Now, addition to the above statement, we can claim its reverse:

$$(x/10 < x) \Rightarrow (x > 0)$$

This means that if we divide a real number by 10 and get a strictly smaller number, the original number *must* be positive.

Logical equivalence

When a logical implication works in both directions, we have *logical equivalence.* This means that the left-hand part of the statement is true *if and only if* the right-hand part is true. The antecedent can also be the consequent, and vice-versa. To symbolize logical equivalence, we use a double-shafted, double-headed arrow, often with extra space on either side (⇔). We can also write the cryptic word "iff." Now we can answer the challenge above with a single statement. We can write either

$$(x/10 < x) \Leftrightarrow (x > 0)$$

or

$$(x/10 < x) \quad \text{iff} \quad (x > 0)$$

How Inequalities Behave

Imagine three variables, a, b, and c. Now suppose that we think of some way to compare their values. Such a "comparison tool" is called a *relation.* In algebra, the variables represent numbers. But in general mathematics, they often represent other things, such as sets or logical statements.

Three properties of relations

Suppose we symbolize a newly dreamed-up relation by a pound sign (#, read as "pound"). Let's define three properties that may (or may not) apply to this relation. Our relation is *reflexive* if and only if, for all possible values of a,

$$a \# a$$

Our relation is *symmetric* if and only if, for all possible values of a and b,

$$(a \# b) \Rightarrow (b \# a)$$

Our relation is *transitive* if and only if, for all possible values of a, b, and c,

$$[(a \# b) \,\&\, (b \# c)] \Rightarrow (a \# c)$$

We can use parentheses, braces, and brackets in logical statements, just as we use them in ordinary mathematical expressions and equations. The ampersand (&) stands for "and."

If we have a relation # that is reflexive, symmetric, and transitive, then # is called an *equivalence relation.* The converse of this is also true: If # is an equivalence relation, then # is reflexive, symmetric, and transitive.

Behavior of the = relation

You may be thinking, "I've seen some of these properties before!" In the solution to Prob. 8 in Chap. 6, we saw that the equality relation (=) is reflexive, symmetric, and transitive, so it is an equivalence relation. In contrast, the various forms of inequality are not equivalence relations.

Behavior of the ≠ relation

"Not equal" fails the test when it comes to the reflexive property. This is trivial; if "not equal" were reflexive, all variables would be different from themselves! The symmetric property, however, does work for the "not equal" relation. If a is not equal to b, then b is not equal to a. Logically, we can write this as

$$(a \neq b) \Rightarrow (b \neq a)$$

"Not equal" can't pass the transitive property test. Suppose we let a be equal to 1, b be equal to 2, and c be equal to 1. Then we have $1 \neq 2$ and $2 \neq 1$, but it is not true that $1 \neq 1$!

Behavior of the > relation

The "strictly larger" relation is neither reflexive nor symmetric. There is no number a such that $a > a$. If $a > b$, we can never say that $b > a$, no matter what a and b are. Therefore, "strictly larger" is not an equivalence relation.

The transitive property does work here. For all numbers a, b, and c, if $a > b$ and $b > c$, then $a > c$. This principle is illustrated in Fig. 11-1.

Mathematicians symbolize the words "for all" by an upside-down capital letter A (∀), and give it the fancy name *universal quantifier.* We can now write

$$(\forall\ a, b, c) : [(a > b)\ \&\ (b > c)] \Rightarrow (a > c)$$

The colon separates the quantifier from the main substance of the statement. We can read the above string of symbols out loud as, "For all a, b, and c: If a is strictly larger than b, and b is strictly larger than c, then a is strictly larger than c."

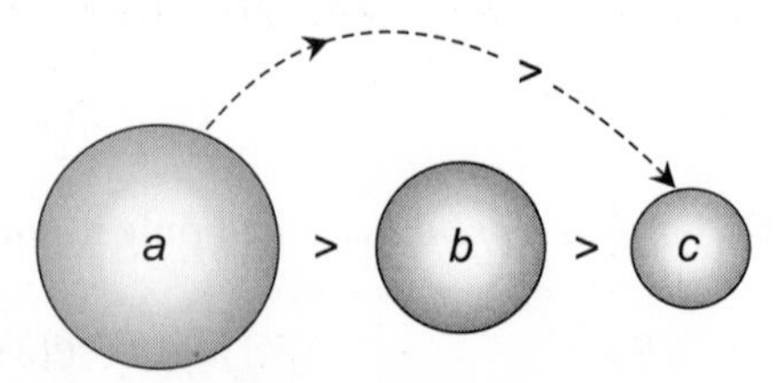

Figure 11-1 The "strictly larger" relation is transitive. If $a > b$ and $b > c$, then $a > c$.

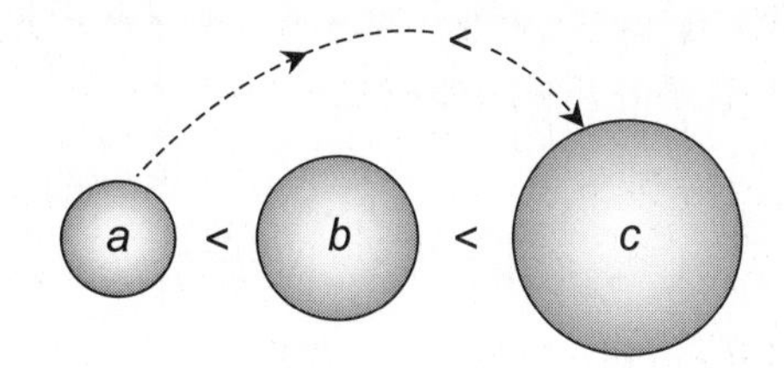

Figure 11-2 The "strictly smaller" relation is transitive. If $a < b$ and $b < c$, then $a < c$.

Behavior of the < relation

The "strictly smaller" relation fails the reflexive and symmetric tests, just as does the "strictly larger" relation. In fact, we can restate the case here by turning all the inequality symbols around. There is no number a such that $a < a$. If $a < b$, we can never say that $b < a$, regardless of the values of a and b. The "strictly smaller" relation is not an equivalence relation.

The "strictly smaller" relation is transitive. For all numbers a, b, and c, if $a < b$ and $b < c$, then $a < c$. This principle is illustrated in Fig. 11-2. We can write

$$(\forall\ a, b, c) : [(a < b)\ \&\ (b < c)] \Rightarrow (a < c)$$

Behavior of the ≥ relation

The "larger than or equal" relation is not symmetric. If $a \geq b$, then $b \geq a$ if a and b happen to be the same, but it never works if a is larger than b. Therefore, the "larger than or equal" relation is not an equivalence relation.

The reflexive and transitive properties hold true for the "larger than or equal" relation. A number a is always larger than or equal to itself, simply because it equals itself, and that's good enough! For all numbers a, b, and c, if $a \geq b$ and $b \geq c$, then $a \geq c$. We can write these two facts in formal terms as

$$(\forall\ a) : a \geq a$$

and

$$(\forall\ a, b, c) : [(a \geq b)\ \&\ (b \geq c)] \Rightarrow (a \geq c)$$

Behavior of the ≤ relation

The case for the "smaller than or equal" relation is similar to the case for the "larger than or equal" relation. If $a \leq b$, then $b \leq a$ if $a = b$, but never if $a < b$. Because of this, the "strictly smaller" relation cannot qualify as an equivalence relation. The reflexive and transitive properties, however, do hold true here. We can logically state these fact as

$$(\forall\ a) : a \leq a$$

and

$$(\forall\ a, b, c) : [(a \leq b)\ \&\ (b \leq c)] \Rightarrow (a \leq c)$$

Are you confused?

Once you know what all these symbols mean, you can read complicated logical statements out loud or write them down in words. You can even break them up and turn them into "mathematical verse." For example, the last logical sentence in the previous paragraph can be written like this.

For all *a*, *b*, and *c*:
If
a is smaller than or equal to *b*,
and
b is smaller than or equal to *c*,
then
a is smaller than or equal to *c*.

Here's a challenge!

Is logical implication an equivalence relation?

Solution

No. To see why, let's make up three statements and give them logical names:

- I'm thinking of a natural number = Tn.
- I'm thinking of a rational number = Tq.
- I'm thinking of a real number = Tr.

Now let's check out the symmetric property. From our knowledge of how the sets of naturals, rationals, and reals (N, Q, and R) are related, we know that

$$Tn \Rightarrow Tq$$

and

$$Tq \Rightarrow Tr$$

These statements translate to the following mathematical verses.

If
I'm thinking of a natural number,
then
I'm thinking of a rational number.
If
I'm thinking of a rational number,
then
I'm thinking of a real number.

We can't reverse either of these implications and still end up with a valid statement. If I'm thinking of a rational number, I'm not necessarily thinking of a natural number. (Suppose it's 1/2.) If I'm thinking of a real number, I'm not necessarily thinking of a rational number. (Suppose it's the positive square root of 2.) Logical implication is not an equivalence relation, because it's not symmetric.

Inequality Morphing

Some of the familiar equation-morphing rules also work for inequalities, but others must be modified, and a few don't work at all. Here are the things we can do with two-part equations, summarized for reference.

- Reverse the order.
- Add the same quantity to both sides.
- Subtract the same quantity from both sides.
- Add one equation to another.
- Multiply both sides by the same quantity.
- Divide both sides by the same nonzero quantity.

Whenever we do one or more of these things to an equation, we get another valid equation. Now let's see how well these rules work for inequalities. (We won't get into formal proofs of these facts.) You can try out some examples if you want to improve your understanding of how they work.

Manipulating ≠ statements

If two quantities a and b are different, we can express that fact in either order. In general, it is always true that if $a \neq b$, then $b \neq a$.

We can add or subtract the same quantity from each side of a "not equal" statement. If two quantities are different to start out with, then they'll still be different if we add or subtract the same quantity from both. If $a \neq b$, then for any number c

$$a + c \neq b + c$$

and

$$a - c \neq b - c$$

We cannot, in general, add two "not equal" statements and get another "not equal" statement. Consider $3 \neq 4$ and $8 \neq 7$. These are both true statements, but when we add them (left-to-left and right-to-right), we get $3 + 8 \neq 4 + 7$. But they are equal!

We can multiply both sides of a "not equal" statement by the same nonzero quantity and get another true statement. If the quantity is 0, then we end up with $0 \neq 0$, which is false.

We can divide both sides of a "not equal" statement by the same nonzero quantity and get another true statement. If the quantity by which we divide through is 0, we get undefined results on both sides of the inequality symbol.

Here's a summary of how we can morph "not equal" statements.

- Can we reverse the order? Yes.
- Can we add the same quantity to both sides? Yes.
- Can we subtract the same quantity from both sides? Yes.
- Can we add one statement to another? Not in general.
- Can we multiply both sides by the same quantity? Only if that quantity is not 0.
- Can we divide both sides by the same nonzero quantity? Yes.

Manipulating > statements

If some quantity a is strictly larger than another quantity b, we cannot reverse the order and still have a valid statement. It is never true that if $a > b$, then $b > a$. However, we can reverse the order if we also reverse the sense of the inequality. If $a > b$, then it is always true that $b < a$.

We can add or subtract the same quantity from each side of a "strictly larger than" statement. If $a > b$, then for any number c

$$a + c > b + c$$

and

$$a - c > b - c$$

We can always add two "strictly larger than" statements (left-to-left and right-to-right) and get another "strictly larger than" statement. For any numbers a, b, c, and d, if we have $a > b$ and $c > d$, then

$$a + c > b + d$$

We can multiply both sides of a "strictly larger than" statement by the same positive quantity and get another valid statement. If $a > b$, then for any positive number p

$$ap > bp$$

If the quantity by which we multiply through is 0, then we end up with $0 > 0$, which is false. If the quantity by which we multiply the statement through happens to be negative, the sense of the inequality is reversed. The "strictly larger than" relation turns into a "strictly smaller than" relation. If $a > b$, then for any negative number n

$$an < bn$$

We can divide both sides of a "strictly larger than" statement by the same positive quantity and get another valid statement. If $a > b$, then for any positive number p

$$a/p > b/p$$

If the quantity by which we divide through is 0, we get undefined results on both sides of the inequality symbol. If the quantity by which we divide through is negative, the sense of the inequality is reversed, just as with multiplication by a negative. If $a > b$, then for any negative number n

$$a/n < b/n$$

Here's a summary of how we can morph "strictly larger than" statements.

- Can we reverse the order? Never, unless we change the inequality to "strictly smaller than."
- Can we add the same quantity to both sides? Yes.
- Can we subtract the same quantity from both sides? Yes.
- Can we add one statement to another? Yes.
- Can we multiply both sides by the same quantity? Only if that quantity is positive.
- Can we divide both sides by the same quantity? Only if that quantity is positive.

Manipulating < statements

If some quantity a is strictly smaller than another quantity b, we cannot reverse the order and still have a valid statement. It is never true that if $a < b$, then $b < a$. But we can reverse the order if we also reverse the sense of the inequality. If $a < b$, then it is always true that $b > a$.

We can add or subtract the same quantity from each side of a "strictly smaller than" statement. If $a < b$, then for any number c

$$a + c < b + c$$

and

$$a - c < b - c$$

We can add two "strictly smaller than" statements (left-to-left and right-to-right) and get another "strictly smaller than" statement. For any numbers a, b, c, and d, if $a < b$ and $c < d$, then

$$a + c < b + d$$

We can multiply both sides of a "strictly smaller than" statement by the same positive quantity and get another valid statement. If $a < b$, then for any positive number p

$$ap < bp$$

If the quantity by which we multiply through is 0, then we end up with $0 < 0$, which is false. If the quantity by which we multiply the statement through is negative, the sense of the inequality is reversed. If $a < b$, then for any negative number n,

$$an > bn$$

We can divide both sides of a "strictly smaller than" statement by the same positive quantity and get another valid statement. If $a < b$, then for any positive number p

$$a/p < b/p$$

If the quantity by which we divide through is 0, we get undefined results on both sides of the inequality symbol. If the quantity by which we divide through is negative, the sense of the inequality is reversed. If $a < b$, then for any negative number n

$$a/n > b/n$$

Here's a summary of how we can morph "strictly smaller than" statements.

- Can we reverse the order? Never, unless we change the inequality to "strictly larger than."
- Can we add the same quantity to both sides? Yes.
- Can we subtract the same quantity from both sides? Yes.
- Can we add one statement to another? Yes.
- Can we multiply both sides by the same quantity? Only if that quantity is positive.
- Can we divide both sides by the same quantity? Only if that quantity is positive.

Manipulating ≥ statements

If some quantity a is larger than or equal to another quantity b, we cannot reverse the order, except in the situation where a happens to equal b. For that reason, it's not generally true that if $a \geq b$, then $b \geq a$. But we can reverse the order if we also reverse the sense of the inequality. If $a \geq b$, then it is always true that $b \leq a$.

We can add or subtract the same quantity from each side of a "larger than or equal" statement. If $a \geq b$, then for any number c

$$a + c \geq b + c$$

and

$$a - c \geq b - c$$

We can always add two "larger than or equal" statements (left-to-left and right-to-right) and get another valid statement. For any numbers a, b, c, and d, if $a \geq b$ and $c \geq d$, then

$$a + c \geq b + d$$

We can multiply both sides of a "larger than or equal" statement by the same nonnegative quantity and get another valid statement. If $a \geq b$, then for any nonnegative number q

$$aq \geq bq$$

If $q = 0$, we end up with $0 \geq 0$, which is true. If the quantity by which we multiply the statement is negative, the sense of the inequality is reversed. If $a \geq b$, then for any negative number n

$$an \leq bn$$

We can divide both sides of a "larger than or equal" statement by the same positive quantity and get another valid statement. If $a \geq b$, then for any positive number p

$$a/p \geq b/p$$

If the quantity by which we divide through is 0, we get undefined results on both sides of the inequality symbol. If the quantity by which we divide through is negative, the sense of the inequality is reversed. If $a \geq b$, then for any negative number n

$$a/n \leq b/n$$

Here's a summary of how we can morph "larger than or equal" statements.

- Can we reverse the order? Not in general, unless we change the inequality to "smaller than or equal."
- Can we add the same quantity to both sides? Yes.
- Can we subtract the same quantity from both sides? Yes.
- Can we add one statement to another? Yes.
- Can we multiply both sides by the same quantity? Only if that quantity is nonnegative.
- Can we divide both sides by the same quantity? Only if that quantity is positive.

Manipulating ≤ statements

If some quantity a is smaller than or equal to another quantity b, we cannot reverse the order, except when a happens to equal b. It's not generally true that if $a \leq b$, then $b \leq a$. But we can reverse the order if we also reverse the sense of the inequality. If $a \leq b$, then it is always true that $b \geq a$.

We can add or subtract the same quantity from each side of a "smaller than or equal" statement. If $a \leq b$, then for any number c

$$a + c \leq b + c$$

and

$$a - c \leq b - c$$

We can always add two "smaller than or equal" statements (left-to-left and right-to-right) and get another valid statement. For any numbers a, b, c, and d, if $a \leq b$ and $c \leq d$, then

$$a + c \leq b + d$$

We can multiply both sides of a "smaller than or equal" statement by the same nonnegative quantity and get another valid statement. If $a \leq b$, then for any nonnegative number q

$$aq \leq bq$$

If $q = 0$, then we end up with $0 \leq 0$, which is true. If the quantity by which we multiply the statement is negative, the sense of the inequality is reversed. If $a \leq b$, then for any negative number n

$$an \geq bn$$

We can divide both sides of a "smaller than or equal" statement by the same positive quantity and get another valid statement. If $a \leq b$, then for any positive number p

$$a/p \leq b/p$$

If the quantity by which we divide through is 0, we get undefined results on both sides of the inequality symbol. If the quantity by which we divide through is negative, the sense of the inequality is reversed. If $a \leq b$, then for any negative number n

$$a/n \geq b/n$$

Here's a summary of how we can morph "smaller than or equal" statements.

- Can we reverse the order? Not in general, unless we change the inequality to "larger than or equal."
- Can we add the same quantity to both sides? Yes.
- Can we subtract the same quantity from both sides? Yes.
- Can we add one statement to another? Yes.
- Can we multiply both sides by the same quantity? Only if that quantity is nonnegative.
- Can we divide both sides by the same quantity? Only if that quantity is positive.

Are you confused?

Some people find it hard to see why multiplying an inequality through by a negative number reverses its sense. Here's an example:

$$3 < 7$$

If you multiply this through by −1 without changing the sense of the inequality, you get

$$3 \times (-1) < 7 \times (-1)$$

which simplifies to

$$-3 < -7$$

This new statement is false! It becomes true if, but only if, you switch the inequality symbol from "strictly smaller than" to "strictly greater than" getting

$$-3 > -7$$

You must be careful if you multiply or divide an inequality through by a variable, or by any expression containing a variable. Suppose you multiply the original inequality in this section through by x. Then you get

$$3x < 7x$$

This changes the situation completely! You no longer have a plain statement of fact. Now you have something that contains an unknown. Keep in mind that x might be positive or negative, or even equal to 0. If x happens to be negative, the sense of the inequality must be reversed if the statement is to remain true. If x happens to be 0, the statement becomes false no matter what. It's best to stick with plain numbers, also called constants, whenever you multiply or divide an inequality through.

Are you still confused?

A whimsical way to state the above warning is to invoke a time-worn proverb known as *Murphy's law*: "If something can go wrong, it will." Whenever you do anything to an equation or inequality, ask yourself, "Is this action completely safe? Can anything bad happen?" If you suspect possible trouble, don't ignore that uneasy feeling. Check out the rules in this chapter to be sure you're "obeying the law"! A single blunder can cause an error that may remain hidden for some time. But eventually, that error will come around and bite you.

Here's a challenge!

Suppose we are given the following inequality, and we are told to derive a statement that clearly indicates all the real numbers x for which it is true:

$$x + 10 < 2x - 24$$

In other words, we must "solve the inequality." How can we do it?

Solution

We can use the rules for inequality morphing to change this statement into something with x on one side and a numeral on the other. First, let's add 24 to each side, getting

$$x + 10 + 24 < 2x - 24 + 24$$

This simplifies to

$$x + 34 < 2x$$

Now let's subtract x from each side. That gives us

$$34 < 2x - x$$

which simplifies to

$$34 < x$$

We can say this in a more intuitive way by turning it around and reversing the sense of the inequality, so we have

$$x > 34$$

This is the standard way to state the solution to any single-variable algebra problem. We put the variable all by itself on the left-hand side of the relation symbol, and a plain numeral all by itself on the right.

Here's a final challenge!

In terms of an inequality statement and set notation, describe how the nonnegative integers relate to the negative real numbers.

Solution

Let's call the set of nonnegative integers Z_{0+}, and the set of negative reals R_{-}. Any negative real number we choose will be smaller than any nonnegative integer we choose. Therefore, if x is an element of Z_{0+} and y is an element of R_{-}, then x is larger than y. In logical form along with set notation, we can write this as

$$[(x \in Z_{0+}) \;\&\; (y \in R_{-})] \Rightarrow x > y$$

Practice Exercises

This is an open-book quiz. You may (and should) refer to the text as you solve these problems. Don't hurry! You'll find worked-out answers in App. B. The solutions in the appendix may not represent the only way a problem can be figured out. If you think you can solve a particular problem in a quicker or better way than you see there, by all means try it!

1. Suppose we see this equation:

   $$7/2 = 14/4 = 21/6$$

   How can we simplify this using the rules for equation morphing, so we get a statement that says a positive integer is equal to itself?

2. How can we morph the equation in Prob. 1 so we get a statement to the effect that a negative integer is equal to itself?

3. What happens if we multiply an equation through by the number 0? What happens if we multiply an equation through by a variable or expression that ultimately turns out to equal 0, although don't know it at the time?

4. In terms of an inequality statement and set notation, describe how the negative integers relate to the natural numbers. Here's a hint: Use the same approach as we did in the final challenge.

5. In terms of an inequality statement and set notation, describe how the nonpositive real numbers relate to the nonnegative real numbers. Here's a hint: Use the same approach as we did in the final challenge.

6. In terms of an inequality statement and set notation, describe how the rational numbers relate to the irrational numbers. Here's a hint: Use the same approach as we did in the final challenge.

7. Write the following statement as a "mathematical verse." Does it represent a valid mathematical law? If not, show a counterexample.

$$(\forall\ a, b, c) : [(a \geq b)\ \&\ (b \leq c)] \Rightarrow (a = c)$$

8. For what real-number values of x is this equation true?

$$x + 4 = 2x$$

9. For what real-number values of y is this inequality true?

$$y/2 \neq 4y + 7$$

10. For what real-number values of z is this inequality true?

$$z/(-3) \leq 6z + 6$$

CHAPTER

12

First-Degree Equations in One Variable

Algebra involves the manipulation of equations or inequalities to find the values of variables, also called *unknowns*. The simplest type of algebraic equation is called a *first-degree equation in one variable*. That means there's only one unknown to solve for, and it's never raised to any power (other than the first power).

Constants, Sums, and Differences

Let's explore what happens when we form equations by adding and subtracting variables and constants on each side of an equality symbol.

Letter constants

A constant can, and often does, take the form of a plain number. Then it appears in an equation as a numeral. We might also see a constant symbolized by a letter such as a. The actual value of a so-called *letter constant* might not be revealed, but we can always be sure that it is fixed.

Letter constants can represent known irrational numbers when those numbers are impossible to write in terms of numerals alone. We've already seen an example in this book: π, the ratio of a circle's circumference to its diameter. We can't write out its exact value as a numeral. Another well-known constant is an irrational number whose first few digits are 2.7182818 ..., and which is known as the *exponential constant*. This constant is symbolized as e. Letter constants abound in physics and engineering. For example, c stands for the speed of light in a vacuum, approximately 186,000 miles per second or 300,000 kilometers per second.

When we see a letter constant in an equation, we must be sure that we know exactly what it means. For example, e and c can represent general mathematical constants, having nothing to do with exponentials or the speed of light. Here's an example:

$$ax + bx - cx - d = e$$

In this equation, c does not stand for the speed of light, and e doesn't mean the exponential constant. All five constants are meant to stand for ordinary numbers. Letters are used to avoid specifying exactly *what* numbers they are. We can, however, place restrictions on them, such as $a > 1$, $b \geq -5$, $d \neq 0$, or $c < e$.

Letter constants are convenient when we want to show an equation in a certain form. So, for example, we can write

$$a + x = b$$

to represent an equation in x, as long as we realize that a and b represent constants, and as long as we know that x is the variable. Once we know that the values of a and b always stay the same no matter what happens to x, we can morph the above equation to solve for x in terms of a and b. In this case, it's easy; we can subtract a from each side to get

$$x = b - a$$

Letter constants usually come from the first half of the English alphabet, and are usually written in lowercase. Greek letters can also represent constants.

Constants plus or minus x

Here are some first-degree equations that contain a variable x with constants added and/or subtracted.

$$x - 4 = 0$$

$$x + 7 = -2$$

$$a - x = 0$$

$$a - 5 + x = 0$$

$$a - x = b$$

These equations can be morphed to get x all by itself on the left sides of the equality symbols, and nothing but constants on the right sides. We can use the rules from Chap. 9 to do this. The above equations then become:

$$x = 4$$

$$x = -9$$

$$x = a$$

$$x = 5 - a$$

$$x = a - b$$

These are the *solutions* to the original equations, because they clearly state the values of x in terms of the constants.

Are you confused?

If you aren't sure how the five solutions come out of the five original equations above, Tables 12-1 through 12-5 show the processes in each case, step-by-step in detail. The processes shown here don't necessarily represent the only ways the equations can be solved. Nevertheless, the solution to any equation should always turn out the same, regardless of the sequence of steps. You might want to try solving some of these equations in two or three different ways.

Table 12-1. Process for solving the equation $x - 4 = 0$.

Statements	Reasons
$x - 4 = 0$	This is the equation we are given
$x - 4 + 4 = 0 + 4$	Add 4 to each side
$x = 4$	Simplify each side

Table 12-2. Process for solving the equation $x + 7 = -2$.

Statements	Reasons
$x + 7 = -2$	This is the equation we are given
$x + 7 - 7 = -2 - 7$	Subtract 7 from each side
$x = -9$	Simplify each side

Table 12-3. Process for solving the equation $a - x = 0$.

Statements	Reasons
$a - x = 0$	This is the equation we are given
$-x = -a$	Subtract a from each side
$-1(-x) = -1(-a)$	Multiply each side by -1
$x = a$	Simplify each side

Table 12-4. Process for solving the equation $a - 5 + x = 0$.

Statements	Reasons
$a - 5 + x = 0$	This is the equation we are given
$a - 5 + x - a = 0 - a$	Subtract a from each side
$-5 + x = -a$	Simplify each side
$-5 + x + 5 = -a + 5$	Add 5 to each side
$x = 5 - a$	Simplify each side

Table 12-5. Process for solving the equation $a - x = b$.

Statements	Reasons
$a - x = b$	This is the equation we are given
$a - x + x = b + x$	Add x to each side
$a = b + x$	Simplify the left side
$b + x = a$	Transpose the left and right sides
$b + x - b = a - b$	Subtract b from each side
$x = a - b$	Simplify the left side

Here's a challenge!

Manipulate the following equation so it contains x all by itself on the left side, and an expression containing the constants without x on the right side.

$$x - a + 5 = -x - b - 7 + c$$

Solution

This can be done in various ways. They'll all produce the same result. To avoid making errors with the signs, let's change all the subtractions to negative additions before we start rearranging things. That gives us

$$x + (-a) + 5 = -x + (-b) + (-7) + c$$

We can add a to each side, and then simplify the left side. This gets one of the constants out of the left side, so we have

$$x + 5 = -x + (-b) + (-7) + c + a$$

Next, let's add -5 to each side, and again simplify the left side. This removes another constant from the left side, so we have

$$x = -x + (-b) + (-7) + c + a + (-5)$$

We can add x to each side, and then simplify both sides. That gets rid of the variable on the right side, leaving only constants there. Now we have

$$2x = -b + (-7) + c + a + (-5)$$

Let's rearrange the right side to get the letter constants in alphabetical order, followed by the numerals. (That's not technically necessary, but it will make things more elegant in the end.) That gives us

$$2x = a + (-b) + c + (-7) + (-5)$$

We can divide through by 2, and then add the two plain numbers in the numerator on the right-hand side, to get

$$x = [a + (-b) + c + (-12)]/2$$

The right-hand distributive law for division over addition, applied to the expression on the right side of the equation, gives us

$$x = a/2 + (-b)/2 + c/2 + (-12)/2$$

We can simplify the right-hand side by changing all the negative additions back to subtractions, and then dividing out the numeral quotient. This gives us

$$x = a/2 - b/2 + c/2 - 6$$

Products and Ratios

Let's see what happens when the quantities on either side of an equation are multiplied by constants, divided by nonzero constants, or both.

Examples

Here are five first-degree equations that contain a variable x multiplied and/or divided by constants.

$$4x = 0$$
$$x/7 = 2$$
$$2x/a = b$$
$$5abx = c$$
$$3x/(4a) = 3$$

Using the rules from Chap. 9, we can manipulate these equations to get x alone on the left side, and the constants all by themselves on the right. That solves the equations. Here are the results.

$$x = 0$$
$$x = 14$$
$$x = ab/2$$
$$x = c/(5ab)$$
$$x = 4a$$

Are you confused?

If you can't see straightaway how these solutions are derived, Tables 12-6 through 12-10 show how the equations can be solved, step-by-step. Note that in the third and fifth original equations above (and in Tables 12-8 and 12-10), we must not let a equal 0. Also, in the fourth solution equation (and in Table 12-9), we must never allow either a or b to equal 0.

Table 12-6. Process for solving the equation $4x = 0$.

Statements	Reasons
$4x = 0$	This is the equation we are given
$4x/4 = 0/4$	Divide each side through by 4
$x = 0$	Simplify each side

Table 12-7. Process for solving the equation $x/7 = 2$.

Statements	Reasons
$x/7 = 2$	This is the equation we are given
$7x/7 = 7 \times 2$	Multiply each side by 7
$x = 14$	Simplify each side

Table 12-8. Process for solving the equation $2x/a = b$, provided $a \neq 0$.

Statements	Reasons
$2x/a = b$	This is the equation we are given
$a(2x/a) = ab$	Multiply each side by a
$2x = ab$	Simplify the left side
$2x/2 = ab/2$	Divide each side by 2
$x = ab/2$	Simplify the left side

Table 12-9. Process for solving the equation $5abx = c$, provided $a \neq 0$ and $b \neq 0$.

Statements	Reasons
$5abx = c$	This is the equation we are given
Require that $a \neq 0$ and $b \neq 0$	We are about to divide through by the product of these constants
Consider $(5ab)$ to be a single constant	This will allow us to solve the equation in a "streamlined" fashion
$5abx/(5ab) = c/(5ab)$	Divide through by the constant $(5ab)$
$x = c/(5ab)$	Simplify the left side

Table 12-10. Process for solving the equation $3x/(4a) = 3$, provided $a \neq 0$.

Statements	Reasons
$3x/(4a) = 3$	This is the equation we are given
Consider $(4a)$ to be a single constant	This will allow us to solve the equation in a "streamlined" fashion
$[3x/(4a)](4a) = 3 \times (4a)$	Multiply through by the constant $(4a)$
$3x = 12a$	Simplify both sides
$(3x)/3 = 12a/3$	Divide each side by 3
$x = 4a$	Simplify each side

Here's a challenge!

Manipulate the following equation so it contains x all by itself on the left side, and an expression containing the constants without x on the right side. Indicate, if applicable, which constants cannot equal 0.

$$3abx/(4cd) = k^2$$

Solution

We must have $c \neq 0$ and $d \neq 0$ because, if either of them are allowed to equal 0, the left-hand side of the equation becomes undefined. Let's multiply the equation through by the quantity $(4cd)$. We get

$$[3abx/(4cd)](4cd) = (4cd)k^2$$

which simplifies to

$$3abx = 4cdk^2$$

Now we can divide the entire equation through by the quantity $(3ab)$. When we do this, we must insist that $a \neq 0$ and $b \neq 0$. That produces

$$3abx/(3ab) = 4cdk^2/(3ab)$$

which simplifies to

$$x = 4cdk^2/(3ab)$$

That does it! We don't have to worry about the fact that one of the constants is squared. The square of a constant is always another constant. The variable, x, is never raised to any power (other than the first power), so the equation is a first-degree equation.

Combinations of Operations

In a first-degree equation that involves a single variable, constants can be added to or subtracted from that variable, and the variable can also be multiplied or divided by nonzero constants.

Examples

Here are some first-degree equations that involve combinations of sums, differences, products, and ratios:

$$8x - 4 = 0$$
$$18x + 7 = -2$$
$$a - 3x = 0$$
$$a - 5 + 15x = 0$$

$$a - 8x = b$$
$$-6a + 3x = 12b$$
$$6a - 3x/(bc) = -24d$$

These seven equations can all be rearranged with the morphing laws we already know, so that x appears alone on the left sides of the equality symbols, and nothing but constants appear on the right sides. Here are the respective solutions:

$$x = 1/2$$
$$x = -1/2$$
$$x = a/3$$
$$x = 1/3 - a/15$$
$$x = a/8 - b/8$$
$$x = 2a + 4b$$
$$x = 2abc + 8bcd$$

Are you confused?

Tables 12-11 through 12-17 break down the solution processes for the above equations. Some of the steps are combined, making the derivations less tedious than those earlier in this chapter. Note that in the last original equation above (and in Table 12-17), it's necessary that $b \neq 0$ and $c \neq 0$. Also note that an attempt has been made to put the solutions in elegant form by avoiding sums or differences in the numerators of fractions, putting fractions in lowest terms, and getting the letters for the constants in alphabetical order.

Table 12-11. Streamlined process for solving the equation $8x - 4 = 0$.

Statements	Reasons
$8x - 4 = 0$	This is the equation we are given
$8x = 4$	Add 4 to each side
$x = 4/8$	Divide each side by 8
$x = 1/2$	Put the fraction into lowest terms

Table 12-12. Streamlined process for solving the equation $18x + 7 = -2$.

Statements	Reasons
$18x + 7 = -2$	This is the equation we are given
$18x = -9$	Subtract 7 from each side
$x = -9/18$	Divide each side by 18
$x = -1/2$	Put the fraction into lowest terms

Table 12-13. Streamlined process for solving the equation $a - 3x = 0$.

Statements	Reasons
$a - 3x = 0$	This is the equation we are given
$-3x = -a$	Subtract a from each side
$3x = a$	Multiply through by -1
$x = a/3$	Divide through by 3

Table 12-14. Streamlined process for solving the equation $a - 5 + 15x = 0$.

Statements	Reasons
$a - 5 + 15x = 0$	This is the equation we are given
$-5 + 15x = -a$	Subtract a from each side
$15x = -a + 5$	Add 5 to each side
$15x = 5 - a$	Commutative law for addition; change negative addition to subtraction
$x = (5 - a)/15$	Divide through by 15
$x = 5/15 - a/15$	Right-hand distributive law for division over subtraction
$x = 1/3 - a/15$	Reduce the numerical fraction to lowest terms

Table 12-15. Streamlined process for solving the equation $a - 8x = b$.

Statements	Reasons
$a - 8x = b$	This is the equation we are given
$-8x = b - a$	Subtract a from each side
$-x = (b - a)/8$	Divide through by 8
$x = -(b - a)/8$	Multiply through by -1
$x = (a - b)/8$	Simplify the right side
$x = a/8 - b/8$	Right-hand distributive law for division over subtraction

Table 12-16. Streamlined process for solving the equation $-6a + 3x = 12b$.

Statements	Reasons
$-6a + 3x = 12b$	This is the equation we are given
$3x = 6a + 12b$	Add $6a$ to each side
$x = (6a + 12b)/3$	Divide through by 3
$x = 2a + 4b$	Right-hand distributive law for division over addition

Table 12-17. Streamlined process for solving the equation $6a - 3x/(bc) = -24d$, provided $b \neq 0$ and $c \neq 0$.

Statements	Reasons
$6a - 3x/(bc) = -24d$	This is the equation we are given
$-3x/(bc) = -6a - 24d$	Subtract $6a$ from each side
$3x/(bc) = 6a + 24d$	Multiply through by -1
$3x = (6a + 24d)(bc)$	Multiply through by (bc)
$3x = 6abc + 24dbc$	Right-hand distributive law for multiplication over addition
$3x = 6abc + 24bcd$	Commutative law for multiplication in second addend on right side
$x = (6abc + 24bcd)/3$	Divide through by 3
$x = 2abc + 8bcd$	Right-hand distributive law for division over addition

Standard form

Whenever you encounter an equation that can be morphed into the following form, then that equation is a first-degree equation:

$$ax + b = 0$$

where x is the variable, and a and b are constants. This is called the *standard form for a first-degree equation in one variable.* It's possible that b can equal 0. If $a = 0$, then x disappears, the statement ends up trivial or false, and it's not a first-degree equation in one variable—because there is no variable! Here are several examples:

$$x = 0$$
$$3x = 0$$
$$-4x = 0$$
$$x + 3 = 0$$
$$x - 4 = 0$$
$$5x + 2 = 0$$
$$5x - 2 = 0$$
$$-5x + 2 = 0$$
$$-5x - 2 = 0$$

Here's a challenge!

Imagine that you are working on a problem in physics, engineering, or some other branch of science and you come across this equation:

$$4/x - 8/k = 0$$

where k is a constant. Can this be rearranged into standard first-degree equation form? If so, how? If not, why not? What is the significance of the result?

Solution

Because k is a constant, the quantity $(8/k)$ must also be a constant, and can be treated as one. However, k cannot be allowed to equal 0, because in the equation as stated, it appears in the denominator of a fraction. With that in mind, you can add the quantity $(8/k)$ to each side of the equation, getting

$$4/x - 8/k + 8/k = 8/k$$

which simplifies to

$$4/x = 8/k$$

Using the rule of cross-multiplication, you know that

$$4k = 8x$$

Now you can add $-8x$ to each side, getting

$$-8x + 4k = -8x + 8x$$

which simplifies to

$$-8x + 4k = 0$$

This is in standard first-degree form, $ax + b = 0$, if you let $a = -8$ and $b = 4k$. Therefore, the original equation is a first-degree equation. At first glance, it might seem that the original equation can't be first-degree, because it looks as if x, which is the denominator of a fraction, is raised to the -1 power. But if a single-variable equation can be converted into standard first-degree form, then it is a first-degree equation.

Here's another challenge!

Suppose you come across this equation in a physics or engineering problem. Note the similarity of this to the equation in the previous challenge:

$$4/x = 0$$

Can this be rearranged into standard first-degree equation form? If so, how? If not, why not? What is the significance of the result?

Solution

You can rewrite this as

$$4/x = 0/1$$

Then you can try using the rule of cross-multiplication, getting

$$4 = 0x$$

If you reverse the sense of the equation, you have

$$0x = 4$$

Now you can subtract 4 from each side, getting

$$0x - 4 = 0$$

It's tempting to think that this is in standard form for a first-degree equation. But the constant by which you multiply x (called the *coefficient* of x) is 0, so the variable x becomes meaningless. No matter what real number you choose for x, you get an absurd result.

Word Problems

Centuries ago, the algebra in this chapter was unknown, even to the best mathematicians. Problems that seem simple to us were difficult for them. When they encountered *word problems* like the ones that follow, they often sought out solutions by making educated guesses until they "got lucky." We have a better way.

Problem A

Imagine a number. We add it to half of itself, and then add the result to 1/5 of itself. The final sum is equal to 51/10. What is the original number?

Solution A

The first step in solving any word problem is to set up an equation representing that problem. Let's call our unknown number x. The statement of the problem tells us that if we take x, then add $x/2$ to that, and then add $x/5$ to that, we get 51/10. The equation is

$$x + x/2 + x/5 = 51/10$$

We can multiply through by 10 to get

$$10x + 5x + 2x = 51$$

Let's apply the right-hand distributive law for multiplication over addition "in reverse" on the left side of the equation. That gives us

$$(10 + 5 + 2)x = 51$$

which simplifies to

$$17x = 51$$

We can divide through by 17 and get

$$x = 3$$

Problem B

The old Widow Johnson sold all her property and put the cash into a savings account. The account contained \$150,000 when she died. She left two children: Jane and Jack. Jane got married and then, like her mother, became a widow. Jane has two children to support now, and Jack is a bachelor living alone. Knowing that Jane would likely need more money than Jack, the old Widow Johnson, in the wisdom of her waning weeks, decided that Jane ought to get three times as much of the inheritance as Jack. How much did each child get?

Solution B

Let's call Jack's share of the inheritance, in dollars, x. Then Jane got $3x$. The total inheritance, in dollars, was 150,000. Therefore,

$$x + 3x = 150{,}000$$

We can simplify this to

$$4x = 150{,}000$$

Then we divide through by 4 to get

$$x = 37{,}500$$

Therefore, Jack received \$37,500. Jane got the other portion of the inheritance. That was $3 \times \$37{,}500$, or \$112,500. We can also figure Jane's share by noting that she got $\$150{,}000 - \$37{,}500$, or \$112,500.

Problem C

Solve the previous problem by letting y represent Jane's share of the inheritance, in dollars, rather than Jack's share.

Solution C

If we call Jane's share of the inheritance y, in dollars, then Jack's share was $y/3$. When we set up the equation on this basis, we get

$$y + y/3 = 150{,}000$$

which can be simplified to

$$(4/3)y = 150{,}000$$

If we multiply through by 3/4, we get

$$(3/4)(4/3)y = 3/4 \times 150{,}000$$

which simplifies to

$$y = 112{,}500$$

Jane's share was therefore \$112,500. Jack got the other portion, which was 1/3 of \$112,500, or \$37,500. We can also figure Jack's share by noting that he got \$150,000 – \$112,500, or \$37,500.

Problem D

The old Widower Jones (who was a good friend of the old Widow Johnson) sold all his property and put the money into a savings account. The account balance was \$130,000 when he died. He left three children: Joann, Jill, and Judy. All three are married today, and well-off. Nevertheless, the old Widower Jones decreed, in the dullness of his demise, that Joann should get \$10,000 less than Jill, and Jill should get \$20,000 less than Judy. How much did each child get?

Solution D

Let's say that x was the amount, in dollars, that Joann received. Then Jill got $x + 10{,}000$ dollars. Judy got $(x + 10{,}000) + 20{,}000$ dollars, or $x + 30{,}000$ dollars. We have

$$x + (x + 10{,}000) + (x + 30{,}000) = 130{,}000$$

Ungrouping these addends and then rearranging them according to the commutative law for addition, we obtain

$$x + x + x + 10{,}000 + 30{,}000 = 130{,}000$$

which simplifies to

$$3x + 40{,}000 = 130{,}000$$

Subtracting 40,000 from each side gives us

$$3x = 90{,}000$$

Dividing through by 3, we get

$$x = 30{,}000$$

That means Joann received \$30,000. Jill got \$10,000 more than Joann, or \$40,000. Judy got \$20,000 more than Jill, or \$60,000.

Are you confused?

Suppose you're trying to solve a word problem, and you set up an equation in one variable to represent it. You hope to get a first-degree equation. If you're lucky, you'll be able to get the equation into the form

$$ax + b = 0$$

where a and b are constants. However, this might not be possible. If you can't get the equation into the standard form for a first-degree equation, you'll have to use more powerful equation-solving techniques than those in this chapter. You'll learn about them in Part 3.

Here's a challenge!

Try playing the following little number game. This example is only one of infinitely many variations on the same theme. Choose any real number you want. Perform the following operations in this order:

- Multiply the number you have chosen by 2.
- Subtract 10 from the previous result.
- Multiply the previous result by 2.
- Add 60 to the previous result.
- Divide the previous result by 4.
- Subtract the number you originally picked.
- You will end up with 10, no matter what number you originally chose.

Explain how this trick works.

Solution

Games like this are fabricated by "reverse engineering." You pick the number that you want the game to end up at (in this case 10), and then perform multiple operations on successive results to obtain x. You can see how this example was put together by going through the above sequence of steps backward, replacing every addition with subtraction, every subtraction with addition, every multiplication with division, and every division with multiplication. You'll get this sequence of operations:

- Start with 10.
- Add x, obtaining $10 + x$.
- Multiply the previous result by 4, getting $40 + 4x$.
- Subtract 60 from the previous result, getting $-20 + 4x$.
- Divide the previous result by 2, getting $-10 + 2x$.
- Add 10 to the previous result, getting $2x$.
- Divide the previous result by 2, getting x.

When you build games of this sort, you must never multiply or divide by any expression containing x. If you do that, you make it possible for someone to defeat the trick by choosing x such that division by 0 occurs somewhere in the sequence of steps.

Practice Exercises

This is an open-book quiz. You may (and should) refer to the text as you solve these problems. Don't hurry! You'll find worked-out answers in App. B. The solutions in the appendix may not represent the only way a problem can be figured out. If you think you can solve a particular problem in a quicker or better way than you see there, by all means try it!

1. Using the S/R table method, morph this into standard form for a first-degree equation in one variable:

$$4x + 4 = 2x - 2$$

2. Using the S/R table method, morph this into standard form for a first-degree equation in one variable:

$$x/3 = 6x + 2$$

3. Using the S/R table method, morph this into standard form for a first-degree equation in one variable:

$$x - 7 = 7x + x/7$$

4. Using the narration method, morph this into standard form for a first-degree equation in one variable:

$$x/3 + x/6 = 12$$

5. Solve, in no more than two steps each, the standard-form equations you derived in Probs. 1 through 4.

6. When you multiply a certain number by 2, then add 8 to that product, and divide that sum by 4, you get −1. Find the number by devising a first-degree equation in x, and then solving for x.

7. Suppose we take a certain number, and then subtract 1/10 of itself. After that, we divide the result by 2, and end up with 135. Find the number by devising a first-degree equation in x, and then solving for x.

8. Bill weighs 10 kilograms (kg) more than Bruce, and Bruce weighs 5 kg more than Bonnie. The combined weight of all three people is 200 kg. What does each person weigh?

9. Imagine that we have a motorboat with a maximum *water speed* (the speed relative to the water it's floating on) of 18 miles per hour (mi/h). We make a trip upstream from our cabin to our cousins' cabin, a distance of 18 miles (mi), running the boat at top speed all the way. The trip takes 1 hour and 12 minutes (1 h 12 min). We would expect it to take exactly 1 h if there were no current in the river. But the current, which we were fighting, slowed us down. How fast was the river flowing? Assume that the river flowed at the same speed all during our journey.

10. Assuming the river keeps flowing at the same speed as we determined when we solved Prob. 9, and our boat's water speed is always 18 mi/h, how long will it take us to travel downstream from our cousins' cabin back to our own? Here's a hint: The travel time (in hours) equals the distance traveled (in miles) divided by the constant speed (in miles per hour).

CHAPTER

13

Mappings, Relations, and Functions

When the elements of two sets are associated with each other in an organized way, the association is called a *mapping*. When the set elements are numbers, a mapping is often called a *relation*. A *function* is a relation that has special characteristics about which you'll learn in this chapter.

Mapping "Territories"

Imagine two sets of points, defined by the shaded rectangles in Fig. 13-1. Suppose you're interested in the subsets shown by the hatched ovals. You want to match the points in the top oval with those in the bottom oval by means of a defined scheme. When you do this, you *map* the elements of one set to the elements of the other set.

Point matching

Think of Fig. 13-1 as portraying two vans carrying people, some of whom are using cell phones to send or receive text messages. The people in one van (call it the upper van) are represented by all the points inside the top rectangle, and the people in the other van (call it the lower van) are represented by all the points inside the bottom rectangle. Within each van, the people involved in cell-phone "texting" are represented by points inside the ovals. The arrows indicate the direction of the mapping—in this case, the direction in which messages are sent.

Domain and range

All the points actively involved in the mapping shown by Fig. 13-1 are inside the ovals. The top oval is called the *domain* of the mapping. Sometimes it's called the *essential domain*. The bottom oval is called the *range*. Here, the domain contains six points, and the range contains five points. In this example, the domain of the mapping is *exactly* the set of people in the upper van who are sending messages. The range is *exactly* the set of people in the lower van who are receiving messages.

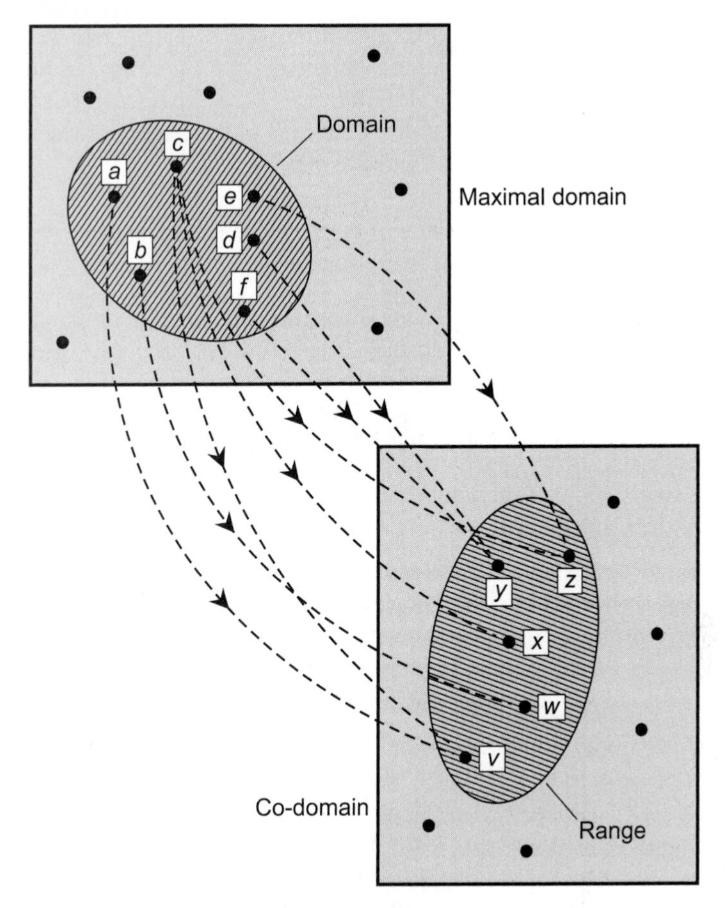

Figure 13-1 A mapping between two sets.

Maximal domain and co-domain

The large rectangles in Fig. 13-1 contain more points than those inside the ovals. The top rectangle is called the *maximal domain* of the mapping, and the bottom rectangle is called the *co-domain*. Some, but not all, of the points inside the maximal domain and the co-domain are actively involved in the mapping. In our example, the maximal domain of the mapping is the set of all people in the upper van, whether they're sending messages or not. The co-domain is the set of all people in the lower van, whether they're receiving messages or not.

The domain is a subset of the maximal domain. The range is a subset of the co-domain. Sometimes, the domain and the maximal domain of a mapping are identical. If that were the case in Fig. 13-1, then the hatched region would fill the top rectangle. Similarly, the co-domain and the range of a mapping might be identical. If that were true in Fig. 13-1, then the hatched region would fill the bottom rectangle.

Ordered pairs

When you have a mapping from the elements of one set to the elements of another set, you can define the mapping in terms of *ordered pairs*. An ordered pair is an expression in parentheses that contains two items separated by a comma. The first item represents an element of the domain. The second element represents an element of the range. In the situation shown by Fig. 13-1, the ordered pairs are (a,v), (b,w), (c,v), (c,x), (c,z), (d,y), (e,z), and (f,y). When you write an ordered pair, you can (but don't have to) put a space after the comma, as you would in an ordinary sequence or a list of set elements.

Are you confused?

In the example of Fig. 13-1, the correspondence between the points in the domain and the points in the range is not one-to-one. Point c in the domain maps to three points in the range. Points v, y, and z in the range are each mapped from two points in the domain. You can imagine that in the upper van, one person is sending messages to three different people in the lower van. In the lower van, three people are receiving messages from two different senders. "Dupes" like this are okay in a general mapping. In some situations, "dupes" are not allowed, as you'll see later in this chapter.

Here's a challenge!

Examine Fig. 13-2. Suppose the upper rectangle represents the set of all positive real numbers, and the lower rectangle represents the set of all negative real numbers. Also imagine that the upper oval represents

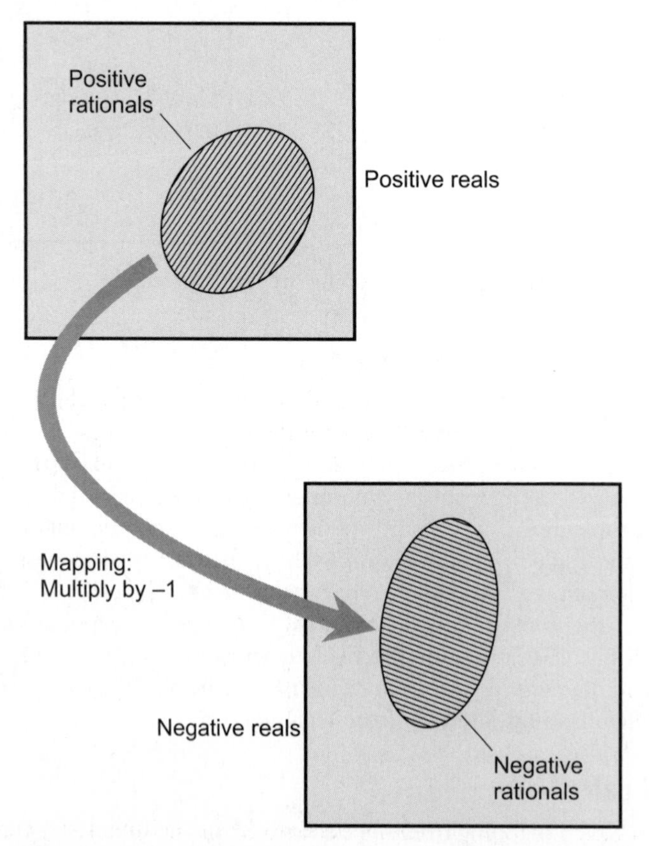

Figure 13-2 A mapping from the positive rational numbers to the negative rational numbers.

the set of all positive rationals, and the lower oval represents the set of all negative rationals. Here's the mapping: any number x in the upper oval is mapped into a number y in the lower oval by taking its additive inverse (multiplying x by -1). How can we define the ordered pairs in this mapping? What is the domain? The maximal domain? The range? The co-domain? What happens in this situation if we want to map a negative real number to something, or if we want to map something to a positive real number?

Solution

We can define the ordered pairs (x,y) as always having the form $(x,-x)$, where x is rational and $x > 0$. The domain is the set of all positive rationals. The maximal domain is the set of all positive reals. The range is the set of all negative rationals. The co-domain is the set of all negative reals. This mapping does not tell us how to map a negative real number to anything. It also fails to tell us how we would map anything to a positive real.

Types of Mappings

Mathematicians have special names for different types of mappings. You should know what these terms mean, even though they may seem strange at first! Imagine two sets of objects, called set X and set Y. Let the variable x represent an element in set X, and let the variable y represent an element in set Y. There are three major ways in which the elements of X can be mapped to the elements of Y.

Injection

Figure 13-3 shows a situation in which elements of set X are mapped to elements of set Y. This mapping has a domain that is a subset of X, and a range that is a subset of Y. Each element x

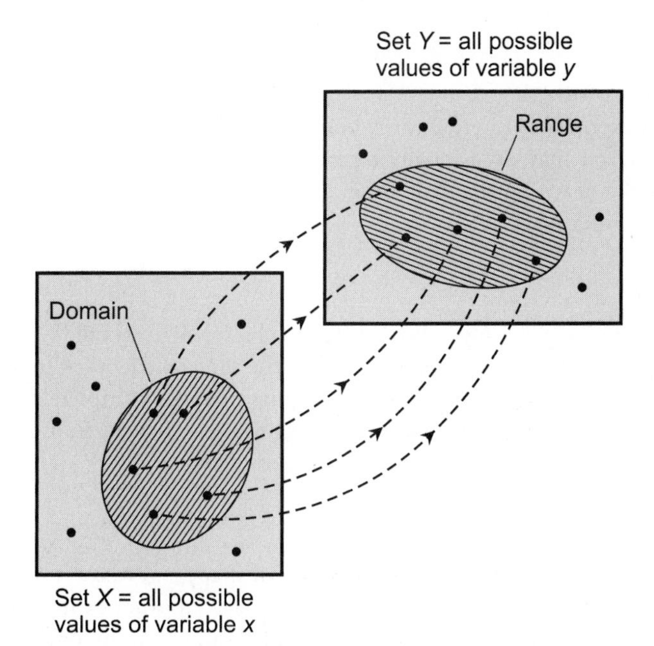

Figure 13-3 An example of an injection. Every element x maps into a single element y, and every element y maps from a single element x.

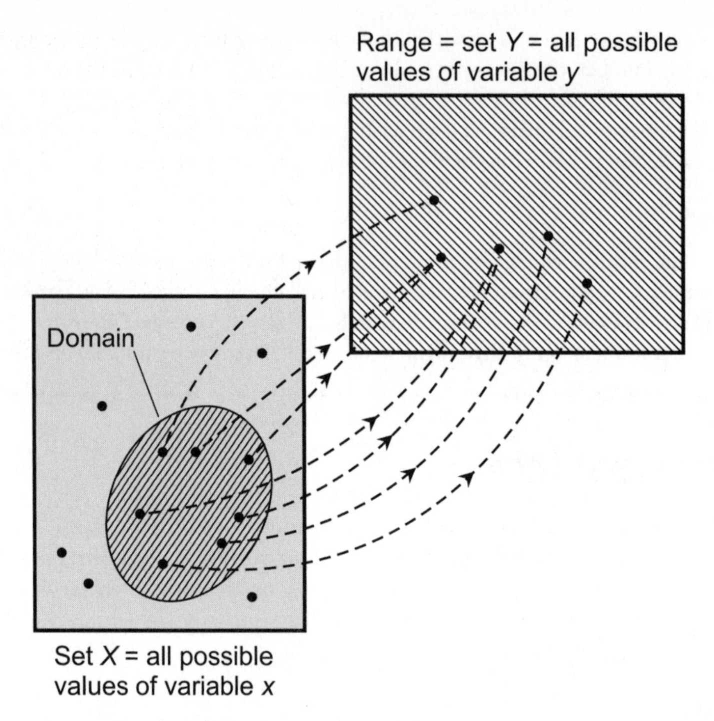

Figure 13-4 An example of a surjection. Every possible element y is accounted for, and is the result of a mapping from at least one element x.

in set X corresponds to one, but only one, element y in set Y. A mapping of this type is called an *injection.* You may occasionally hear an injection called a *one-to-one mapping,* or simply *one-to-one.* But that's a little misleading, because an injection doesn't necessarily involve all the elements of either set X or set Y.

Surjection

The mapping in Fig. 13-4 is a little different. Here, elements of X are mapped to *all* the elements of Y. The domain is a subset of X, but the range is identical to Y. This type of mapping is called a *surjection.* Because it maps elements of set X completely onto set Y, a surjection is sometimes called an *onto mapping,* or simply *onto.* A surjection can be one-to-one, but it doesn't have to be, and in this example, it clearly isn't!

Bijection

Figure 13-5 shows an example of a third type of mapping, called a *bijection,* between two sets X and Y. This is an injection that is also a surjection. Another expression you might hear is, "A bijection is both one-to-one and onto." The old-fashioned term for a bijection is *one-to-one correspondence.* That term is rarely used nowadays, because it sounds too much like "one-to-one," which some people call an injection.

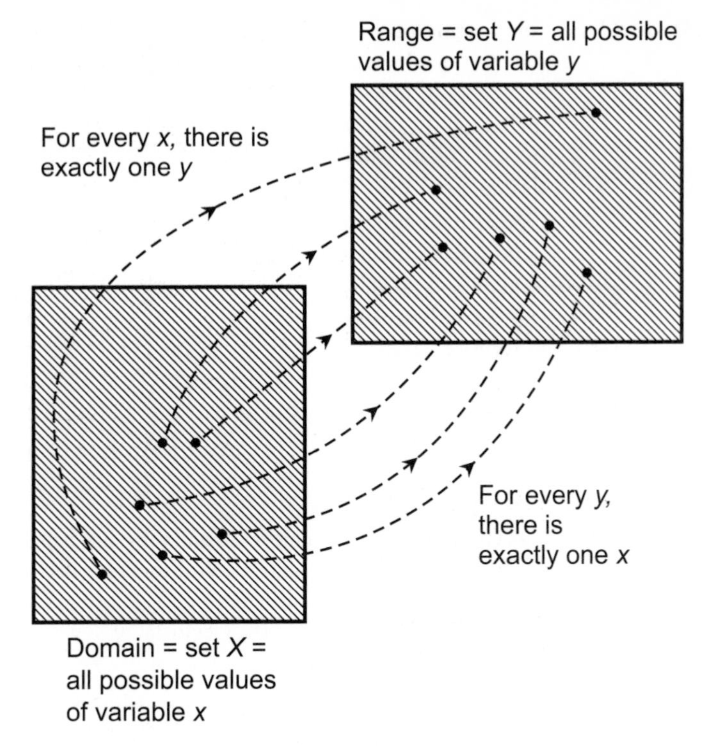

Figure 13-5 An example of a bijection. It is both an injection and a surjection.

Are you confused?

In Fig. 13-3, the domain is shown as a proper subset of set X, and the range is shown as a proper subset of Y. However, the domain could be the entire set X, or the range could be the entire set Y, or both. (If both were true, we'd have a bijection, which is a special sort of injection!) In Fig. 13-4, the domain is shown as a proper subset of X, but the range is shown as the entire set Y. Again, the domain could contain all of the elements in set X.

Here's a challenge!

Let X be the set of all real numbers x larger than 0 but smaller than 1. Let Y be the set of all real numbers y strictly larger than 1. Give an example of an injection from X into Y. Give an example of a bijection between X and Y.

Solution

If we add 1 to any number x in set X, we get a number y in set Y that's larger than 1 but smaller than 2. We can write

$$y = x + 1$$

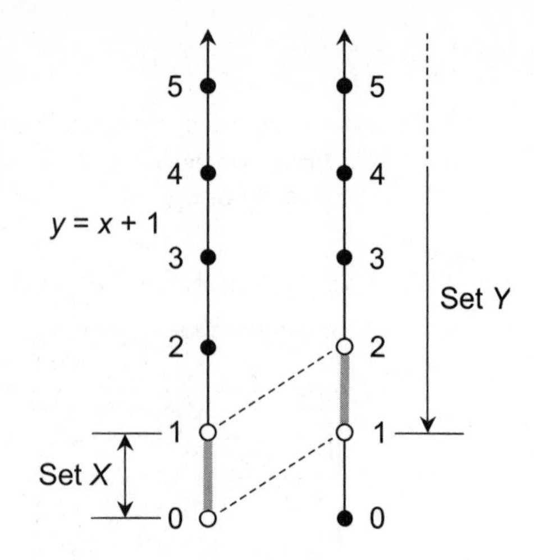

Figure 13-6 An injection from the set X of all reals between, but not including, 0 and 1 to the set Y of all reals strictly larger than 1. Open circles indicate points not in the domain and range, which are shown as heavy gray lines.

This mapping is an injection. The domain is the whole set X, and the range happens to be a proper subset of Y, as shown in Fig. 13-6. This mapping is one-to-one, because for any x in the domain, there's a single y in the range, and vice-versa. But it's not onto the entire set Y.

Now let's consider a different mapping. If we take the reciprocal of any number x in set X, we get a number y in set Y that's larger than 1. We can write

$$y = 1/x$$

This mapping is a bijection. No matter what x between 0 and 1 we choose, the reciprocal is always a *unique* (*one and only one*, or *exactly one*) real number y larger than 1. Conversely, no matter what real number y larger than 1 we choose, we can always find a unique real number x between 0 and 1 that has y as its reciprocal. This mapping, shown in Fig. 13-7, is a bijection, because every element in set Y is accounted for.

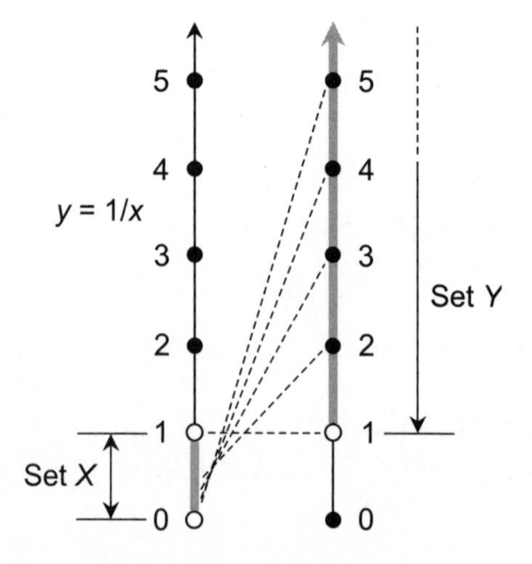

Figure 13-7 A bijection between the set X of all reals between 0 and 1 and the set Y of all reals strictly larger than 1. Open circles indicate points not included in the domain and range, which are shown as heavy gray lines.

Examples of Relations

Whenever you can express a mapping in terms of ordered pairs, then that mapping is a *relation*. The examples in the "challenges" you've seen so far in this chapter are all relations. This section will give you some more examples.

Independent vs. dependent variable

In a relation, the elements of the domain and the range can be represented by variables. If we say that x is a nonspecific element of the domain and y is a nonspecific element of the range, then x is the *independent variable* and y is the *dependent variable*. A relation therefore maps values of the independent variable to values of the dependent variable. We can also call x the "input variable" and y the "output variable," as computer scientists sometimes do.

An injective relation

Relations between sets of numbers are often represented by equations. We write the dependent variable all by itself on the left side of the equality symbol, and then write an expression containing the independent variable on the right side. Ordered pairs are produced by putting values for x into the equation, and then calculating the values for y. Here is an example:

$$y = x + 2$$

for all real numbers x. When we put specific values of x into this, we get results such as:

- If $x = -5$, then $(x,y) = (-5,-3)$
- If $x = -1$, then $(x,y) = (-1,1)$
- If $x = 0$, then $(x,y) = (0,2)$
- If $x = 3/2$, then $(x,y) = (3/2,7/2)$
- If $x = 4$, then $(x,y) = (4.6)$
- If $x = 25$, then $(x,y) = (25,27)$

This mapping is one-to-one, because for every value of x, there is exactly one value of y, and vice-versa. By definition, therefore, the mapping is an injection. We can call this relation an *injective relation.*

A surjective relation

Suppose both the maximal domain X and the co-domain Y of a particular mapping include all real numbers. Let the essential domain be the set of all nonnegative real numbers, that is, the set of all x such that $x \geq 0$. Let the range be the set of all real numbers y, so it is the same as the co-domain. Now consider this equation:

$$y = \pm(x^{1/2})$$

When we plug specific values of the independent variable x into this equation, we get results such as:

- If $x = 1/9$, then $(x,y) = (1/9,1/3)$ or $(1/9,-1/3)$
- If $x = 1/4$, then $(x,y) = (1/4,1/2)$ or $(1/4,-1/2)$
- If $x = 1$, then $(x,y) = (1,1)$ or $(1,-1)$
- If $x = 4$, then $(x,y) = (4,2)$ or $(4,-2)$
- If $x = 9$, then $(x,y) = (9,3)$ or $(9,-3)$
- If $x = 0$, then $(x,y) = (0,0)$

This mapping is clearly not an injection! For every nonzero value of x, there are two values of y. But the mapping is onto the entire co-domain. No matter what real number y we choose, we can square it and get a nonnegative real number x. That means the mapping is a surjection, so we can call this relation a *surjective relation.*

A bijective relation

Let's modify the relation in the preceding section by restricting the co-domain and range Y to the set of nonnegative reals. Then we get this equation to represent it:

$$y = x^{1/2}$$

When there is no sign in front of an expression raised to the 1/2 power, then by convention, the 1/2 power indicates the nonnegative square root. Now there's only one output value y for every input value x. We've simply declared that all negative output values are invalid! Here are some of the ordered pairs in this relation:

- If $x = 1/9$, then $(x,y) = (1/9,1/3)$
- If $x = 1/4$, then $(x,y) = (1/4,1/2)$
- If $x = 1$, then $(x,y) = (1,1)$
- If $x = 4$, then $(x,y) = (4,2)$
- If $x = 9$, then $(x,y) = (9,3)$
- If $x = 0$, then $(x,y) = (0,0)$

We now have a one-to-one mapping, and it's also onto the entire co-domain. That means it's both injective and surjective. The equation

$$y = x^{1/2}$$

represents a *bijective relation* within the set of nonnegative reals. No matter what nonnegative real number x we plug into this relation, we get a unique nonnegative real number y out of it. It also works the opposite way: No matter what nonnegative real y we want to get out of this relation, we can find a unique nonnegative real x to plug in that will give it to us.

Are you confused?

Graphs can make relations easier to understand. In the next chapter, you'll learn about one of the most common graphing schemes in mathematics, and you'll see how the above relations look when illustrated that way.

Here's a challenge!

Using Fig. 13-8 as a guide, define a relation that maps the set of natural numbers onto the set of rational numbers. List the first 13 ordered pairs in this relation. Is this relation injective? Is it surjective? Is it bijective?

Solution

Let's represent natural numbers by the independent variable n, and corresponding rational numbers by the dependent variable r. We can start with 0 for n, and map it into the rational number 0 for r. Then we

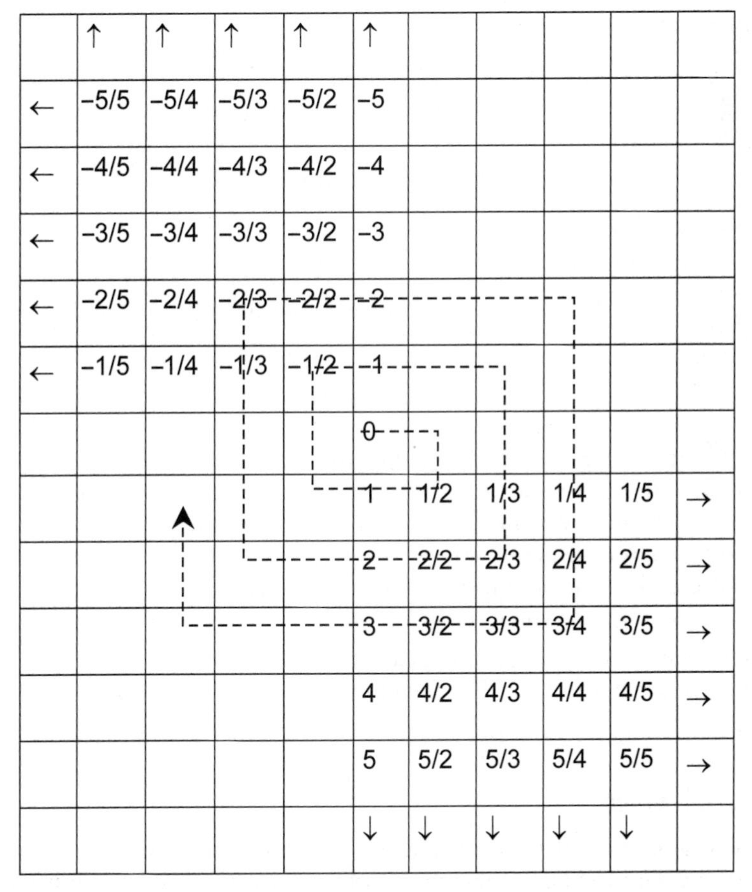

	↑	↑	↑	↑	↑					
←	−5/5	−5/4	−5/3	−5/2	−5					
←	−4/5	−4/4	−4/3	−4/2	−4					
←	−3/5	−3/4	−3/3	−3/2	−3					
←	−2/5	−2/4	−2/3	−2/2	−2					
←	−1/5	−1/4	−1/3	−1/2	−1					
					0					
					1	1/2	1/3	1/4	1/5	→
					2	2/2	2/3	2/4	2/5	→
					3	3/2	3/3	3/4	3/5	→
					4	4/2	4/3	4/4	4/5	→
					5	5/2	5/3	5/4	5/5	→
					↓	↓	↓	↓	↓	

Figure 13-8 This two-dimensional list of all rationals (duplicated from Chap. 9) suggests a way to define a relation between the set of naturals and the set of rationals.

can proceed in Fig. 13-8 along the dashed "square spiral," increasing n by 1 each time we move to the next block, and letting r be equal to the rational number in the block. When we follow that pattern, the ordered pairs in the form (n,r) are (0,0), (1,1/2), (2,1), (3,–1/2), (4,–1), (5,1/3), (6,2/3), (7,2/2), (8,2), (9,–1/3), (10,–2/3), (11,–2/2), (12,–2), and so on forever!

Some of the r values in the range of this relation have more than one n counterpart in the domain. For example, if $r = 1$, then $n = 2$. But when $r = 2/2$ (which is equal to 1), we have $n = 7$. The same thing happens with many other rationals. For example, the ordered pairs (4,–1) and (11,–2/2) have the same value of r, because $-1 = -2/2$. If we keep writing out the list of ordered pairs for a long time, we'll keep coming across "dupes" such as this.

An injection must be one-to-one, but the relation we've defined here is not of that sort! Because it's not an injection, this relation can't be a bijection. But it's a surjection. For every rational number r, we can always find at least one natural number n that maps to it. Remember that in Chap. 9, we deliberately engineered this two-dimensional list so it would account for every possible rational number.

Examples of Functions

A *function* is a relation in which every element in the domain has *at most* one element in the range. That is, for every value of the independent variable, there can never be more than one value for the dependent variable. Even so, a single value of the dependent variable might be mapped from two, three, four, or more values of the independent variable—even infinitely many. Let's look at some examples of functions.

Add 1 to the input

Consider a relation in which the independent variable is x and the dependent variable is y, and for which the domain and range are both the entire set of real numbers. Our relation is defined as follows:

$$y = x + 1$$

This is a function between x and y, because there's never more than one value of y for any value of x. In fact, for every value of x, there is exactly one value of y. This function is bijective. It maps values of x onto the entire set R, and it is one-to-one.

Mathematicians name functions by giving them letters of the alphabet such as f, g, and h. In this notation, the dependent variable is replaced by the function letter followed by the independent variable in parentheses. We might write

$$f(x) = x + 1$$

to represent the above equation, and then we can say, "f of x equals x plus 1."

Square the input

Now let's look at another simple relation. The independent variable is v and the dependent variable is w. The domain is the entire set of reals, and the range is the set of nonnegative reals. Here's the equation that represents the relation:

$$w = v^2$$

This is another example of a function. If we call it g, we can write

$$g(v) = v^2$$

For every value of v in the domain of g, there is exactly one value of w, which we can also call $g(v)$, in the range. But the reverse is not true. For every nonzero value of w in the range of g, there are two values of v in the domain. These two values are additive inverses (negatives) of each other. For example, if $w = 49$, then $v = 7$ or $v = -7$. This is not a problem; a relation can be many-to-one and still be a function. The trouble happens when a relation is one-to-many. Then it can't be a function.

Our function g is not injective because it's two-to-one, not one-to-one (except when $v = 0$). Therefore, it cannot be bijective. The function g is surjective, however, because every possible value in its range (the set of nonnegative reals) is accounted for. In formal language we say, "For any nonnegative real number w in the range of g, there is at least one v in the domain such that $g(v) = w$."

Cube the input

Here's another relation. Let's call the independent variable t and the dependent variable u. The domain and range are both the entire set of reals. The equation is

$$u = t^3$$

This is a function. If we call it h, then

$$h(t) = t^3$$

For every value of t in the domain of h, there is exactly one value of u in the range. The reverse is also true. For every value of u in the range of h, there is exactly one t in the domain.

This function is one-to-one, so it's injective. It maps onto the entire range, so it's surjective as well. That means h is a bijection.

Are you confused?

As with relations, graphing can be useful when you want to see how functions map values of an independent variable into values of a dependent variable. Graphs of the above three functions are shown in the next chapter.

Here's a challenge!

With any relation, you can transpose the values of the independent and dependent variables while leaving their names the same. You can also transpose the domain and the range. When you do these things, you get another relation, which is called the *inverse relation* (or simply the *inverse* if the context is clear). The inverse of a relation is denoted by writing a superscript −1 after the name of the relation. If you have $h(z)$, its inverse is written $h^{-1}(z)$. The inverse of a function, however, is not necessarily another function! Look again at the three functions f, g, and h above. The inverse of one of these functions is not a function. Which one?

Solution

The inverse of g is not a function. Here's the function again. Remember that the domain is the entire set of reals, and the range is the set of nonnegative reals:

$$w = g(v) = v^2$$

If you take the equation $w = v^2$ and transpose the positions of the independent and dependent variables, you get

$$v = w^2$$

This is the same as

$$w = \pm(v^{1/2})$$

The plus-or-minus symbol is important here! It indicates that for every value of the independent variable v you plug into this relation, you'll get two values of w, one positive and the other negative. You can also write

$$g^{-1}(v) = \pm(v^{1/2})$$

The function g is two-to-one (except when $v = 0$), and that's okay. But the inverse is one-to-two (except when $w = 0$). That makes g^{-1} a legitimate relation, but prevents it from being a function.

The other two functions above have inverses that are also functions. Both f and h are one-to-one, so their inverses must also be one-to-one. We have

$$f(x) = x + 1$$

and

$$f^{-1}(x) = x - 1$$

We also have

$$h(t) = t^3$$

and

$$h^{-1}(t) = t^{1/3}$$

If a function is one-to-one over a certain domain and range, then you can transpose the values of the independent and dependent variables while leaving their names the same, and you can transpose the domain and range. The resulting inverse is a function. If a function is many-to-one, then its inverse is one-to-many, and is therefore not a function.

Practice Exercises

This is an open-book quiz. You may (and should) refer to the text as you solve these problems. Don't hurry! You'll find worked-out answers in App. B. The solutions in the appendix may not represent the only way a problem can be figured out. If you think you can solve a particular problem in a quicker or better way than you see there, by all means try it!

1. Imagine a dance at which there are 15 boys and 15 girls. Every boy writes his name on a slip of paper, and then puts the slip into a jar. (No two boys have the same name.) The girls pick slips out of the jar, one at a time and one for each girl, to determine who their dance partner will be. If we think of this action as a mapping from the set of boys to the set of girls, what type of mapping is this?

2. Imagine another dance at which there are 10 boys and 15 girls. Every girl writes her name on a slip of paper, and then puts the slip into a jar. (No two girls have the same name.) The boys pick slips out of the jar, with every other boy taking two slips instead of only one, so all the slips get taken. This determines how the dance partners will be assigned. Five of the boys end up dancing with one girl, but the other five have to contend with two girls! If we think of this action as a mapping from the set of boys to the set of girls, what type of mapping is it?

3. If we think of the mapping in Prob. 2 in reverse (that is, from the set of girls to the set of boys) what type of mapping is it?

4. Imagine that you and I were once members of Internet Network Alpha, which had 60,000 members. We were not totally honorable characters, you and I. We conspired to send a mass e-mail message (also called *spam*) to every single one of the 175,000 members of Internet Network Beta. Let A be the set of all members of Alpha at that time, and let B be the set of all members of Beta. Suppose that we had the latest programs to defeat antispam software in other people's computers, so we succeeded in our dubious quest. Every single member of Beta got our message. What sort of mapping did we carry out from set A to set B? What was the maximal domain? What was the essential domain? What was the co-domain? What was the range? (Note: As a result of our behavior, we were kicked out of Internet Network Alpha, a punishment which, we realize today, was well deserved.)

5. Consider the set Q of all rational numbers and the set R of all real numbers. Suppose we devise a relation from Q to R that takes every integer q in Q and doubles it to get an even integer r in R. The maximal domain is Q, and the essential domain is Z, the set of integers, which is a proper subset of Q. The co-domain is R, and the range is Z_{even}, the set of all even integers, which is a proper subset of R. What type of relation from Q to R have we devised?

6. Suppose that we create a relation between the set Q of rational numbers and the set Z of integers. To generate an integer z from a rational q, we find the fractional equivalent of q in lowest terms, and then chop off the denominator. If we let q be the independent variable and z be the dependent variable, what type of relation is this? Is it injective? Is it surjective? Is it bijective?

7. Suppose that we create a relation between the set Z of integers and the set Q of rationals. To generate a rational q from a nonzero integer z, we find the reciprocal of

z (that is, we divide 1 by z). If $z = 0$, we say that q is not defined. If we let z be the independent variable and q be the dependent variable, is this relation injective? Is it surjective? Is it bijective?

8. Again, consider a relation between the set Z of integers and the set Q of rationals. Imagine that this relation works in the same way as the relation in Prob. 7, but with one exception. If $z = 0$, we say that $q = 0$ by default. If we let z be the independent variable and q be the dependent variable, is this relation injective? Surjective? Bijective?
9. Consider the relation $y = x^4$, where the domain is the entire set of reals and the range is the set of nonnegative reals. Is this relation a function? If so, why? If not, why not? Is the relation injective? Is it surjective? Is it bijective?
10. Imagine that the values of the independent and dependent variables in Prob. 9 are transposed while leaving their names the same. Also suppose that the domain of the new relation is the set of nonnegative reals, and the range of the new relation is the entire set of reals. Is this inverse relation a function? If so, why? If not, why not? Is the relation injective? Is it surjective? Is it bijective?

CHAPTER

14

The Cartesian Plane

In the 1600s, the French mathematician *Rene Descartes* (pronounced "re-NAY day-CART") invented a way to illustrate relations and functions. It became a graphing scheme now known as *Cartesian* (pronounced "car-TEE-zhun") *coordinates* or the *Cartesian plane.*

Two Number Lines

The Cartesian plane is put together by placing two real-number lines so they intersect at a right angle. The number lines usually intersect at their 0 points. The point where the axes intersect is called the *origin.*

Variables and ordered pairs

Figure 14-1 shows a simple set of Cartesian coordinates. The independent variable is portrayed along a horizontal line, and the dependent variable is portrayed along a vertical line. The number-line scales are graduated in increments of the same size.

Figure 14-2 shows how several ordered pairs of the form (x, y) are plotted as points on the Cartesian plane. Here, x is the independent variable and y is the dependent variable.

The quadrants

Any pair of intersecting lines divides a plane into four parts. In the Cartesian system, these parts are called *quadrants* (Fig. 14-3):

- In the *first quadrant,* both variables are positive.
- In the *second quadrant,* the independent variable is negative and the dependent variable is positive.
- In the *third quadrant,* both variables are negative.
- In the *fourth quadrant,* the independent variable is positive and the dependent variable is negative.

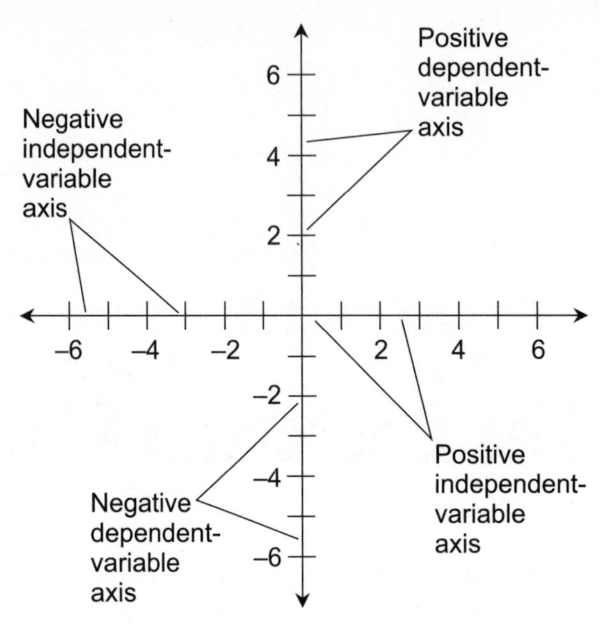

Figure 14-1 The Cartesian plane consists of two real-number lines intersecting at a right angle, forming axes for the variables.

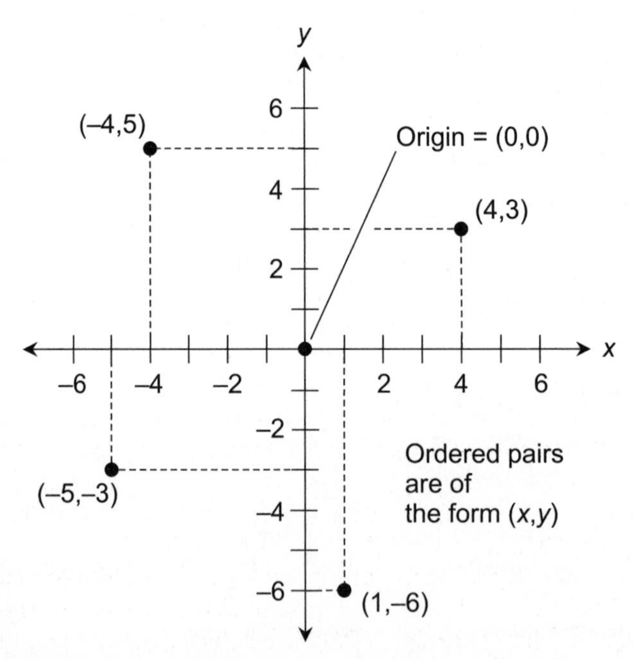

Figure 14-2 Five ordered pairs plotted as points on the Cartesian plane. The dashed lines are for axis location reference only.

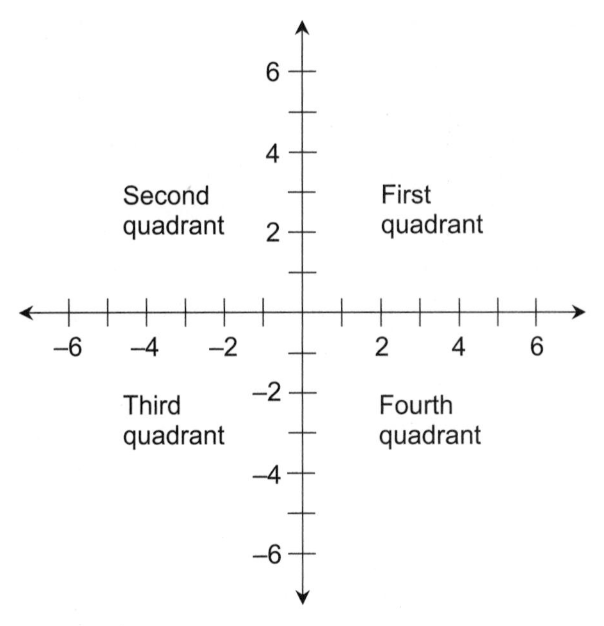

Figure 14-3 The Cartesian plane is divided into quadrants. The first, second, third, and fourth quadrants are sometimes labeled I, II, III, and IV respectively.

If a point lies exactly on one of the axes or at the origin, then it is not in any quadrant. The quadrants are sometimes labeled with Roman numerals. In most Cartesian graphs, they're located like this:

- Quadrant I is at the upper right
- Quadrant II is at the upper left
- Quadrant III is at the lower left
- Quadrant IV is at the lower right

Axis increments

In a true Cartesian coordinate plane, both axes are *linear,* and both axes are graduated in increments of the same size. This means that for any given axis, the change in value is always directly proportional to the physical displacement. If we move 1/4 of an inch along an axis and the value changes by 1 unit, then that fact is true everywhere along that axis and it is also true everywhere along the other axis.

In a more generalized system called *rectangular coordinates* or the *rectangular coordinate plane,* the two axes do not have to be graduated in the same increments. The value on one axis might change by 1 unit for every 1/4 of an inch, while the value on the other axis changes by 25 units for every 1/4 of an inch. The increments we select for each axis depend on what sort of relation or function we want to graph. It's best to choose the increments so a graph is easy to read.

Are you confused?

The coordinate planes in Figs. 14-1, 14-2, and 14-3 show values only to up to ±6 for each variable. If we want to show graphs "far out," we can increase the numbers on one or both scales. Instead of going from −6 to 6 in increments of 1 unit per division, we can go from −60 to 60 in increments of 10 units per division, or from −3,000 to 3,000 in increments of 500 units per division. If we want to graph something "close in," we can make the numbers on the scales smaller. We might go from −0.6 to 0.6 in increments of 0.1 unit per division, or from −0.0012 to 0.0012 in increments of 0.0002 unit per division! Our use of 6 increments on each of the four scales is arbitrary. We can have more or fewer, as long as we draw the coordinate system so it's easy to read.

Here's a challenge!

Imagine an ordered pair (x, y). You have plotted its point on the Cartesian plane. Neither x nor y is equal to 0, so the point does not fall on either axis. What will happen to the location of the point if you multiply both x and y by −1?

Solution

The point will move diagonally to the opposite quadrant. In other words, it will go "kitty-corner" across the coordinate plane, as follows:

- If it starts out in the first quadrant, it will move to the third.
- If it starts out in the second quadrant, it will move to the fourth.
- If it starts out in the third quadrant, it will move to the first.
- If it starts out in the fourth quadrant, it will move to the second.

If you have trouble envisioning this, draw a Cartesian plane on a piece of graph paper. Then plot a specific point or two in each quadrant. Calculate how the x and y values change when you multiply both of them by −1, and then plot the new points.

Three Relations

To graph a relation in Cartesian coordinates, we can pick a few numerical values for the independent variable, calculate the resulting values for the dependent variable, and plot the ordered pairs as points. When we've plotted enough points so we're reasonably sure we know what the graph will look like, we can connect the points with a smooth line or curve. This line or curve is the actual graph.

Add 2 to the input

Figure 14-4 shows some points plotted in Cartesian coordinates for the following relation, which was the first one we evaluated in Chap. 13:

$$y = x + 2$$

These points lie along a straight line. We can plot more points for the relation, and they will always lie along the same straight line. The line is the graph of the relation.

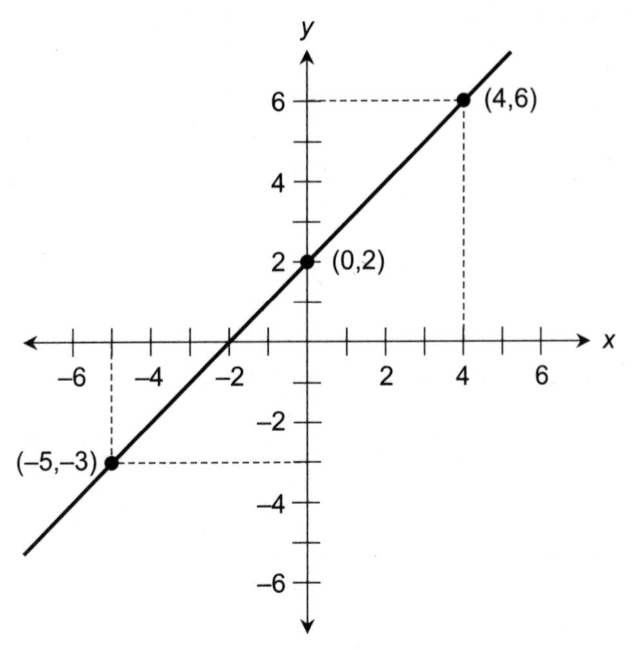

Figure 14-4 Cartesian graph of the relation $y = x + 2$.

Positive/negative square root

The second of the three relations we evaluated in Chap. 13 took the positive or negative square root of *x*, as follows:

$$y = \pm(x^{1/2})$$

Figure 14-5 shows some points for this relation, along with the curve that connects them. This curve is called a *parabola*. It is characteristic of *quadratic equations*, which we'll study in Part 3.

Nonnegative square root

The third relation we looked at in Chap. 13 was the same as the second one, but without the negative values of *y*. It involved taking the nonnegative square root:

$$y = x^{1/2}$$

Some points for this relation are plotted in Fig. 14-6, along with the curve that connects them. It appears as the nonnegative half of the curve in Fig. 14-5.

Are you confused?

How many points must you plot before you truly know what the graph of a relation looks like? The best answer is, "It depends." With simple relations such as those in this chapter, a few points are enough. With more complicated relations, you might have to plot many points before the complete graph can be

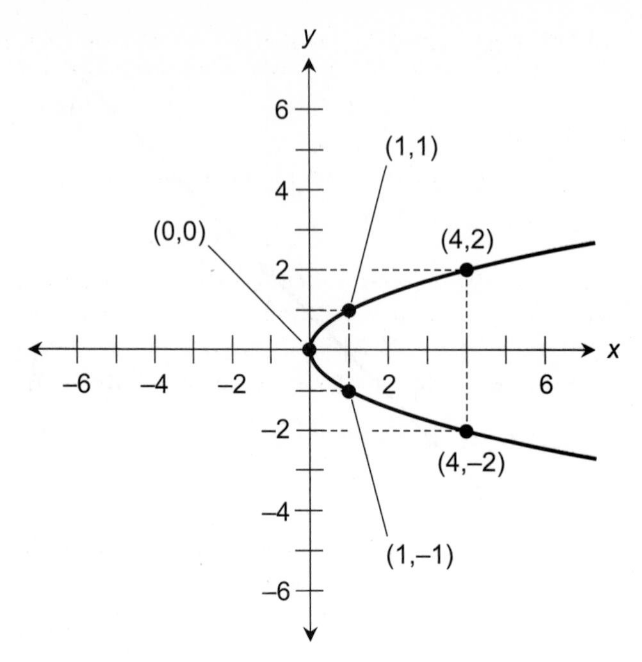

Figure 14-5 Cartesian graph of the relation $y = \pm(x^{1/2})$.

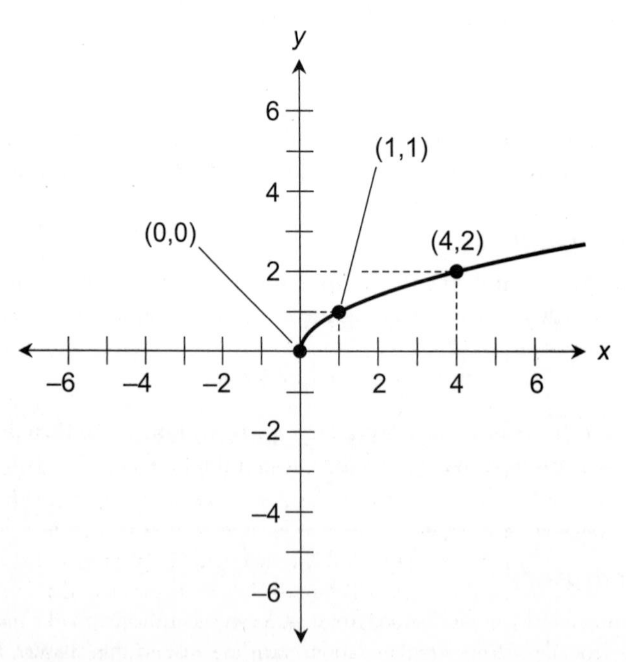

Figure 14-6 Cartesian graph of the relation $y = x^{1/2}$.

determined. There are computer programs that generate detailed graphs of relations by plotting millions of points and connecting them by means of a scheme called *curve fitting*.

Here's a challenge!

Draw graphs of the inverses of the three relations in this section.

Solution

First, let's figure out the equations for these inverse relations. To do that, we must transpose the values of the variables without changing their names. We must also transpose the domain and the range. The three original relations are

$$y = x + 2$$
$$y = \pm(x^{1/2})$$
$$y = x^{1/2}$$

In Chap. 13, we were given the domains and ranges for these relations. In the first case, the domain and range are both the entire set of reals. In the second case, the domain is the set of nonnegative reals, and the range is the set of all reals. In the third case, the domain and range are both the set of nonnegative reals. Switching the values of the variables by reversing their positions in the original equation, we get

$$x = y + 2$$
$$x = \pm(y^{1/2})$$
$$x = y^{1/2}$$

When we manipulate these equations to get x in terms of y, the results are

$$y = x - 2$$
$$y = x^2$$
$$y = (+x)^2$$

When we transpose the domain and the range from the original relations in the first case, they both remain the entire set of reals. In the second case, the domain becomes the set of all reals, and the range becomes the set of nonnegative reals. In the third case, the domain and range both remain the set of nonnegative reals. The plus sign in the last equation means that we consider it only for nonnegative values of x, because negative values of x aren't part of the domain.

Now that we know the equations for the inverse relations, we're ready to graph them. A simple trick makes it easy to graph the inverse of any relation. We draw the line representing all points where the independent and dependent variables have the same value. We can imagine this line, represented by the equation $y = x$ in these examples, as a "point reflector." It works like the "number reflector" for generating negative numbers from positive numbers on the number line. (We devised that gimmick in Chap. 3. Now we're operating in two dimensions instead of one.) For any point that's part of the graph of the original relation, we can find its counterpart in the graph of the inverse relation by going to the opposite side of the "point reflector," exactly the same distance away. Figure 14-7 shows how this works. The line connecting a point in the original graph and its "mate" in the inverse graph is perpendicular to the "point reflector." In addition, the "point reflector" intersects every point-connecting line exactly in the middle. Technically, we say that the "point reflector" is a *perpendicular bisector* of any line connecting a point with its inverse point.

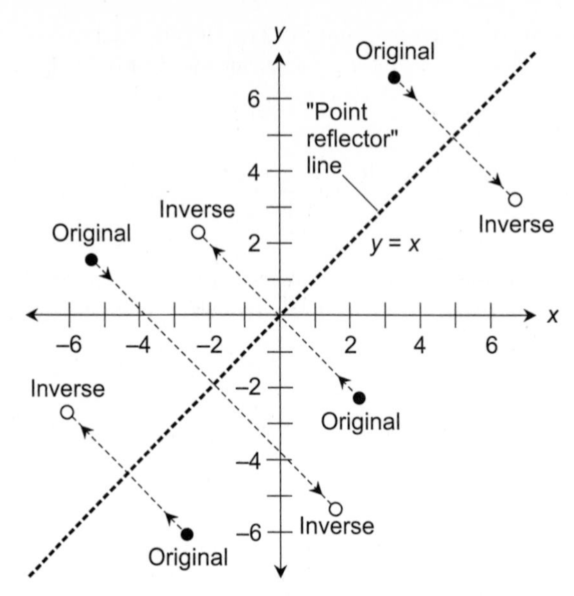

Figure 14-7 Any point in the graph of the inverse of a relation can be located on the basis of its "mate" in the graph of the original relation.

When we want to graph the inverse of a relation, we "flip the whole graph over" along a "hinge" corresponding to the "point reflector." That moves every point in the graph of the original relation to its new position in the graph of the inverse. When we do this to the graphs from Figs. 14-4, 14-5, and 14-6, we get the graphs in Figs.14-8, 14-9, and 14-10 respectively. These graphs show the inverses of the original three relations. Note that the positions of the x and y axes have not been switched, but the values of the variables, as well as the domain and range, have been!

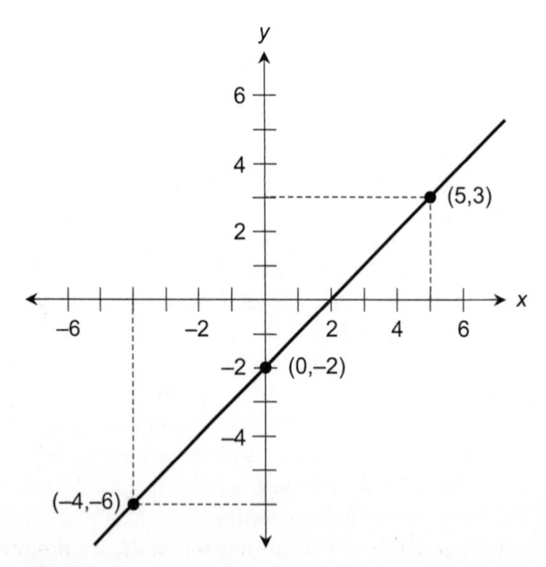

Figure 14-8 Cartesian graph of the relation $y = x - 2$.

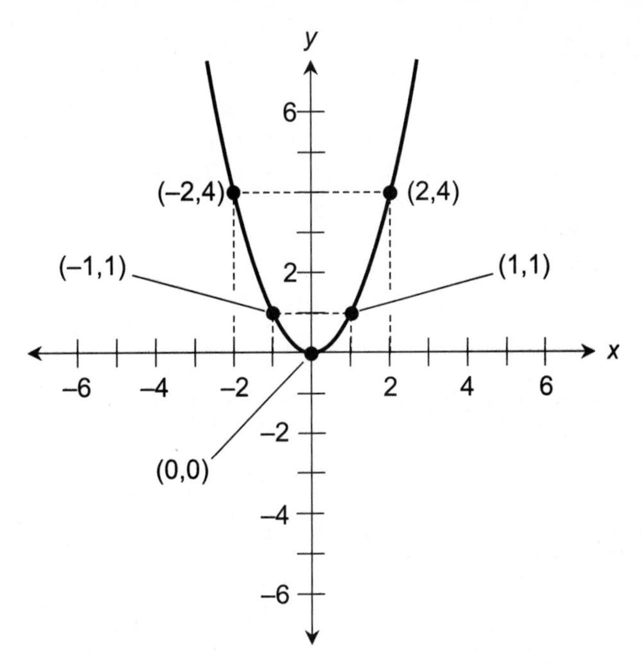

Figure 14-9 Cartesian graph of the relation $y = x^2$.

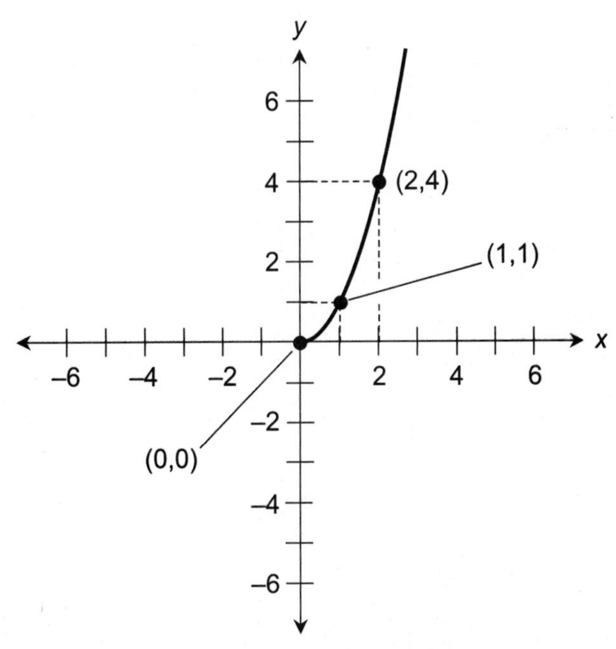

Figure 14-10 Cartesian graph of the relation $y = (+x)^2$.

Three Functions

The process for graphing a function is the same as it is for graphing a relation. Remember, a function is nothing more than a relation with special properties! In this section, we'll look at the graphs of the three functions we evaluated in Chap. 13.

Add 1 to the input

Figure 14-11 shows some points plotted in Cartesian coordinates, along with the straight line that connects them, for the following function:

$$y = x + 1$$

These points lie along a straight line. Note the similarity between this graph and the one shown in Fig. 14-4. The only difference here is that the line is exactly 1 unit lower on the coordinate plane.

Square the input

In Fig. 14-12, several points are plotted, and the curve connecting them is drawn, for the function

$$w = v^2$$

This graph is the same curve as the one shown in Fig. 14-9, but the variable names are different.

Cube the input

Figure 14-13 shows what happens when the independent variable is cubed rather than squared. Several points, along with the curve connecting them, are plotted for the function

$$u = t^3$$

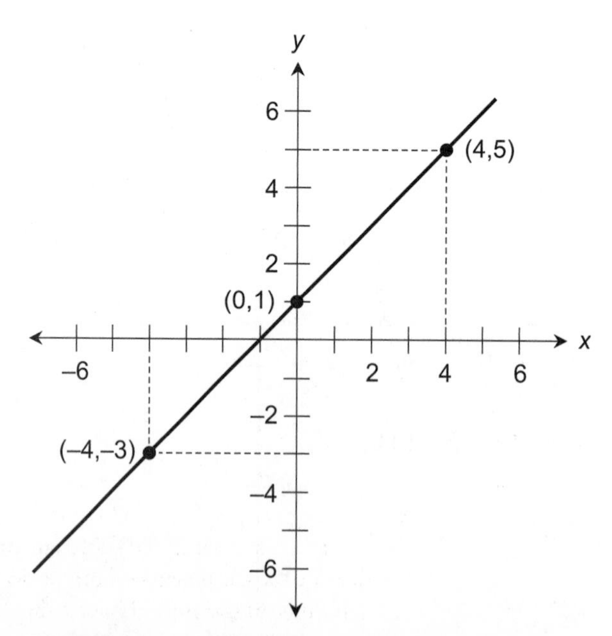

Figure 14-11 Cartesian graph of the function $y = x + 1$.

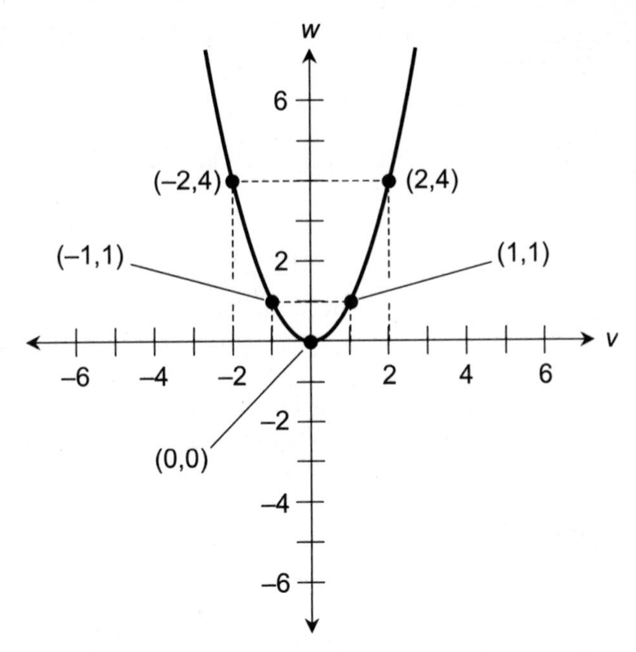

Figure 14-12 Cartesian graph of the function $w = v^2$.

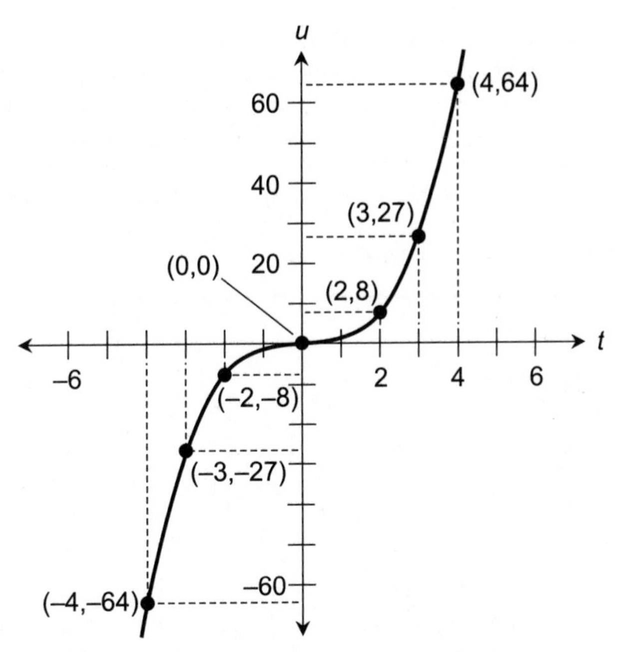

Figure 14-13 Rectangular graph of the function $u = t^3$. This is not a true Cartesian graph because the increments on the vertical axis differ in size from those on the horizontal axis.

Note the difference in "magnification" between the t and u axes. This difference makes the graph fit nicely into the available space. Even though an increment on the u axis represents 10 times the numerical change as an increment of the same length on the t axis, both axes are linear.

Are you confused?

In Chap. 13, you learned that a function never maps a single value of the independent variable to more than one value of the dependent variable. You can use this fact to determine whether or not a given relation is a function by looking at its graph. Draw a vertical line somewhere on the graph. "Vertical" in this context means "parallel to the dependent-variable axis." Imagine moving this vertical line to the right and left. Sometimes—maybe all the time—this vertical line will intersect the graph of the relation. For the relation to qualify as a function, the movable vertical line must never intersect the graph at more than one point. (It's okay if there are places, or even large regions, where the vertical line doesn't intersect the graph at all.) This trick can be called the *vertical-line test.*

Here's a challenge!

How can you tell, merely by looking at their graphs, which of the three relations in this section have inverses that are functions? Don't actually graph the inverses. You'll get a chance to do that in the last three Practice Exercises.

Solution

You can conduct a *horizontal-line test* on the graph of a relation to see if its inverse is a function. Draw a horizontal line parallel to the independent-variable axis. Imagine moving this horizontal line up and down. For the inverse to qualify as a function, this movable line must never intersect the graph of the original relation at more than one point. (It's okay if there are places or regions where the horizontal line doesn't intersect the graph at all.) Now conduct this test on the graphs shown in Figs. 14-11, 14-12, and 14-13. You'll see that the line in Fig. 14-11 checks out, so the inverse of this relation is a function. The same is true for the relation graphed in Fig. 14-13. But the curve in Fig. 14-12 fails the horizontal-line test! That means that the inverse of that relation is not a function.

Practice Exercises

This is an open-book quiz. You may (and should) refer to the text as you solve these problems. Don't hurry! You'll find worked-out answers in App. B. The solutions in the appendix may not represent the only way a problem can be figured out. If you think you can solve a particular problem in a quicker or better way than you see there, by all means try it!

1. Imagine an ordered pair (x, y), and suppose we have plotted its point on the Cartesian plane. Neither x nor y is equal to 0, so the point does not fall on either axis. What happens to the location of the point if we multiply x by -1 and leave y the same?
2. Imagine an ordered pair (x, y), and suppose we have plotted its point on the Cartesian plane. Neither x nor y is equal to 0, so the point does not fall on either axis. What happens to the location of the point if we multiply y by -1 and leave x the same?

3. Imagine an ordered pair (x, y), and suppose we have plotted its point on the Cartesian plane. Where, in relation to the point for (x, y), will we find the point for $(6x, 6y)$? Where, in relation to the point for (x, y), will we find the point for $(x/4, y/4)$?
4. The vertical-line test can be used to see whether or not a graph portrays a function. How can we use the same test on a graph to determine whether or not a given numerical value is in the domain?
5. How can we use the horizontal-line test on a graph to determine whether or not a given numerical value is in the range of a function or relation?
6. Sketch a graph of the equation $y = |x|$ for all real numbers x. Does this equation represent a function of x?
7. Sketch a graph of the equation $y = |x + 1|$ for all real numbers x. Does this equation represent a function of x?
8. Sketch a graph of the inverse of $y = x + 1$. Do this by applying the "point reflector" scheme to Fig. 14-11.
9. Sketch a graph of the inverse of $w = v^2$. Do this by applying the "point reflector" scheme to Fig. 14-12.
10. Sketch a graph of the inverse of $u = t^3$. Don't use the "point reflector" scheme from Fig. 14-13. Derive the inverse using algebra, and then plot the graph from scratch.

CHAPTER

15

Graphs of Linear Relations

We've seen the graphs of some relations and functions. Now it's time to focus on the graphs of *linear relations.* These always appear as straight lines in the Cartesian plane. In particular, we're interested in the equations and graphs of *linear functions*—linear relations where the straight-line graph is not vertical (that is, not parallel to the dependent-variable axis).

Slope-Intercept Form

One of the best known ways to relate the graph of a linear function with its equation defines the *slope* of the line and the point where it crosses the dependent-variable axis. A *two-variable linear equation* of this sort is said to be in *slope-intercept form.* Let's call it the *SI form* for short.

What is slope?

The slope of a straight line in the Cartesian plane is an expression of the steepness with which the line ramps upward or downward as we move to the right. A horizontal line has a slope of 0. A line that ramps upward as we move to the right has positive slope that increases without limit as the slant angle approaches 90° (vertical and going straight up). If the line ramps down as we move to the right, the slope decreases from 0, becoming more negative without limit as the slant angle approaches −90° (vertical and going straight down).

To figure out the exact slope of a line in the Cartesian plane, we must know the coordinates of two points on that line. These can be any two points, as long as they're different. The slope of a line passing through two points is equal to the difference in the y values divided by the difference in the x values for the points. In this context, mathematicians abbreviate "the difference in" by writing the uppercase Greek letter delta (Δ). These differences are often called *increments.* The slope of a line is usually symbolized as m. Therefore,

$$m = \Delta y / \Delta x$$

Sometimes the slope of a straight line is informally called *rise over run*. This notion works as long as the independent variable is on the horizontal axis, the dependent variable is on the vertical axis, and we move to the right.

Try two points!

Suppose we see a line in the Cartesian plane, and are able to locate two points on it and determine their exact coordinates:

$$(x_1, y_1) = (-2, 4)$$

and

$$(x_2, y_2) = (3, 5)$$

The slope is $(y_2 - y_1)$, which we call Δy, divided by $(x_2 - x_1)$, which we call Δx:

$$\begin{aligned} m &= \Delta y/\Delta x \\ &= (y_2 - y_1)/(x_2 - x_1) \\ &= (5 - 4)/[3 - (-2)] \\ &= 1/(3 + 2) \\ &= 1/5 \end{aligned}$$

This situation is illustrated in Fig. 15-1.

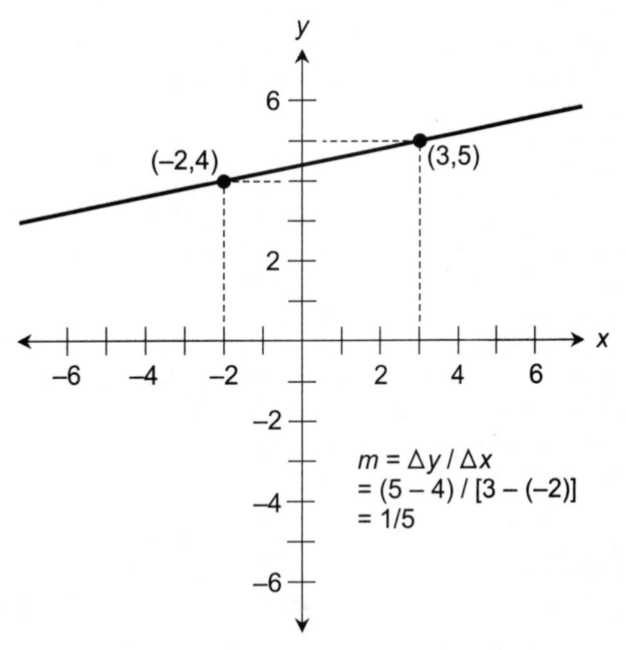

Figure 15-1 The slope of a line can be calculated from the coordinates of two points on that line.

Switching the order

We can switch the points (x_1, y_1) and (x_2, y_2) and still get the same slope when we calculate it as above. That's because both the numerator and the denominator end up being additive inverses (exact negatives) of what they were before. Let's take

$$(x_1, y_1) = (3, 5)$$

and

$$(x_2, y_2) = (-2, 4)$$

The slope is again equal to $\Delta y/\Delta x$. Calculating, we get

$$\begin{aligned} m &= \Delta y/\Delta x \\ &= (y_2 - y_1)/(x_2 - x_1) \\ &= (4 - 5)/(-2 - 3) \\ &= -1/(-5) \\ &= 1/5 \end{aligned}$$

When we know the coordinates of two points on a line, we can figure the slope going from the first point to the second, or going from the second point to the first; it doesn't matter. But we must be careful not to confuse the coordinates. We can reverse the *external* sequence in which we work with the points, but we can't reverse the *internal* sequence of either of the ordered pairs defining those points!

What is the intercept?

When we talk about the SI form of a straight line in the Cartesian plane, the term *intercept* refers to the value of a variable at the point where the line crosses the axis for that variable. If y is the dependent variable, then we often talk about the *y-intercept.* That's what is usually meant when we work with the SI form of an equation when graphing it in the xy-plane. Two examples are shown in Fig. 15-2.

An intercept can be thought of as an ordered pair where one of the values (the one on the axis not being intercepted) is 0. We can plug 0 into a linear equation for one of the variables, and solve for the remaining variable to get its intercept. This method can be more convenient than rearranging everything into SI form or drawing a graph, but all by itself it doesn't give us any visual reinforcement of the situation.

Putting it together

In Chap. 12, you learned the standard form for a first-degree equation in one variable. If the variable is x, the standard form is

$$ax + b = 0$$

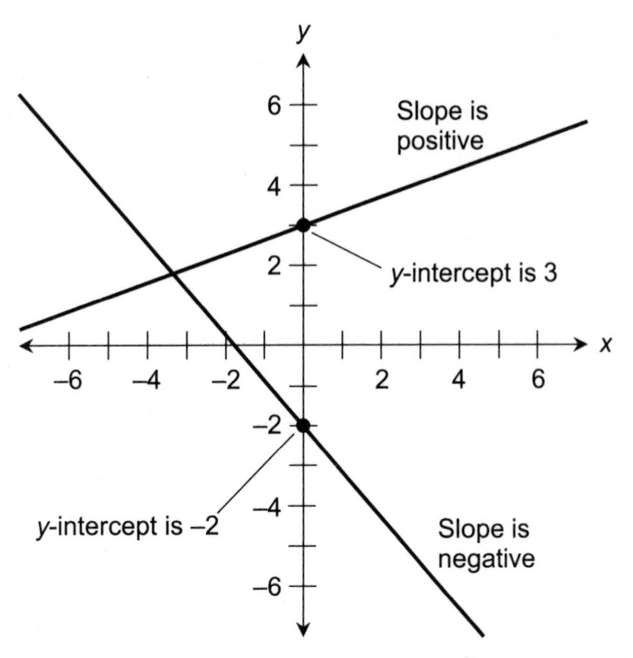

Figure 15-2 Two examples of *y*-intercept points for straight lines. The line that ramps upward as we move to the right has positive slope; the line that ramps downward as we move to the right has negative slope.

where a and b are constants. If you substitute y for 0 and then transpose the left and right sides, you get an equation for a linear function where y is the dependent variable and x is the independent variable:

$$y = ax + b$$

As things work out, the constant a is the slope of the graph, and the constant b is the y-intercept. Because the slope is usually symbolized by m instead of a, you can write

$$y = mx + b$$

This is the classical expression of the SI form for a linear function.

Are you confused?

If the graph of a linear relation is a vertical line, then the slope is undefined, and the relation is not a function. The graph of a linear function *must* be a nonvertical line; otherwise it would fail the vertical-line test in the worst possible way! Whenever you see a linear relation that simply says x is equal to some constant, then you know that relation is not a function of x. Figure 15-3 shows some examples. Note that all the lines are vertical; they are parallel to the dependent-variable axis.

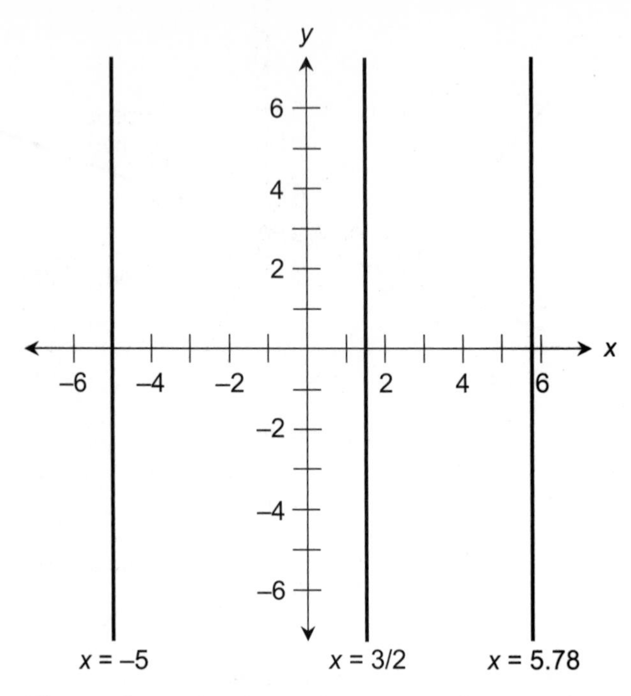

Figure 15-3 These linear relations are not functions of x. The slopes of the graphs are undefined, because they are all straight lines parallel to the y axis.

Here's a challenge!

Put the following equation into SI form as a linear function of x, and graph it on that basis:

$$8x + 4y = 12$$

Solution

We must rearrange this equation to get y all by itself on the left side of the equality symbol, and an expression containing only x and one or more constants on the right side. Subtracting $8x$ from both sides gives us

$$4y = -8x + 12$$

Dividing each side by 4 puts it into SI form:

$$y = (-8x)/4 + 12/4$$
$$= -2x + 3$$

The slope is −2, and the y-intercept is 3. Figure 15-4 shows the graphing process. We plot the y-intercept point on the y axis at the mark for 3 units. That gives us a point with coordinates (0, 3). To plot the line,

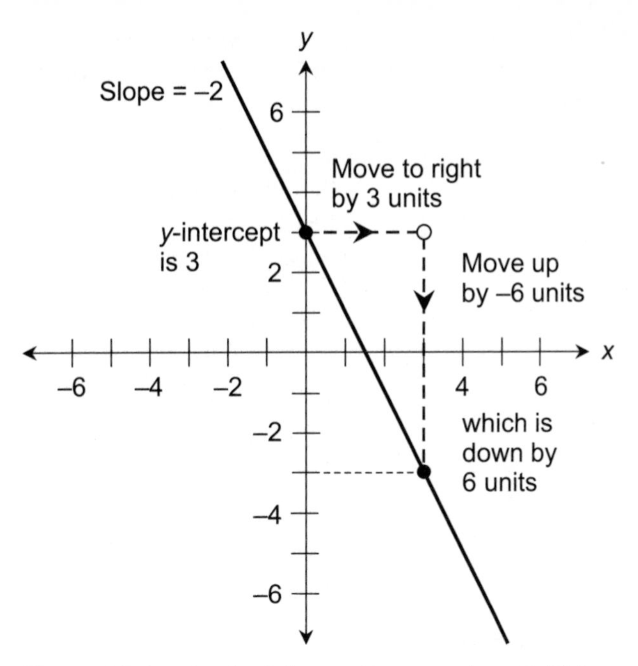

Figure 15-4 Graph of the equation $8x + 4y = 12$. This graph can be drawn easily when we morph the equation into its SI form $y = -2x + 3$.

we must know the coordinates of another point. We can find one by moving horizontally to the right by any number of units we want (call it n units), and then moving straight up from there by mn units, where m is the slope. We should move far enough to the right so the two points will be well separated. That will make it easy to draw the graph accurately. Let's move to the right from (0, 3) by 3 units. That gives us the point (3, 3), shown as an open circle to indicate that it's not actually part of the graph. Then we move straight up by $mn = -2 \times 3 = -6$ units, which is equivalent to moving straight down by 6 units. We have now found a second point on the graph. Our new point is 3 units to the right and 6 units below the y-intercept point, so its coordinates are (3, −3). We plot this point, and then we connect the two points with a solid line to obtain the graph. We extend the line somewhat beyond the points in either direction, keeping in mind that the true, complete line (in the mathematical cosmos) extends infinitely far in each direction!

Here's another challenge!

Just as there is a standard form for a first-degree equation in one variable, there's a standard form for a two-variable linear equation. Here it is:

$$ax + by + c = 0$$

where x is the independent variable, y is the dependent variable, and a, b, and c are constants. Show how this equation can be rearranged into the SI form, expressing y as a function of x (as long as b, the coefficient of y, is not equal to 0).

Table 15-1. Conversion of a general two variable linear equation to SI form. This only works if the constant b (the coefficient of y) is not equal to 0. In this result, the slope is equal to $-a/b$, and the y-intercept is equal to $-c/b$.

Statements	Reasons
$ax + by + c = 0$	This is the equation we are given
$ax + by = -c$	Subtract c from each side
$by = -ax - c$	Subtract ax from each side
Require that $b \neq 0$	We're about to divide through by b
$y = (-ax - c)/b$	Divide through by b
$y = -ax/b - c/b$	Right-hand distributive law for division over subtraction
$y = (-a/b)x - c/b$	Rearrange to define the coefficient for x
$y = (-a/b)x + (-c/b)$	Change subtraction to negative addition, putting the equation into strict SI form

Solution

Table 15-1 is an S/R derivation of a SI equation from the standard form of a two-variable linear equation. Here, the familiar m for slope is replaced by $-a/b$, and the familiar b for slope is replaced by $-c/b$. The result is in the correct form; that's the important thing! The coefficient of y cannot equal 0 in the original equation; that would cause both the slope and the y-intercept to be undefined. The graph in such a case would exist, but it would be a vertical line, so it would not represent a function of x.

Point-Slope Form

Another common way to express a linear function is known as the *point-slope form.* As the name suggests, we can draw the graph of a function if we know the coordinates of any single point on the line, and if we also know the slope of the line. Let's call this the *PS form.*

The form

Here is the standard PS form for a linear function. Later in this chapter, we'll figure out how this form is derived:

$$y - y_0 = m(x - x_0)$$

where x is the independent variable, y is the dependent variable, m is the slope, and (x_0, y_0) are the coordinates of a known point on the graph.

An example

Suppose we're told that there's a linear function whose graph contains the point $(-1, 2)$. We are also told that the slope of the graph is 2. The independent variable is x, and the dependent variable is y. Our task is to draw a graph of the function.

Let's begin by assigning $x_0 = -1$ and $y_0 = 2$. When we plug these numbers into the standard PS equation, we get

$$y - 2 = 2[x - (-1)]$$

which simplifies to

$$y - 2 = 2(x + 1)$$

We can now find another point on the graph by plugging in a value for x and solving for y. We only need one more point to determine the straight line in the Cartesian plane that represents this function. Let's try $x = 1$. We solve for y in steps:

$$\begin{aligned} y - 2 &= 2(1 + 1) \\ y - 2 &= 2 \times 2 \\ y - 2 &= 4 \\ y &= 6 \end{aligned}$$

This tells us that (1, 6) represents a point on the graph. We already know that (−1, 2) is on it. When we plot these two points on the plane and draw a straight line through them both, we get the graph shown in Fig. 15-5.

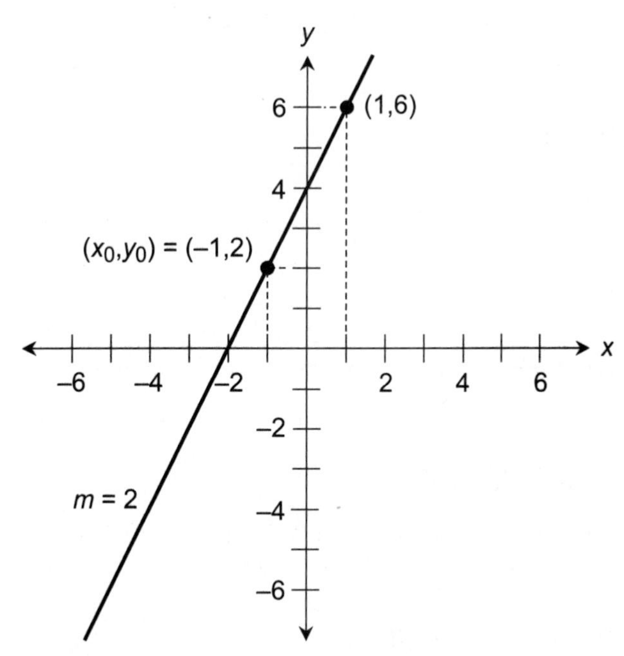

Figure 15-5 Graph of a linear function based on the knowledge that the point (−1, 2) is on the line, and the slope is equal to 2.

Table 15-2 Conversion of a linear equation from PS to SI form. Remember that m, x_0, and y_0 are constants. The y-intercept, called b in the classical expression of the SI form, turns out to be the quantity $(y_0 - mx_0)$.

Statements	Reasons
$y - y_0 = m(x - x_0)$	This is the equation we are given
$y - y_0 = mx - mx_0$	Distributive law of multiplication over subtraction
$y = mx - mx_0 + y_0$	Add y_0 to each side
$y = mx + (-mx_0) + y_0$	Change subtraction to negative addition
$y = mx + (-mx_0 + y_0)$	Grouping of addends
$y = mx + (y_0 - mx_0)$	Simplify second addend on right side

Are you confused?

The PS form of a linear function is actually a generalized version of the SI form. The PS form is handy when we don't know the y-intercept of a graph, but we do know the coordinates of some point in one of the quadrants. If we are told only those coordinates and the slope, we can easily write down an equation representing the function using the PS form. We can then draw the graph by finding another point using that equation, and connecting the two points with a straight line.

Here's a challenge!

Look again at the general PS form for a linear equation, whose graph has a known point with coordinates $(x_0 y_0)$ and a slope m, and where x is the independent variable and y is the dependent variable:

$$y - y_0 = m(x - x_0)$$

Convert this equation into SI form.

Solution

Table 15-2 is an S/R derivation that shows how this can be done.

Equations from Graphs

Let's derive the SI and PS forms of linear equations by looking at how their graphs behave generally. Then we'll derive a standard form for a linear equation based on two known points.

Known slope and y-intercept

Imagine a line in Cartesian coordinates that has slope m and crosses the y axis at the point $(0, b)$, as shown in Fig. 15-6. If we move away from $(0, b)$ on the line, the slope is always equal to the difference in the y value divided by the difference in the x value, or $\Delta y / \Delta x$.

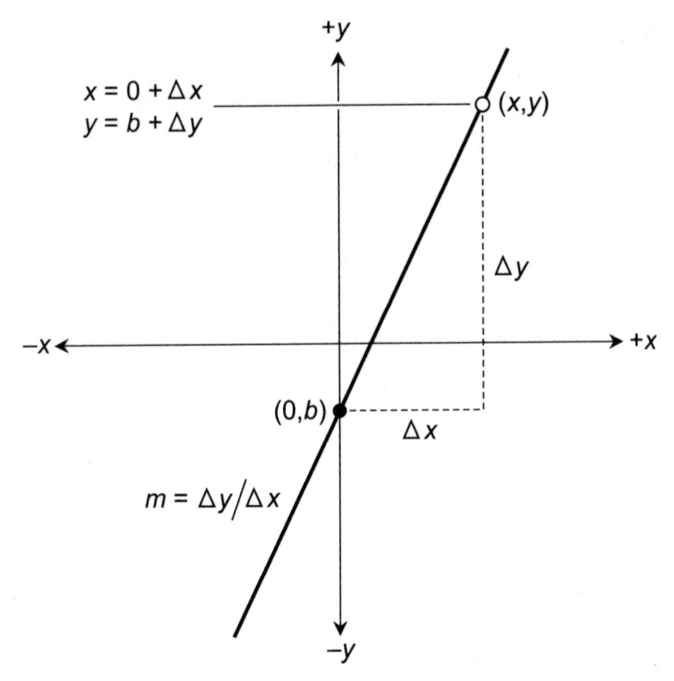

Figure 15-6 The SI form of a linear equation can be derived from this generic graph.

Suppose we move from (0, b) to some point (x, y) on the line by going Δx units to the right and Δy units upward. The x coordinate of the point (x, y) will be $0 + \Delta x$, because we have moved Δx units horizontally from a point where $x = 0$. The y coordinate of the point (x, y) will be $b + \Delta y$, because we have moved Δy units vertically from a point where $y = b$.

If we can get an equation that allows us to calculate y in terms of x for the arbitrary point (x, y), then we will have demonstrated how y is a function of x. As things turn out, we'll also get the SI form of the equation for the line.

We can express Δy in terms of the slope m and the increment Δx by morphing the formula that defines slope. That formula, once again, is

$$m = \Delta y / \Delta x$$

Multiplying through by Δx, we get

$$m\Delta x = \Delta y$$

Now remember that

$$y = b + \Delta y$$

We can substitute $m\Delta x$ for Δy in this equation, getting

$$y = b + m\Delta x$$

But in this situation, Δx is exactly equal to x! That's because, by traversing the increment Δx, we have moved from the y axis (where $x = 0$) horizontally by x units. Because of this lucky coincidence, we can substitute x for Δx in the above equation, getting

$$y = b + mx$$

If we want to be picayune, we can reverse the order of the addends to state it as

$$y = mx + b$$

Known point and slope

Imagine a line in the Cartesian plane that passes through a point whose coordinates are $(x_0 y_0)$, where x_0 and y_0 are known constants. Suppose the line has slope m as shown in Fig. 15-7. If we move away from (x_0, y_0) along the line, the slope is always equal to $\Delta y/\Delta x$.

Let's go from (x_0, y_0) to some arbitrary point (x, y) on the line, just as we did when we derived the SI equation. The x coordinate of (x, y) will be $x_0 + \Delta x$, because we have moved Δx units horizontally from a point where $x = x_0$. The y coordinate of the point (x, y) will be $y_0 + \Delta y$, because we have moved Δy units vertically from a point where $y = y_0$. Now remember, once again, how slope is defined:

$$m = \Delta y/\Delta x$$

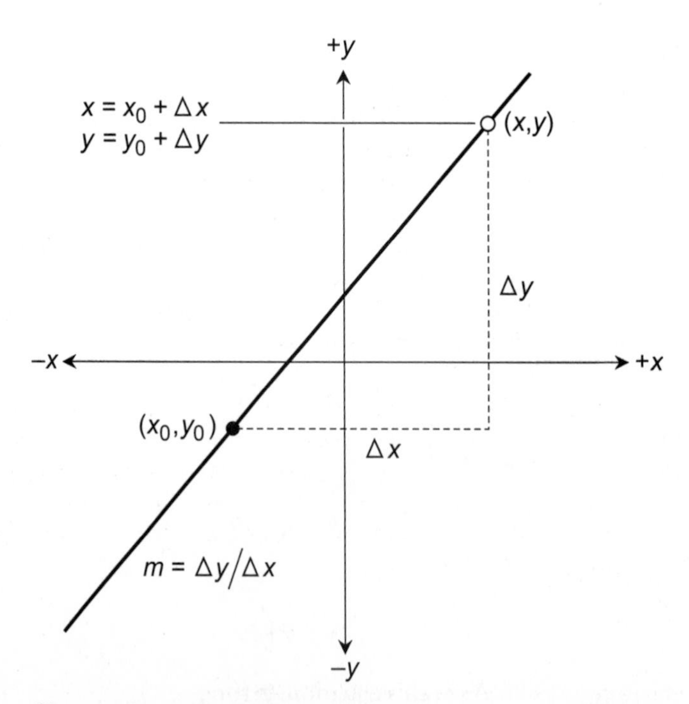

Figure 15-7 The PS form of a linear equation can be derived from this generic graph.

As before, we have

$$m\Delta x = \Delta y$$

Observe that in Fig. 15-7,

$$y = y_0 + \Delta y$$

Let's substitute $m\Delta x$ for Δy here. That gives us

$$y = y_0 + m\Delta x$$

Now we can see from Fig. 15-7 that

$$x = x_0 + \Delta x$$

Subtracting x_0 from each side, we obtain

$$x - x_0 = \Delta x$$

Substituting $(x - x_0)$ for Δx in the equation for y in terms of y_0 and $m\Delta x$, we get

$$y = y_0 + m(x - x_0)$$

Subtracting y_0 from each side gets us to the PS form

$$y - y_0 = m(x - x_0)$$

Are you confused?

The preceding two examples show lines with positive slope. There's good reason to wonder, "What happens when the slope of the line is negative?" The answer is, "Nothing special, as long as we're careful."

Think of what happens when we move to the right in the Cartesian plane. If the slope of a line is positive, then Δy is always positive as we move to the right. If the slope is negative, then Δy is always negative as we move to the right. Either way, we add Δy when we move to the right. If we keep adding a positive Δy, we go higher and higher. If we keep adding a negative Δy, we go lower and lower. In this context, "higher" means "in the positive y direction," and "lower" means "in the negative y direction."

Similar sign-related confusion can occur when we work with the SI form of a linear equation. The standard form, once again, is

$$y = mx + b$$

This equation has a plus sign whether b is positive or negative. If $m = 3$ and $b = -2$, for example,

$$y = 3x + (-2)$$

which is the same as

$$y = 3x - 2$$

When we work with the PS form, we come across another "sign-rigid" situation, but with minus signs instead of a plus sign! The general form of the equation is always

$$y - y_0 = m(x - x_0)$$

It contains two minus signs, and pays no heed to whether y_0 or x_0 happen to be positive or negative. For example, if $m = 5$, $x_0 = -4$, and $y_0 = -8$, then

$$y - (-8) = 5[x - (-4)]$$

which is the same as

$$y + 8 = 5(x + 4)$$

We must always pay close attention to signs when working with the standard forms of linear equations. It's easy to get them wrong! If we see, for example,

$$y = 2x - 4$$

then the y-intercept is $b = -4$. If we see

$$y - 3 = -4(x + 5)$$

then the graph contains a point whose coordinates are $(x_0, y_0) = (-5, 3)$.

Here's a challenge!

Imagine a straight line that passes through two points whose Cartesian coordinates are (x_1, y_1) and (x_2, y_2). Derive an equation for this line in terms of the independent variable x and the dependent variable y. Call this the *two-point form* of a linear equation. Consider x_1, x_2, y_1, and y_2 to be constants.

Solution

Figure 15-8 shows a generic example of this situation. The line has a negative slope, but that doesn't make any difference in the way things will turn out. Let's start by calculating the slope of the line. It is $\Delta y/\Delta x$. Let's move to the right, from the point (x_1, y_1) to the point (x_2, y_2). Then

$$\Delta y = y_2 - y_1$$

and

$$\Delta x = x_2 - x_1$$

The slope is

$$\begin{aligned} m &= \Delta y/\Delta x \\ &= (y_2 - y_1)/(x_2 - x_1) \end{aligned}$$

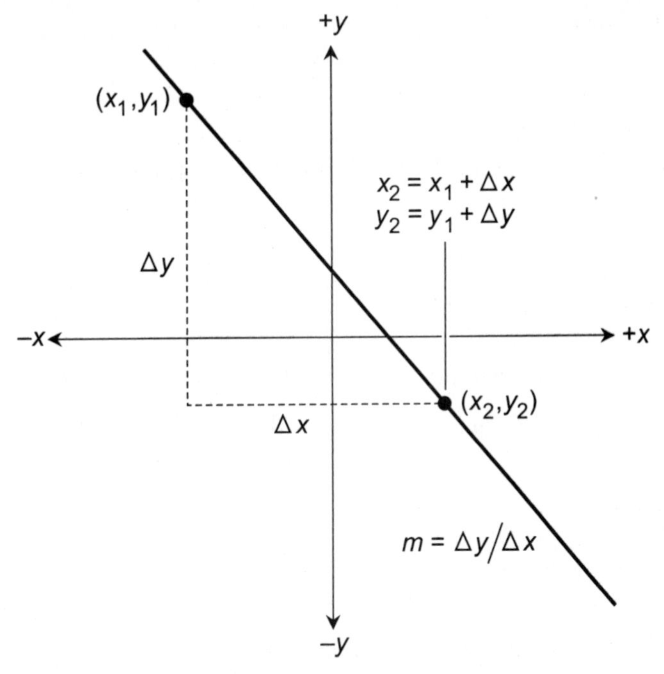

Figure 15-8 A two-point form of a linear equation can be derived from this generic graph.

Now let's use the PS form to derive an equation for the line. We have two points to choose from. Either point will work. Let's use (x_1, y_1). We can substitute x_1 for x_0, and y_1 for y_0 in the classical PS equation to get

$$y - y_1 = m(x - x_1)$$

Substituting $(y_2 - y_1)/(x_2 - x_1)$ for m, we obtain

$$y - y_1 = [(y_2 - y_1)/(x_2 - x_1)]\ (x - x_1)$$

which can be rearranged to

$$y - y_1 = (x - x_1)(y_2 - y_1)/(x_2 - x_1)$$

This is a mess, but it's the best we can do when we aren't given the slope directly. The good news is that most of the values in this equation are constants. When the coordinates for the points are given as numbers, we can plug them in and get the equation by means of straightforward arithmetic.

Practice Exercises

This is an open-book quiz. You may (and should) refer to the text as you solve these problems. Don't hurry! You'll find worked-out answers in App. B. The solutions in the appendix may not represent the only way a problem can be figured out. If you think you can solve a particular problem in a quicker or better way than you see there, by all means try it!

1. Imagine two points P and Q plotted on the Cartesian plane, where the independent variable is u and the dependent variable is v. Point P is defined by $(-1, -6)$, and Q is defined $(2, 2)$. What is the slope of the line connecting these two points if we go in the direction from P to Q?
2. Calculate the slope of the line in Prob. 1 on the basis of going in the direction from Q to P, showing that the slope doesn't depend on which way we move along the line.
3. Derive an equation of the line described in Probs. 1 and 2 in PS form.
4. Derive an equation of the line described in Probs. 1, 2, and 3 in SI form.
5. Sketch a graph of the linear equation discussed in Probs. 1 through 4 using the simplest possible method. Label the slope as m and the v-intercept as b, and indicate their values.
6. Suppose we see two equations where s is the independent variable and t is the dependent variable:

   $$t = s + 5$$

   and

   $$t = 5 - s$$

   Someone says, "When graphed, these equations will produce lines oriented at a 90° angle with respect to each other." How can she say this without drawing the graphs? Under what circumstances will she be right? Under what circumstances will she be wrong?
7. Our advisor, who introduced herself in Prob. 6, goes on to make the claim, "The two lines we talked about will intersect somewhere on the t axis." She's right! What is the exact point of intersection?
8. Graph the two lines we discussed in Probs. 6 and 7, and label the point of intersection.
9. Find an equation for the line in Cartesian (x, y) coordinates that passes through the two points $(2, 8)$ and $(0, -4)$. Use the results of the last challenge. Put the equation into PS form.
10. Find an equation for the line in Cartesian (x, y) coordinates that passes through the two points $(-6, -10)$ and $(6, -12)$. Use the results of the last challenge. Put the equation into SI form.

CHAPTER

16

Two-by-Two Linear Systems

In Chap. 12, we saw how we can solve first-degree equations in one variable. Now it's time to work with pairs of linear equations in two variables, also known as *two-by-two linear systems.* Solving a linear system involves reducing it to first-degree equations, one for each of the variables.

Morph and Mix

When we want to solve a two-by-two linear system, we can put both equations into slope-intercept (SI) form. Then we can take the right sides of the resulting equations and mix them to get a first-degree equation in one variable. We can solve that equation, and finally plug the result into either of the SI equations to solve for the other variable.

Morph both equations into SI form

Suppose we are told to find the values of x and y such that both of the following equations are true:

$$8x + 4y = 16$$

and

$$7x - y = 41$$

Let's put these equations into SI form. In the first case, we start with

$$8x + 4y = 16$$

Subtracting $8x$ from each side gives us

$$4y = -8x + 16$$

Dividing through by 4, we get

$$y = -2x + 4$$

Now for the second equation. We begin with

$$7x - y = 41$$

Subtracting $7x$ from each side, we get

$$-y = -7x + 41$$

Multiplying through by -1 gives us

$$y = 7x - 41$$

Mix the right sides and solve

Now let's put the right side of the first SI equation on the left side of an equals sign, and the right side of the second SI equation on the right side of the same equals sign. This produces a first-degree equation in one variable:

$$-2x + 4 = 7x - 41$$

Now let's solve this for x. When we subtract $7x$ from each side, we get

$$-9x + 4 = -41$$

We can subtract 4 from each side to obtain

$$-9x = -45$$

Finally, we divide through by -9 to get

$$x = 5$$

Substitute and solve again

To solve for y, we can take either of the SI equations and plug in 5 for x. Let's use the first one. That gives us

$$\begin{aligned} y &= -2 \times 5 + 4 \\ &= -10 + 4 \\ &= -6 \end{aligned}$$

We have used algebra to find that $x = 5$ and $y = -6$. We can express this as the ordered pair $(5,-6)$ if we imagine x as the independent variable and y as the dependent variable.

Two versions of the SI form

In a two-by-two system, it often doesn't matter which variable we consider independent and which one we consider dependent. But a true SI equation always has the dependent variable alone on the left side of the equals sign, and the independent variable on the right side along with constants that represent characteristics of a graph. For example, if we see the equation

$$x - y = 10$$

then we can say that both of the following are SI equivalents of it:

$$y = x - 10$$

and

$$x = y + 10$$

In the first case, we treat y as the dependent variable. In the second case, we treat x as the dependent variable.

Are you confused?

Once you've solved a two-by-two linear system (or think you have), you should consider your solutions *tentative* until you've plugged them into both of the original equations and worked out the arithmetic to be sure that they're correct. The solution to the system originally stated on page 251 appears to be $x = 5$ and $y = -6$. Is it, really? Check the first original equation:

$$8x + 4y = 16$$
$$8 \times 5 + 4 \times (-6) = 16$$
$$40 - 24 = 16$$
$$16 = 16$$

That checks out fine! Now for the second original equation:

$$7x - y = 41$$
$$7 \times 5 - (-6) = 41$$
$$35 + 6 = 41$$
$$41 = 41$$

It works out here, too! You can now be confident that the solutions are right.

Here's a challenge!

While flying directly into a high-altitude wind, an airplane has a *groundspeed* (speed measured with respect to the earth) of 750 kilometers per hour (km/h). When flying right along with that same wind at the same *airspeed* (speed measured with respect to the surrounding air), the plane has a groundspeed of 990 km/h. What is the airspeed of the plane? What is the speed of the wind relative to the earth?

Solution

Let x represent the airspeed of the plane, and let y represent the speed of the wind, both in kilometers per hour. When the plane flies against the wind, the wind takes away from its groundspeed. Therefore

$$x - y = 750$$

Let's put this equation into SI form. Subtracting x from each side gives us

$$-y = -x + 750$$

Multiplying through by -1 tells us that

$$y = x - 750$$

When the plane flies with the wind, the wind adds to its groundspeed. That means

$$x + y = 990$$

To morph this into SI form, we can subtract x from each side, getting

$$y = -x + 990$$

Now let's mix the right sides of these two SI equations:

$$x - 750 = -x + 990$$

When we add 750 to each side, we obtain

$$x = -x + 1{,}740$$

We can add x to each side to get

$$2x = 1{,}740$$

Finally we divide through by 2, discovering that

$$x = 870$$

Now we know that the airspeed of the plane is 870 km/h. Let's plug this value into one of the SI equations. We can use the first one, getting

$$\begin{aligned} y &= x - 750 \\ &= 870 - 750 \\ &= 120 \end{aligned}$$

This tells us that the wind is blowing at 120 km/h with respect to the earth.

These values should be checked by plugging them into both of the original equations to be sure they're correct. Here we go:

$$x + y = 990$$
$$870 + 120 = 990$$
$$990 = 990$$

and

$$x - y = 750$$
$$870 - 120 = 750$$
$$750 = 750$$

We can now state these solutions with confidence, at least until the wind changes or the airplane alters its cruising speed!

Double Elimination

There's another way to solve two-by-two systems of linear equations, which I like to call *double elimination*. It's also called the *addition method*. We can morph one or both equations so that when we add them in their entirety, one of the variables disappears. That allows us to solve for the other variable. We can then do the same thing to make the other variable disappear, and solve for the one!

Get in the same form

Let's solve the following two-by-two linear system using double elimination, first for u (by eliminating t) and then for t (by eliminating u):

$$2t + 5u = -7$$

and

$$u = 4t - 3$$

Before we go any farther, we must get both equations in the same form. Let's use the form of the first equation. The second equation can be morphed into that form by subtracting $4t$ from each side, getting

$$-4t + u = -3$$

Eliminate the first variable

Our first objective is to make t vanish when we add multiples of the equations. We can multiply the first original equation through by 2, getting

$$4t + 10u = -14$$

Adding the two morphed equations in their entirety causes the variable t to vanish:

$$\begin{array}{r} -4t + u = -3 \\ 4t + 10u = -14 \\ \hline 11u = -17 \end{array}$$

When we divide this result through by 11, we see that $u = -17/11$.

Eliminate the second variable

Now let's look at the first original equation and the second morphed equation again. They are

$$2t + 5u = -7$$

and

$$-4t + u = -3$$

Our goal this time is to find a way to make u vanish when we add multiples of the equations. Let's multiply the second equation through by −5. That morphs it into

$$20t - 5u = 15$$

We can add the two equations in their entirety:

$$\begin{array}{r} 2t + 5u = -7 \\ 20t - 5u = 15 \\ \hline 22t = 8 \end{array}$$

This is easily solved to get $t = 8/22$, which reduces to 4/11. We can now state the solution to this system as an ordered pair $(t,u) = (4/11,-17/11)$.

Are you confused?

Let's plug these results into the original equations to be sure they're accurate. The first equation figures out this way:

$$2t + 5u = -7$$

$$2 \times 4/11 + 5 \times (-17/11) = -7$$

$$8/11 + (-85/11) = -7$$

$$-77/11 = -7$$

$$-7 = -7$$

That works! The second original equation comes out like this:

$$u = 4t - 3$$
$$-17/11 = 4 \times 4/11 - 3$$
$$-17/11 = 16/11 - 3$$
$$-17/11 = 16/11 - 33/11$$
$$-17/11 = (16 - 33)/11$$
$$-17/11 = -17/11$$

That checks out as well. We can be confident that we've found the correct solution to the original two-by-two linear system.

Here's a challenge!

Derive a general formula using double elimination that solves the following two-by-two linear system for the variables x and y in terms of the constants a through f. Here are the equations:

$$ax + by = c$$

and

$$dx + ey = f$$

Solution

First, let's cause x to disappear so we can solve for y. To do this, we must get coefficients for x that have the same absolute value but opposite sign in the two equations. Let's multiply the first equation through by d, and multiply the second equation through by $-a$. When we add the resulting equations in their entirety, we get

$$\begin{array}{r} dax + dby = dc \\ -adx - aey = -af \\ \hline dby - aey = dc - af \end{array}$$

The terms dax and $-adx$ are additive inverses, so when we add them, they vanish. Now, we can invoke the distributive law "backward" in the left side of this result to get

$$(db - ae)y = dc - af$$

We can solve for y if we divide through by $(db - ae)$, assuming that $(db - ae) \neq 0$:

$$y = (dc - af)/(db - ae)$$

Now let's cause y to disappear so we can solve for x. We multiply the first original equation through by e, and multiply the second equation through by $-b$. When we add the resulting equations in their entirety, we get

$$\begin{array}{r} eax + eby = ec \\ -bdx - bey = -bf \\ \hline eax - bdx = ec - bf \end{array}$$

The terms *eby* and $-bey$ are additive inverses, so they disappear from the sum. Next, we can apply the distributive law "backward" in the left side, obtaining

$$(ea - bd)x = ec - bf$$

We can solve for x if we divide through by $(ea - bd)$, assuming that $(ea - bd) \neq 0$:

$$x = (ec - bf)/(ea - bd)$$

Are you astute?

Have you noticed that the "taboos" in the above derivations are actually two different ways of saying the same thing? They're stated like this:

$$(db - ae) \neq 0$$

and

$$(ea - bd) \neq 0$$

Both of these inequalities are equivalent to the statement $ae \neq bd$. Can you see why? What do you think will happen if a linear system has coefficients in the above form such that $ae = bd$? You'll get a chance to explore a situation like that in exercises 5, 6, and 7 at the end of this chapter!

Rename and Replace

A two-by-two linear system can be unraveled by renaming one variable in terms of the other, and then creating a single-variable, first-degree equation from the result. We solve that equation, and then plug the number into a strategic spot to solve for the other variable. This process is usually called the *substitution method.* I like to call it *rename and replace.*

Rename one variable

When we want to solve a two-by-two linear system by rename-and-replace, we begin by morphing one of the equations into SI form. Consider this pair of equations:

$$-7v + w + 10 = 0$$

and

$$4v + 8w = -40$$

We can take the first equation and add $7v$ to each side, getting

$$w + 10 = 7v$$

Then we can subtract 10 from each side to obtain this SI equation with w playing the role of the dependent variable:

$$w = 7v - 10$$

Make a first-degree equation

The second step involves substituting our "new name" for w into the equation we haven't touched yet, which in this case is the second original. That gives us

$$4v + 8(7v - 10) = -40$$

The distributive law of multiplication over subtraction can be applied to the second addend on the left side of the equals sign to get

$$4v + 56v - 80 = -40$$

Summing the first two addends in the left side of this equation gives us

$$60v - 80 = -40$$

Adding 80 to each side, we obtain

$$60v = 40$$

This tells us that $v = 40/60 = 2/3$.

Plug the number into the best place

Now that we have solved for one of the variables, we can replace the resolved unknown with its solution in any relevant equation containing both variables. The simplest approach is to use is the SI equation we derived in the first step:

$$w = 7v - 10$$

When we replace v by 2/3 here, we get

$$w = 7 \times 2/3 - 10$$

Taking the product on the right side of the equals sign, and changing 10 into 30/3 to obtain a common denominator, we come up with

$$w = 14/3 - 30/3$$

Now it's a matter of mere arithmetic:

$$\begin{aligned} w &= (14 - 30)/3 \\ &= -16/3 \end{aligned}$$

We've derived the solution to this system: $v = 2/3$ and $w = -16/3$.

Are you confused?

As always, we had better check our work to be sure the solutions we obtained satisfy both of the original equations. Here's the first check:

$$-7v + w + 10 = 0$$
$$-7 \times (2/3) + (-16/3) + 10 = 0$$
$$-14/3 - 16/3 + 10 = 0$$
$$-30/3 + 10 = 0$$
$$-10 + 10 = 0$$
$$0 = 0$$

All right! Here's the second check:

$$4v + 8w = -40$$
$$4 \times 2/3 + 8 \times (-16/3) = -40$$
$$8/3 - 128/3 = -40$$
$$(8 - 128)/3 = -40$$
$$-120/3 = -40$$
$$-40 = -40$$

All right again! Our solutions are correct.

Here's a challenge!

Solve the following pair of equations as a two-by-two linear system using the substitution method:

$$3x - \pi y = -1$$

and

$$-8x + 2y = 4$$

This process is going to be messy! If we remember that π is a plain old real number, it won't be too bad. The signs will be tricky, though.

Solution

Let's tackle the second equation and get it into a form that expresses y in terms of x. First, we can add $8x$ to each side, getting

$$2y = 8x + 4$$

When we divide through by 2, we get

$$y = 4x + 2$$

Now we can substitute $(4x + 2)$ for y in the first original equation, obtaining

$$3x - \pi(4x + 2) = -1$$

When we apply the distributive law on the left side of the equals sign, we get

$$3x - (4\pi x + 2\pi) = -1$$

This is the equivalent of

$$3x + [-1(4\pi x + 2\pi)] = -1$$

which simplifies to

$$3x - 4\pi x - 2\pi = -1$$

When we add 2π to each side, we get

$$3x - 4\pi x = -1 + 2\pi$$

We can use the distributive law "backward" to morph the left side of this equation, obtaining

$$(3 - 4\pi)x = -1 + 2\pi$$

Now we can divide through by $(3 - 4\pi)$ to get

$$x = (-1 + 2\pi)/(3 - 4\pi)$$

It's a mess, all right! But we've found a real number that's equal to x. We can plug this number into the SI equation we derived earlier, getting

$$\begin{aligned} y &= 4[(-1 + 2\pi)/(3 - 4\pi)] + 2 \\ &= (-4 + 8\pi)/(3 - 4\pi) + 2 \end{aligned}$$

Believe it or not, this can be simplified. But we must take a step back, and then we can take two steps forward. Let's "complexify" the number 2 and write it as twice the denominator in the fraction above, divided by that denominator. The idea is to get a common denominator, add some fractions, and get a simpler expression as a result. In mathematical terms,

$$2 = 2(3 - 4\pi)/(3 - 4\pi)$$

It takes some intuition to see, in advance, how a scheme like this will work. (With practice, you'll develop this "sixth sense.") Our solution for y can now be rewritten as

$$y = (-4 + 8\pi)/(3 - 4\pi) + 2(3 - 4\pi)/(3 - 4\pi)$$

This gives us a sum of two fractions with the common denominator $(3 - 4\pi)$. Therefore:

$$y = [(-4 + 8\pi) + 2(3 - 4\pi)]/(3 - 4\pi)$$

Applying the distributive law, we get

$$y = (-4 + 8\pi + 6 - 8\pi)/(3 - 4\pi)$$

When we add up the terms in the numerator here, we get our reward:

$$y = 2/(3 - 4\pi)$$

We've arrived at our solutions! They are:

$$x = (-1 + 2\pi)/(3 - 4\pi)$$

and

$$y = 2/(3 - 4\pi)$$

How about some extra credit?

Plug the above numbers into the original equations for x and y, and verify that the answers we got are correct. You're on your own! Here's a hint: $(3 - 4\pi)$ divided by itself is equal to 1.

Practice Exercises

This is an open-book quiz. You may (and should) refer to the text as you solve these problems. Don't hurry! You'll find worked-out answers in App. B. The solutions in the appendix may not represent the only way a problem can be figured out. If you think you can solve a particular problem in a quicker or better way than you see there, by all means try it!

1. The sum of two numbers is 44. Their difference is 10. What are the two numbers? Use the morph-and-mix method.
2. The sum of two numbers is 100. One of them is 6 times the other. What are the two numbers? Use the morph-and-mix method.
3. Imagine that you and I are traveling in a car on a level highway at constant speed. I'm the driver. There are no other vehicles or living things in sight. You throw a baseball straight out in front of the car. The ball strikes the pavement at 135 miles per hour (mi/h). Then you throw another baseball directly backward, exactly as hard as the first one. The second ball hits the highway, moving opposite to the direction of the car, at 15 mi/h. (The ball is not only moving backward relative to the car; it's also moving backward relative to the pavement!) How fast am I driving? How fast do you hurl the baseballs relative to the car? Forget about the possible effects of wind and gravity. Here's a hint: Feel free to draw diagrams.
4. The sum of two numbers is −83. Their difference is 13. What are the two numbers? Use double elimination.

5. When we attempt to solve the following two-by-two linear system, we will fail. Try it using the double-elimination method and see what happens. Here are the equations:

$$2x + y = 3$$

and

$$6x + 3y = 12$$

Is something wrong with one or both of these equations? Why can't we solve this system?

6. Put the two original equations from Prob. 5 into SI form. What does this tell us about their graphs? Would drawing the graphs provide any clues about why the system can't be solved?

7. Look again at the general derivation for solving a two-by-two system by double elimination. Remember the "taboo" concerning the constants. Does this have anything to do with why the system described in Probs. 5 and 6 can't be solved? Plug in the numbers and see what happens.

8. Solve the two-by-two linear system given in Prob. 1 using the rename-and-replace (substitution) method.

9. Solve the two-by-two linear system given in Prob. 4 using the rename-and-replace method.

10. Attempt to solve the following pair of equations as a two-by-two linear system by substitution. You'll get a meaningless result. Why?

$$s = 2r - 3$$

and

$$-10r + 5s + 15 = 0$$

CHAPTER

17

Two-by-Two Linear Graphs

Let's graph the two-by-two linear systems we looked at in Chap. 16. This will give you a chance to see what's going on when a system has a single solution (is *consistent*), has no solution (is *inconsistent*), or has infinitely many solutions (is *redundant* or *dependent*). Any linear system must fall into one of these three categories.

We Morphed, We Mixed, We Can Graph

In the preceding chapter, we used the morph-and-mix method to solve the following linear system for x and y:

$$8x + 4y = 16$$

and

$$7x - y = 41$$

We found that $x = 5$ and $y = -6$. Now we'll illustrate this situation graphically, calling x the independent variable and y the dependent variable.

Find two points for each line

When we want to physically draw a graph of a line on graph paper, we must know the coordinates of two points that lie on that line. We can then connect the two points, imagining the line extending forever in either direction. When we want to graph a two-by-two linear system, the most efficient approach is to find the points where the two lines cross one of the axes, and then also find the point where the two lines intersect (if there is such a point).

When we morphed the above two equations before mixing them on our way to a solution, we put them into SI form. Those SI equations, once again, are

$$y = -2x + 4$$

and

$$y = 7x - 41$$

The y-intercepts are at 4 and −41. The ordered pairs for those points are (0,4) and (0,−41). The lines intersect at the solution point where $x = 5$ and $y = -6$, corresponding to the ordered pair (5,−6).

Connect the points

We have determined that one line passes through the points (0, 4) and (5, −6), while the other line passes through the points (0, −41) and (5, −6). Figure 17-1 shows the graph of this system on the Cartesian plane.

Are you confused?

In Fig. 17-1, two of the points we found are close to the origin, while the third point is far away. It's difficult to plot points accurately when they're so diverse, because the increments must be large (in this case 10 units per division). It's also difficult to draw a line based on two points that are close together. If we want to

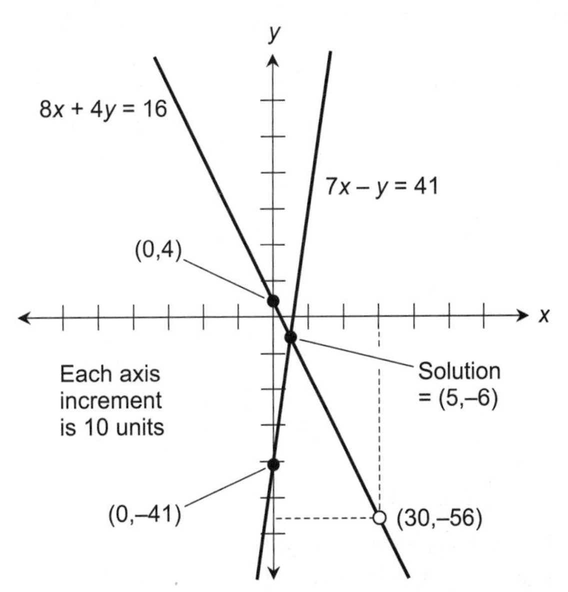

Figure 17-1 Graphs of $8x + 4y = 16$ and $7x - y = 41$ as a two-by-two linear system where the independent variable is x and the dependent variable is y. On both axes, each increment represents 10 units.

precisely draw the line through two points that are close together, we can find a third point farther out on that line and use it. For example, in the equation $8x + 4y = 16$ for the line through (0, 4) and (5, −6), we can plug in $x = 30$ and easily solve for y using the SI form:

$$y = -2x + 4$$
$$= -2 \times 30 + 4$$
$$= -60 + 4$$
$$= -56$$

That tells us that the point (30, −56) is on the line. It is shown as an open circle in Fig. 17-1. This point is far enough away from (0, 4) so we can easily draw the line.

Here's a challenge!

Revisit the airplane challenge from Chap. 16. While flying straight into the wind, an airplane has a groundspeed of 750 km/h. When flying straight downwind at the same airspeed, the plane has a groundspeed of 990 km/h. Let x represent the airspeed of the plane, and let y represent the speed of the wind, both in kilometers per hour. Then

$$x - y = 750$$

and

$$x + y = 990$$

Graph these two equations, showing that the airspeed of the plane is 870 km/h, and the wind is blowing at 120 km/h.

Solution

Let's write down the SI forms of the equations that we derived before we mixed them. They are

$$y = x - 750$$

and

$$y = -x + 990$$

The y-intercepts are −750 and 990, so we can plot the points (0, −750) and (0, 990) on a Cartesian plane where x is the independent variable and y is the dependent variable. The solution we obtained in Chap. 16 was $x = 870$ and $y = 120$, giving us another point with the ordered pair (870,120). Figure 17-2 shows these three points and the lines through them. That's how this two-by-two linear system looks when graphed. The reference points we found are well separated, so the lines are easy to draw.

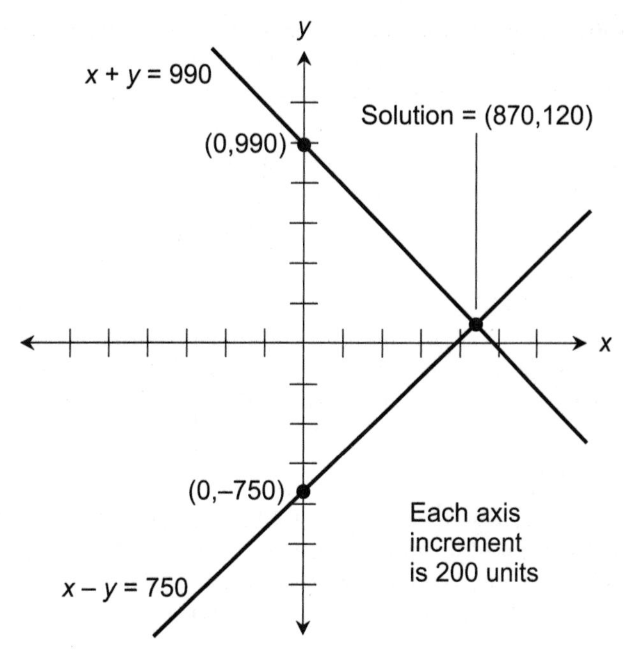

Figure 17-2 Graphs of $x - y = 750$ and $x + y = 990$ as a two-by-two linear system where the independent variable is x and the dependent variable is y. On both axes, each increment represents 200 units.

We Added, We Eliminated, We Can Graph

In Chap. 16, we solved the following two-by-two linear system for t and u using double elimination:

$$2t + 5u = -7$$

and

$$u = 4t - 3$$

We found that $t = 4/11$ and $u = -17/11$. Let's graph this situation, calling t the independent variable and u the dependent variable.

Find two points for each line

As before, let's find the u-intercepts and the solution, using these points as the basis for drawing our lines. The second equation is already in SI form. The u-intercept for its graph is -3, so we know that $(0, -3)$ is on the line. We have to work on the first equation a little. We start with

$$2t + 5u = -7$$

When we subtract $2t$ from each side, we get

$$5u = -2t - 7$$

We can divide through by 5 to obtain

$$u = (-2/5)t - 7/5$$

The u-intercept for this line is $-7/5$, so the point $(0, -7/5)$ is on it. Note that the ordered pairs here are always of the form (t, u), because it's customary to list the independent variable first and the dependent variable after it. In Chap. 16, we found that the solution for the linear system was $t = 4/11$ and $u = -17/11$. Therefore, the point $(4/11, -17/11)$ is where the lines intersect.

Connect the points

We have the points we need to graph the lines. Our first line passes through $(0, -3)$ and $(4/11, -17/11)$. The second line goes through $(0, -7/5)$ and $(4/11, -17/11)$. Unfortunately, two of these points are so close together that it's hard to draw the line through them precisely, as can be seen by looking at the graph of this system (Fig. 17-3).

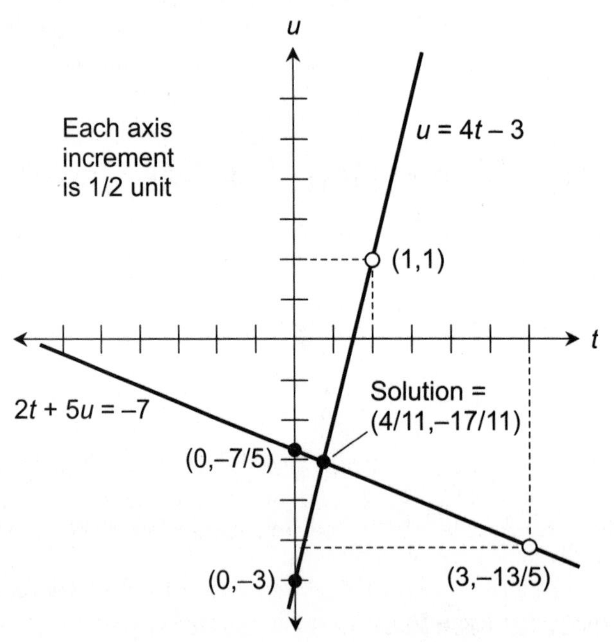

Figure 17-3 Graphs of $2t + 5u = -7$ and $u = 4t - 3$ as a two-by-two linear system where the independent variable is t and the dependent variable is u. On both axes, each increment represents 1/2 unit.

Let's find another point on each line to make our line-drawing task easier. If we plug in $t = 3$ to the SI version of the first equation, we get

$$
\begin{aligned}
u &= (-2/5)t - 7/5 \\
&= (-2/5) \times 3 - 7/5 \\
&= -6/5 - 7/5 \\
&= -13/5
\end{aligned}
$$

That gives us the point $(3, -13/5)$, shown as a small open circle on the line for the first equation. If we plug in $t = 1$ to the second equation, we get

$$
\begin{aligned}
u &= 4t - 3 \\
&= 4 \times 1 - 3 \\
&= 4 - 3 \\
&= 1
\end{aligned}
$$

That gives us the point $(1, 1)$, shown as a small open circle on the line for the second equation.

Are you confused?

When we want to find well-spaced points to draw lines in situations like this, a little common sense is a big help. We don't want any of the points to land off the scale on either axis, but we have to get them far enough away from the other points so we can easily draw the lines. The calculations are usually simple if we use integers for the "plug-ins."

Here's a challenge!

Draw a graph of the above system with the variables transposed. That is, make u the independent variable and t the dependent variable.

Solution

Let's tackle this problem from scratch. To begin, we must get the two original equations into SI form with t as the dependent variable. That will make it easy to find the t-intercepts. Then we'll make up a coordinate system with a horizontal u axis and a vertical t axis and plot the points. Finally, we'll draw the lines through the points.

Here are the original equations, each followed by a step-by-step manipulation to get it into SI form with t as the dependent variable. By now, you can figure out the reasoning behind each step, so we don't have to drag ourselves through the justifications!

$$
\begin{aligned}
2t + 5u &= -7 \\
2t &= -5u - 7 \\
t &= (-5/2)u - 7/2
\end{aligned}
$$

and

$$u = 4t - 3$$
$$u + 3 = 4t$$
$$4t = u + 3$$
$$t = (1/4)u + 3/4$$

Now we know that the t-intercepts are −7/2 and 3/4, so we can plot the points corresponding to the ordered pairs (0,−7/2) and (0,3/4). The solution is still $t = 4/11$ and $u = -17/11$, but when we write this as an ordered pair, it must be of the form (u,t), so we plot the intersection point as (−17/11, 4/11). Figure 17-4 is a Cartesian graph of this situation. We've extended the negative (downward) t axis so we can plot the t-intercept for the first line and have a little extra room to extend the line past the point.

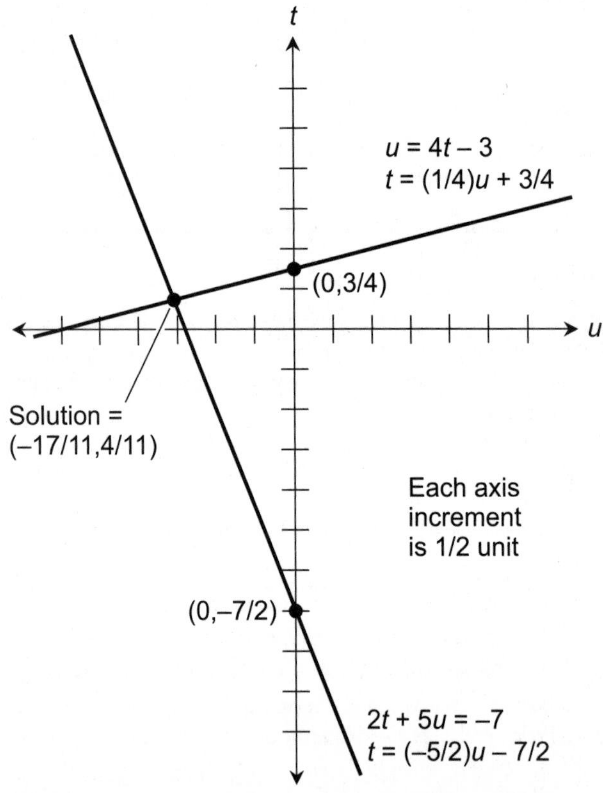

Figure 17-4 Graphs of $2t + 5u = -7$ and $u = 4t - 3$ as a two-by-two linear system where the independent variable is u and the dependent variable is t. On both axes, each increment represents 1/2 unit. The SI forms of the equations are shown below the originals.

We Renamed, We Replaced, We Can Graph

In Chap. 16, we solved the following two-by-two linear system for v and w using the rename-and-replace scheme, also called the substitution method:

$$-7v + w + 10 = 0$$

and

$$4v + 8w = -40$$

We found that $v = 2/3$ and $w = -16/3$. Let's graph this system, calling v the independent variable and w the dependent variable.

Find two points for each line

Our first step is to get both of the original equations into SI form with w as the dependent variable. For the first equation:

$$-7v + w + 10 = 0$$
$$w + 10 = 7v$$
$$w = 7v - 10$$

This tells us that the w-intercept for one of the lines is -10, so we can plot the point $(0, -10)$ on the Cartesian plane. For the second equation:

$$4v + 8w = -40$$
$$8w = -4v - 40$$
$$w = (-4/8)v - 40/8$$
$$w = (-1/2)v - 5$$

The w-intercept for the other line is -5, so we can plot $(0, -5)$ on the coordinate plane. We know that the lines intersect at the solution point where $v = 2/3$ and $w = -16/3$, so we can plot $(2/3, -16/3)$. These three known points are plotted in Fig. 17-5.

Connect the points

One line passes through $(0, -10)$ and $(2/3, -16/3)$. The other line passes through $(0, -5)$ and $(2/3, -16/3)$. The points here are badly "bunched up." Let's find new points on both lines so we can obtain accurate graphs. Look again at the SI equation

$$w = 7v - 10$$

The slope of the line is 7. It ramps steeply upward as we move to the right. If we go 1 unit to the right, for example, we'll go 7 units upward. If we go 2 units to the right, we'll go 14 units

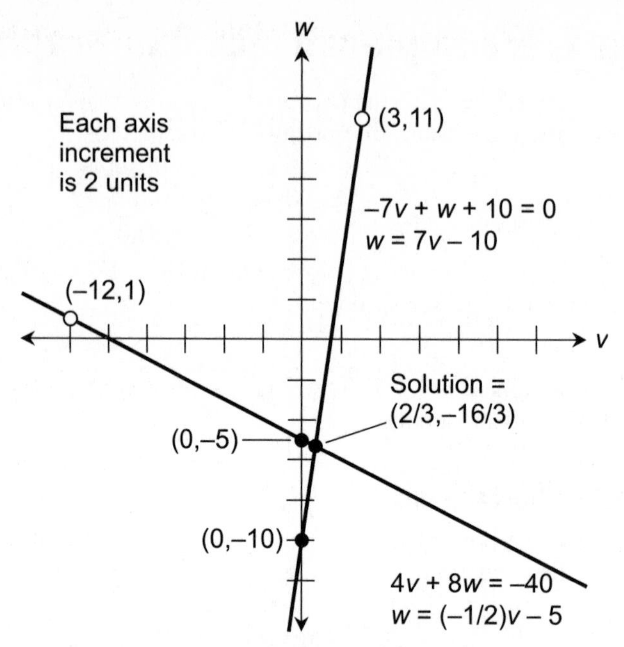

Figure 17-5 Graphs of $-7v + w + 10 = 0$ and $4v + 8w = -40$ as a two-by-two linear system where the independent variable is v and the dependent variable is w. On both axes, each increment represents 2 units. The SI forms of the equations are shown below the originals.

upward. How about 3 units to the right? Will the w value go off the scale? Let's plug in $v = 3$ and see what we get:

$$\begin{aligned} w &= 7 \times 3 - 10 \\ &= 21 - 10 \\ &= 11 \end{aligned}$$

It's still in the field of view, so we can plot (3, 11) on the plane as a small open circle, and draw the line for the first equation through it and (0, −10).

There's no doubt that we need to find a more distant point on the line for the second equation. In SI form, that equation is

$$w = (-1/2)v - 5$$

so the slope is −1/2. Knowing this and the w-intercept, we can get a good idea of the position and orientation of the line. It passes through the point (0, −5), ramps gradually downward as we move to the right, and ramps gradually upward as we move to the left. Let's plug in $v = -12$ and calculate:

$$\begin{aligned} w &= (-1/2) \times (-12) - 5 \\ &= 6 - 5 \\ &= 1 \end{aligned}$$

This is in the field of view, so we can plot the point (−12, 1) as a small open circle, and draw the line for the second equation through it and (0, −5).

Are you confused?

What happens when one of the lines in a graph has an undefined slope? How can you solve and graph a linear system such as this, for example?

$$y = 2x - 2$$

and

$$x = 3$$

You can't reduce the second equation to SI form. Its graph is a vertical line passing through (3, 0). The first equation has a slope of 2. Because this sloping line extends forever both ways, upward and to the right as well as downward and to the left, it *must* cross the vertical line $x = 3$, which extends forever straight upward and straight downward, at some point. This is a perfectly decent two-by-two linear system. For extra credit, you can solve it and graph it.

Here's a trick!

Look again at the system we graphed in Fig. 17-5. The original two equations are:

$$-7v + w + 10 = 0$$

and

$$4v + 8w = -40$$

Suppose we want to draw a graph of this system with w as the independent variable and v as the dependent variable. Instead of deriving the SI versions all over again with the variables reversed, we can do a trick with the coordinate system and the lines drawn on it.

For a minute, imagine Fig. 17-5 as an unbreakable unit with the axes, points, and lines "all glued together." The positive v axis goes toward the right and the positive w axis goes upward. Now think: How can we morph this figure so the positive w axis goes toward the right and the positive v axis goes upward? If we can do that, we'll end up with a graph of the two-by-two linear system with w as the independent variable and v as the dependent variable.

Actually, this process takes only three steps. We *rotate* the drawing in Fig. 17-5 by 90° (one-quarter of a turn) counterclockwise. Then we *mirror* the drawing left-to-right. Finally, we relabel the points by

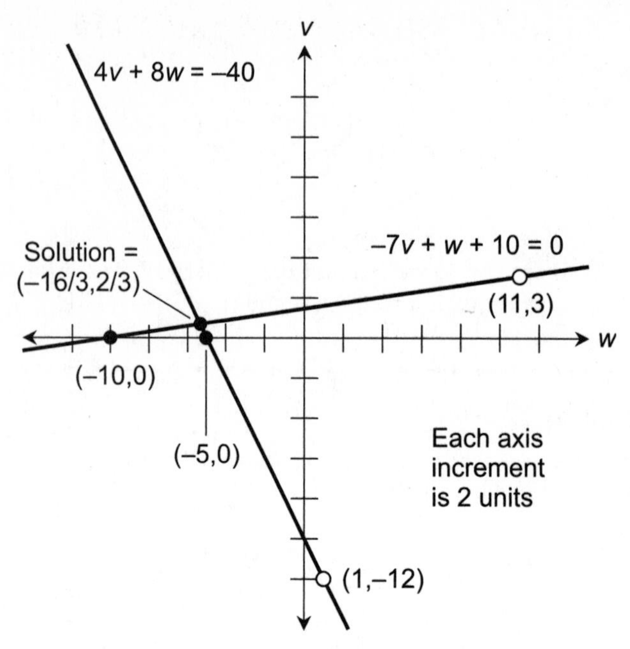

Figure 17-6 Graphs of $-7v + w + 10 = 0$ and $4v + 8w = -40$ as a two-by-two linear system where the independent variable is w and the dependent variable is v. This graph was obtained by rotating Fig. 17-5 by 90° counterclockwise, mirroring it right-to-left, and relabeling the points.

transposing the numbers in the ordered pairs. Once we've done these three things, we have a graph of the two-by-two system with the variables transposed. Figure 17-6 shows the result.

Are you astute?

Do you notice something familiar about the transformation we just performed, as if we've done it before? We have! These maneuvers are the equivalent of mirroring the whole system along the axis corresponding to the line where the values of the independent variable and the values of the dependent variable are identical. (In this case, that's the line $w = v$.) That transformation produces the graph of the inverse of a relation, as you learned in Chap. 14.

Look at the SI versions of the equations we graphed in Fig. 17-5. For reference, here they are again:

$$w = 7v - 10$$

and

$$w = (-1/2)v - 5$$

These are both functions of v. If we put these equations into SI form with v rather than w as the dependent variable, we obtain the inverse relations. Here are the manipulations, starting with the original equations:

$$-7v + w + 10 = 0$$
$$-7v + 10 = -w$$
$$-7v = -w - 10$$
$$7v = w + 10$$
$$v = (1/7)w + 10/7$$

and

$$4v + 8w = -40$$
$$4v = -8w - 40$$
$$v = (-8/4)w - 40/4$$
$$v = -2w - 10$$

These inverse relations are both functions of w. Can you see why?

We Couldn't Solve, but We Can Graph

In exercises 5, 6, and 7 at the end of Chap. 16, we scrutinized a two-by-two linear system. We could not solve it. Here's the pair of equations:

$$2x + y = 3$$

and

$$6x + 3y = 12$$

Let's graph these equations. This will help us see why no ordered pair (x,y) satisfies them as a system.

Find two points for each line

If we consider x as the independent variable and y as the dependent variable, then the SI version of the first equation is

$$y = -2x + 3$$

and the SI version of the second equation is

$$y = -2x + 4$$

The y-intercepts are 3 and 4, respectively. That means the point (0, 3) lies on the line for the first equation, and the point (0, 4) lies on the line for the second equation.

When taken together as a two-by-two linear system, these equations have no common solution. That means their graphs do not intersect, so we can't use their point of intersection as a graph-plotting aid. But there's another way. Both lines have slopes of −2. If we move n units to the right along either line (where n is any number), we must move $2n$ units downward to stay on the line. Let's move to the right by 4 units along each line. That means we must go downward by 8 units. Starting at (0, 3), we'll end up at (4, −5). Starting at (0, 4), we'll end up at (4, −4).

Connect the points

Figure 17-7 shows the four points we've found, and the two lines connecting them. Note that the lines are parallel, so they have no intersection point. These two lines, taken together, form the graph of the two-by-two linear system represented by the inconsistent equations

$$2x + y = 3$$

and

$$6x + 3y = 12$$

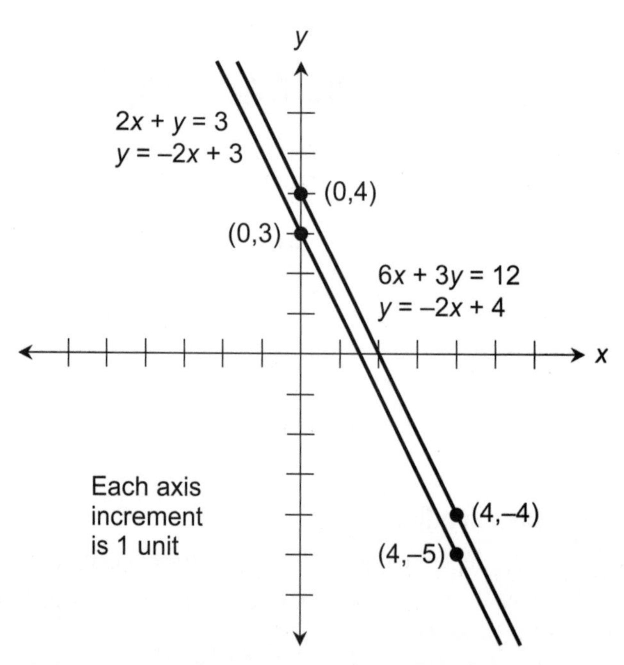

Figure 17-7 Graphs of $2x + y = 3$ and $6x + 3y = 12$ as a two-by-two linear system where the independent variable is x and the dependent variable is y. The lines are parallel and distinct. On both axes, each increment represents 1 unit. The SI forms of the equations are shown below the originals.

Are you confused?

Suppose the graphs of a two-by-two linear system show up as parallel, vertical lines, both with undefined slope. If you call x the independent variable and y the dependent variable, then neither of the equations is a function of x, although they are both relations between x and y. This sort of system can always be reduced to the form

$$x = a$$

and

$$x = b$$

where a and b are constants, and $a \neq b$.

If the graphs of a two-by-two linear system show up as parallel, horizontal lines in the same coordinate system, then they both have slopes of 0, and they both represent functions. Such a system can always be reduced to the form

$$y = c$$

and

$$y = d$$

where c and d are constants, and $c \neq d$.

Here's a challenge!

In exercise 10 at the end of Chap. 16, we tried to solve the following pair of equations as a linear system by substitution, but failed when we got the meaningless result $0 = 0$:

$$s = 2r - 3$$

and

$$-10r + 5s + 15 = 0$$

We showed that these equations are equivalent to a single function of r. State and graph that function, letting r be the independent variable and s be the dependent variable. Label a few of the infinitely many ordered pairs that satisfy the system.

Solution

The function can be stated as the first equation above. It tells us that the s-intercept is -3, so we can plot $(0, -3)$ on the Cartesian plane. Additional points can be found by moving to the right or left from $(0, -3)$ and moving upward or downward by twice that distance. The slope is 2, so if we start at any point on the line and move to the right by 1 unit (add 1 to r), we must move upward by 2 units (add 2 to s) to stay on the line. If we move to the left by 1 unit (subtract 1 from r), we must move downward by 2 units

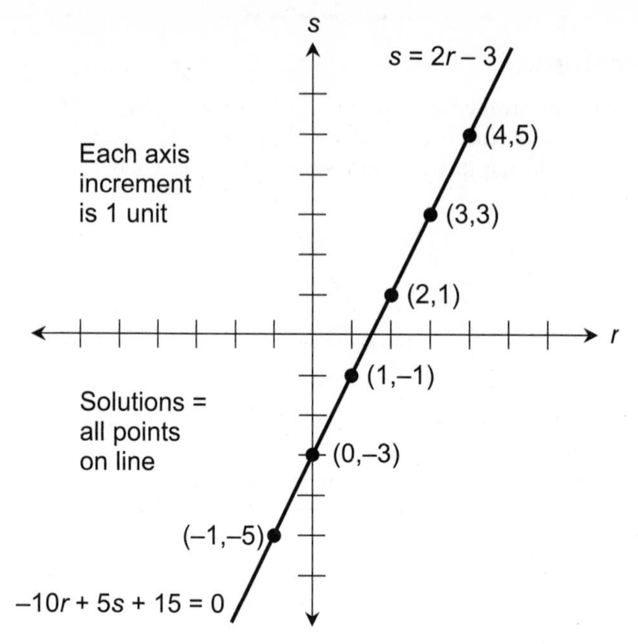

Figure 17-8 Graphs of $s = 2r - 3$ and $-10r + 5s + 15 = 0$ as a two-by-two linear system where the independent variable is r and the dependent variable is s. The lines coincide because the two equations are equivalent. On both axes, each increment represents 1 unit.

(subtract 2 from s). Some of the resulting points are plotted in Fig. 17-8, along with the line connecting them. The line is the graph of the function

$$s = 2r - 3$$

Any point on the line (not only the ones plotted in the figure) can be called a solution to this redundant two-by-two linear system. Because the domain and the range are both the entire set of real numbers, the *solution set* (the set containing all the ordered pairs that satisfy the system) is not only infinite, but nondenumerable!

Practice Exercises

This is an open-book quiz. You may (and should) refer to the text as you solve these problems. Don't hurry! You'll find worked-out answers in App. B. The solutions in the appendix may not represent the only way a problem can be figured out. If you think you can solve a particular problem in a quicker or better way than you see there, by all means try it!

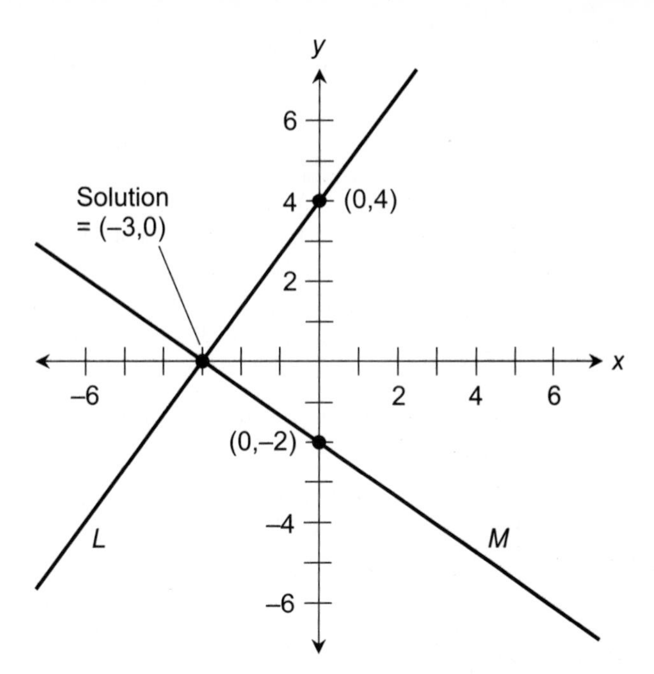

Figure 17-9 Illustration for Practice Exercises 1 and 2.

1. Refer to Fig. 17-9. What is the equation of line L in SI form?
2. Refer to Fig. 17-9. What is the equation of line M in SI form?
3. Using the morph-and-mix method, show that the solution to the equations derived in Probs. 1 and 2 is $x = -3$ and $y = 0$, which shows up as the point $(-3, 0)$.
4. Rotate and mirror Fig. 17-9, obtaining a new graph that shows the system with y as the independent variable and x as the dependent variable. Call the transposed line L by the new name L^*, and call the transposed line M by the new name M^*.
5. Using the graph derived in Prob. 4, what is the equation of line L^* in SI form? Remember that x is now the dependent variable.
6. Using the graph derived in Prob. 4, what is the equation of line M^* in SI form? Remember that x is now the dependent variable.
7. Using the morph-and-mix method, show that the solution to the equations derived in Probs. 5 and 6 is $y = 0$, and $x = -3$, which shows up as the point $(0, -3)$.
8. Suppose we see a linear function where x is the independent variable and y is the dependent variable. Let's call the function f and state it like this:

$$f(x) = mx + b$$

where m is the slope and b is the y-intercept of the graph. In this context, $f(x)$ is just another name for y. Now imagine that we derive the inverse of this function so y

becomes the independent variable and x becomes the dependent variable. We call this inverse function f^{-1}, and state it as follows:

$$f^{-1}(y) = ny + c$$

where n is the slope and c is the x-intercept. Now if f is a function, f^{-1} is *almost always* a function. But sometimes it isn't. (Some texts will say, under such conditions, that f has no inverse, or that f^{-1} does not exist. Others, such as this book, will say that if we consider f as a relation, then f always has an inverse relation, but that relation might not be a function.) Under what conditions is the inverse relation of a linear function *not* another function? Here's a hint: A straight line in Cartesian coordinates represents a function if and only if its slope is defined.

9. Draw graphs of three linear functions in Cartesian coordinates whose inverses are relations, but not functions.

10. Derive a version of f^{-1} as we defined it in Prob. 8, but that contains the constants m and b instead of the constants n and c. Here's a hint: state f as

$$y = mx + b$$

and morph this into SI form with x alone on the left side of the equals sign. Then, in place of the isolated x, write $f^{-1}(y)$.

CHAPTER

18

Larger Linear Systems

In this chapter we'll solve a *three-by-three linear system.* That's a triplet of linear equations in three variables. There are numerous ways to tackle problems of this sort. We'll look at only one example and one method. The scheme presented here can be broken down into three major steps:

- Eliminate one variable to get a two-by-two system
- Solve that two-by-two system for both of its variables
- Solve for the original eliminated variable by substitution

Eliminate One Variable

Consider three linear equations, each having three variables: x, y, and z. Our mission: Find the numbers for x, y, and z that satisfy all three equations. Here is the system we'll solve in this chapter:

$$-4x + 2y - 3z = 5$$
$$2x - 5y = z - 1$$
$$3x = -6y + 7z$$

Choose the vanishing variable

Let's decide which variable we want to eliminate. It doesn't make any difference whether it's x, y, or z. If we do all the calculations right, we'll get the same answer in the end, no matter which variable we choose at this stage. Let's get rid of z, so we are left with two equations in x and y.

Get the equations into form

Now that we've chosen the variable to eliminate, we must get all three equations into the same form. The following form is as good as any:

$$ax + by + cz = d$$

where a, b, c, and d are constants. The first equation is already in this form. The second equation is

$$2x - 5y = z - 1$$

Subtracting z from each side will put it into the form we want:

$$2x - 5y - z = -1$$

The third equation is

$$3x = -6y + 7z$$

Adding $6y$ to each side, we get

$$3x + 6y = 7z$$

We can subtract $7z$ from each side and it comes into the sought-after form:

$$3x + 6y - 7z = 0$$

We now have the three-by-three system in this uniform condition:

$$-4x + 2y - 3z = 5$$

$$2x - 5y - z = -1$$

$$3x + 6y - 7z = 0$$

Make z vanish once

Here are the first two revised equations again, for reference:

$$-4x + 2y - 3z = 5$$

and

$$2x - 5y - z = -1$$

Let's multiply the second equation through by −3, getting

$$-6x + 15y + 3z = 3$$

If we add this to the first equation, we obtain the sum

$$\begin{array}{r} -4x + 2y - 3z = 5 \\ -6x + 15y + 3z = 3 \\ \hline -10x + 17y = 8 \end{array}$$

Make z vanish again

Now let's scrutinize the second two revised equations. Here they are again:

$$2x - 5y - z = -1$$

and

$$3x + 6y - 7z = 0$$

We can multiply the first equation through by −7, getting

$$-14x + 35y + 7z = 7$$

When we add the second equation to this, we get

$$\begin{array}{r} -14x + 35y + 7z = 7 \\ 3x + 6y - 7z = 0 \\ \hline -11x + 41y = 7 \end{array}$$

State the two-by-two

We have now derived two different equations in the variables x and y. This is a two-by-two linear system, which we know how to solve:

$$-10x + 17y = 8$$

and

$$-11x + 41y = 7$$

Are you confused?

When we set out to get rid of the variable z, we decided to work on the first two equations in the revised three-by-three system, and then work on the second two equations. You might ask, "Why can't we eliminate z between the first equation and the second one, and then between the first equation and the third one? Or between the first and the third, and then between the second and the third?" The answer is, "We can! Either of those alternative schemes will work just as well as the one we chose. The intermediate equations will differ, but the ultimate solution to the three-by-three system will turn out the same."

Whenever you want to solve a three-by-three system of equations, you must somehow involve *all three* of the equations in the solution process. As you continue to study algebra and take more advanced courses, you'll learn other ways to solve three-by-three systems than the methods presented in this book. You might also learn techniques to solve four-by-four or larger systems. In any case, if you want to find the solution to a system of equations, you must give every one of those equations some "say" in the outcome.

Here's a challenge!

Put the original equations in our three-by-three linear system into a form where z appears all by itself on the left sides of the equals signs, and expressions containing only constants, x, and y appear on the right sides. This defines two independent variables (in this case x and y) and a single dependent variable (z). Here are the original equations, for reference:

$$-4x + 2y - 3z = 5$$
$$2x - 5y = z - 1$$
$$3x = -6y + 7z$$

Solution

To morph these equations, we can use the same sort of algebra that gets two-variable linear equations into SI form. Here are the processes, step-by-step. You have learned enough algebra so you can follow along without detailed explanations.

$$-4x + 2y - 3z = 5$$
$$2y - 3z = 4x + 5$$
$$-3z = 4x - 2y + 5$$
$$-z = (4/3)x - (2/3)y + 5/3$$
$$z = (-4/3)x + (2/3)y - 5/3$$

$$2x - 5y = z - 1$$
$$2x - 5y + 1 = z$$
$$z = 2x - 5y + 1$$

$$3x = -6y + 7z$$
$$3x + 6y = 7z$$
$$(3/7)x + (6/7)y = z$$
$$z = (3/7)x + (6/7)y$$

These three equations represent relations that map pairs of variables (in this case x and y) into a single variable (in this case z). If we call the relations f, g, and h, then we can write:

$$f(x, y) = (-4/3)x + (2/3)y - 5/3$$
$$g(x, y) = 2x - 5y + 1$$
$$h(x, y) = (3/7)x + (6/7)y$$

Solve the Two-by-Two

Let's continue our quest to solve the three-by-three linear system. We have obtained a two-by-two system in x and y. Here it is again:

$$-10x + 17y = 8$$

and

$$-11x + 41y = 7$$

We can use any of the methods described in Chap. 16 to tackle this. Let's use double elimination.

Eliminate x, solve for y

Let's multiply the first equation through by −11 and the second equation through by 10. When we do that, we get

$$110x - 187y = -88$$

and

$$-110x + 410y = 70$$

Now let's add these two equations in their entirety:

$$\begin{array}{r} 110x - 187y = -88 \\ -110x + 410y = 70 \\ \hline 223y = -18 \end{array}$$

Dividing through by 223 tells us that $y = -18/223$. This fraction happens to be in lowest terms (a fact that you can verify if you like).

Eliminate y, solve for x

Let's multiply the first equation in our two-by-two system through by −41 and the second equation through by 17. That gives us

$$410x - 697y = -328$$

and

$$-187x + 697y = 119$$

Adding these equations, we get

$$\begin{array}{r} 410x - 697y = -328 \\ -187x + 697y = 119 \\ \hline 223x = -209 \end{array}$$

Dividing through by 223 tells us that $x = -209/223$. This fraction, like the previous one with the same denominator, is in lowest terms.

Two down, one to go!

We've now solved for two of the three unknowns in our three-by-three system. We have the values for x and y:

$$x = -209/223$$

and

$$y = -18/223$$

In the next section, we'll substitute these values back into one of the original three equations and solve for z. Then we'll check our work. Something tells me that z is going to be a fraction with a denominator of 223. What do you think?

Are you confused?

You might again question the choice of solution processes. "Why," you might ask, "do we use the double-elimination method to solve the two-by-two system here? Why not use morph-and-mix or rename-and-replace?" The answer is, of course, "We can use either of those methods."

Here's a challenge!

Solve the preceding two-by-two system using the morph-and-mix method. Consider x the dependent variable.

Solution

Once again, here's our pair of equations:

$$-10x + 17y = 8$$

and

$$-11x + 41y = 7$$

We must get both of these into SI form, with x all by itself on the left sides of the equals signs. Step-by-step, the first equation morphs like this:

$$-10x + 17y = 8$$

$$-10x = -17y + 8$$

$$10x = 17y - 8$$

$$x = (17/10)y - 8/10$$

The second equation morphs as follows:

$$-11x + 41y = 7$$

$$-11x = -41y + 7$$

$$11x = 41y - 7$$

$$x = (41/11)y - 7/11$$

Now we mix the right-hand sides, getting a first-degree equation in y:

$$(17/10)y - 8/10 = (41/11)y - 7/11$$

Let's get a common denominator here. We can multiply the numerators and denominators on the left side of the equals sign by 11, and multiply the numerators and denominators on the right side of the equals sign by 10. That gives us

$$(187/110)y - 88/110 = (410/110)y - 70/110$$

Multiplying the whole equation through by 110 to get rid of the fractions, we obtain

$$187y - 88 = 410y - 70$$

Adding 88 to each side produces

$$187y = 410y + 18$$

Subtracting $410y$ from each side, we have

$$-223y = 18$$

We can divide through by −223 to get $y = 18/(-223) = -18/223$. This agrees with the result we obtained by double elimination. We can plug this into either of the SI equations we derived above, solving for x. The first equation will do. Step-by-step:

$$\begin{aligned} x &= (17/10)y - 8/10 \\ &= (17/10)(-18/223) - 8/10 \\ &= -306/2{,}230 - 8/10 \\ &= -306/2{,}230 - 1{,}784/2{,}230 \\ &= (-306 - 1{,}784)/2{,}230 \\ &= -2{,}090/2{,}230 \\ &= -209/223 \end{aligned}$$

It agrees again! This practically guarantees that our answers are correct. We've found x and y by two different routes, and they've come out the same both times. If we made an error somewhere, the answers would almost certainly disagree.

Substitute Back

Now we have the values for x and y, and we're confident that they're correct because we've arrived at them from two different directions. Here they are again:

$$x = -209/223$$

and

$$y = -18/223$$

Plug them in

We can use any of the original equations or their revisions to solve for z. Let's use the third original equation:

$$3x = -6y + 7z$$

Plugging in the numbers for x and y, and proceeding step-by-step, we get

$$3 \times (-209/223) = -6 \times (-18/223) + 7z$$

$$-627/223 = 108/223 + 7z$$

$$-735/223 = 7z$$

$$7z = -735/223$$

$$z = -105/223$$

Are you confused?

If the last step in the above calculation confuses you, note that $-735/7 = -105$. That's the numerator in the fraction. I had a feeling that -735 would cleanly divide by 7, because I acted on my hunch that z would be a fraction with a denominator of 223. By now, you have probably noticed that in linear systems, fractional solutions have a way of coming out with identical denominators. There's a good reason for that, but it would be a distraction to delve into the "why" of it right now. The important thing is that we have all three solutions to our original three-by-three linear system, at least tentatively:

$$x = -209/223$$

$$y = -18/223$$

$$z = -105/223$$

Here's a challenge!

In a problem as messy and lengthy as this, it's *vital* to check the derived solutions. Here are the original equations once again, for reference:

$$-4x + 2y - 3z = 5$$
$$2x - 5y = z - 1$$
$$3x = -6y + 7z$$

"So," you ask, "What's the challenge here? It's only arithmetic!" The answer: "It's a challenge to force ourselves through the tedium. But it has to be done if we want to be sure our answers are correct."

Solution

Let's start with the first equation. Plugging in the values for x, y, and z, and then doing the arithmetic carefully, we pass through the following steps:

$$-4x + 2y - 3z = 5$$
$$-4 \times (-209/223) + 2 \times (-18/223) - 3 \times (-105/223) = 5$$
$$836/223 - 36/223 + 315/223 = 5$$
$$(836 - 36 + 315)/223 = 5$$
$$1{,}115/223 = 5$$
$$5 = 5$$

So far, we're doing okay. Now for the second equation check:

$$2x - 5y = z - 1$$
$$2 \times (-209/223) - 5 \times (-18/223) = -105/223 - 1$$
$$-418/223 + 90/223 = -105/223 - 223/223$$
$$(-418 + 90)/223 = (-105 - 223)/223$$
$$-328/223 = -328/223$$

Two checks are done, and one remains. Here we go:

$$3x = -6y + 7z$$
$$3 \times (-209/223) = -6 \times (-18/223) + 7 \times (-105/223)$$
$$-627/223 = 108/223 - 735/223$$
$$-627/223 = (108 - 735)/223$$
$$-627/223 = -627/223$$

Mission accomplished!

General Linear Systems

Now that you've seen two-by-two and three-by-three linear systems, you might wonder, "What does a three-by-three graph look like?" or "What happens when there are many linear equations in many variables?" or "What happens when the number of equations is not the same as the number of variables?"

Two-by-two geometry

The graph of a linear equation in two variables shows up as a straight line in the Cartesian plane, as you saw in Chap. 17. When you have two such equations, their graphs always appear in one of three ways:

- Two different lines that intersect at a single, unique point
- Two different lines that are parallel
- Two lines that precisely coincide

In the first case, the system has a single solution, corresponding to the point where the lines intersect. In the second case, there is no solution. In the third situation, there are infinitely many solutions.

Cartesian three-space

The graph of a linear equation in three variables can't be drawn on a Cartesian plane. Instead, we need to use a system that can portray all of space. The most common such system is called *Cartesian three-space.* It makes use of three coordinate axes, all of which intersect at their zero points, and in such a way that each axis is perpendicular to the other two.

Cartesian three-space is sometimes drawn in perspective, as in Fig. 18-1. In this example, the variables are x, y, and z. If the drawings were literal, the x axis would appear horizontal on the page, the y axis would appear vertical on the page, and the z axis would be perpendicular to the page. Note that the positive x axis goes to the right, the positive y axis goes upward, and the positive z axis comes toward us.

Figure 18-2 shows two specific points, P and Q, plotted in Cartesian three-space. The coordinates of point P are $(-5, -4, 3)$, and the coordinates of point Q are $(3, 5, -2)$. Points are denoted as *ordered triples* in the form (x, y, z), where the first number represents the value on the x axis, the second number represents the value on the y axis, and the third number represents the value on the z axis.

Three-by-three geometry

The graph of a linear equation in three variables appears as a flat plane (not a line!) in Cartesian three-space. When we have three linear equations in three variables, their graphs can show up in any of the following ways:

- Three different planes that all intersect at a single, unique point.
- Three different planes, two of which are parallel, and the third of which intersects the other two in a pair of parallel lines.

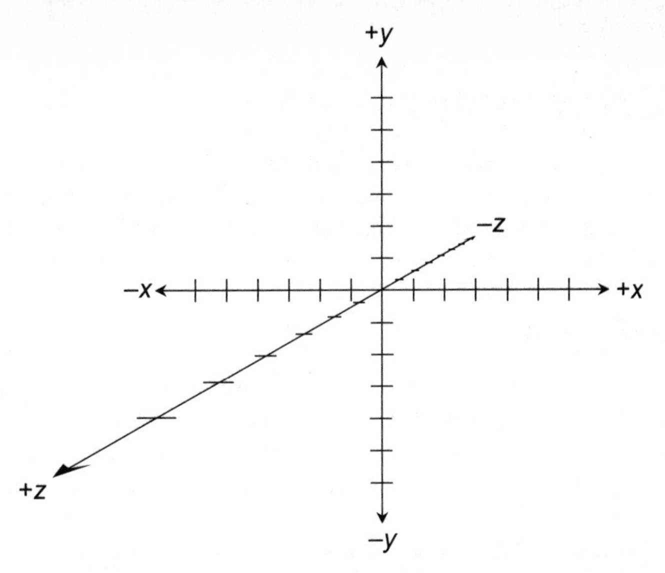

Figure 18-1 Cartesian three-space. The *x* axis increases positively from left to right, the *y* axis increases positively from the bottom up, and the *z* axis increases positively from far to near. All three axes intersect at the origin and are mutually perpendicular.

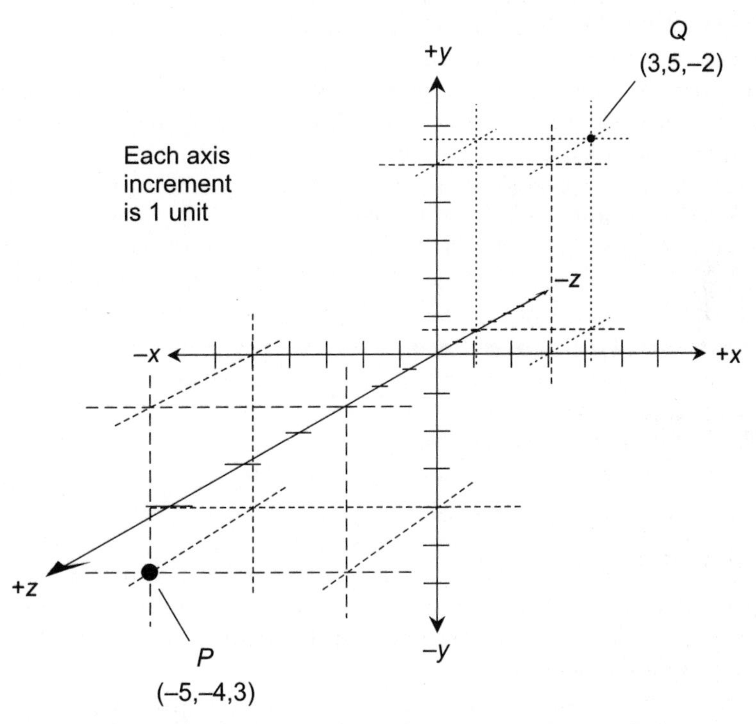

Figure 18-2 Two points in Cartesian three-space, along with the corresponding ordered triples of the form (x,y,z). On all three axes, each increment represents 1 unit.

- Three different planes, such that each pair of planes intersects in a different line, and all three of those lines are mutually parallel.
- Three different planes that are mutually parallel.
- Two planes that precisely coincide, and a third plane parallel to them both.
- Two planes that precisely coincide, and a third plane that intersects them both in a single line.
- Three different planes that all intersect in a single line.
- Three planes that all precisely coincide.

In the first case, the system has a unique solution, corresponding to the point where all three planes intersect. In each of the second through fifth cases, there is no solution. In each of the last three cases, there are infinitely many solutions. Try to envision all these situations. Note that two planes are parallel in space if and only if they do not intersect.

Are you confused?

Most people have trouble envisioning the graphs of three-by-three linear systems. Computer programs have been developed to portray systems like these, and such programs can be a big help. But they, too, give only a limited perspective. A true view would require a three-dimensional hologram that we could walk around in! When it comes to *n-by-n linear systems* where n is a natural number larger than 3, even a walk-through hologram can't give us a complete picture.

It's not easy to verbally describe what happens in n-by-n linear systems when n is large. Often, a unique solution exists in this type of system, but not always. A unique solution always comes down to a single point in *Cartesian n-space.* When n is large, there are many ways that an n-by-n linear system can fall short of a unique solution. Even so, if we write up a system of n linear equations in n variables "at random," the chance is good that it will have a single solution.

As you can imagine, the process of solving an n-by-n system of linear equations where $n > 3$ is bound to be time-consuming and tedious. See how much longer it took us to solve the three-by-three system in this chapter than it took us to solve the two-by-two systems in Chap. 16! As n increases, so does the time it will take us to solve the system, unless we have access to a computer. The process of solving an n-by-n linear system is ideally suited to computer applications, which grind out solutions by brute force.

Fewer equations than variables

All of the linear systems we've examined so far have the same number of equations as variables. Occasionally, a linear system has fewer equations than variables. Whenever that happens, there is no single solution to the whole system.

Think of a linear system with two variables but only one equation. If we consider this as a one-by-two linear system (one equation, two variables), then it has infinitely many solutions. It is the same thing as a redundant two-by-two system. In the Cartesian plane, it shows up as a single straight line.

Now imagine a one-by-three linear system (a single equation in x, y, and z). Its graph in Cartesian three-space is a single flat plane. The solutions are all the *ordered triples* (x,y,z) that represent points on the plane. There are infinitely many such points, so this type of system has infinitely many solutions. This type of system can never have a unique solution.

Things get more interesting with two-by-three linear systems (two equations and three variables x, y, and z). The graph appears as two planes in Cartesian three-space. The planes might intersect in a straight line, in which case there are infinitely many solutions: all the ordered triples (x,y,z) that correspond to points on that line. A second possibility is that the two planes are parallel. Then they don't intersect anywhere, and there are no solutions to the system. A third possibility is that the planes coincide. Then, again, there are infinitely many solutions: all the ordered triples (x,y,z) that correspond to points on that plane. Two flat planes in space can never intersect in a single point. That means a two-by-three linear system can't have a unique solution.

More equations than variables

When a linear system in two variables has more than two equations, or a linear system in three variables has more than three equations, we have "extra data." We can imagine such a system as a two-by-two or three-by-three linear system with one or more extra equations thrown in.

If a two-by-two or three-by-three linear system is inconsistent, then adding one or more extra equations *cannot* make it consistent. If a two-by-two or three-by-three linear system is consistent, then adding one or more extra equations *might* make it inconsistent or redundant, but not necessarily.

These statements can be generalized to any number of linear equations in any number of variables. If we encounter 27 equations in 28 variables, we can be certain that the system has no unique solution. But if we see 28 or more equations in 28 variables, we can't know if the system has a unique solution until we try to solve it, preferably with the help of a computer. If the system has no unique solution, the algebra will lead us into absurd or useless statements.

Here's a challenge!

Draw a graph of the following three-by-two linear system to illustrate why it does not have a unique solution:

$$y = x + 1$$

$$y = 3x - 1$$

$$y = -2x - 3$$

Solution

These three equations are in SI form. The y-intercepts are 1, −1, and −3, respectively. The slopes are 1, 3, and −2 respectively; they can be used to find second points as shown in Fig. 18-3, which portrays the graphs of the equations as a single system. Each line intersects both of the others, but there is no single point common to all three lines. For any linear system to have a solution, no matter how many variables there are, the graphs of all the equations must have a single point in common. If there are two variables, that point can be represented by a unique ordered pair. If there are three variables, the point can be represented by a unique ordered triple. If there are n variables, the point can be represented by a unique *ordered n-tuple.*

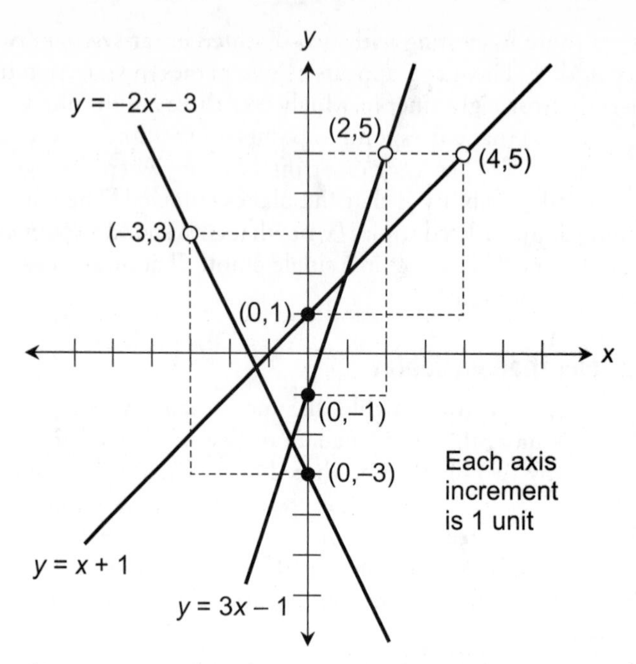

Figure 18-3 Graphs of three equations in two variables, considered as a linear system. There is no solution, because there is no single point common to all three lines. On both axes, each increment represents 1 unit.

Practice Exercises

This is an open-book quiz. You may (and should) refer to the text as you solve these problems. Don't hurry! You'll find worked-out answers in App. B. The solutions in the appendix may not represent the only way a problem can be figured out. If you think you can solve a particular problem in a quicker or better way than you see there, by all means try it!

1. Here are the three revised original equations for the three-by-three system we tackled in this chapter:

$$-4x + 2y - 3z = 5$$

$$2x - 5y - z = -1$$

$$3x + 6y - 7z = 0$$

In the section "Eliminate One Variable," we got rid of z between the first two of these equations, and then between the second two. Now eliminate z between the first and third equations.

2. Derive a two-by-two linear system in x and y from the solution to Prob. 1, along with the equation in x and y that we derived from the second and third three-variable equations in the section "Eliminate One Variable."

3. Solve the two-by-two linear system obtained in the solution to Prob. 2. Use the double-elimination method.

4. Solve for z by substituting the values for x and y (solution 3) into the first equation stated in Prob. 1.

5. Derive a two-by-two linear system in x and y from the solution to Prob. 1, along with the equation in x and y that we derived from the first two three-variable equations in the section "Eliminate One Variable."

6. Solve the two-by-two linear system obtained in the solution to Prob. 5. Use the morph-and-mix method, treating y as the independent variable and x as the dependent variable.

7. Solve for z by substituting the values for x and y (solution 6) into the second equation stated in Prob. 1.

8. The following four-by-two linear system has a unique solution, even though there are more equations than variables. How can we know this without doing any algebra or graphing the equations? What is that solution?

$$y = -x + 1$$

$$y = -2x + 1$$

$$y = 3x + 1$$

$$y = 4x + 1$$

9. Plug in the values for x and y that appear to solve the set of equations in Prob. 8, based on the reasoning in the solution to Prob. 8. Verify that these values satisfy all four equations.

10. Graph all four of the lines presented in Prob. 8. On the basis of this graph, explain why any pair or triplet of these equations, taken as a two-by-two or three-by-two system, has the same unique solution as any other pair or triplet of the equations.

CHAPTER

19

The Matrix Morphing Game

There's a neat way to solve three-by-three linear systems by removing the variables from the notation. Only the coefficients remain, and these can be arranged into columns and rows to form a table-like array called a *matrix*. In this chapter, you'll learn how to manipulate *matrixes* (or *matrices*) to solve three-by-three linear systems.

How to Build a Matrix

When you encounter a three-by-three linear system, each variable is multiplied by a constant in each equation. Constants can also appear by themselves without variables. The form can vary from one equation to another. Matrices aren't so flexible. They have to be written in a specific, standard form. Everything must go into a preassigned "cubbyhole."

The equation form

Before we can write any three-by-three linear system as a matrix, we have to get the triplet of equations into the following form:

$$a_1x + b_1y + c_1z = d_1$$

$$a_2x + b_2y + c_2z = d_2$$

$$a_3x + b_3y + c_3z = d_3$$

Where x, y, and z are the variables, and the letters a, b, c, and d with numeric subscripts are the coefficients. A coefficient can be any real number including 0. When we want to get a system into matrix form, we must write all the coefficients, even the ones that are equal to 0!

The matrix form

Once we have morphed all the equations into the proper form, we can remove the variables and equals signs. Then we can put the numbers into an array having three rows, each with four numbers:

a_1	b_1	c_1	d_1
a_2	b_2	c_2	d_2
a_3	b_3	c_3	d_3

Some texts enclose matrices in huge parentheses or brackets. But they aren't really necessary, and they can clutter things up. Let's not use them.

Are you confused?

You might ask, "Can matrices represent two-by-two systems? Can they represent systems larger than three-by-three? Can they represent asymmetrical systems such as five-by-four?" The answers are "Yes," "Yes," and "Yes." In this chapter, we'll look at three-by-three systems only, but matrix techniques can be applied to any linear system.

Here's a challenge!

Put the following three-by-three linear system into matrix form:

$$7y = 3z + 3$$
$$8z = -2x - 7$$
$$12x = 7y$$

Solution

None of these three equations is in the proper form for conversion to matrix notation. We'll have to manipulate them. Here are the processes, step-by-step. For the first equation:

$$7y = 3z + 3$$
$$0x + 7y = 3z + 3$$
$$0x + 7y - 3z = 3$$

For the second equation:

$$8z = -2x - 7$$
$$2x + 8z = -7$$
$$2x + 0y + 8z = -7$$

For the third equation:

$$12x = 7y$$
$$12x - 7y = 0$$
$$12x - 7y + 0z = 0$$

Now we have these three equations that make up the linear system:

$$0x + 7y - 3z = 3$$
$$2x + 0y + 8z = -7$$
$$12x - 7y + 0z = 0$$

We may want to write the above equations like this, so we are sure to get the signs of the coefficients right:

$$0x + 7y + (-3z) = 3$$
$$2x + 0y + 8z = -7$$
$$12x + (-7y) + 0z = 0$$

We can write this system in matrix form by removing the variables and equals signs, and then aligning the coefficients into neat rows and columns:

0	7	–3	3
2	0	8	–7
12	–7	0	0

Matrix Operations

Imagine the matrix for a three-by-three linear system as a game board with 12 positions, arranged in three horizontal rows and four vertical columns. Let's invent a *matrix morphing game*. There are three types of moves in this game: *swap*, *multiply*, and *add*. We can make as many of these moves as we want.

Swap

We may interchange all the elements between two rows in a matrix, while keeping the elements of both rows in the same order from left to right. For example, if we start with

a_1	b_1	c_1	d_1
a_2	b_2	c_2	d_2
a_3	b_3	c_3	d_3

we can change it to

a_3	b_3	c_3	d_3
a_2	b_2	c_2	d_2
a_1	b_1	c_1	d_1

In this case, the first and third rows have been swapped. Note that we cannot swap individual elements or vertical columns! The swap maneuver is only allowed between entire rows.

Multiply

We may multiply all the elements in any row by a nonzero constant, keeping the elements in the same order from left to right. For example, if we have

a_1	b_1	c_1	d_1
a_2	b_2	c_2	d_2
a_3	b_3	c_3	d_3

we can change this to

a_1	b_1	c_1	d_1
ka_2	kb_2	kc_2	kd_2
a_3	b_3	c_3	d_3

In this case, all the elements in the second row have been multiplied by k. Because the absolute value of k can be smaller than 1, we can extrapolate this rule to let us multiply or divide all the elements in any row by a nonzero constant. As with the swap move, we can operate only on entire rows. We can't do this maneuver with individual elements or with columns.

Add

We may add all the elements in any row to all the elements in another row, and then replace the elements in either row by the sum, taking care to keep the elements of both rows in the same order from left to right. Suppose we start with this matrix:

a_1	b_1	c_1	d_1
a_2	b_2	c_2	d_2
a_3	b_3	c_3	d_3

We can change it to either of the following:

a_1	b_1	c_1	d_1
$a_1 + a_2$	$b_1 + b_2$	$c_1 + c_2$	$d_1 + d_2$
a_3	b_3	c_3	d_3

or

$a_1 + a_2$	$b_1 + b_2$	$c_1 + c_2$	$d_1 + d_2$
a_2	b_2	c_2	d_2
a_3	b_3	c_3	d_3

Note that the replaced row must be one of the two involved in the sum. In this example, we aren't allowed to replace the third row with the sum of the first and second rows.

The final goal

The matrix morphing game, like any sensible game, has an ultimate objective. Our goal is to get a matrix representing a three-by-three linear system into this form:

1	0	0	x
0	1	0	y
0	0	1	z

where x, y, and z are real numbers. This is called the *unit diagonal form.*

Now imagine that we start with a three-by-three linear system, make it into a matrix, and then play the matrix morphing game until we get the unit diagonal form. Do you suspect that the values x, y, and z, which appear in the far right column, will represent the solution to the linear system? If so, you are right, provided the system is consistent (has a unique solution).

Are you confused?

Do you wonder why we can't multiply all the elements in a row by 0? Think about this for a minute. Doing that would wipe out one of the equations in the system, leaving us with a two-equation system having three variables. Such a system doesn't contain enough information to define a unique solution.

Are you still confused?

You might also ask, "Why can't we add two rows in a matrix and then replace the remaining row (the one not involved in the sum) with that sum?" Well, we could try it, and for awhile it might seem to work. But such a move would delete the information in the equation represented by the replaced row, turning it into a mere hybrid of the other two rows. We would be left with the equivalent of a two-equation system in three variables. We might end up thinking that the original system was redundant, when in fact it was consistent.

A Sample Problem

In this section, we'll solve a three-by-three linear system using matrix operations. The game can be played in many ways. The process in this section doesn't represent the only avenue by which the final result can be reached. But if we manage to avoid making mistakes, we'll always get to the same destination, no matter what road we take!

The game plan

Anyone who plays a game needs a plan. Here's a strategy for the matrix morphing game that works well for me. First, we get equations into form:

$$a_1x + b_1y + c_1z = d_1$$
$$a_2x + b_2y + c_2z = d_2$$
$$a_3x + b_3y + c_3z = d_3$$

Then we make the matrix:

a_1	b_1	c_1	d_1
a_2	b_2	c_2	d_2
a_3	b_3	c_3	d_3

Next, we get the matrix into *echelon form,* which looks like this:

#	#	#	#
0	#	#	#
0	0	#	#

where a pound sign (#) can represent any real number. Then we go for the *diagonal form:*

@	0	0	#
0	@	0	#
0	0	@	#

where a pound sign can represent any real number, and an at sign (@) can represent any *non-zero* real number. Our final goal is the unit diagonal form:

1	0	0	x
0	1	0	y
0	0	1	z

Once we've put a matrix into the unit diagonal form, we may find that one or more of the solutions x, y, or z is a fraction that can be reduced. We ought to reduce all solutions to their lowest forms in the interest of elegance.

What if we can't "win"?

If we find it impossible to get a matrix into the unit diagonal form, our failure indicates one of four things:

- We didn't try hard enough
- We made a mistake somewhere
- The original system is inconsistent
- The original system is redundant

The system to be solved

Now let's methodically tackle a three-by-three linear system and solve it using the matrix morphing game. Here are the equations:

$$3x + z = 2y + 11$$
$$4y + 2z = x$$
$$-5x + y = 3z - 20$$

Formatting the equations

None of these equations is in the proper form for assembling a matrix. Let's get them that way! Fortunately, the maneuvers are simple. With the first equation, we can subtract $2y$ from each side to get

$$3x - 2y + z = 11$$

With the second equation, we can subtract x from each side to obtain

$$-x + 4y + 2z = 0$$

With the third equation, we can subtract $3z$ from each side, getting

$$-5x + y - 3z = -20$$

We now have the equations in form, and can state the whole system like this:

$$\begin{aligned} 3x - 2y + z &= 11 \\ -x + 4y + 2z &= 0 \\ -5x + y - 3z &= -20 \end{aligned}$$

Building the matrix

To construct the matrix, we remove all the variables and arrange the remaining numbers into an orderly array, paying close attention to the signs:

3	−2	1	11
−1	4	2	0
−5	1	−3	−20

Deriving the echelon form

There are many different routes by which we can arrive at an echelon form of this matrix. Let's start by getting 0 at the extreme left in the bottom row. We can multiply the second row by −5 to obtain

3	−2	1	11
5	−20	−10	0
−5	1	−3	−20

We can add the second and third rows and then replace the third row with the sum to get

3	−2	1	11
5	−20	−10	0
0	−19	−13	−20

Now let's get 0 at the extreme left in the second row. If we multiply the second row by −3/5, we come up with

3	−2	1	11
−3	12	6	0
0	−19	−13	−20

Adding the first and second rows and then replacing the second row with the sum, we have

3	−2	1	11
0	10	7	11
0	−19	−13	−20

Now the number −19 in the third row must somehow be made to vanish, which means we must turn it into 0. Let's use the second row to "attack" it. If we multiply the second row through by 19 and the third row through by 10 (combining two moves), we get

3	−2	1	11
0	190	133	209
0	−190	−130	−200

Adding the second and third rows and then replacing the third row with the sum, we obtain a matrix in echelon form:

3	−2	1	11
0	190	133	209
0	0	3	9

Deriving the diagonal form

To morph the echelon matrix into diagonal form, we simply keep playing the game. In this case, we want to get 0s in the places that now contain −2, 1, and 133. Let's start with the third element in the first row, currently equal to 1. We can use the third row to "attack" it. Let's divide the third row by −3 to get

3	−2	1	11
0	190	133	209
0	0	−1	−3

Adding the first and third rows and then replacing the first row with the sum, we get

3	−2	0	8
0	190	133	209
0	0	−1	−3

Let's make the number 133 vanish. We can use the third row as the "weapon" here. If we multiply the third row by 133, we get

3	−2	0	8
0	190	133	209
0	0	−133	−399

Adding the second two rows and then replacing the second row with the sum, we obtain

3	−2	0	8
0	190	0	−190
0	0	−133	−399

We now have an opportunity to reduce the sizes of the numbers in the second and third rows. Let's divide the second row through by 190, and divide the third row through by −133. That gives us

3	−2	0	8
0	1	0	−1
0	0	1	3

We have only to make the element −2 in the first row vanish, and we'll have the matrix in diagonal form. We can multiply the second row by 2 to obtain

3	−2	0	8
0	2	0	−2
0	0	1	3

Adding the first and second rows and then replacing the first row with the sum, we get

3	0	0	6
0	2	0	−2
0	0	1	3

Deriving the unit diagonal form

Our remaining task is simple and clean. We can divide the top row by 3 and the middle row by 2, obtaining the unit diagonal matrix

1	0	0	2
0	1	0	−1
0	0	1	3

Stating and reducing the solution

The solution to the system is now clear. We have no fractions to reduce, because we kept the sizes of the numbers down as we went along. (Otherwise we'd have fractions with large numerators and denominators, although they would divide out cleanly and leave us with integers.) Evidently,

$$x = 2$$

$$y = -1$$

$$z = 3$$

Are you confused?

If you think we've finished solving this problem, you're indeed confused. We must check our work! Only then can we be totally confident that our solution is correct, because it's easy to make mistakes in the matrix morphing game. When we plug the values for *x*, *y*, and *z* into the first original equation, here's what happens:

$$3x + z = 2y + 11$$

$$3 \times 2 + 3 = 2 \times (-1) + 11$$

$$6 + 3 = -2 + 11$$

$$9 = 9$$

Check one. Now for the second equation:

$$4y + 2z = x$$

$$4 \times (-1) + 2 \times 3 = 2$$

$$-4 + 6 = 2$$

$$2 = 2$$

Check two. Finally, the third equation:

$$-5x + y = 3z - 20$$

$$-5 \times 2 + (-1) = 3 \times 3 - 20$$

$$-10 - 1 = 9 - 20$$

$$-11 = -11$$

Check three. Mission accomplished.

Here's an extra-credit challenge!

Play the matrix morphing game with the linear system we just got done solving, but take a different route this time. Don't reduce the sizes of any of the integers. Let the absolute values grow as large as they "want"!

When you get the unit diagonal form, you should end up with sloppy fractions in the far-right column. When you reduce these fractions, you should get

1	0	0	2
0	1	0	−1
0	0	1	3

Solution

You're on your own. Have fun!

Practice Exercises

This is an open-book quiz. You may (and should) refer to the text as you solve these problems. Don't hurry! You'll find worked-out answers in App. B. The solutions in the appendix may not represent the only way a problem can be figured out. If you think you can solve a particular problem in a quicker or better way than you see there, by all means try it!

1. Put the following three-by-three linear system into the proper form for conversion to matrix notation:

$$x = y - z - 7$$

$$y = 2x + 2z + 2$$

$$z = 3x - 5y + 4$$

2. Write the set of equations from the solution to Prob. 1 in the form of a matrix.
3. Write the set of equations represented by the following matrix:

0	4	−1	−2
5	−3/2	8	1
1	1	1	1

4. Put the matrix of Prob. 3 into echelon form.
5. Put the matrix derived in the solution to Prob. 4 into diagonal form.
6. Reduce the matrix derived in the solution to Prob. 5 to a form with the smallest possible absolute values in each row, such that all the numbers in the matrix are integers.
7. Reduce the matrix derived in the solution to Prob. 6 to unit diagonal form. Then state the tentative solution to the three-by-three linear system we derived from the matrix in Prob. 3 and stated in solution 3.
8. Check the values for x, y, and z derived in the solution to Prob. 7 to be sure they're correct. To do this, plug the numbers into the equations stated in the solution to Prob. 3.

9. Put the following three-by-three linear system into matrix form. Then play the matrix morphing game for awhile, trying to get the unit diagonal form. It can't be done! Why?

$$x + y + z = 1$$

$$x + y + z = 2$$

$$x + y + z = 3$$

10. Put the following three-by-three linear system into matrix form. Then play the matrix morphing game for awhile, trying to get the unit diagonal form. It can't be done! Why?

$$x + y + z = 1$$

$$2x + 2y + 2z = 2$$

$$3x + 3y + 3z = 3$$

CHAPTER

20

Review Questions and Answers

Part Two

This is not a test! It's a review of important general concepts you learned in the previous nine chapters. Read it though slowly and let it "sink in." If you're confused about anything here, or about anything in the section you've just finished, go back and study that material some more.

Chapter 11

Question 11-1

What things can we do with the equation if we want to change its form, but be sure that the result is equivalent to the original equation and still valid?

Answer 11-1

We can do any of these things:

- Switch the left and right sides.
- Add a quantity to the left side, and add the same quantity to the right side.
- Subtract a quantity from the left side, and subtract the same quantity from the right side.
- Multiply the left side by a nonzero quantity, and multiply the right side by the same nonzero quantity.
- Divide the left side by a nonzero quantity, and multiply the right side by the same nonzero quantity.

Question 11-2

What are the five types of inequalities that can exist between two numbers, variables, or mathematical expressions p and q? How are these inequalities symbolized?

Answer 11-2

If we have two numbers, variables, or mathematical expressions p and q, then five types inequalities can exist:

- If p is strictly smaller than q, we write $p < q$.
- If p is smaller than or equal to q, we write $p \leq q$.
- If p is not equal to q but we don't know which is smaller, we write $p \neq q$.
- If p is larger than or equal to q, we write $p \geq q$.
- If p is strictly larger than q, we write $p > q$.

Question 11-3

Suppose we multiply an inequality through by a nonzero real number. Under what circumstances will the sense of the inequality be reversed? Under what circumstances will the sense of the inequality stay the same?

Answer 11-3

The sense of the inequality is reversed if we multiply through by a negative real number, and it remains the same if we multiply through by a positive real number.

Question 11-4

Suppose we divide an inequality through by a nonzero real number. Under what circumstances will the sense of the inequality be reversed? Under what circumstances will the sense of the inequality stay the same?

Answer 11-4

The sense of the inequality is reversed if we divide through by a negative real number, and it remains the same if we divide through by a positive real number.

Question 11-5

If a quantity x is strictly smaller than another quantity y, then we can also say x is smaller than or equal to y. How would we write this fact entirely in logical symbols?

Answer 11-5

Remember the symbols for "strictly smaller than," "smaller than or equal to," and "logically implies." The above statement can be written symbolically as

$$(x < y) \Rightarrow (x \leq y)$$

Question 11-6

We can't square both sides of a "strictly smaller than" inequality and be sure that we'll get another valid statement. Provide an example that shows why.

Answer 11-6

Consider the fact that $-3 < 2$. If we square both sides, we get $9 < 6$, which is false. There are infinitely many other examples.

Question 11-7

Is it possible to multiply a quantity by a positive real number and have the product be smaller than the original quantity? If so, give an example.

Answer 11-7

Yes, this is possible. If the original quantity is negative, then when we multiply it by a positive real number, it gets more negative. That means it gets smaller. Remember that numbers get smaller, not larger, as they grow more negative!

Question 11-8

Is it possible to divide a positive real number by another positive real number, and have the quotient turn out strictly larger than the original? If so, give an example.

Answer 11-8

Yes, this can happen. Suppose we start with a positive real number, and then divide it by some positive number smaller than 1. That will give us a quotient larger than the original. For example, if we divide 3 by 1/4, we get 12, which is larger than 3.

Question 11-9

Suppose we have a "strictly larger than" inequality. What can we do to both sides, and be sure that the result will be another valid "strictly larger than" statement?

Answer 11-9

We can do any of the following things:

- Add the same quantity to both sides.
- Subtract the same quantity from both sides.
- Multiply both sides by the same positive quantity.
- Divide both sides by the same positive quantity.

Question 11-10

Imagine a hypothetical relation called *clobber* (symbolized by ©) and three variables a, b, and c. How do we define the reflexive, symmetric, and transitive properties for the clobber relation? If clobber has all three of these properties, what sort of relation is it?

Answer 11-10

Clobber is reflexive if and only if, for all possible values of a,

$$a \copyright a$$

Clobber is symmetric if and only if, for all possible values of a and b,

$$(a \copyright b) \Rightarrow (b \copyright a)$$

Clobber is transitive if and only if, for all possible values of a, b, and c,

$$[(a \copyright b) \text{ and } (b \copyright c)] \Rightarrow (a \copyright c)$$

If clobber has all three of these properties, then clobber is an equivalence relation.

Chapter 12

Question 12-1

Suppose a is a constant and x is a variable for which we want to solve. We see the following equation:

$$x - 2a = 5a$$

How can we get an equation with x alone on the left side of the equals sign and a multiple of a alone on the right?

Answer 12-1

We can morph the equation as follows:

$$\begin{aligned} x - 2a &= 5a \\ (x - 2a) + 2a &= 5a + 2a \\ x &= 7a \end{aligned}$$

Question 12-2

Suppose b is a constant and z is a variable for which we want to solve. We see:

$$z + 3b = b/2$$

How can we get an equation with z alone on the left side of the equals sign and a multiple of b alone on the right?

Answer 12-2

We can morph the equation as follows:

$$\begin{aligned} z + 3b &= b/2 \\ 2(z + 3b) &= 2(b/2) \\ 2z + 6b &= b \\ (2z + 6b) - 6b &= b - 6b \\ 2z &= -5b \\ z &= (-5/2)b \end{aligned}$$

Question 12-3

When we divide both sides of a first-degree equation by any expression containing the variable, there's a hidden risk. What is that risk?

Answer 12-3

The expression containing the variable must not be equal to 0, or we'll run into trouble of some sort. The fact that the expression contains an unknown means that we can't be sure it's nonzero until we know the solution to the equation! When trying to solve a first-degree equation, therefore, it's best to avoid dividing through by any expression that contains the variable.

Question 12-4

Suppose we see the following equation where a, b, c, and d are constants and x is the variable:

$$3abx = 7x/(c + d)$$

To be sure this equation makes sense, we must restrict the values of the constants. How?

Answer 12-4

We must require that $c \neq -d$. That is, we can't let c and d be additive inverses. If c and d happen to be additive inverses, then we will find ourselves dividing by 0 on the right side of the equation.

Question 12-5

Which of the following equations are first-degree equations in one variable? Which are not? Letters a, b, and c represent constants. Letters x, y, and z represent variables.

$$x + 3a + 2b = c$$

$$(c + a)x = b$$

$$(x + y)a = b$$

$$y^2 + ay + b = c$$

$$x + y + z = 0$$

Answer 12-5

The first and second equations are first-degree equations in one variable. The third and fifth equations aren't, because they contain more than one variable. The fourth equation isn't, because it contains the square of the variable.

Question 12-6

When using letters to represent constants, or when reading texts in which letters are used to represent constants, we have to be clear and careful with the context. Why?

Answer 12-6

Some letters are widely used to represent specific physical, chemical, or mathematical constants. In physics, c represents the speed of light in a vacuum; N is often used to represent a chemical constant called the Avogadro constant; e is commonly used to represent the exponential constant. We don't want to let c be a general constant if we're writing about relativity theory, or N stand for a general constant if the subject happens to be chemistry, or e indicate a general constant if the discussion involves exponential functions.

Question 12-7

Suppose a, b, and c are constants, and x is a variable. How can we manipulate the following equation so it contains x all by itself on the left side, and an expression containing the constants without x on the right side, showing and justifying every step in an S/R table?

$$x - 2c + 7 = 4x + 2ax + b + c$$

Answer 12-7

We can morph the equation as described in Table 20-1. The end result is

$$x = (b + 3c - 7) / (-3 - 2a)$$

Question 12-8

We had better be careful about something in the last step of the process shown by Table 20-1. What's that?

Answer 12-8

We must require that $2a \neq -3$. If we do not impose that restriction, we allow for the possibility of division by 0 in the last step.

Question 12-9

What is the standard form for a first-degree equation in one variable?

Table 20-1. Equation morphing process for Answer 12-7. We solve for x in terms of the constants. But there's a catch, which is addressed in Question and Answer 12-8.

Statements	Reasons
$x - 2c + 7 = 4x + 2ax + b + c$	This is the equation we are given
$x + 7 = 4x + 2ax + b + 3c$	Add $2c$ to each side
$x = 4x + 2ax + b + 3c - 7$	Subtract 7 from each side
$x - 4x = 2ax + b + 3c - 7$	Subtract $4x$ from each side
$x - 4x - 2ax = b + 3c - 7$	Subtract $2ax$ from each side
$x + (-4x) + (-2a)x = b + 3c - 7$	Change subtractions to negative additions on left side of equation
$[1 + (-4) + (-2a)]x = b + 3c - 7$	Right-hand distributive law on left side of equation
$(-3 - 2a)x = b + 3c - 7$	Simplify left side of equation
$x = (b + 3c - 7)/(-3 - 2a)$	Divide each side by $(-3 - 2a)$

Answer 12-9

If a and b are constants and x is the variable, then the standard form is

$$ax + b = 0$$

Any first-degree equation in one variable can always be put into this form. The variable might be called something other than x, and the constants a and b might be expressed in terms of other constants and numbers, but this basic form can always be derived.

Question 12-10

Put the equation derived in Answer 12-7, and shown in the last line of Table 20-1, into the standard form for a first-degree equation in one variable.

Answer 12-10

Here's the equation in the form we got when we solved for x in terms of the constants:

$$x = (b + 3c - 7)/(-3 - 2a)$$

We can multiply each side by $(-3 - 2a)$ to obtain

$$(-3 - 2a)x = b + 3c - 7$$

Next, we can subtract the quantity $(b + 3c - 7)$ from each side, getting

$$(-3 - 2a)x - (b + 3c - 7) = 0$$

Technically, this equation is in the standard form for a first-degree equation in one variable. But we might want to rename the expressions made up of a, b, c, and numerals. Let's use single letters p and q, as follows:

$$(-3 - 2a) = p$$

and

$$-(b + 3c - 7) = q$$

Now we have

$$px + q = 0$$

Chapter 13

Question 13-1

Figure 20-1 shows a mapping between sets. Four sets are identified, labeled A, B, C, and D. The five points in set B represent all the elements in that set, and the five elements in set C represent all the elements of that set. The dashed curves represent the entire mapping; they "tell the whole story." Furthermore, $B \subset A$ and $C \subset D$. Which set is the maximal domain? Which set is the co-domain? Which set is the essential domain? Which set is the range?

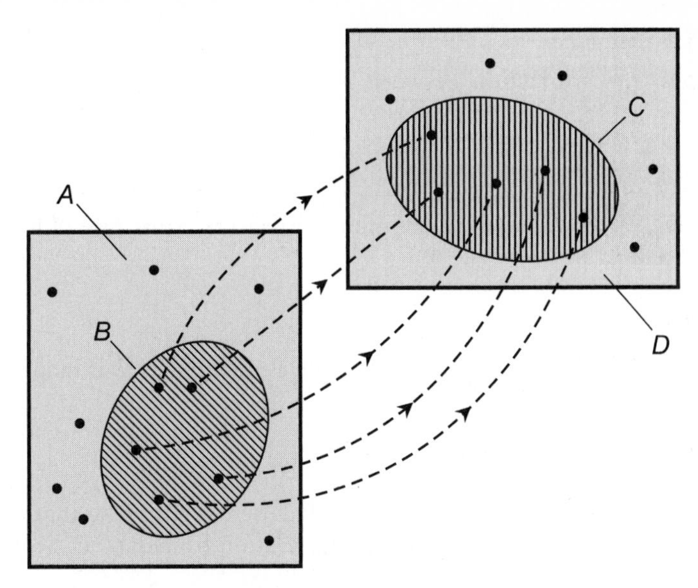

Figure 20-1 Illustration for Questions and Answers 13-1 through 13-8.

Answer 13-1

The maximal domain is set *A*. The co-domain is set *D*. The essential domain is set *B*. The range is set *C*.

Question 13-2

In Fig 20-1, the five points within set *C* represent the entire set. On that basis, what can we say about the mapping?

Answer 13-2

The mapping is onto set *C*. That is, it is a surjection.

Question 13-3

In Fig 20-1, note that each of the points in set *B* maps to a single point in set *C*. What can we say about the mapping on this basis?

Answer 13-3

The mapping is one-to-one. That is, it is an injection.

Question 13-4

In Fig 20-1, suppose that each point in set *B* maps to a single point in set *C*, and vice-versa. On this basis, and according to what we know so far about the points and the sets, what type of mapping is this?

Answer 13-4

The mapping is a bijection between *B* and *C*, because it is both a surjection and an injection.

Question 13-5

Based on Fig. 20-1, and on the descriptions given so far, what type of relation is the mapping between *B* and *C*?

Answer 13-5

The mapping is a function, because no single element in set *B* is mapped to more than one element in set *C*.

Question 13-6

Suppose the sense of the mapping in Fig. 20-1 were reversed, so that it went from set *C* to set *B*. This would give us the inverse of the relation from *B* to *C*. Would that inverse be a function?

Answer 13-6

Yes, because no single element in set *C* would map to more than one element in set *B*.

Question 13-7

Let's call the five points in set *B* of Fig. 20-1 by the names b_1 through b_5, and the five points in set *C* by the names c_1 through c_5. Suppose that b_1 maps to c_5, b_2 maps to c_4, b_3 maps to c_3, b_4 maps to c_2, and b_5 maps to c_1. How can we state this mapping as a set of ordered pairs?

Answer 13-7

We can state the mapping as the set

$$\{(b_1,c_5), (b_2,c_4), (b_3,c_3), (b_4,c_2), (b_5,c_1)\}$$

Question 13-8

In the ordered pairs given in Answer 13-7 relevant to Fig. 20-1, which elements are values of the independent variable? Which elements are values of the dependent variable?

Answer 13-8

The elements b_1 through b_5 are values of the independent variable, and the elements c_1 through c_5 are values of the dependent variable.

Question 13-9

Figure 20-2 shows a mapping that's almost the same as the mapping of Fig. 20-1. The only difference is that there is an extra element in set *B*, and it maps to one of the existing elements

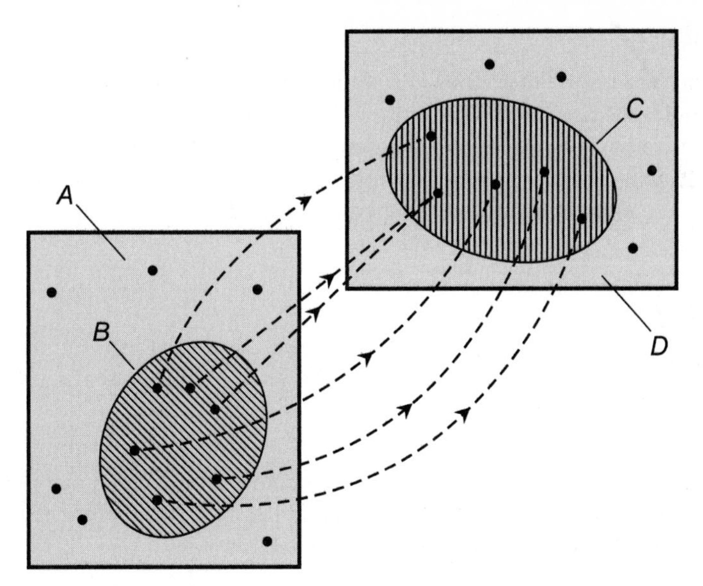

Figure 20-2 Illustration for Questions and Answers 13-9 and 13-10.

in set *C*. The six points in set *B* represent all the elements in that set, and the five elements in set *C* represent all the elements of that set. As before, the dashed curves represent the entire mapping. Is this new mapping a surjection? Is it an injection? Is it a bijection? Is it a relation? Is it a function?

Answer 13-9

The mapping of Fig. 20-2 is a surjection, because it maps to every element of set *C*. It's not an injection, however, because it's not one-to-one. That means it is not a bijection. This mapping is a relation. It's also a function, because no single element in set *B* maps to more than one element in set *C*.

Question 13-10

Consider the mapping of Fig. 20-2 as a relation. Now suppose the sense were reversed, so that the mapping went from set *C* to set *B*, giving us the inverse of the relation. Would that inverse be a function?

Answer 13-10

No, because one of the elements of set *C* would map to two elements of set *B*.

Chapter 14

Question 14-1

The Cartesian plane is divided into four quadrants. How are those quadrants normally oriented? How are the positive and negative values of the variables portrayed in those quadrants?

Answer 14-1

The first quadrant of the Cartesian plane is usually the one at the upper right, and both variables are positive. In the second quadrant, which is usually at the upper left, the independent variable is negative and the dependent variable is positive. In the third quadrant, which is usually at the lower left, both variables are negative. In the fourth quadrant, which is usually at the lower right, the independent variable is positive and the dependent variable is negative.

Question 14-2

Figure 20-3 shows three points, called *P*, *Q*, and *R*, plotted in the Cartesian plane. In which quadrants do these points lie? What are the ordered pairs representing these points?

Answer 14-2

Point *P* is in the third quadrant, and is represented by the ordered pair (−5,−3). Point *Q* is not in any quadrant because it is directly on one of the axes; it is represented by (0,−1). Point *R* is in the first quadrant, and is represented by (2,4).

Question 14-3

Imagine horizontal lines (that is, lines parallel to the *x* axis) running through each of the three points in Fig. 20-3. What are the equations of these lines? Do any of them represent functions of *x*?

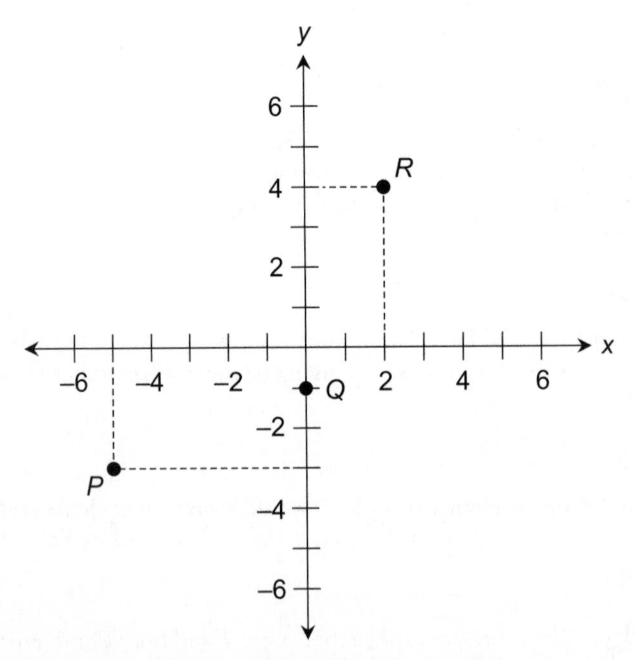

Figure 20-3 Illustration for Questions and Answers 14-2 through 14-6.

Answer 14-3

A horizontal line through point P has the equation $y = -3$. A horizontal line running through point Q has the equation $y = -1$. A horizontal line through point R has the equation $y = 4$. All three of these lines represent functions of x, because they all pass the vertical-line test. A movable vertical line (that is, a line parallel to the y axis) will never intersect any one of those three horizontal lines at more than one point.

Question 14-4

Imagine vertical lines running through each of the three points in Fig. 20-3. What are the equations of these lines? Do any of them represent functions of x?

Answer 14-4

A vertical line through point P has the equation $x = -5$. A vertical line running through point Q has the equation $x = 0$. A vertical line through point R has the equation $x = 2$. None of these lines represent functions of x, because they all fail the vertical-line test. A movable vertical line that intersects any of those three lines at one point will intersect it at infinitely many other points as well.

Question 14-5

Suppose the coordinates of each of the points in Fig. 20-3 are transposed; that is, the order of each ordered pair is reversed. Call the new points P^*, Q^*, and R^*. In which quadrants will these new points appear?

Answer 14-5

The coordinates of P^* will be $(-3, -5)$, so P^* will be in the third quadrant, just as is P. The coordinates of Q^* will be $(-1, 0)$, so it will not be in any quadrant, but on the negative x axis. The coordinates of R^* will be $(4, 2)$, so it will be in the first quadrant, just as is R.

Question 14-6

Suppose both of the coordinates of each of the points in Fig. 20-3 are multiplied by -1. Call the new points $-P$, $-Q$, and $-R$. In which quadrants will these new points appear?

Answer 14-6

The coordinates of $-P$ will be $(5, 3)$, so $-P$ will be in the first quadrant. The coordinates of $-Q$ will be $(0, 1)$, so it will not be in any quadrant, but on the positive y axis. The coordinates of $-R$ will be $(-2, -4)$, so it will be in the third quadrant.

Question 14-7

Figure 20-4 shows the graphs of four different relations between x and y in Cartesian coordinates. Which of these relations are functions of x? How can we tell?

Answer 14-7

Using the vertical-line test, we can see that curve E and line G both represent functions of x, at least within the viewing region of this graph. They both pass the test. However, neither curve F nor curve H are functions of x, because they both fail the vertical-line test.

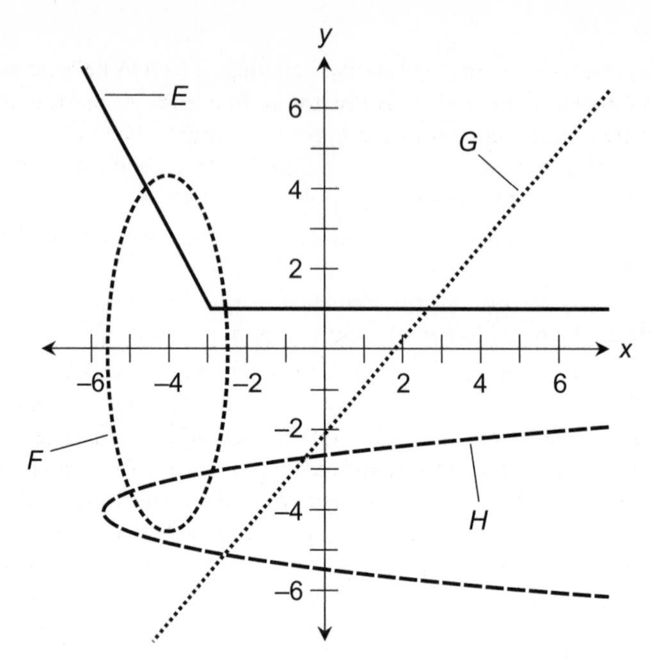

Figure 20-4 Illustration for Questions and Answers 14-7 through 14-10.

Question 14-8

Can we restrict the domains of relations shown by curves *F* or *H* to make either of them represent a nontrivial function of *x*? If so, how?

Answer 14-8

Imagine "taking slices" of curve *F* or curve *H* by considering only that portion of the curve that falls between two movable vertical lines. This limits the values of *x* that apply to the curves, thereby restricting the domain. No matter how we "slice it," we always get two values of *y* for each value of *x* when we do this with either *F* or *H*. The only way we can get a function out of curve *F* is to bring the two vertical lines together so they both intersect *F* at the extreme left point or the extreme right point of the ellipse. That restricts the domain so severely that we get mappings of one point onto one other point, but those are trivial functions! In the case of curve *H*, we can bring the two vertical lines together so they both intersect the curve at the extreme left point. Again, that's a trivial result.

Question 14-9

Consider the inverses of the relations shown in Fig. 20-4. A convenient way to imagine these inverses is to let *y* be the independent variable and let *x* be the dependent variable, so we have mappings from values of *y* to values of *x*. Which of the inverses, defined in this way, are functions?

Answer 14-9

Using the horizontal-line test, we can see that G and H represent functions of y, at least within the viewing region. They both pass the test. But neither E nor F represents functions of y, because they fail the horizontal-line test. (Part of curve E is itself a horizontal line.)

Question 14-10

Can we restrict the domains of the inverses of relations E or F to make either of them into a nontrivial function of y? If so, how?

Answer 14-10

In the case of the inverse of the "bent line" curve E, we can make it into a function if we confine the domain to values of y strictly larger than 1. This is based on the assumption that the lines that make up curve E continue straight for infinite distances in both directions. The inverse of curve F is not so cooperative. Imagine "taking slices" of by considering only that portion of the curve that falls between two movable horizontal lines. This limits the values of y that apply to the curve, thereby restricting the domain. No matter how we "slice it," we always get two values of x for each value of y. The only way we can get a function out of the inverse of curve F is to bring the two horizontal lines together so they both intersect the graph at the extreme top point or the extreme bottom point. Those are trivial results.

Chapter 15

Question 15-1

Figure 20-5 is a Cartesian graph showing the same three points P, Q, and R as we saw in Fig. 20-3. Three lines, each extending indefinitely in either direction, pass through pairs of these points. Let's call the lines PQ, QR, and PR. (It's visually apparent which is which!) Assume that the ordered pairs for the points are all pairs of integers. In other words, assume that the points are *exactly* where they appear to be. Based on this information, how can we determine the slope of line PQ? How can we determine the y-intercept of line PQ?

Answer 15-1

The slope of a line is equal to the change in the y-value (Δy) divided by the change in the x-value (Δx) as we move from one point on the line to another point. We know the ordered pairs for the two points as $P = (-5,-3)$ and $Q = (0,-1)$. Therefore,

$$\begin{aligned}\Delta y &= -1 - (-3)\\ &= -1 + 3\\ &= 2\end{aligned}$$

and

$$\begin{aligned}\Delta x &= 0 - (-5)\\ &= 0 + 5\\ &= 5\end{aligned}$$

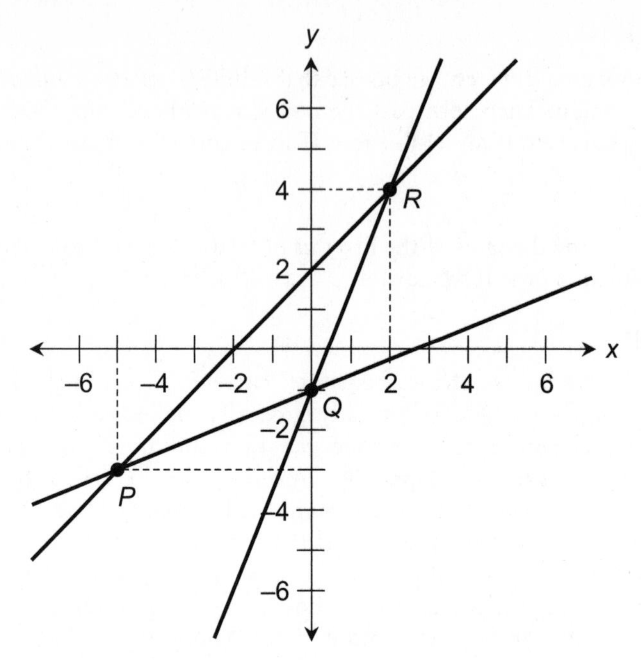

Figure 20-5 Illustration for Questions and Answers 15-1 through 15-10.

That means $\Delta y/\Delta x = 2/5$, which is the slope of the line. The y-intercept is the y-value at Q, because Q happens to lie on the y axis. That value is −1.

Question 15-2

In Fig. 20-5, how can we determine the slope of line QR? The y-intercept?

Answer 15-2

We know the ordered pairs for the two points as $Q = (0,-1)$ and $R = (2,4)$. Therefore,

$$\begin{aligned}\Delta y &= 4 - (-1)\\ &= 4 + 1\\ &= 5\end{aligned}$$

and

$$\begin{aligned}\Delta x &= 2 - 0\\ &= 2\end{aligned}$$

That means $\Delta y/\Delta x = 5/2$, which is the slope of the line. As with line PQ, the y-intercept is −1, which is the y-value at point Q.

Question 15-3

In Fig. 20-5, how can we determine the slope of line *PR*? The *y*-intercept?

Answer 15-3

We know the ordered pairs for the two points as $P = (-5,-3)$ and $R = (2,4)$. Therefore,

$$\begin{aligned}\Delta y &= 4 - (-3) \\ &= 4 + 3 \\ &= 7\end{aligned}$$

and

$$\begin{aligned}\Delta x &= 2 - (-5) \\ &= 2 + 5 \\ &= 7\end{aligned}$$

That means $\Delta y / \Delta x = 7/7 = 1$, which is the slope of the line. The *y*-intercept can be inferred. Note that if we move to the right from point *P* by Δx units, we must go up by the same number of units to stay on the line. If we increase the *x*-value of point *P* by 5 units, we arrive at the *y* axis, and we'll be at a point 5 units above the *y*-value of *P*. The *y*-value of *P* is −3, so 5 more than that is 2. The *y*-intercept of line *PR* is therefore equal to 2.

Question 15-4

Based on Answers 15-1, 15-2, and 15-3, what are the slope-intercept forms of the equations for lines *PQ*, *QR*, and *PR*?

Answer 15-4

Now that we know the slopes and the *y*-intercepts of all three lines, we can write the slope-intercept equations straightaway. Remember the general slope-intercept form for a line in Cartesian coordinates:

$$y = mx + b$$

where x is the independent variable, y is the dependent variable, m is the slope, and b is the *y*-intercept. For line *PQ*, we have

$$y = (2/5)x - 1$$

For line *QR*, we have

$$y = (5/2)x - 1$$

For line *PR*, we have

$$y = x + 2$$

Question 15-5

How can we determine the point-slope form of the equation for line *PQ*, based on the coordinates of point *P* and the slope of the line?

Answer 15-5

Remember the point-slope form for a straight line in Cartesian coordinates:

$$y - y_0 = m(x - x_0)$$

where x is the independent variable, y is the dependent variable, (x_0, y_0) are the coordinates of a point on the line, and m is the slope of the line. We know that $P = (-5, -3)$, so $x_0 = -5$ and $y_0 = -3$. We also know that for line *PQ*, the slope m is equal to 2/5. Therefore, the point-slope equation for line *PQ* is

$$y - (-3) = (2/5)[x - (-5)]$$

which can be simplified to

$$y + 3 = (2/5)(x + 5)$$

Question 15-6

How can we determine the point-slope form of the equation for line *QR*, based on the coordinates of point *R* and the slope of the line?

Answer 15-6

We know that $R = (2, 4)$, so $x_0 = 2$ and $y_0 = 4$. We also know that for line *QR*, the slope m is equal to 5/2. Therefore, the point-slope equation for line *QR* is

$$y - 4 = (5/2)(x - 2)$$

Question 15-7

How can we determine the point-slope form of the equation for line *PR*, based on the coordinates of point *P* and the slope of the line?

Answer 15-7

We know that $P = (-5, -3)$, so $x_0 = -5$ and $y_0 = -3$. We also know that for line *PR*, the slope m is equal to 1. Therefore, the point-slope equation for line *PR* is

$$y - (-3) = x - (-5)$$

which can be simplified to

$$y + 3 = x + 5$$

Question 15-8

How can we determine the point-slope form of the equation for line *PR*, based on the coordinates of point *R* and the slope of the line?

Answer 15-8

We know that $R = (2, 4)$, so $x_0 = 2$ and $y_0 = 4$. We also know that for line *PR*, the slope m is equal to 1. Therefore, the point-slope equation for line *PR* is

$$y - 4 = x - 2$$

Question 15-9

It's intuitively obvious that the equations we derived in Answers 15-7 and 15-8 must represent the same line. How can we prove it by showing that the equations are equivalent?

Answer 15-9

If we can convert one of the equations into the other using the rules for equation morphing, it will prove that the equations are equivalent. Let's start with

$$y + 3 = x + 5$$

We can subtract 7 from each side, getting

$$y - 4 = x - 2$$

That's all there is to it!

Question 15-10

Starting with the slope-intercept forms, how can we morph the equations for lines *PQ*, *QR*, and *PR* in Fig. 20-5 into the form

$$ax + by = c$$

where a, b, and c are *integer* constants?

Answer 15-10

From Answer 15-4, the slope-intercept form of the equation for line *PQ* is

$$y = (2/5)x - 1$$

We can multiply through by 5 to obtain

$$5y = 2x - 5$$

Subtracting $2x$ from each side gives us

$$-2x + 5y = -5$$

That's in the form we want! Again from Answer 15-4, the slope-intercept form of the equation for line QR is

$$y = (5/2)x - 1$$

When we multiply each side by 2, we get

$$2y = 5x - 2$$

Subtracting $5x$ from each side, we obtain

$$-5x + 2y = -2$$

That's in the form we want! Once again referring to Answer 15-4, the slope-intercept form of the equation for line PR is

$$y = x + 2$$

Subtracting x from each side gives us

$$-x + y = 2$$

That's in the form we want!

Chapter 16

Question 16-1

In Chap. 16, we learned how a two-by-two linear system in variables x and y can be solved by the following process:

- Morph both equations into SI form with y all by itself on the left side of the equals sign.
- Mix the two equations to get a first-degree equation in x.
- Solve the first-degree equation for x.
- Substitute that solution back into one of the SI equations to solve for y.

How can we solve such a system by morphing and mixing alone, without substituting either variable for the other?

Answer 16-1

We can go through the morph-and-mix process twice, first for one variable and then for the other. We proceed like this:

- Morph both equations into SI form with y all by itself on the left side of the equals sign.
- Mix the two equations to get a first-degree equation in x.

- Solve the first-degree equation for x.
- Morph both equations into SI form with x all by itself on the left side of the equals sign.
- Mix the two equations to get a first-degree equation in y.
- Solve the first-degree equation for y.

Question 16-2

How can we put the following two-by-two linear system into a pair of SI equations with y all by itself on the left side of the equals sign?

$$2x - y + 8 = 0$$

and

$$x - 3y + 9 = 0$$

Answer 16-2

By now, we're good enough at equation manipulation to write down the steps one after another, without having to justify everything. For the first original equation, we can do this:

$$2x - y + 8 = 0$$

$$-y + 8 = -2x$$

$$-y = -2x - 8$$

$$y = 2x + 8$$

and for the second original equation, we can do this:

$$x - 3y + 9 = 0$$

$$-3y + 9 = -x$$

$$-3y = -x - 9$$

$$3y = x + 9$$

$$y = (1/3)x + 3$$

Question 16-3

How can we combine the two equations from Answer 16-2 to get a first-degree equation and solve the original system for x?

Answer 16-3

We can mix the right sides of the two SI equations together and then solve the resulting first-degree equation in x by manipulation. Here it goes, one step at a time:

$$2x + 8 = (1/3)x + 3$$
$$6x + 24 = x + 9$$
$$6x + 15 = x$$
$$5x + 15 = 0$$
$$5x = -15$$
$$x = -3$$

Question 16-4

How can we put the two-by-two linear system from Question 16-2 into a pair of SI equations with x all by itself on the left sides of the equals signs?

Answer 16-4

For the first original equation, we can do this:

$$2x - y + 8 = 0$$
$$2x + 8 = y$$
$$2x = y - 8$$
$$x = (1/2)y - 4$$

and for the second original equation, we can do this:

$$x - 3y + 9 = 0$$
$$x + 9 = 3y$$
$$x = 3y - 9$$

Question 16-5

How can we combine the two equations from Answer 16-4 to get a first-degree equation and solve the original system for y?

Answer 16-5

We can mix the right sides of the two SI equations together and then solve the resulting first-degree equation in y by manipulation, as follows:

$$(1/2)y - 4 = 3y - 9$$
$$y - 8 = 6y - 18$$
$$y + 10 = 6y$$
$$10 = 5y$$
$$y = 2$$

Question 16-6

How can we be sure the solution we obtained in Answers 16-3 and 16-5 is in fact the correct solution to the original two-by-two linear system?

Answer 16-6

The solution we have obtained is $x = -3$ and $y = 2$. We must plug these values into both of the original equations to be certain we've gotten the right solution. For the first original equation, we proceed like this:

$$\begin{aligned} 2x - y + 8 &= 0 \\ 2 \times (-3) - 2 + 8 &= 0 \\ -6 - 2 + 8 &= 0 \\ 0 &= 0 \end{aligned}$$

It checks out! For the second original equation, we do this:

$$\begin{aligned} x - 3y + 9 &= 0 \\ -3 - (3 \times 2) + 9 &= 0 \\ -3 - 6 + 9 &= 0 \\ 0 &= 0 \end{aligned}$$

It checks again! Now we know the solution we obtained is correct.

Question 16-7

Do you suspect that I concocted the above problem so it would come out with a pair of "clean integers" for the solution? If so, you are right! How can we compose a two-by-two linear system as a test problem (for someone else to solve), and be sure the solution will turn out to be a pair of integers?

Answer 16-7

We can choose a point where the graphs of two lines intersect, and assign different slopes to those lines. Then we can write down the equations in point-slope form, using the solution point as the reference for both lines. Finally, we can convert the point-slope equations to some other form to get the test problem. For extra credit, you can try this and then solve the test problem you've created.

Question 16-8

How can we add multiples of the two original equations stated in Question 16-2 to solve the linear system for x? For reference, here are the equations again:

$$\begin{aligned} 2x - y + 8 &= 0 \\ x - 3y + 9 &= 0 \end{aligned}$$

Answer 16-8

We can multiply the first equation through by -3 and then add it to the second equation, getting the sum

$$\begin{aligned} -6x + 3y - 24 &= 0 \\ x - 3y + 9 &= 0 \\ -5x - 15 &= 0 \end{aligned}$$

To solve this, we add 15 to each side and then divide through by −5, as follows:

$$-5x - 15 = 0$$
$$-5x = 15$$
$$x = -3$$

Question 16-9

How can we add multiples of the two original equations stated in Question 16-2 to solve the linear system for y?

Answer 16-9

We can multiply the second equation through by −2 and then add it to the first equation, getting the sum

$$2x - y + 8 = 0$$
$$-2x + 6y - 18 = 0$$
$$5y - 10 = 0$$

To solve this, we add 10 to each side and then divide through by 5, like this:

$$5y - 10 = 0$$
$$5y = 10$$
$$y = 2$$

Question 16-10

How can we solve the two-by-two linear system stated in Question 16-2 by the rename-and-replace (substitution) method? Here are the original equations again, for reference:

$$2x - y + 8 = 0$$

and

$$x - 3y + 9 = 0$$

Answer 16-10

We start by converting either of the two equations to SI form, so one of the variables appears all alone on the left side of the equals sign. Let's use the first equation and isolate y on the left side. To manipulate the equation, we proceed just as we did in Answer 16-2:

$$2x - y + 8 = 0$$
$$-y + 8 = -2x$$
$$-y = -2x - 8$$
$$y = 2x + 8$$

Next, we substitute the quantity $(2x + 8)$ for y in the second equation and solve the result for x, as follows:

$$x - 3y + 9 = 0$$

$$x - 3(2x + 8) + 9 = 0$$

$$x - 6x - 24 + 9 = 0$$

$$-5x - 15 = 0$$

$$-5x = 15$$

$$x = -3$$

Now that we know the value of x, we can plug it into either equation and solve for y. Let's use the first equation. Then we proceed as follows:

$$2x - y + 8 = 0$$

$$2 \times (-3) - y + 8 = 0$$

$$-6 - y + 8 = 0$$

$$-6 + 8 = y$$

$$y = 2$$

Chapter 17

Question 17-1

Let's consider again the two-by-two system we saw in Question 16-2. How can we graph this system in Cartesian coordinates, with x as the independent variable and y as the dependent variable? Here are the original equations:

$$2x - y + 8 = 0$$

and

$$x - 3y + 9 = 0$$

Answer 17-1

We can use the SI forms of the equations to find their y-intercepts, and the solution of the system to find a third point that lies on both lines. We're lucky here, because the intersection point is fairly far away from the y axis. The SI forms of the equations were derived in Answer 16-2. Respectively, they are:

$$y = 2x + 8$$

and

$$y = (1/3)x + 3$$

We know from these equations that the y-intercepts are 8 and 3, so the point (0, 8) is on the first line and the point (0, 3) is on the second line. The solution to the system, as we've already determined, is (−3, 2). That point lies on both lines. Figure 20-6 shows plots of these three points, along with plots of the lines connecting the points and representing the equations.

Question 17-2

How can we determine the x-intercept of the line representing the equation $y = 2x + 8$ on the basis of the known slopes and the point data in Fig. 20-6? (We can set $y = 0$ and then calculate x by arithmetic, but the purpose of this exercise is to demonstrate the geometric principles of slope and intercept.)

Answer 17-2

The slope of the line is 2. Therefore, if we start from any point on the line and move in the positive x direction by n units, we must move in the positive y direction by $2n$ units to stay on the line. In the opposite sense, if we start from any point on the line and move in the negative y direction by p units, we must move in the negative x direction by $p/2$ units to stay on the line. Let's start at the point (0, 8), which is the y-intercept. If we move in the negative y direction

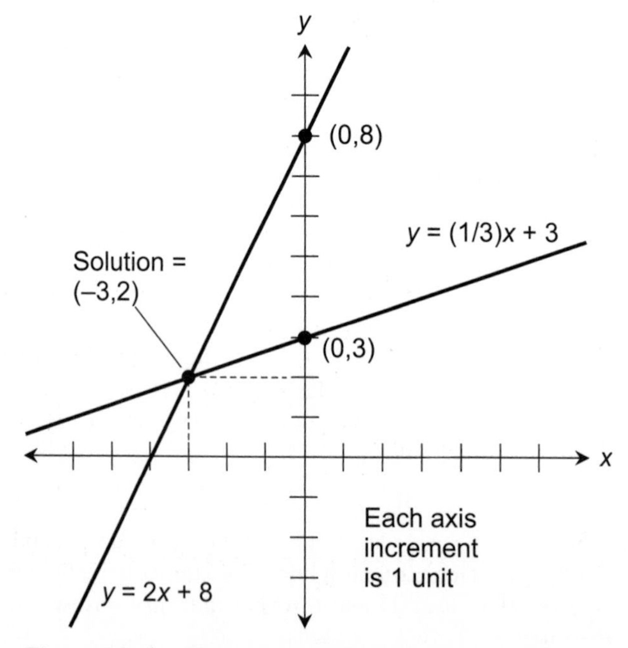

Figure 20-6 Illustration for Questions and Answers 17-1 through 17-6.

along the line until $y = 0$, we'll end up on the x axis, and we'll have gone 8/2, or 4, units in the negative x direction from the y axis. That means the x-intercept is -4. It looks that way in the figure; now we know it's really true.

Question 17-3

How can we verify that the preceding answer is correct by manipulating the equation for the line?

Answer 17-3

We can start with either the original equation or the SI form with y as the dependent variable, and manipulate things until we get the SI form with x all alone on the left side of the equals sign. In Answer 16-4, that was done starting with the original equation. If we start with the SI form with y as the dependent variable, we can proceed like this:

$$y = 2x + 8$$

$$y - 8 = 2x$$

$$(1/2)y - 4 = x$$

$$x = (1/2)y - 4$$

This SI equation tells us that the x-intercept is equal to -4.

Question 17-4

How can we determine the x-intercept of the line representing the equation $y = (1/3)x + 3$ on the basis of the known slopes and the point data in Fig. 20-6? (As in Question 17-2, we can set $y = 0$ and then solve for x; but again, this exercise is meant to show how slope and intercept are related geometrically.)

Answer 17-4

The slope of the line is 1/3. Therefore, if we start from any point on the line and move in the positive x direction by n units, we must move in the positive y direction by $n/3$ units to stay on the line. In the opposite sense, if we start from any point on the line and move in the negative y direction by p units, we must move in the negative x direction by $3p$ units to stay on the line. Let's start at the point (0, 3), which is the y-intercept. If we move in the negative y direction along the line until $y = 0$, we'll end up on the x axis, and we'll have gone 3×3, or 9, units in the negative x direction from the y axis. That means the x-intercept is -9. This is outside the field of view in Fig. 20-6.

Question 17-5

How can we verify that the preceding answer is correct by manipulating the equation for the line?

Answer 17-5

As we did in Answer 17-3, we can start with either the original equation or the SI form with y as the dependent variable, and manipulate things until we get the SI form with x all alone on the left side of the equals sign. In Answer 16-4, that was done starting with the original equation. If we start with the SI form with y as the dependent variable, we can do this:

$$y = (1/3)x + 3$$
$$y - 3 = (1/3)x$$
$$3y - 9 = x$$
$$x = 3y - 9$$

This SI equation tells us that the x-intercept is equal to −9.

Question 17-6

How can we morph Fig. 20-6, the graph of the system from Question and Answer 17-1 and Fig. 20-6, to get a graph that will show the same system with y as the independent variable and x as the dependent variable?

Answer 17-6

We can use the rotate-and-mirror method. We start with Fig. 20-6 and rotate the entire assembly as a single mass—axes, lines, and points—counterclockwise by a quarter-turn (90°). Then we mirror the whole thing left-to-right. Finally, we reverse the ordered pairs to obtain the new points. The result is shown in Fig. 20-7.

Question 17-7

What are the SI forms of the equations for the lines in Fig. 20-7?

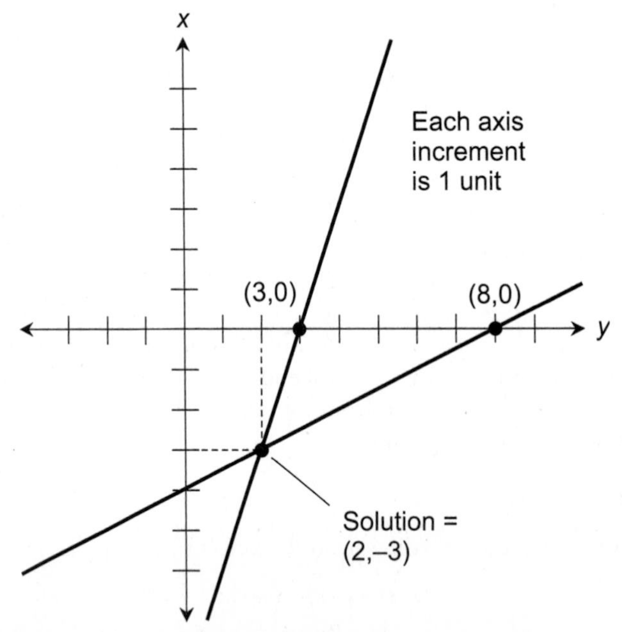

Figure 20-7 Illustration for Questions and Answers 17-6 through 17-10.

Answer 17-7

They are the equations we derived in Answers 17-3 and 17-5:

$$x = (1/2)y - 4$$

and

$$x = 3y - 9$$

Question 17-8

What is the *x*-intercept of the line representing the equation $x = (1/2)y - 4$?

Answer 17-8

We can answer this straightaway, because the equation is in SI form with x as the dependent variable. It's −4.

Question 17-9

What is the *x*-intercept of the line representing the equation $x = 3y - 9$?

Answer 17-9

Again, we can infer this from the SI equation having x as the dependent variable. It's −9.

Question 17-10

Based the known slopes of the lines, and on the point data shown in Fig. 20-7, at what points would the lines representing the two-by-two linear system intersect the graph of the equation $y = -2$?

Answer 17-10

The graph of the equation $y = -2$ would appear as a vertical line in Fig. 20-7, parallel to the x axis and running through the point (−2, 0). To reach this line from the point (2, −3), we can travel in the negative y direction by 4 units along either of our existing lines. First, let's move along the line for $x = (1/2)y - 4$. The slope is 1/2. That means if we go 4 units in the negative y direction, we must go 4/2, or 2, units in the negative x direction to stay on the line. That will put us at the point (−2, −5). Now let's move along the line for the equation $x = 3y - 9$. The slope is 3. Therefore, if we go 4 units in the negative y direction, we must go 4×3, or 12, units in the negative x direction to stay on the line. That will get us to the point (−2, −15).

Chapter 18

Question 18-1

Here is a set of equations that forms a three-by-three linear system:

$$4x = 8 + 4y + 4z$$
$$2y = 5 + x - 5z$$
$$4z = 13 - 2x + y$$

How can we put the first of these equations into the form $ax + by + c = d$, where a, b, c, and d are constants?

Answer 18-1

Here are the steps we can take, one at a time, starting with the original equation:

$$4x = 8 + 4y + 4z$$
$$4x - 4y = 8 + 4z$$
$$4x - 4y - 4z = 8$$

Question 18-2

How can we put the second equation in Question 18-1 into the form $ax + by + c = d$, where a, b, c, and d are constants?

Answer 18-2

Here are the steps we can take, one at a time, starting with the original equation:

$$2y = 5 + x - 5z$$
$$-x + 2y = 5 - 5z$$
$$-x + 2y + 5z = 5$$

Question 18-3

How can we put the third equation in Question 18-1 into the form $ax + by + c = d$, where a, b, c, and d are constants?

Answer 18-3

Here are the steps we can take, one at a time, starting with the original equation:

$$4z = 13 - 2x + y$$
$$2x + 4z = 13 + y$$
$$2x - y + 4z = 13$$

Question 18-4

Based on the rearrangements in Answers 18-1 through 18-3, how can we state the three-by-three linear system from Question 18-1 now? What strategies can we use to solve it?

Answer 18-4

We can state the system by combining the final equations from Answers 18-1 through 18-3, like this:

$$4x - 4y - 4z = 8$$
$$-x + 2y + 5z = 5$$
$$2x - y + 4z = 13$$

We can solve this system in many different ways. In Chap. 18, we learned to solve systems of this type by getting rid of one variable, resulting in a two-by-two linear system, solving that system, and then substituting back to solve for the variable we eliminated. That's the method we'll use here. We can get rid of any of the three variables by morphing and adding any two pairs of the three-variable equations.

Question 18-5

How can we obtain a two-by-two linear system in x and z from the three-by-three system as stated in Answer 18-4, using the first two equations and then the second two?

Answer 18-5

Here are the first two equations from the three-by-three linear system as stated in Answer 18-4:

$$4x - 4y - 4z = 8$$
$$-x + 2y + 5z = 5$$

We can divide the top equation through by 2 and then add the bottom equation, getting the sum

$$2x - 2y - 2z = 4$$
$$-x + 2y + 5z = 5$$
$$x + 3z = 9$$

That's the first equation in our two-by-two system. To get the second equation, let's look at the second two equations from the three-by-three system as stated in Answer 18-4:

$$-x + 2y + 5z = 5$$
$$2x - y + 4z = 13$$

We can multiply the bottom equation through by 2 and then add it to the top equation, getting

$$-x + 2y + 5z = 5$$
$$4x - 2y + 8z = 26$$
$$3x + 13z = 31$$

Now we have the following two-by-two linear system in the variables x and z:

$$x + 3z = 9$$
$$3x + 13z = 31$$

Question 18-6

How can we solve the above two-by-two system for z using the addition method?

Answer 18-6

Let's multiply the top equation through by −3 and then add the bottom equation to it. This gives us the sum

$$-3x - 9z = -27$$
$$3x + 13z = 31$$
$$4z = 4$$

This simplifies to the tentative solution $z = 1$.

Question 18-7

How can we solve the above two-by-two system for x by substituting in the solution for z we obtained in Answer 18-6?

Answer 18-7

We can plug in the value 1 for z in either of the equations in the two-by-two system as stated at the end of Answer 18-5. Let's use the top equation. We have

$$x + 3z = 9$$
$$x + 3 \times 1 = 9$$
$$x + 3 = 9$$
$$x = 6$$

Now we have the tentative solutions $x = 6$ and $z = 1$.

Question 18-8

Now that we know the values of x and z in the original three-by-three system, how can we find the value of y?

Answer 18-8

We can plug in the value 6 for x and the value 1 for z in any of the original three equations, as they are stated in Question 18-1. Let's use the first one:

$$4x = 8 + 4y + 4z$$
$$4 \times 6 = 8 + 4y + 4 \times 1$$
$$24 = 8 + 4y + 4$$
$$20 = 8 + 4y$$
$$12 = 4y$$
$$3 = y$$

Now we have the complete, but still tentative, solution to the original three-by-three system:

$$x = 6$$
$$y = 3$$
$$z = 1$$

Question 18-9

How can we be sure that the solution we have obtained for the three-by-three system presented in Question 18-1 is correct?

Answer 18-9

We must substitute our solutions into all three of the original equations. We did it indirectly for the first equation in Answer 18-8, but if we want to be completely rigorous, we must plug all three values into that equation along with the other two. (What if we made a mistake in Answer 18-8? Don't laugh. Things like that can and do happen!) Here are the original three equations again, for reference:

$$4x = 8 + 4y + 4z$$
$$2y = 5 + x - 5z$$
$$4z = 13 - 2x + y$$

Grinding out the numbers in the first equation, we get

$$4x = 8 + 4y + 4z$$
$$4 \times 6 = 8 + 4 \times 3 + 4 \times 1$$
$$24 = 8 + 12 + 4$$
$$24 = 24$$

Check one. In the second equation, we get

$$2y = 5 + x - 5z$$
$$2 \times 3 = 5 + 6 - 5 \times 1$$
$$6 = 5 + 6 - 5$$
$$6 = 6$$

Check two. In the third equation, we get

$$4z = 13 - 2x + y$$
$$4 \times 1 = 13 - 2 \times 6 + 3$$
$$4 = 13 - 12 + 3$$
$$4 = 4$$

Check three. Mission accomplished! Note that in the mixed addition/subtraction here, we proceed directly from left to right after the equals sign. We subtract 12, and then we add 3. We don't subtract the quantity (12 + 3)! If we have any doubts when we come across a situation like this, we can change the subtraction to negative addition. If we do that in the above calculation, we get an extra step, so the whole sequence goes like this:

$$4z = 13 - 2x + y$$
$$4 \times 1 = 13 - 2 \times 6 + 3$$
$$4 = 13 - 12 + 3$$
$$4 = 13 + (-12) + 3$$
$$4 = 4$$

Question 18-10

Suppose that when we checked our solutions in the preceding answer, we found that they did not work out. What could we do then?

Answer 18-10

Whenever we check solutions and discover that they don't work out, it means we've made a mistake somewhere. In that case, we must go back through each step in our solution process and find the mistake. We could also start all over and approach the problem in another way. For example, we could eliminate a different variable in the beginning, and/or use different equations to get the two-by-two linear system in the intermediate phase. After that, of course, we'd have to check our solutions again for correctness.

Chapter 19

Question 19-1

Here is the three-by-three linear system taken from Answer 18-4:

$$4x - 4y - 4z = 8$$
$$-x + 2y + 5z = 5$$
$$2x - y + 4z = 13$$

How can we arrange these equations into a matrix that can be manipulated to solve this system?

Answer 19-1

We take the coefficients, remove the variables, and place the remaining numerals neatly in the cells of a table. The first equation is represented by the top row of the matrix, the second equation is represented by the middle row, and the third equation is represented by the bottom row. The result looks like this:

4	−4	−4	8
−1	2	5	5
2	−1	4	13

Question 19-2

What general procedure can we use to solve the three-by-three system, starting with this matrix?

Answer 19-2

We can play the matrix morphing game to get the matrix into echelon form, which looks like this:

#	#	#	#
0	#	#	#
0	0	#	#

where a pound sign (#) can represent any real number. Then we continue the game and go for the diagonal form, which looks like this:

@	0	0	#
0	@	0	#
0	0	@	#

where a pound sign can represent any real number, and an at sign (@) can represent any *nonzero* real number. Then we must get the matrix into the unit diagonal form, which looks like this:

1	0	0	x
0	1	0	y
0	0	1	z

Once we have the matrix in the unit diagonal form, assuming we haven't made any mistakes, the values x, y, and z represent the solution to the original three-by-three linear system. We can reduce the values we have found, if they are fractions, to their lowest forms. Finally, we should plug our answers into the original equations to be sure we've arrived at the correct solution to the system.

Question 19-3

How is the matrix morphing game played?

Answer 19-3

There are three types of moves:

- Swap two rows, while keeping the elements of both rows in the same order from left to right.
- Multiply or divide all the elements in any row by a nonzero constant, keeping the elements in the same order from left to right.
- Add all the elements in any row to all the elements in another row, and then replace the elements in either row by the sum, keeping the elements of both rows in the same order from left to right.

Question 19-4

Using the rules outlined above, how can we get the matrix from Answer 19-1 into echelon form, and then reduce the sizes of the elements to make the matrix easier to work with later?

Answer 19-4

Here's the matrix again, for reference:

4	−4	−4	8
−1	2	5	5
2	−1	4	13

We can multiply the second row by 2 to get

4	−4	−4	8
−2	4	10	10
2	−1	4	13

Now we can add the second and third rows and then replace the third row with the sum, obtaining

4	−4	−4	8
−2	4	10	10
0	3	14	23

If we divide the first row by 2, we get

2	−2	−2	4
−2	4	10	10
0	3	14	23

Adding the first two rows and then replacing the second row with the sum, we obtain

2	−2	−2	4
0	2	8	14
0	3	14	23

We can multiply the second row by −3 and the third row by 2 to get

2	−2	−2	4
0	−6	−24	−42
0	6	28	46

Adding the second row to the third and then replacing the third row with the sum, we get

2	−2	−2	4
0	−6	−24	−42
0	0	4	4

This matrix is in echelon form. We can reduce the sizes of the numbers in this matrix, making it easier to work with as we continue the game. Let's divide the first row by 2, the second row by 6, and the third row by 4, getting

1	−1	−1	2
0	−1	−4	−7
0	0	1	1

Question 19-5

Using the rules of the matrix morphing game, how can we get the last matrix in Answer 19-4 into diagonal form?

Answer 19-5

Multiplying the second row of the echelon-form matrix we just derived by −1, we get

1	−1	−1	2
0	1	4	7
0	0	1	1

Adding the first row to the second, and then replacing the first row with the sum, we obtain

1	0	3	9
0	1	4	7
0	0	1	1

If we multiply the third row by −3, we get

1	0	3	9
0	1	4	7
0	0	−3	−3

Adding the first and third rows and then replacing the first row with the sum gives us

1	0	0	6
0	1	4	7
0	0	–3	–3

Now let's multiply the third row by 4/3. We get

1	0	0	6
0	1	4	7
0	0	–4	–4

Adding the second and third rows and then replacing the second row with the sum, we obtain the diagonal matrix

1	0	0	6
0	1	0	3
0	0	–4	–4

Question 19-6

How can we get the last matrix in Answer 19-5 into unit diagonal form?

Answer 19-6

This is easy, because we reduced our numbers along the way. We can divide the third row of the last matrix in Answer 19-5 by –4 to get

1	0	0	6
0	1	0	3
0	0	1	1

Question 19-7

What is the solution to the original three-by-three system based on Answer 19-6?

Answer 19-7

The solution can be read down the last column for x, y, and z in order:

$$x = 6$$
$$y = 3$$
$$z = 1$$

We know these values are correct, because they are the same values we found and verified in the previous section for the same original three-by-three linear system.

Question 19-8

Suppose that we play the matrix morphing game in an attempt to solve a three-by-three linear system, and we come up with a matrix that looks like this:

0	0	0	−5
1	1	1	2
0	0	0	7

At first, we suspect we made a mistake somewhere. But when we try the game again, we get the same matrix. What is the reason for this absurd result?

Answer 19-8

The original system of equations is inconsistent. There is no unique solution.

Question 19-9

Suppose that we play the matrix morphing game in an attempt to solve another three-by-three linear system, and we come up with this elegant but useless matrix:

1	1	1	1
1	1	1	1
1	1	1	1

What does this tell us?

Answer 19-9

The original system of equations is redundant. There are infinitely many solutions.

Question 19-10

Can the matrix morphing game be used to solve larger systems, such as four-by-four, five-by-five, and so on?

Answer 19-10

Yes. The matrix morphing game can be applied to linear systems of any finite dimension. That's the good news. There's bad news, too: The number of steps in the process increases much faster than the dimension. If carried out "manually" on a large system, matrix morphing is no game. It's more like slow torture! But there's some more good news: Computers can be programmed to play the matrix morphing game quickly and without pain, even for gigantic linear systems.

PART 3

Nonlinear Equations and Relations

CHAPTER

21

Imaginary and Complex Numbers

All the algebra we've done so far has been of the first degree. Variables were never squared, cubed, or raised to any higher power. Before we start dealing with equations of higher degree, let's study numbers a little more. This chapter introduces the *imaginary numbers* and the *complex numbers.*

The Square Root of −1

Early mathematicians couldn't represent fractions or ratios. Then the rational numbers were developed. At one time, there was no such thing as the number zero. Then the idea was conceived, and zero "found its place." Negative numbers evolved. The distinction between the rationals and irrationals followed. Finally, someone defined quantities that, when squared, would produce negative numbers.

A mystery number

If you square a positive real number, you get a positive real. If you square 0, you get 0. If you square a negative real, you get a positive real. You can't square any real number and get a negative real. Within the set of real numbers, expressions such as $(-1)^{1/2}$ or $(-3)^{1/2}$ or $(-1/5)^{1/2}$ or $(-\pi)^{1/2}$ are undefined.

Before square roots of negative numbers were defined more than one mathematician must have asked, "What if there *actually are* numbers that can be squared to produce negative reals, but no one has found them yet?" They *imagined* that such numbers existed, and then they explored how those numbers would behave in arithmetic. The outcome of their "mind experiments" was the discovery of a new realm of numbers that they called "imaginary." That term has been used ever since.

Many mathematics texts use the lowercase English letter i to stand for the 1/2 power of −1, which means the positive square root of −1. The choice of i is reasonable enough; it stands for "imaginary." In science, engineering, and applied mathematics, the letter j is often used, because i plays other roles, notably in the expressions for sequences and series. In this book, we'll call the unit imaginary number j, not i. This choice is based on my assumption that you're more likely to go into science or engineering than pure mathematics, so you should get used to the notation they prefer.

Positive and negative j

The square root of -1 can have either of two values, just as can the square root of any positive real number. One of these is j. The other is $-j$, the product of j and -1. These two numbers are not the same, just as the positive and negative square roots of 1 are not the same!

Are you confused?

Some people have trouble envisioning this *unit imaginary number,* also called the *j operator.* Does the idea escape your "mind's eye"? If so, don't worry about it. Recall from Chap. 3 that the natural numbers—the simplest ones—are built up from a set containing nothing! All numbers are abstract in the literal sense, so j isn't any more bizarre than 0, or -1, or any other number.

Here's a challenge!

All of the laws of real-number arithmetic also apply to the unit imaginary number. Based on that fact, figure out what happens as j is raised to increasing integer powers starting with the 1st power.

Solution

Keep in mind that j is the positive square root of -1, which is $(-1)^{1/2}$. The parentheses are important in this expression. If we leave them out, someone might get the idea that we're discussing the quantity $-(1^{1/2})$, which is equal to -1. Because all the laws of the reals also apply to j, we can be sure that $j^1 = j$. By definition, $j^2 = -1$. From this we can calculate

$$\begin{aligned} j^3 &= j^2 \times j \\ &= -1 \times j \\ &= -j \end{aligned}$$

Now for the 4th power:

$$\begin{aligned} j^4 &= j^3 \times j \\ &= -j \times j \\ &= -1 \times j \times j \\ &= -1 \times j^2 \\ &= -1 \times (-1) \\ &= 1 \end{aligned}$$

And the 5th power:

$$\begin{aligned} j^5 &= j^4 \times j \\ &= 1 \times j \\ &= j \end{aligned}$$

And the 6th:

$$\begin{aligned} j^6 &= j^5 \times j \\ &= j \times j \\ &= j^2 \\ &= -1 \end{aligned}$$

Can you see what will happen if we keep going like this, increasing the integer power by 1 over and over? We'll go in a four-way cycle. If you grind things out, you'll see for yourself that $j^7 = -j$, $j^8 = 1$, $j^9 = j$, $j^{10} = -1$, and so on. In general, if n is a positive integer,

$$j^n = j^{n+4}$$

The Imaginary Number Line

The unit imaginary number j can be multiplied by any real number to get the positive square root of some negative real number. Conversely, the positive square root of any negative real number is equal to some positive-real multiple of j. If we want to multiply j by a positive real number b, we write jb. If we want to multiply j by a negative real number $-b$, we write $-jb$, putting the minus sign in front of j rather than between j and b. For example,

$$\begin{aligned} j5 &= (-1)^{1/2} \times 25^{1/2} \\ &= (-1 \times 5)^{1/2} \\ &= (-5)^{1/2} \end{aligned}$$

and

$$\begin{aligned} (-4)^{1/2} &= (-1 \times 4)^{1/2} \\ &= (-1)^{1/2} \times 4^{1/2} \\ &= j2 \end{aligned}$$

If we take the real number line and multiply the value of every point by j, the result is the *imaginary number line* (Fig. 21-1).

Are you confused?

"Why," you might ask, "do we write j before the real-number numeral and not after it?" It's a matter of preference. Engineers usually write the j before the real number. If you see other notations for imaginary numbers such as $2j$, $2i$ (the way most pure mathematicians write it), or even $i2$, keep in mind that they all refer to the same quantity, which we would call $j2$.

Be careful!

In the "challenge" calculation at the end of the previous section, j was raised to integer powers. If we're not careful, we can confuse expressions like these with integer multiples of j. We must pay close attention to whether that real number is meant to be a multiple of j (as in $j4$), or a power of j (as in j^4).

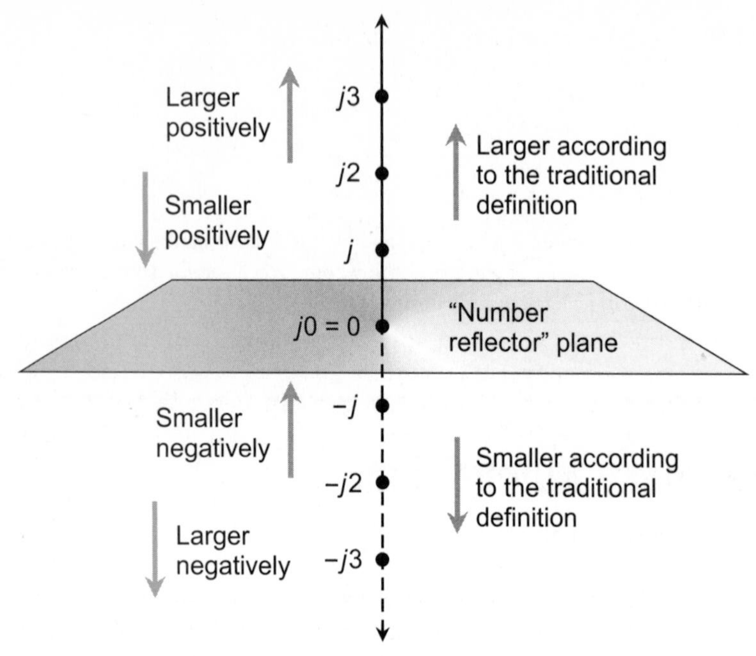

Figure 21-1 The imaginary number line. The imaginary values are defined according to the values of the real-number multiples of j.

Relative and absolute values

Imagine a "number reflector" plane, like a mathematical mirror, perpendicular to the imaginary number line and passing through the point for $j0$, which is identical to the real number zero. (Zero is the only real number that's also imaginary.) The definitions of the positive and negative imaginary numbers, as well as the definitions for "larger than" and "smaller than," are analogous to the definitions for the real numbers. If you go upward on the line, the value of the imaginary number increases. If you go downward, the value decreases.

In Fig. 21-1, the distance of an imaginary number from the point for $j0$ is defined as its *absolute value*. The absolute value of an imaginary number is equal to the nonnegative real number you get if you remove the j, and also remove the minus sign (if there is one). To denote the absolute value of an imaginary number or imaginary-number expression, we enclose it between vertical lines, just as we do with real numbers and real-number expressions. For example,

$$|j3| = 3$$

and

$$|-j3| = 3$$

In general, if b is any nonnegative real number, then

$$|jb| = b$$

and

$$|-jb| = b$$

To add, move upward

Think of upward distances on the imaginary number line as positive imaginary displacements, and downward distances as negative imaginary displacements. If we have an imaginary number jb_1 and we want to add another imaginary number jb_2 to it, we first find the point on the number line representing jb_1. Then we move up along the line by b_2 units. That will get us to the point representing the sum of the two numbers, $jb_1 + jb_2$.

As an example, suppose $b_1 = -3$ and $b_2 = 2$. We start at the point for $-j3$ and move up 2 units. That gets us to the point for $-j3 + j2$. It happens to be $-j$, as shown on the left side of Fig. 21-2.

Now suppose that we start with $j2$ and travel upward by -3 units. We're talking about displacement here, not simple distance, so negatives can make sense! An upward displacement

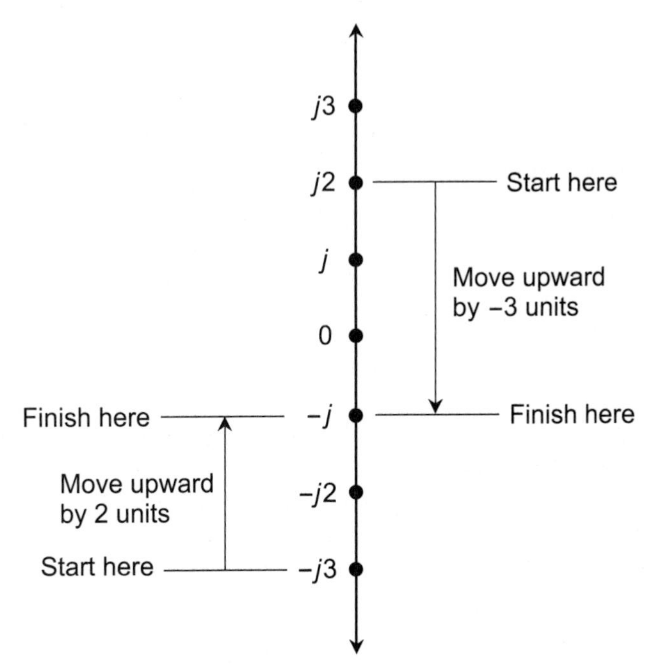

Figure 21-2 On the left, a number-line rendition of $-j3 + j2$. On the right, a number-line rendition of $j2 + (-j3)$. When we move negatively upward, we move downward by the equivalent distance.

of −3 units is the same as a downward displacement of 3 units. This process is shown on the right in Fig. 21-2. When we add the imaginary numbers $-j3$ and $j2$ in either order as shown, we end up at the same point, which corresponds to $-j$. We have geometrically analyzed these two facts:

$$-j3 + j2 = -j$$

and

$$j2 + (-j3) = -j$$

To subtract, move downward

Look again at the right-hand side of Fig. 21-2. We add a negative imaginary number to some other imaginary number. Adding a negative imaginary number is the same thing as subtracting the product of −1 and that number. If we have an imaginary number jb_1 and we want to subtract another imaginary number jb_2 from it, we must first find the point on the number line representing jb_1. Then we travel downward by b_2 units along the imaginary number line. That will get us to the point representing $jb_1 - jb_2$.

Are you confused?

Do you suspect that the laws of real-number arithmetic apply to all imaginary numbers, and not just to j itself? If so, you're right! Look back at the end of Chap. 9 if you want to review those laws.

Let's see how the distributive law, familiar with respect to multiplication and addition of real numbers, can be used to scrutinize the two imaginary-number sums we just analyzed. We can separate the real-number multiples, called the *real coefficients,* from j and then find the sums like this:

$$\begin{aligned} -j3 + j2 &= j(-3 + 2) \\ &= j(-1) \\ &= -j \end{aligned}$$

and

$$\begin{aligned} j2 + (-j3) &= j\,[2 + (-3)] \\ &= j(2 - 3) \\ &= j(-1) \\ &= -j \end{aligned}$$

Here's a challenge!

In terms of the imaginary number line, express the fact that when we subtract $-j5$ from $-j3$, we get $j2$. Write it down in the simplest possible form.

Solution

When we subtract a negative imaginary number, we move negatively downward along the imaginary number line, meaning that we actually travel upward. Figure 21-3 shows how this works. We start at $-j3$ and

Figure 21-3 Here, we start with $-j3$ and then subtract $-j5$, ending up with $j2$. When we go negatively downward, we go upward by the equivalent distance.

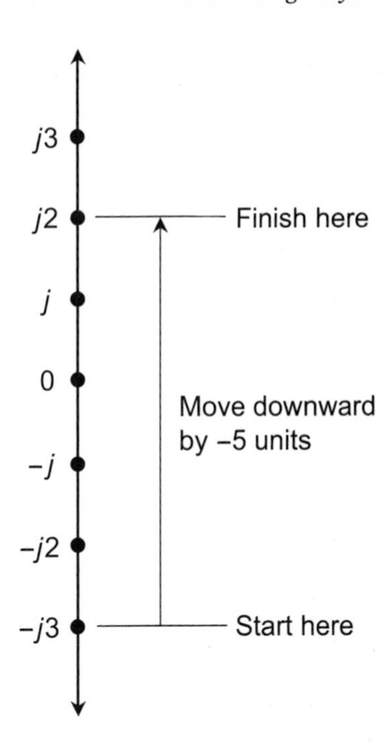

move down by −5 units, which means we really move up by 5 units. We finish at the point corresponding to $j2$. We can write

$$-j3 - (-j5) = j2$$

Simplifying, we can write

$$-j3 + j5 = j2$$

Real + Imaginary = Complex

When you add a real number and an imaginary number, you get a *complex number.* In this context, the term "complex" does not mean "complicated." A better word would be "composite," but that term has already been taken! (A composite number is a natural number that can be factored into a product of two or more primes.) All the rules of arithmetic you learned in Chap. 9 apply to complex numbers, complex-number variables, and expressions containing complex numbers.

How they are written

When we write a complex number, we put down the real-number part first, then a plus or minus sign, and the imaginary-number part. Here are some examples:

$$4 + j3$$
$$-4 + j5$$

$$-5 - j3$$
$$1 - j6$$
$$0 + j0$$

When a complex number is written as a difference between a real number and an imaginary number, we can rewrite it as a sum. The third and fourth of the above complex numbers can be converted to the sums

$$-5 + (-j3)$$

and

$$1 + (-j6)$$

The complex numbers $0 + j0$ and $0 - j0$ are the same as the real number 0. They are also identical to the imaginary numbers $j0$ and $-j0$.

The complex-number plane

The set of complex numbers needs two dimensions—a *plane*—to be graphically defined. The set of coordinates shown in Fig. 21-4 is the *complex-number plane,* in which we can plot

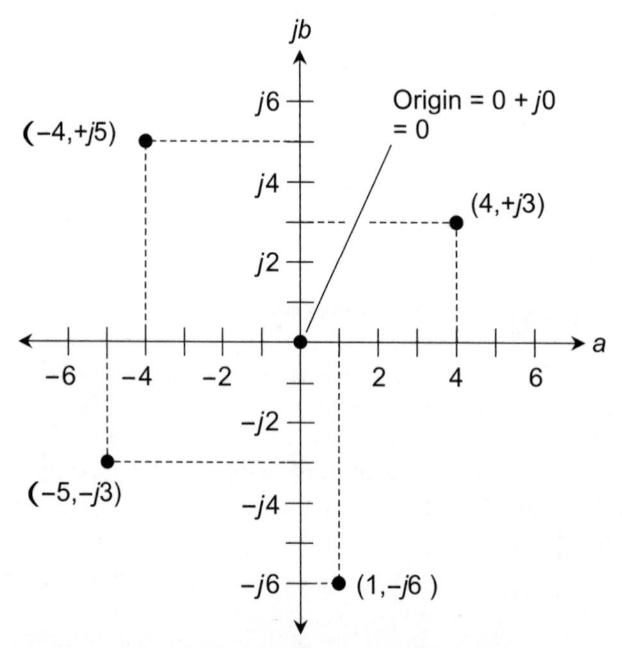

Figure 21-4 The complex-number plane, showing five values plotted as points. The dashed reference lines help to show the coordinates of the points on the axes.

quantities that are part real and part imaginary. The real part is expressed toward the right for positive and toward the left for negative. The imaginary part goes upward for positive and downward for negative. Any point in the plane, representing a unique complex number, can be expressed as an ordered pair (a,jb) or written as $a + jb$, where a and b are real numbers and j is the unit imaginary number.

If $a = 0$ and $b \neq 0$, a complex number $a + jb$ is called *pure imaginary.* If $a \neq 0$ and $b = 0$, a complex number $a + jb$ is called *pure real.* If both a and b are positive, the point representing a complex number is in the first quadrant of the plane. If a is negative and b is positive, the point is in the second quadrant. If both a and b are negative, the point is in the third quadrant. If a is positive and b is negative, the point is in the fourth quadrant.

Adding and subtracting complex numbers

When we want to add two complex numbers, we must add the real parts and the complex parts separately. For example, the sum of $4 + j7$ and $45 - j83$ works out like this:

$$\begin{aligned}(4 + j7) + (45 - j83) &= (4 + 45) + j(7 - 83) \\ &= 49 + j(-76) \\ &= 49 - j76\end{aligned}$$

Subtracting complex numbers is a little more involved; it's best to convert a difference to a sum. For example, we can find $(4 + j7) - (45 - j83)$ by multiplying the second complex number by -1 and then adding the two complex numbers, like this:

$$\begin{aligned}(4 + j7) - (45 - j83) &= (4 + j7) + [-1(45 - j83)] \\ &= (4 + j7) + (-45 + j83) \\ &= -41 + j90\end{aligned}$$

Multiplying complex numbers

When you multiply one complex number by another, you should treat both of the numbers as sums called *binomials,* which means "expression with two names." Any sum with two addends is a binomial. You'll work with them a lot in the coming chapters. If a, b, c, and d are real numbers, then

$$\begin{aligned}(a + jb)(c + jd) &= ac + jad + jbc + j^2 bd \\ &= (ac - bd) + j(ad + bc)\end{aligned}$$

Remember that $j^2 = -1$! That's why you get a minus sign between ac and bd. This rule is an adaptation of the product of sums rule you learned in Chap. 9.

Dividing complex numbers

When you want to divide a complex number by another complex number, things get a little messy. You won't have to do this very often, but if you ever find yourself faced with a complex-number division problem, you can use the following formula. If a, b, c, and d are real numbers, and if c and d are not both equal to 0, then

$$(a + jb)/(c + jd) = (ac + bd)/(c^2 + d^2) + j(bc - ad)/(c^2 + d^2)$$

Complex conjugates

Every complex number has a sort of "mirror image." Imagine two complex numbers that have the same real coefficients, but where one is a sum and the other is a difference, like this:

$$a + jb$$

and

$$a - jb$$

These two numbers are called *complex conjugates.* When you add $a + jb$ to its conjugate, you get a pure real number:

$$\begin{aligned}(a + jb) + (a - jb) &= (a + a) + (jb - jb)\\ &= 2a + j0\\ &= 2a\end{aligned}$$

When you multiply $a + jb$ by its conjugate, you get another pure real number:

$$\begin{aligned}(a + jb)(a - jb) &= a^2 - jab + jba - j^2b^2\\ &= a^2 - jab + jab + b^2\\ &= a^2 + b^2\end{aligned}$$

Complex conjugates show up together when you solve certain equations. You'll see some examples in Chap. 23.

Absolute value of a complex number

The *absolute value* of a complex number $a + jb$ is the distance from the origin $(0, j0)$ to the point (a, jb) on the complex-number plane. For $a + j0$ when a is positive, the absolute value is a. For $a + j0$ when a is negative, the absolute value is $-a$. For a pure imaginary number $0 + jb$ where b is positive, the absolute value is b. If b is negative, the absolute value of $0 + jb$ is equal to $-b$.

Suppose we want to find the absolute value of $-22 - j0$. This is a pure real number. It is the same as $-22 + j0$, because $j0 = 0$. Therefore, the absolute value of this complex number is $-(-22) = 22$. What about the absolute value of $0 - j34$? This is a pure imaginary number. The value of b is -34, because $0 - j34 = 0 + j(-34)$. Therefore, the absolute value is $-(-34) = 34$.

If a complex number $a + jb$ is neither pure real or pure imaginary, the absolute value must be found by going through a little arithmetic. First, we square both a and b separately. Then we add the squares. Finally, we take the positive square root of that sum. We can write this as a formula:

$$|a + jb| = (a^2 + b^2)^{1/2}$$

Let's find the absolute value of $3 - j4$. In this case, $a = 3$ and $b = -4$. Squaring both of these and adding the results gives us

$$3^2 + (-4)^2 = 9 + 16$$
$$= 25$$

The positive square root of 25 is 5. Therefore, $|3 - j4| = 5$.

Are you confused?

Do you wonder how the set of complex numbers relates to the sets of natural numbers, integers, rational numbers, irrational numbers, imaginary numbers, and real numbers? The answer is that every one of those other sets is a *proper subset* of the set of complex numbers.

The set of complex numbers, sometimes denoted *C*, is the "grandmother of all number sets" as far as most algebra is concerned. It's as far into the universe of numbers as we'll go. Someday, you might take courses that take you "farther out." Who knows? Maybe you'll discover or invent a new realm of numbers that nobody has worked with before.

Here's a challenge!

Find the sum $(2 + j3)$ and $(3 + j)$. Then plot both of these numbers, as well as their sum, in the complex-number plane. These three points, along with the origin, form the vertices of a quadrilateral (four-sided geometric figure). It's a special sort of quadrilateral. What sort? Why?

Solution

First, we should note that when j is multiplied by 1 in a complex number, there's no need to write down the numeral 1 after the j. That's why the j is all alone in the second addend above. To find the sum of these two complex numbers, we add their pure real and pure imaginary parts separately and then put the results back together, like this:

$$(2 + j3) + (3 + j) = (2 + 3) + j(3 + 1)$$
$$= 5 + j4$$

Figure 21-5 shows the two complex numbers, along with their sum and the origin, as ordered pairs in the complex-number plane.

The four points lie at the vertices of a *parallelogram.* Remember from geometry: A parallelogram is a quadrilateral in which the pairs of opposite sides are parallel. To prove that the quadrilateral in Fig. 21-5 is a parallelogram, we can show that the pairs of line segments forming opposite sides have identical slopes.

The slope m_1 of the line segment connecting $(0, j0)$ and $(2, j3)$ can be found by taking the ratio of the difference in jb to the difference in a, like this:

$$m_1 = \Delta jb / \Delta a$$
$$= (j3 - j0)/(2 - 0)$$
$$= j3/2$$

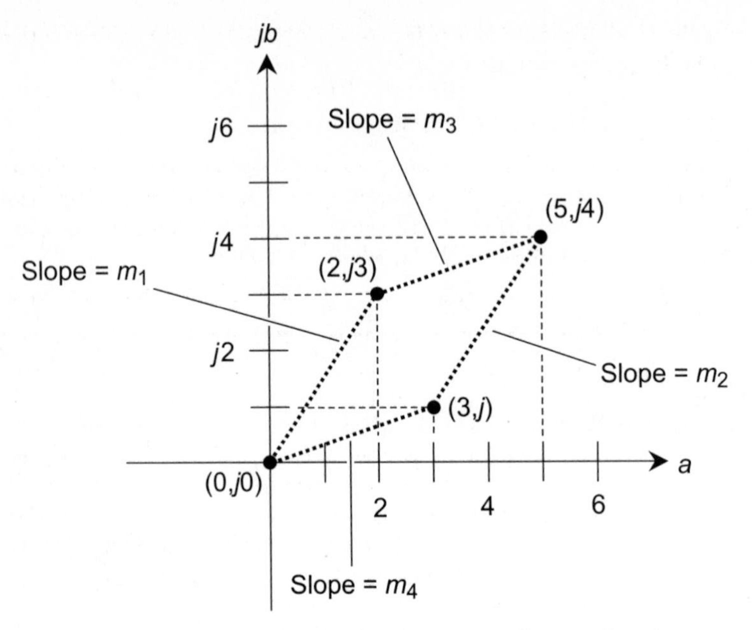

Figure 21-5 Addition of $(2 + j3)$ and $(3 + j)$, illustrated in the complex-number plane.

The line segment connecting $(3, j)$ and $(5, j4)$ is opposite from the side with slope m_1. Let's call its slope m_2. Then

$$\begin{aligned} m_2 &= \Delta jb/\Delta a \\ &= (j4 - j)/(5 - 3) \\ &= j3/2 \end{aligned}$$

These two opposite sides are parallel. We're halfway there! Now let's find the slope m_3 of the line segment connecting $(2, j3)$ and $(5, j4)$. It is

$$\begin{aligned} m_3 &= \Delta jb/\Delta a \\ &= (j4 - j3)/(5 - 2) \\ &= j/3 \end{aligned}$$

Finally, let's find the slope m_4 of the line segment connecting $(0, j0)$ and $(3, j)$. This line segment is opposite from the side with slope m_3. We have

$$\begin{aligned} m_4 &= \Delta jb/\Delta a \\ &= (j - 0)/(3 - 0) \\ &= j/3 \end{aligned}$$

Those two opposite sides are also parallel. This proves that the quadrilateral in Fig. 21-5 is a parallelogram.

Here's an extra-credit challenge!

Whenever you add two complex numbers and diagram the process after the fashion of Fig. 21-5, you'll get a parallelogram, or else all four points will lie along a single straight line (a "squashed parallelogram"). If you're ambitious, prove this. You're on your own. Here's a hint: Call the two complex numbers $a + jb$ and $c + jd$, where a, b, c, and d are real numbers.

Practice Exercises

This is an open-book quiz. You may (and should) refer to the text as you solve these problems. Don't hurry! You'll find worked-out answers in App. C. The solutions in the appendix may not represent the only way a problem can be figured out. If you think you can solve a particular problem in a quicker or better way than you see there, by all means try it!

1. The laws of arithmetic for real numbers also apply to imaginary numbers. On that basis, how can we determine the value of j^0?
2. What is the value of j^{-2}? The value of j^{-4}? The value of j^{-6}? The value of j^{-8}? What happens as this trend continues?
3. Determine the value of j^{-1} in two ways. First, use the difference of powers law. Here's a hint: Note that $j^{-1} = j^{3-4}$. Second, use the law of cross multiplication. Again, here's a hint: Find the value of an unknown (call it z) when $1/j = z/1$.
4. Using the difference of powers law and all the other things we've learned, determine the values of j^{-3}, j^{-5}, and j^{-7}. Here are some hints:

$$j^{-3} = j^{1-4}$$
$$j^{-5} = j^{-1-4}$$
$$j^{-7} = j^{-3-4}$$

   What happens as this trend continues?
5. Using what we've learned in the chapter text and so far in this set of exercises, create a table that shows what happens when j is raised to any integer power.
6. Find the following:
   (a) $(4 + j5) + (3 - j8)$
   (b) $(4 + j5) - (3 - j8)$
   (c) $(4 + j5)(3 - j8)$
   (d) $(4 + j5)/(3 - j8)$
7. Find the difference between the complex conjugates $(a + jb)$ and $(a - jb)$. First, subtract the second from the first. Then subtract the first from the second.

8. Find the quotient $(a + jb)/(a - jb)$, assuming that a and b are not both equal to 0.

9. Find the quotient $(a - jb)/(a + jb)$, assuming that a and b are not both equal to 0. How does this compare with the answer to Exercise 8?

10. Suppose k is a positive real-number constant. How many pure real numbers have absolute values equal to k? How many pure imaginary numbers have absolute values equal to k? How many complex numbers have absolute values equal to k? Draw a diagram in the complex-number plane that shows these situations.

CHAPTER

22

Quadratic Equations with Real Roots

In this chapter, you'll learn about *single-variable quadratic equations*, often called *quadratics* for short. First, you'll see how you can recognize one. Then you'll learn how you can solve it if there's at least one solution, called a *root*, in the set of real numbers. Sometimes a quadratic equation has two real roots, sometimes it has only one real root, and sometimes it has no real roots.

Second-Degree Polynomials

All quadratics are *second-degree equations*. That means they can be written in a form that contains a variable raised to the second power (squared), but no power higher than that.

Polynomials

A quadratic equation can always be portrayed as a *polynomial*, which means "expression with multiple names," on the left side of the equals sign, with 0 on the right side. Here are some examples of polynomials. Only the second of these can form a quadratic if we set it equal to 0.

$$x - 3$$
$$-x^2 + 3x - 6$$
$$x - y^3 + 7z^2$$
$$2x - 2y^5 - 2z^7$$
$$ax^4y - bxz^3 - cy^2z^2$$

where a, b, and c are constants, and x, y, and z are variables. Polynomials contain parts called *terms* that are added together. The terms in a polynomial are also known as *monomials*. Sometimes the monomials are complicated; they can even be polynomials themselves. A polynomial can contain any finite number of monomials—hundreds, thousands, or millions perhaps—but you'll rarely see a polynomial with more than a dozen terms.

As you know, all cases of subtraction are sums in disguise. The above expressions can be written as pure sums like this:

$$x + (-3)$$
$$-x^2 + 3x + (-6)$$
$$x + (-y^3) + 7z^2$$
$$2x + (-2y^5) + (-2z^7)$$
$$ax^4y + (-bxz^3) + (-cy^2z^2)$$

Polynomial standard form

Any single-variable quadratic, no matter how complicated it looks when you first see it, can be rewritten so it appears in *polynomial standard form.* When the equation is in this form, the left side of the equals sign contains a constant multiple of the variable squared, added to a constant multiple of the variable itself, added to a constant. On the right side of the equals sign, you find 0 all by itself. Here's the general equation:

$$ax^2 + bx + c = 0$$

where a, b, and c are constants, and x is the variable. All of the following are quadratic equations in polynomial standard form:

$$3x^2 + 2x + 5 = 0$$
$$3x^2 - 4x = 0$$
$$-7x^2 - 5 = 0$$
$$-4x^2 = 0$$

In all but the first of these, some of the constants equal 0. But the first constant, the one by which x^2 is multiplied, can never be 0 in a quadratic equation. If you set $a = 0$ in the standard form of a single-variable quadratic equation, you'll get the standard form for single-variable first-degree equation, because the term containing x^2 vanishes:

$$bx + c = 0$$

Mutant quadratics

Quadratic equations frequently appear in disguise. I call them *mutant quadratics.* Every such equation has two things in common. First, it isn't in polynomial standard form. Second, it can be morphed into that form without changing the set of roots we get when we solve it. Here are some examples:

$$x^2 = 2x + 3$$
$$x = 4x^2 - 7$$
$$x^2 + 4x = 7 + x$$
$$x - 2 = -8x^2 - 22$$
$$3 + x = 2x^2$$

Tables 22-1 through 22-5 show how we can convert the above equations to polynomial standard form. Note the last step in Table 22-5. To be "true to form," the polynomial should show the terms by descending powers of the variable. The term containing x^2 should come first, then the term containing x, and finally the constant.

Table 22-1. Conversion of $x^2 = 2x + 3$ to polynomial standard form.

Statements	Reasons
$x^2 = 2x + 3$	This is the equation we are given
$x^2 - 2x = 3$	Subtract $2x$ from each side
$x^2 - 2x - 3 = 0$	Subtract 3 from each side

Table 22-2. Conversion of $x = 4x^2 - 7$ to polynomial standard form.

Statements	Reasons
$x = 4x^2 - 7$	This is the equation we are given
$-4x^2 + x = -7$	Subtract $4x^2$ from each side
$-4x^2 + x + 7 = 0$	Add 7 to each side

Table 22-3. Conversion of $x^2 + 4x = 7 + x$ to polynomial standard form.

Statements	Reasons
$x^2 + 4x = 7 + x$	This is the equation we are given
$x^2 + 4x - 7 = x$	Subtract 7 from each side
$x^2 + 3x - 7 = 0$	Subtract x from each side

Table 22-4. Conversion of $x - 2 = -8x^2 - 22$ to polynomial standard form.

Statements	Reasons
$x - 2 = -8x^2 - 22$	This is the equation we are given
$8x^2 + x - 2 = -22$	Add $8x^2$ to each side
$8x^2 + x + 20 = 0$	Add 22 to each side

Table 22-5. Conversion of $3 + x = 2x^2$ to polynomial standard form.

Statements	Reasons
$3 + x = 2x^2$	This is the equation we are given
$-2x^2 + 3 + x = 0$	Subtract $2x^2$ from each side
$-2x^2 + x + 3 = 0$	Commutative law for addition

Are you confused?

Suppose you see an equation in one variable. You think it's a quadratic, but you aren't sure. Here's an example:

$$3x^2 + x - 8 = 4x^2 + 7x + 4$$

If you can convert this equation to polynomial standard form, you can be certain that it's a quadratic. If you can't convert it, then two things are possible: you didn't try hard enough, or it isn't a quadratic. If it isn't a quadratic and you want to prove that it isn't, you must morph the equation into a form that's obviously not convertible into polynomial standard form. That can be tricky. The next "challenge" will show you an example.

As things work out, you can convert the above equation to polynomial standard form. If you subtract $4x^2$ from each side, then subtract $7x$ from each side, and finally subtract 4 from each side, you get

$$-x^2 - 6x - 12 = 0$$

Here's a challenge!

Show that the following is the equivalent of a first-degree equation, not a quadratic:

$$7x^2/2 + 7x - 5 = -23x^2/(-10) + 6x^2/5 + 3x$$

Solution

It takes a little intuition to solve this, but nothing beyond the mathematical skill we've acquired by now! Let's look closely at the right side of this equation. It contains two terms with x^2. These terms are:

$$-23x^2/(-10)$$

and

$$6x^2/5$$

We can add these two terms to get a single term in x^2. If we multiply both the numerator and denominator of the first of the above expressions by -1, and if we multiply both the numerator and denominator of the second expression by 2, we get

$$23x^2/10$$

and

$$12x^2/10$$

We now have a common denominator, so we can find the sum

$$23x^2/10 + 12x^2/10 = 35x^2/10$$

which reduces to $7x^2/2$. This allows us to rewrite the right side of the original equation to obtain

$$7x^2/2 + 7x - 5 = 7x^2/2 + 3x$$

Subtracting $7x^2/2$ from each side, we get

$$7x - 5 = 3x$$

When we subtract $3x$ from each side now, we get an equation in the standard single-variable, first-degree form:

$$4x - 5 = 0$$

Binomial Factor Form

There's another way to express a quadratic equation: as a product of two binomials that is equal to 0. In some ways, this form is simpler than the polynomial standard form. As you'll soon see, the *binomial factor form* of a quadratic tells you the roots directly.

Binomials in quadratics

When a left side of a quadratic is expressed as a product of binomials, both of the binomials must be in a specific form. The first term in each binomial is a multiple of the variable. The multiplicand in that term is the *coefficient* of the variable. The second term is a constant, sometimes called the stand-alone constant. Here are some examples:

$$x + 1$$
$$x - 5$$
$$3x + 5$$
$$-17x + 24$$
$$8x - 13$$
$$-7x - 11$$

Multiplying two binomials

If we multiply two binomials of the above sort where both binomials contain the same variable, and if we then set the product equal to 0, we get a quadratic equation. Here's the general binomial factor form for a quadratic:

$$(px + q)(rx + s) = 0$$

where p and r are the coefficients, q and s are the constants, and x is the variable.

To produce a quadratic, neither of the coefficients p nor r can be equal to 0. Other than that, there's no restriction on the values of the coefficients and constants. As long as p, q, r, and s are all real numbers, then the resulting equation will have real roots.

The fact that the above general equation is a quadratic might not appear obvious, but when we multiply out the left side of the equation using the product of sums rule, the truth is revealed. Start with the expression

$$(px + q)(rx + s)$$

Multiplying this out according to the product of sums rule gives us

$$pxrx + pxs + qrx + qs$$

Using the commutative law for multiplication in the first two terms and then simplifying the first term, we get

$$prx^2 + psx + qrx + qs$$

The distributive law "in reverse" allows us to rewrite this as

$$prx^2 + (ps + qr)x + qs$$

When we set this equal to 0, we get an equation in polynomial standard form:

$$prx^2 + (ps + qr)x + qs = 0$$

Are you confused?

If you have trouble seeing why the above equation is in polynomial standard form, you can rename the coefficients and constants like this:

$$pr = a$$

$$ps + qr = b$$

$$qs = c$$

By substitution, you get

$$ax^2 + bx + c = 0$$

If you wonder about the requirement that $p \neq 0$ and $r \neq 0$ in the binomial factor form, the reason for this restriction should be clear now. If $p = 0$ or $r = 0$, then $pr = 0$, the term containing x^2 vanishes, and you have the one-variable first-degree equation

$$(ps + qr)x + qs = 0$$

What are the roots?

If we find a quadratic equation in binomial factor form and we want to find the roots, we're in luck! Here is the binomial factor form again, for a quadratic in the variable x:

$$(px + q)(rx + s) = 0$$

where neither p nor r equals 0. If either factor (that is, either binomial) happens to equal 0, then the entire expression on the left side of the equals sign becomes 0. The equation then reduces to $0 = 0$, indicating a root! The roots of the above quadratic can therefore be found by solving these two first-degree equations:

$$px + q = 0$$

and

$$rx + s = 0$$

In the first equation, we can subtract q from each side, getting

$$px = -q$$

Dividing through by p, which we have said is nonzero, we get

$$x = -q/p$$

In the second equation, we can subtract s from each side to obtain

$$rx = -s$$

Then we divide through by r, which we have restricted to nonzero values, getting

$$x = -s/r$$

The roots of the quadratic are $x = -q/p$ or $x = -s/r$. If X is the set of solutions, called the *solution set*, then

$$X = \{-q/p, -s/r\}$$

The solution set for an equation with multiple roots is the set containing all of those roots. As you continue studying algebra, you'll come across equations with solution sets containing three, four, or more elements. The solution set for a true quadratic, however, never has more than two elements.

Here's a challenge!

Consider the following quadratic equation in standard form:

$$x^2 - 2x - 15 = 0$$

Put this equation into binomial factor form. Then find the solution set X. Here's a hint: The coefficients and constants in the binomial factor form are integers.

Solution

The process of putting a quadratic into binomial factor form is called, not surprisingly, *factoring*. Learning how to factor a quadratic takes some practice. Some quadratics are easy to factor, others are difficult, and some can't be factored at all—at least, not into binomials with real-number constants and coefficients. Remember the general binomial factor form:

$$(px + q)(rx + s) = 0$$

We must find the values p, q, r, and s. We're told that they're all integers. The coefficient of the x^2 term in the original equation is equal to 1. That means $pr = 1$ in the binomial factor form. The last constant in the original equation is -15. This tells us that $qs = -15$. Because $pr = 1$, we have two possibilities:

- $p = 1$ and $r = 1$
- $p = -1$ and $r = -1$

Because $qs = -15$, we have eight possibilities:

- $q = 15$ and $s = -1$
- $q = -15$ and $s = 1$
- $q = 1$ and $s = -15$
- $q = -1$ and $s = 15$
- $q = 3$ and $s = -5$
- $q = -3$ and $s = 5$
- $q = 5$ and $s = -3$
- $q = -5$ and $s = 3$

We can trim this down to four possibilities by eliminating duplicate scenarios. That leaves us with these choices:

- $q = 15$ and $s = -1$
- $q = -15$ and $s = 1$
- $q = 3$ and $s = -5$
- $q = -3$ and $s = 5$

Now let's look at the coefficient of x in the original equation. It's -2. This tells us that

$$ps + qr = -2$$

If we "play around" with our choices for awhile, we'll see that if we let $p = 1$, $q = 3$, $r = 1$, and $s = -5$, we get

$$\begin{aligned} ps + qr &= 1 \times (-5) + 3 \times 1 \\ &= -5 + 3 \\ &= -2 \end{aligned}$$

That's the right coefficient for x in the original! Let's try those numbers in the binomial factors for the left side of the equation. We get

$$\begin{aligned} (px + q)(rx + s) &= (x + 3)(x - 5) \\ &= x^2 - 5x + 3x - 15 \\ &= x^2 - 2x - 15 \end{aligned}$$

That's the left side of the original quadratic. Now we know that it can be written as

$$(x+3)(x-5)=0$$

The roots are easy to find. If $x = -3$ or $x = 5$, the left side of the equation becomes equal to 0. The solution set X is therefore

$$X = \{-3,5\}$$

Let's check to be sure these solutions work in the original equation. For $x = -3$, we have

$$x^2 - 2x - 15 = 0$$
$$(-3)^2 - 2 \times (-3) - 15 = 0$$
$$9 - (-6) - 15 = 0$$
$$9 + 6 - 15 = 0$$
$$15 - 15 = 0$$
$$0 = 0$$

For $x = 5$, we have

$$x^2 - 2x - 15 = 0$$
$$5^2 - 2 \times 5 - 15 = 0$$
$$25 - 10 - 15 = 0$$
$$15 - 15 = 0$$
$$0 = 0$$

Completing the Square

There are other ways to look for the roots of a quadratic. One of these methods is called *completing the square.*

Perfect squares

Suppose that we come across a quadratic equation whose left side breaks down into two identical factors, equivalent to a single factor multiplied by itself. A polynomial of this type is called a *perfect square.* Here are some examples:

- The polynomial $x^2 + 2x + 1$ factors into $(x + 1)^2$
- The polynomial $x^2 - 2x + 1$ factors into $(x - 1)^2$
- The polynomial $x^2 + 4x + 4$ factors into $(x + 2)^2$
- The polynomial $9x^2 + 12x + 4$ factors into $(3x + 2)^2$

Quadratics built from perfect squares (by setting the right side of the equation equal to 0) are easy to solve. The two roots are identical, so in effect there's only one root, "done twice over."

Such a root is said to have *multiplicity* 2. If the above polynomial expressions are placed on the left sides of equations with 0 on the right, we get these quadratics:

$$x^2 + 2x + 1 = 0$$
$$x^2 - 2x + 1 = 0$$
$$x^2 + 4x + 4 = 0$$
$$9x^2 + 12x + 4 = 0$$

Respectively, they factor into:

$$(x + 1)^2 = 0$$
$$(x - 1)^2 = 0$$
$$(x + 2)^2 = 0$$
$$(3x + 2)^2 = 0$$

Because the expressions on the left sides of the above equations are equal to 0, we can take the square roots of both sides in each case without "plus-or-minus" ambiguity. This gives us the following first-degree equations, respectively:

$$x + 1 = 0$$
$$x - 1 = 0$$
$$x + 2 = 0$$
$$3x + 2 = 0$$

The solutions to these first-degree equations are easy to find:

$$x = -1$$
$$x = 1$$
$$x = -2$$
$$x = -2/3$$

These are the roots of the original quadratics. We can write down the solution sets as the single-element sets $\{-1\}$, $\{1\}$, $\{-2\}$, and $\{-2/3\}$.

Positive numbers on the right

Now we'll depart slightly from the polynomial standard form. Let's use the same perfect squares as above, but set the right sides to positive numbers rather than 0. Here are the new equations:

$$x^2 + 2x + 1 = 1$$
$$x^2 - 2x + 1 = 4$$
$$x^2 + 4x + 4 = 16$$
$$9x^2 + 12x + 4 = 25$$

The left sides can be factored exactly as before, getting

$$(x + 1)^2 = 1$$
$$(x - 1)^2 = 4$$
$$(x + 2)^2 = 16$$
$$(3x + 2)^2 = 25$$

If we take the square roots of both sides in each case here, remembering to include both the positive and negative results, we obtain

$$x + 1 = \pm(1)^{1/2}$$
$$x - 1 = \pm(4)^{1/2}$$
$$x + 2 = \pm(16)^{1/2}$$
$$3x + 2 = \pm(25)^{1/2}$$

These simplify to the following "double-barreled" first-degree equations:

$$x + 1 = \pm 1$$
$$x - 1 = \pm 2$$
$$x + 2 = \pm 4$$
$$3x + 2 = \pm 5$$

Let's think of these as equation pairs:

$$x + 1 = 1 \text{ or } x + 1 = -1$$
$$x - 1 = 2 \text{ or } x - 1 = -2$$
$$x + 2 = 4 \text{ or } x + 2 = -4$$
$$3x + 2 = 5 \text{ or } 3x + 2 = -5$$

When we solve these four pairs of first-degree equations, we get the following results, which are the roots of the original quadratics:

$$x = 0 \text{ or } x = -2$$
$$x = 3 \text{ or } x = -1$$
$$x = 2 \text{ or } x = -6$$
$$x = 1 \text{ or } x = -7/3$$

We can write the solution sets respectively as $\{0,-2\}$, $\{3,-1\}$, $\{2,-6\}$, and $\{1,-7/3\}$. I'll let you plug these values into the original quadratics to be sure we've found the right roots! Here are the original quadratics once again, in order:

$$x^2 + 2x + 1 = 1$$
$$x^2 - 2x + 1 = 4$$

$$x^2 + 4x + 4 = 16$$
$$9x^2 + 12x + 4 = 25$$

Morphing into a perfect square

When we come across a quadratic equation in polynomial standard form, we can sometimes add a positive real number to both sides, getting a perfect square on the left side. That will leave us with a nonzero value on the right, but as long as the left side is a perfect square, we can solve the equation just as we did in the four cases above. We can take the first quadratic from the above list and subtract 1 from each side, getting

$$x^2 + 2x = 0$$

We can take the second equation and subtract 4 from each side, getting

$$x^2 - 2x - 3 = 0$$

In the third equation, we can subtract 16 from each side to obtain

$$x^2 + 4x - 12 = 0$$

Finally, in the fourth equation, we can take 25 away from each side and get

$$9x^2 + 12x - 21 = 0$$

If you add the right constants to both sides of each of these equations, you'll get quadratics with perfect squares on their left-hand sides.

Are you confused?

Now that we've taken four solutions and manufactured four problems from them, let's retrace our steps and get the solutions back. In this way, we can get a good "feel" for how completing the square actually works. Imagine that we're confronted with the following four quadratics in polynomial standard form:

$$x^2 + 2x = 0$$
$$x^2 - 2x - 3 = 0$$
$$x^2 + 4x - 12 = 0$$
$$9x^2 + 12x - 21 = 0$$

We can take the first of these equations and add 1 to each side, getting

$$x^2 + 2x + 1 = 1$$

That gives us a perfect square on the left side. (Recognizing perfect squares when they appear in polynomial form is a "sixth sense" that evolves over time, and it takes practice to develop it.) Factoring, we obtain

$$(x + 1)^2 = 1$$

We can take the square root of both sides and get

$$x + 1 = \pm 1$$

which can be expressed as the pair

$$x + 1 = 1 \text{ or } x + 1 = -1$$

The solutions are found to be $x = 0$ or $x = -2$, so the solution set is $\{0,-2\}$. The other three equations can be worked out in similar fashion.

Are you still confused?

Do you wonder what happens if, in order to complete the square in a quadratic, you must *subtract* a positive number from both sides, getting a *negative* number on the right side? That's a good question. In that case, the roots turn out to be imaginary or complex. We'll deal with such equations in Chap. 23.

Here's a challenge!

Go through maneuvers similar to those we just completed, but with the second, third, and fourth quadratics from above:

$$x^2 - 2x - 3 = 0$$

$$x^2 + 4x - 12 = 0$$

$$9x^2 + 12x - 21 = 0$$

Solution

You're on your own! Start with perfect squares on the left sides of the equals signs and positive numbers on the right, and then take away those positive numbers from both sides to "unsquare" the equations.

The Quadratic Formula

The technique of completing the square can be applied to the general polynomial standard form of a quadratic equation. This gives us a tool for solving quadratics by "brute force": the so-called *quadratic formula.*

Deriving the formula

Remember the polynomial standard form where x is the variable, and a, b, and c are real-number constants with $a \neq 0$. The general formula is

$$ax^2 + bx + c = 0$$

Let's rewrite this as

$$ax^2 + bx = -c$$

Because $a \neq 0$ in any quadratic equation, we can divide each side by a, getting

$$x^2 + (b/a)x = -c/a$$

It's tempting to think that there must be some constant that we can add to both sides of this equation to get a perfect square on the left side of the equals sign. It takes some searching, but that constant does exist. It is $b^2/(4a^2)$. When we add it to both sides of the above equation, we obtain

$$x^2 + (b/a)x + b^2/(4a^2) = -c/a + b^2/(4a^2)$$

We can now factor the left side into the square of a binomial to get

$$[x + b/(2a)]^2 = -c/a + b^2/(4a^2)$$

The two terms in the right side of this equation can be added using the sum of quotients rule from Chap. 9 to obtain

$$[x + b/(2a)]^2 = (-4a^2c + ab^2) / (4a^3)$$

Canceling out the extra factors of a in the numerator and denominator on the right side, we get

$$[x + b/(2a)]^2 = (-4ac + b^2) / (4a^2)$$

Let's rewrite the numerator on the right side as a difference, so the equation becomes

$$[x + b/(2a)]^2 = (b^2 - 4ac) / (4a^2)$$

If we take the square root of both sides here, remembering the negative as well as the positive, we get

$$x + b/(2a) = \pm[(b^2 - 4ac) / (4a^2)]^{1/2}$$

The denominator in the right side is a perfect square; it's equal to $(2a)^2$. Therefore, we can simplify the expression on that side of the equals sign a little bit, considering it as a ratio of square roots rather than the square root of a ratio. We obtain

$$x + b/(2a) = \pm(b^2 - 4ac)^{1/2} / (2a)$$

If we subtract $b/(2a)$ from both sides, we get

$$x = \pm(b^2 - 4ac)^{1/2} / (2a) - b/(2a)$$

which expresses x in terms of the constants a, b, and c (finally!). An equation that states the general solution to an unknown is called a *formula*.

We're not quite done yet, because this formula can be simplified. We have a common denominator, $2a$, in the difference of the two ratios on the right side. We can therefore rewrite the above formula as

$$x = [\pm(b^2 - 4ac)^{1/2} - b] / (2a)$$

In most texts, the numerator is written with $-b$ first, like this:

$$x = [-b \pm (b^2 - 4ac)^{1/2}] / (2a)$$

An example

Now it's time to solve a specific equation using the quadratic formula. Let's try this:

$$9x^2 + 12x - 21 = 0$$

In this case, we have $a = 9$, $b = 12$, and $c = -21$. We can plug these numbers into the quadratic formula and grind it out:

$$\begin{aligned} x &= [-b \pm (b^2 - 4ac)^{1/2}] / (2a) \\ &= \{-12 \pm [12^2 - 4 \times 9 \times (-21)]^{1/2}\} / (2 \times 9) \\ &= \{-12 \pm [144 - (-756)]^{1/2}\} / 18 \\ &= (-12 \pm 900^{1/2}) / 18 \\ &= (-12 \pm 30) / 18 \\ &= (-12 + 30) / 18 \text{ or } (-12 - 30) / 18 \\ &= 18/18 \text{ or } -42/18 \\ &= 1 \text{ or } -7/3 \end{aligned}$$

The solution set is therefore $\{1,-7/3\}$. You can check these roots by plugging them back into the original equation.

Another example

Now let's see what happens when we solve this equation with the quadratic formula:

$$4x^2 - 24x + 36 = 0$$

Here, we have $a = 4$, $b = -24$, and $c = 36$. Plugging in and grinding out, we obtain

$$\begin{aligned} x &= [-b \pm (b^2 - 4ac)^{1/2}] / (2a) \\ &= \{-(-24) \pm [(-24)^2 - 4 \times 4 \times 36]^{1/2}\} / (2 \times 4) \\ &= [24 \pm (576 - 576)^{1/2}\}/ 8 \\ &= (24 \pm 0^{1/2}\}/ 8 \\ &= 24/8 \\ &= 3 \end{aligned}$$

The solution set is $\{3\}$. There is only one root. Feel free to check it!

The discriminant

In the second example above, $b^2 - 4ac = 0$. In that case, it doesn't matter whether we add $(b^2 - 4ac)^{1/2}$ to $-b$ or subtract $(b^2 - 4ac)^{1/2}$ from $-b$ in the formula. This gives us a quick way to tell whether a quadratic equation has two roots or only one. The quantity $b^2 - 4ac$ is called the *discriminant* of the general quadratic equation

$$ax^2 + bx + c = 0$$

If the discriminant is a positive real number, then the associated quadratic has two real roots. If the discriminant is equal to 0, then the quadratic has one real root with multiplicity 2.

There are plenty of quadratic equations in which the discriminant is a negative real number. Here's an example:

$$4x^2 - 4x + 36 = 0$$

This is almost exactly the same equation as we solved in the second example above. The only difference is that the second coefficient is 4 rather than 24. The discriminant here is

$$\begin{aligned} b^2 - 4ac &= 4^2 - 4 \times 4 \times 36 \\ &= 16 - 576 \\ &= -560 \end{aligned}$$

When we apply the quadratic formula, we must take the square root of -560. That's an imaginary number! A negative discriminant gives us imaginary or complex roots. We'll explore situations like this in the next chapter. We'll also see what happens when one or more of the coefficients in a quadratic equation are imaginary or complex.

Are you confused?

Do you wonder how we came up with the constant $b^2/(4a^2)$ to make a perfect square in the process of deriving the quadratic formula? Again, this is the "sixth sense" at work, a form of intuition that you can develop only with practice.

Here's a challenge!

In the derivation of the quadratic formula, we made a "quantum leap" when we claimed that the polynomial

$$x^2 + (b/a)x + b^2/(4a^2)$$

is a perfect square that can be factored into

$$[x + b/(2a)]^2$$

Show that this is actually true.

Solution

We can work it the other way, multiplying the factors out to get the polynomial. Let's rewrite the above squared binomial as a product of sums, like this:

$$[x + b/(2a)]\ [x + b/(2a)]$$

When we apply the product of sums and product of quotients rules from Chap. 9, this becomes

$$x^2 + xb/(2a) + xb/(2a) + b^2/(4a^2)$$

We can add the two middle terms together, getting

$$x^2 + 2xb/(2a) + b^2/(4a^2)$$

Canceling out the 2 in the numerator and denominator of the middle term, we obtain

$$x^2 + xb/a + b^2/(4a^2)$$

which can also be written as

$$x^2 + (b/a)x + b^2/(4a^2)$$

That's the original polynomial.

Practice Exercises

This is an open-book quiz. You may (and should) refer to the text as you solve these problems. Don't hurry! You'll find worked-out answers in App. C. The solutions in the appendix may not represent the only way a problem can be figured out. If you think you can solve a particular problem in a quicker or better way than you see there, by all means try it!

1. Multiply out the following equation, putting it into polynomial standard form.

$$(-7x - 5)(-2x + 9) = 0$$

2. Factor the following quadratic. Then find the roots and state the solution set. Here's a hint: The coefficients and constants in the factors are all integers.

$$x^2 + 10x + 25 = 0$$

3. Factor the following quadratic. Then find the roots and state the solution set. Here's a hint: The coefficients and constants in the factors are all integers.

$$2x^2 + 8x - 10 = 0$$

4. Factor the following quadratic. Then find the roots and state the solution set. Here's a hint: The coefficients and constants in the factors are all integers.

$$12x^2 + 7x - 10 = 0$$

5. Manipulate the following equation so it has the square of a binomial on the left side of the equals sign and 0 on the right side. Here's a hint: The coefficient and constant in the binomial are both integers.

$$16x^2 - 40x + 25 = 0$$

6. What is the single root of the quadratic stated in Prob. 5? What is the solution set?

7. Manipulate the following equation so it has the square of a binomial on the left side of the equals sign and a positive real number on the right side. Here's a hint: The coefficient and constant in the binomial are both integers.

$$x^2 + 6x - 7 = 0$$

8. What are the roots of the quadratic stated in Prob. 7? What is the solution set?

9. How many real roots does the following quadratic have? Find the real root or roots, if any exist, using the quadratic formula. What is the real-number solution set?

$$-2x^2 + 3x + 35 = 0$$

10. How many real roots does the following quadratic have? Find the real root or roots, if any exist, using the quadratic formula. What is the real-number solution set?

$$4x^2 + x + 3 = 0$$

CHAPTER

23

Quadratic Equations with Complex Roots

Now that we know a little bit about what to expect when the discriminant of a quadratic equation is negative, let's explore this territory in more detail. A negative discriminant indicates that the roots of a quadratic are imaginary or complex.

Complex Roots by Formula

The quadratic formula, like basic addition facts and multiplication tables, is worth committing to memory. (That's why I keep repeating it. Here it is again!) When we have a quadratic equation of the form

$$ax^2 + bx + c = 0$$

then the roots can be found using the formula

$$x = [-b \pm (b^2 - 4ac)^{1/2}] / (2a)$$

Square root of the discriminant

The discriminant of a quadratic equation is equal to the square of the coefficient of x, minus 4 times the product of the coefficient of x^2 and the stand-alone constant. In the quadratic formula as stated above, the discriminant d is

$$d = b^2 - 4ac$$

When the coefficients and constant in a quadratic are real numbers, then the discriminant is always a real number. If $d > 0$, then the square root of d can be either a positive real or its additive inverse. If $d = 0$, then the square root of d is 0. If $d < 0$, then the square root of d can be either of two values, one positive imaginary and the other negative imaginary.

Let's take an example. Suppose that we work out the discriminant d for a quadratic, and we find that $d = -16$. Then the positive square root of d is equal to $j4$, and the negative square root of d is equal to $-j4$.

Here's another example. Let's revisit the quadratic stated in Practice Exercise 10 at the end of Chap. 22:

$$4x^2 + x + 3 = 0$$

In this case, $a = 4$, $b = 1$, and $c = 3$, so

$$\begin{aligned} d &= b^2 - 4ac \\ &= 1^2 - 4 \times 4 \times 3 \\ &= 1 - 48 \\ &= -47 \end{aligned}$$

The positive square root of -47 is $j(47^{1/2})$, and the negative square root is $-j(47^{1/2})$.

Now let's look at the general case. If $d < 0$, then $|d| > 0$. (As you know, it also happens to be true that if $d < 0$, then $|d| = -d$, a fact that we'll use later in this chapter.) We can express the positive square root of d as

$$d^{1/2} = j(|d|^{1/2})$$

and we can express the negative square root of d as

$$-(d^{1/2}) = -j(|d|^{1/2})$$

Stated as a "plus-or-minus" expression, we have

$$\pm(d^{1/2}) = \pm j(|d|^{1/2})$$

Substituting $\pm j(|d|^{1/2})$ in place of $\pm(b^2 - 4ac)^{1/2}$ in the quadratic formula, we get

$$x = [-b \pm j(|d|^{1/2})] \;/\; (2a)$$

This equation can be used *if and only if* the real-number discriminant, d, is negative. It's important to remember what "if and only if" means in this context! We can *always* use this formula when $d < 0$. But we must *never* use it when $d = 0$ or when $d > 0$, because in those cases, the j operator does not belong there.

Imaginary roots: a specific case

Consider the following quadratic equation in which the coefficient of x^2 is positive, the coefficient of x is equal to 0, and the stand-alone constant is positive:

$$3x^2 + 75 = 0$$

In the general polynomial standard form, we have $a = 3$, $b = 0$, and $c = 75$. The discriminant is therefore

$$\begin{aligned} d &= b^2 - 4ac \\ &= 0^2 - (4 \times 3 \times 75) \\ &= -900 \end{aligned}$$

The positive-or-negative square root of the discriminant is

$$\begin{aligned} \pm(d^{1/2}) &= \pm j(|-900|^{1/2}) \\ &= \pm j(900^{1/2}) \\ &= \pm j30 \end{aligned}$$

The roots can now be found as

$$\begin{aligned} x &= [-b \pm j(|d|^{1/2})] / (2a) \\ &= (0 \pm j30) / (2 \times 3) \\ &= \pm j30/6 \\ &= \pm j5 \end{aligned}$$

Are you astute?

Do you suspect that this particular equation can be solved more easily without resorting to the quadratic formula? If so, you're right! If we subtract 75 from each side, we get

$$3x^2 = -75$$

Dividing through by 3 gives us

$$x^2 = -25$$

When we take the positive-or-negative square root of both sides, we obtain

$$\begin{aligned} x &= \pm[(-25)^{1/2}] \\ &= \pm j5 \end{aligned}$$

Imaginary roots: the general case

Let's see what happens with a general quadratic when the coefficient of x is 0. If we write it in polynomial standard form, we get

$$ax^2 + 0x + c = 0$$

which simplifies to

$$ax^2 + c = 0$$

Subtracting c from each side gives us

$$ax^2 = -c$$

We can divide through by a because, in a quadratic, a is never 0. Doing that, we get

$$x^2 = -c/a$$

Taking the positive-or-negative square root of each side, we obtain

$$x = \pm[(-c/a)^{1/2}]$$

If a and c have opposite signs, then $-c/a$ is positive. Therefore, the roots are both real and are additive inverses of each other:

$$x = (-c/a)^{1/2} \text{ or } x = -[(-c/a)^{1/2}]$$

If a and c have the same sign, then $-c/a$ is negative. That means the roots are pure imaginary and are additive inverses of each other:

$$x = j(|-c/a|^{1/2}) \text{ or } x = -j(|-c/a|^{1/2})$$

In this case, the discriminant is

$$\begin{aligned} d &= b^2 - 4ac \\ &= 0^2 - 4ac \\ &= -4ac \end{aligned}$$

Because a and c have the same sign, $-4ac < 0$. Therefore, $d < 0$.

Conjugate roots

The discriminant in a quadratic can be negative even when b, the coefficient of x, is not equal to 0. The only requirement is that $4ac$ be larger than b^2. Here's an example:

$$(45/2)x^2 + 3x + 1 = 0$$

In this case, $a = 45/2$, $b = 3$, and $c = 1$, so we have

$$\begin{aligned} 4ac &= 4 \times (45/2) \times 1 \\ &= 90 \end{aligned}$$

and

$$b^2 = 3^2$$
$$= 9$$

Therefore,

$$d = b^2 - 4ac$$
$$= 9 - 90$$
$$= -81$$

This tells us that the roots of the quadratic are not real numbers. To find the roots, we can plug in the value −81 for d in the "abbreviated discriminant" form of the quadratic formula:

$$x = [-b \pm j(|d|^{1/2})] / (2a)$$
$$= [-3 \pm j(|-81|)^{1/2}] / [2 \times (45/2)]$$
$$= [-3 \pm j(81^{1/2})] / 45$$
$$= (-3 \pm j9) / 45$$
$$= -3/45 \pm j(9/45)$$
$$= -1/15 \pm j(1/5)$$

The roots are

$$x = -1/15 + j(1/5) \text{ or } x = -1/15 - j(1/5)$$

These are complex conjugates. As things work out, the roots are always complex conjugates in a quadratic where $d < 0$, even when $b \neq 0$. The solution set X in this example is

$$X = \{-1/15 + j(1/5), -1/15 - j(1/5)\}$$

Are you ambitious?

For complementary credit, plug the roots we've just found into the original quadratic to be sure that they work. You're on your own. Here's a hint: This is a messy process, but if you're careful and patient, all the "garbage" will drop out in the end.

Are you confused?

The above derivations are abstract, but it's important that you follow through them so you understand the reasoning behind each step. Here are the results, wrapped up into two statements. In any quadratic:

- If the discriminant is a negative real number and the coefficient of x is 0, then the roots are pure imaginary, and are additive inverses.
- If the discriminant is a negative real number and the coefficient of x is a nonzero real number, then the roots are not pure imaginary, but are complex conjugates.

Here's a challenge!

Prove the second of the above statements.

Solution

Consider once again the general polynomial equation

$$ax^2 + bx + c = 0$$

The discriminant is

$$d = b^2 - 4ac$$

Suppose that $d < 0$ and $b \neq 0$. We can use the "abbreviated discriminant" version of the quadratic formula for complex roots:

$$x = [-b \pm j(|d|^{1/2})] / (2a)$$

Applying the right-hand distributive rule for addition and subtraction "in reverse" to the right side, we can split that expression into a sum or difference of two ratios with a common denominator, like this:

$$x = -b/(2a) \pm j(|d|^{1/2})/(2a)$$

Therefore, we can write the two roots as

$$x = -b/(2a) + j(|d|^{1/2})/(2a) \text{ or } x = -b/(2\text{a}) - j(|d|^{1/2})/(2a)$$

The fact that these are complex conjugates is obscure because the expressions are messy. To bring things into focus, we can change a couple of names. Let

$$p = -b/(2a)$$

and

$$q = (|d|^{1/2})/(2a)$$

Now we can rewrite the roots as

$$x = p + jq \text{ or } x = p - jq$$

These are complex conjugates (but not pure imaginary numbers) for all nonzero real numbers p and q. You might ask, "Are p and q really both nonzero?" The answer is "Yes." We can be sure that $p \neq 0$ because $b \neq 0$ and $a \neq 0$, so $-b/(2a)$ can't be 0. We can be sure that $q \neq 0$ because $d \neq 0$ and $a \neq 0$, so $(|d|^{1/2})/(2a)$ can't be 0.

Imaginary Roots in Factors

Now that we've found a way to solve quadratics that have real coefficients, a real constant, and a negative real discriminant, let's see what the factors of such equations look like. We'll start with situations where the coefficient of x is 0, so the roots are pure imaginary.

A specific case

Consider again the quadratic equation we solved earlier, in which the roots are pure imaginary and additive inverses:

$$3x^2 + 75 = 0$$

We simplified this equation in the second version of the solution when we divided both sides by 3, obtaining

$$x^2 + 25 = 0$$

We found that the roots are $j5$ and $-j5$. Knowing these roots, it's reasonable to think that we should be able to figure out what this equation looks like in binomial factor form. In one case, we have $x = j5$. If we take that equation and subtract $j5$ from each side, we get

$$x - j5 = 0$$

In the other case, we have $x = -j5$. We can add $j5$ to each side of that, getting

$$x + j5 = 0$$

This suggests that the binomial factor form of the quadratic is

$$(x - j5)(x + j5) = 0$$

The left side of this equation is 0 if and only if $x = j5$ or $x = -j5$. Let's multiply it out and see what we get. To avoid confusion with the signs, we can rewrite the first term as a sum:

$$[x + (-j5)](x + j5) = 0$$

Applying the product of sums rule gives us

$$x^2 + xj5 + (-j5x) + (-j5)(j5) = 0$$

Using the commutative law for multiplication in the third and fourth terms, we obtain

$$x^2 + xj5 + (-xj5) + (-j \times j) \times 25 = 0$$

Note that $-j \times j = 1$. Also, the second and third terms of the polynomial add up to 0. We can therefore simplify to get

$$x^2 + 25 = 0$$

Multiplying through be 3 gives us

$$3x^2 + 75 = 0$$

That's the original equation.

The general case

Consider a general quadratic in which the coefficient of x^2 is a positive real number a, the coefficient of x is 0, and the stand-alone constant is a positive real number c. Then we have

$$ax^2 + c = 0$$

Subtracting c from each side, we get

$$ax^2 = -c$$

Dividing through by a, which we know is not 0 because we've stated that it's positive, we obtain

$$x^2 = -c/a$$

which can be rewritten as

$$x^2 = -1(c/a)$$

Because a and c are both positive, we know that the ratio c/a is positive as well. Its positive and negative square roots are therefore both real numbers. We can take the square root of both sides of the above equation, getting

$$\begin{aligned} x &= \pm[-1(c/a)]^{1/2} \\ &= \pm(-1)^{1/2}\ [\pm(c/a)^{1/2}] \\ &= \pm j[(c/a)^{1/2}] \\ &= j[(c/a)^{1/2}] \text{ or } -j[(c/a)^{1/2}] \end{aligned}$$

Knowing these roots, we can get the binomial factor form of the original quadratic. In one case,

$$x = j[(c/a)^{1/2}]$$

If we subtract $j[(c/a)^{1/2}]$ from each side, we get

$$x - j[(c/a)^{1/2}] = 0$$

In the other case,

$$x = -j[(c/a)^{1/2}]$$

When we add $j[(c/a)^{1/2}]$ to each side, we get

$$x + j[(c/a)^{1/2}] = 0$$

The binomial factor form of the original quadratic is therefore

$$\{x - j[(c/a)^{1/2}]\}\{x + j[(c/a)^{1/2}]\} = 0$$

To be sure this is the correct equation in binomial factor form, (that is, to be sure we haven't made any mistakes!), we had better multiply out the left side. To avoid confusion with the signs, let's rewrite the first term as a sum:

$$\{x + (-j)[(c/a)^{1/2}]\}\{x + j[(c/a)^{1/2}]\} = 0$$

Using the product of sums rule, we can expand to get

$$x^2 + xj[(c/a)^{1/2}] + (-j)[(c/a)^{1/2}]x + \{-j[(c/a)^{1/2}]\}\{j[(c/a)^{1/2}]\} = 0$$

Taking advantage of the commutative law for multiplication in the third and fourth terms, we can rewrite it as

$$x^2 + xj[(c/a)^{1/2}] + \{-xj[(c/a)^{1/2}]\} + (-j \times j)(c/a) = 0$$

The second and third terms are additive inverses and $-j \times j = 1$, so we can simplify this to

$$x^2 + c/a = 0$$

Multiplying through by *a*, we get

$$ax^2 + c = 0$$

which is the original quadratic.

Are you confused?

We've seen what happens in a quadratic when the coefficient of x^2 and the stand-alone constant are positive while the coefficient of x is equal to 0. You might ask, "What happens if they are both negative?" The answer is simple. If $a < 0$ and $c < 0$, then we can multiply the equation through by -1, and we'll have the case where $a > 0$ and $c > 0$. For example, if we see

$$-2x^2 - 79 = 0$$

we can multiply each side by -1, getting

$$(-1) \times (-2x^2 - 79) = -1 \times 0$$

which simplifies to

$$2x^2 + 79 = 0$$

Here's a challenge!

Investigate what happens in the general case if a is positive and c is negative in the quadratic equation

$$ax^2 + c = 0$$

Solution

We have $a > 0$ and $c < 0$. That means $-c > 0$. Let's rewrite the above equation as

$$ax^2 - (-c) = 0$$

We can add $-c$ to each side, getting

$$ax^2 = -c$$

Dividing through by a, we obtain

$$x^2 = -c/a$$

Because $-c > 0$ and $a > 0$, we know that $-c/a > 0$. We can take the positive-negative square root of both sides to get

$$x = \pm(-c/a)^{1/2}$$

Stated separately, the roots are

$$x = (-c/a)^{1/2} \text{ or } x = -[(-c/a)^{1/2}]$$

These are both real numbers, and are additive inverses.

Here's another challenge!

Investigate what happens in the general case if a is negative and c is positive in the quadratic equation

$$ax^2 + c = 0$$

Solution

We have $a < 0$ and $c > 0$. That means $-a > 0$. We can subtract c from each side, getting

$$ax^2 = -c$$

Multiplying through by -1 gives us

$$(-a)x^2 = c$$

Dividing through by $(-a)$, we obtain

$$x^2 = c/(-a)$$

Because $c > 0$ and $-a > 0$, we know that $c/(-a) > 0$. We can take the positive-negative square root of both sides to get

$$x = \pm[c/(-a)]^{1/2}$$

which can be rewritten as

$$x = \pm(-c/a)^{1/2}$$

Stated separately, the roots are

$$x = (-c/a)^{1/2} \text{ or } x = -[(-c/a)^{1/2}]$$

These are the same roots we get if $a > 0$ and $c < 0$. Both roots are real, and they are additive inverses of each other.

Conjugate Roots in Factors

We've seen the binomial factor forms of quadratics where the roots are pure real or pure imaginary. Now let's look at the factors when the roots are complex conjugates, but are not pure imaginary numbers.

An example

We can use the quadratic formula to find the roots of a quadratic equation that has complex conjugate roots. Then we can generate the binomial factor form of the quadratic from those roots. Let's try this:

$$5x^2 + 6x + 5 = 0$$

We have $a = 5$, $b = 6$, and $c = 5$ in the polynomial standard form. The discriminant is

$$\begin{aligned} d &= b^2 - 4ac \\ &= 6^2 - 4 \times 5 \times 5 \\ &= 36 - 100 \\ &= -64 \end{aligned}$$

The square root of the discriminant is pure imaginary:

$$\begin{aligned} \pm(d^{1/2}) &= \pm[(-8)^{1/2}] \\ &= \pm j8 \end{aligned}$$

The roots are therefore

$$\begin{aligned} x &= [-b \pm j(|d|^{1/2})] \,/\, (2a) \\ &= (-6 \pm j8) \,/\, (2 \times 5) \\ &= (-6 \pm j8) \,/\, 10 \\ &= -6/10 \pm j(8/10) \\ &= -3/5 \pm j(4/5) \end{aligned}$$

Stated individually,

$$x = -3/5 + j(4/5) \text{ or } x = -3/5 - j(4/5)$$

The binomial factor form of the original quadratic must therefore look like this:

$$\{x - [-3/5 + j(4/5)]\}\{x - [-3/5 - j(4/5)]\} = 0$$

We should multiply this out to be sure we have found the correct binomial factors. First, let's convert the subtractions into negative additions, obtaining

$$\{x + (-1)[-3/5 + j(4/5)]\}\{x + (-1)[-3/5 + (-j)(4/5)]\} = 0$$

The distributive law for multiplication over addition lets us rewrite this as

$$\{x + 3/5 + [-j(4/5)]\}[x + 3/5 + j(4/5)] = 0$$

We now have two *trinomial factors* on the left side! We can multiply them out to get the equation

$$\begin{aligned} &x^2 + (3/5)x + j(4/5)x \\ &+ (3/5)x + (3/5)^2 + j(3/5)(4/5) \\ &+ (-j)(4/5)x + (-j)(4/5)(3/5) + (-j)(j)(4/5)^2 \\ &= 0 \end{aligned}$$

Fortunately, we have some pairs of terms that cancel out, so we can simplify to

$$x^2 + (6/5)x + 9/25 + 16/25 = 0$$

and further to

$$x^2 + (6/5)x + 1 = 0$$

Multiplying both sides of this equation by 5 gives us the original quadratic in polynomial standard form:

$$5x^2 + 6x + 5 = 0$$

Are you confused?

If you wonder how the two trinomials in the above situation can be multiplied out, remember the rule you derived when you solved Practice Exercise 10 at the end of Chap. 9. If you've forgotten how that rule works, review it now. It can be extrapolated to all real, imaginary, and complex numbers.

Here's a challenge!

Consider a general quadratic equation in polynomial standard form:

$$ax^2 + bx + c = 0$$

Suppose that b is not equal to 0. Also suppose that the discriminant, d, is negative, where

$$d = b^2 - 4ac$$

Write the general quadratic equation in factored form.

Solution

Let's begin with the quadratic formula we obtained earlier in this chapter for cases of this sort:

$$x = [-b \pm j(|d|^{1/2})] \;/\; (2a)$$

We can break the numerator apart and write this as the sum of two fractions with a common denominator:

$$x = -b/(2a) \pm j(|d|^{1/2})/(2a)$$

The roots can be expressed separately like this:

$$x = -b/(2a) + j(|d|^{1/2})/(2a) \text{ or } x = -b/(2a) - j(|d|^{1/2})/(2a)$$

This suggests that the factored form of the equation ought to be

$$\{x - [-b/(2a) + j(|d|^{1/2})/(2a)]\}\{x - [-b/(2a) - j(|d|^{1/2})/(2a)]\} = 0$$

Here's another challenge!

Multiply out the factored equation we've just obtained, verifying that it's equivalent to the original general quadratic in polynomial standard form.

Solution

We can simplify things if we temporarily rename two of the terms in the binomial. We'll give them their "legitimate" names back later. Let

$$p = -b/(2a)$$

and

$$q = (|d|^{1/2})/(2a)$$

This makes the above factored equation look like this:

$$[x - (p + jq)][x - (p - jq)] = 0$$

To avoid making mistakes with signs, we can rewrite this as

$$[x + (-1)(p + jq)][x + (-1)(p - jq)] = 0$$

and then as

$$[x + (-p) + (-jq)][x + (-p) + jq] = 0$$

Using the multiplication rule for trinomials along with the commutative law for multiplication in some of the terms, we can expand this to

$$\begin{gathered} x^2 + (-px) + jqx \\ + (-px) + (-p)^2 + (-jqp) \\ + (-jqx) + jqp + (-j)(j)q^2 = 0 \end{gathered}$$

which simplifies to

$$x^2 + (-2px) + p^2 + q^2 = 0$$

Now let's substitute the "legitimate" names for p and q back into this equation. That gives us

$$x^2 + \{-2[-b/(2a)]x\} + [-b/(2a)]^2 + [(|d|^{1/2})/(2a)]^2 = 0$$

which simplifies to

$$x^2 + (b/a)x + b^2/(4a^2) + |d|/(4a^2) = 0$$

Now remember that $d = b^2 - 4ac$. Because d is negative, its absolute value is equal to its additive inverse. The additive inverse of a difference (subtraction) can be obtained by switching the order of the terms on either side of the minus sign. Therefore,

$$|d| = 4ac - b^2$$

Let's substitute the quantity $(4ac - b^2)$ for $|d|$ in the previous equation, getting

$$x^2 + (b/a)x + b^2/(4a^2) + (4ac - b^2)/(4a^2) = 0$$

We can combine the last two terms of this polynomial into a single fraction because we have the common denominator $4a^2$. That gives us

$$x^2 + (b/a)x + (b^2 + 4ac - b^2)/(4a^2) = 0$$

which simplifies to

$$x^2 + (b/a)x + (4ac)/(4a^2) = 0$$

and further to

$$x^2 + (b/a)x + c/a = 0$$

Multiplying through by *a*, we obtain

$$ax^2 + bx + c = 0$$

which is the original general quadratic in polynomial standard form.

Practice Exercises

This is an open-book quiz. You may (and should) refer to the text as you solve these problems. Don't hurry! You'll find worked-out answers in App. C. The solutions in the appendix may not represent the only way a problem can be figured out. If you think you can solve a particular problem in a quicker or better way than you see there, by all means try it! Note: Some of these problems take you a little beyond the material covered directly in the text of this chapter. Nevertheless, you should be able to solve them using the rules and techniques you've been taught so far. Be patient, and be careful with the signs and *j* operators!

1. What are the roots of the following quadratic equation? What is the solution set *X*?

$$(x - j7)(x + j7) = 0$$

2. Convert the equation given in Prob. 1 into polynomial standard form.
3. Use the quadratic formula to find the roots of the polynomial equation as it is expressed in the solution to Prob. 2.
4. We haven't dealt with quadratic equations in which the roots are pure imaginary but are *not* additive inverses. However, we can "manufacture" such an equation. Suppose we want the roots to be

$$x = j7 \text{ or } x = -j3$$

Write down the binomial factor form of a quadratic with these two roots.

5. Convert the equation from the solution to Prob. 4 into polynomial standard form.
6. Use the quadratic formula to find the roots of the polynomial equation as it is expressed in the solution to Prob. 5.
7. What are the roots of the following quadratic equation? What is the solution set *X*?

$$(x + 2 + j3)(x - 2 - j3) = 0$$

8. Convert the equation given in Prob. 7 into polynomial standard form.
9. Plug the roots from the solution to Prob. 7 into the polynomial standard equation from the solution to Prob. 8, showing that those roots actually work.
10. Convert the following quadratic into polynomial standard form:

$$(j2x + 2 + j3)(-j5x + 4 - j5) = 0$$

CHAPTER

24

Graphs of Quadratic Functions

When a quadratic equation is in polynomial standard form but the 0 on the right is replaced by another variable and then the sides are transposed, we have a *quadratic function*. For example,

$$x^2 + 2x + 1 = 0$$

is a quadratic equation in x, but

$$y = x^2 + 2x + 1$$

is a quadratic function with the independent variable x and the dependent variable y.

The real roots (if any) of a quadratic equation are called *zeros* when we talk about the associated quadratic function. The zeros represent the values of the independent variable for which the dependent variable equals 0.

Two Real Zeros

A quadratic equation with real coefficients and a real constant can have two real roots, one real root, or no real roots. When a quadratic equation has two real roots, its associated quadratic function has two real zeros. When we graph such a function, it crosses the independent-variable axis at two distinct points.

The parabola

When a quadratic function has real coefficients and a real constant, its graph is a curve called a *parabola*. Figure 24-1 shows several such graphs. All parabolas have a characteristic shape that's easy to recognize. Some are "narrow" and others are "broad," but the general contour is the same in them all. Some "hold water"; they are said to *open upward* or be *concave upward*. Others "spill water"; they are said to *open downward* or be *concave downward*. The graph of a quadratic function always passes the "vertical-line test."

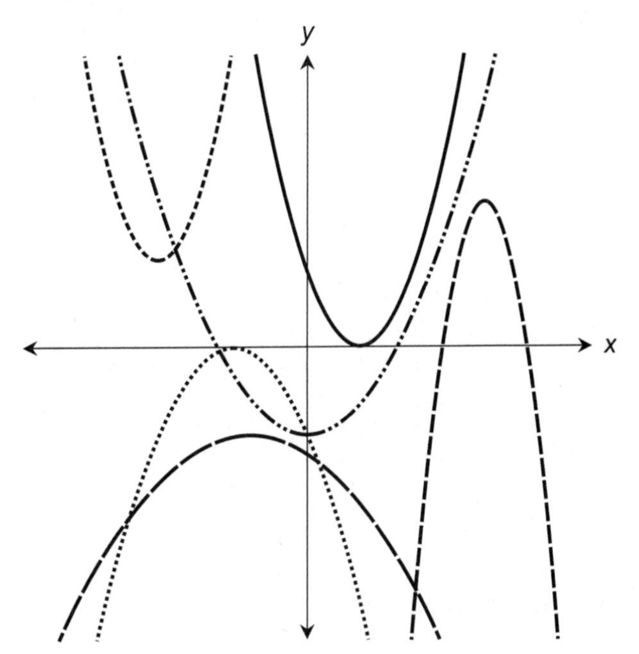

Figure 24-1 The graphs of quadratic functions with real coefficients and a real constant are always parabolas, and they always pass the "vertical-line test" for a function.

Parabola opens upward

Figure 24-2 shows a generic graph of a quadratic function of x with two real zeros, which we call r and s. At the points on the x axis where $x = r$ and $x = s$, the parabola crosses. This parabola opens upward. Imagine that its quadratic function is

$$y = ax^2 + bx + c$$

The x-intercepts are r and s. They represent the roots of the equation

$$ax^2 + bx + c = 0$$

so they can be found using the quadratic formula. If r is the smaller of the two zeros and s is the larger, then

$$r = [-b - (b^2 - 4ac)^{1/2}] / (2a)$$

and

$$s = [-b + (b^2 - 4ac)^{1/2}] / (2a)$$

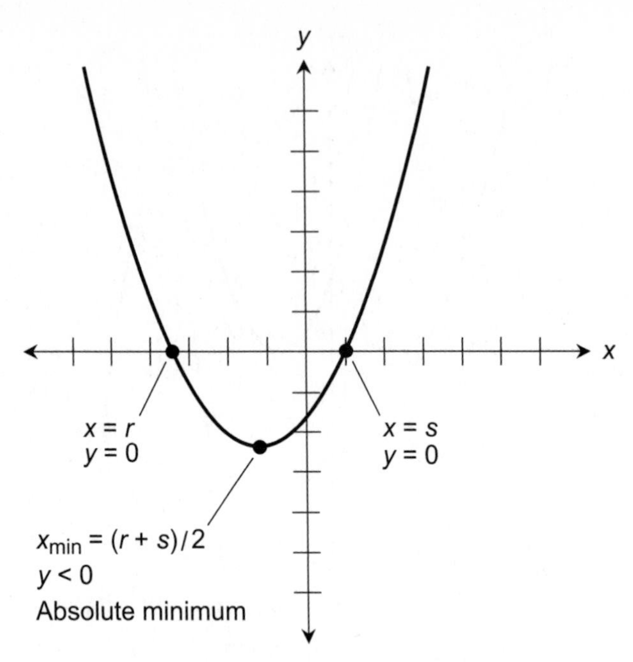

Figure 24-2 Graph of a quadratic function with two real zeros when the coefficient of x^2 is positive. The parabola opens upward, crosses the x axis twice, and has an absolute minimum with $y < 0$.

Whenever a parabola opens upward, it has a single point at which it "bottoms out." This point is called the *absolute minimum* of the graph. In a few moments, we'll discover how this point can be located.

Parabola opens downward

Figure 24-3 shows a generic graph of a quadratic function of x with two real zeros, again called r and s. But this parabola opens downward. If r is the smaller of the two zeros and s is the larger, then

$$r = [-b + (b^2 - 4ac)^{1/2}] / (2a)$$

and

$$s = [-b - (b^2 - 4ac)^{1/2}] / (2a)$$

Note the subtle difference between these two equations and those for the case where the parabola opens upward! Because of this, the values of r and s are reversed, as compared to their

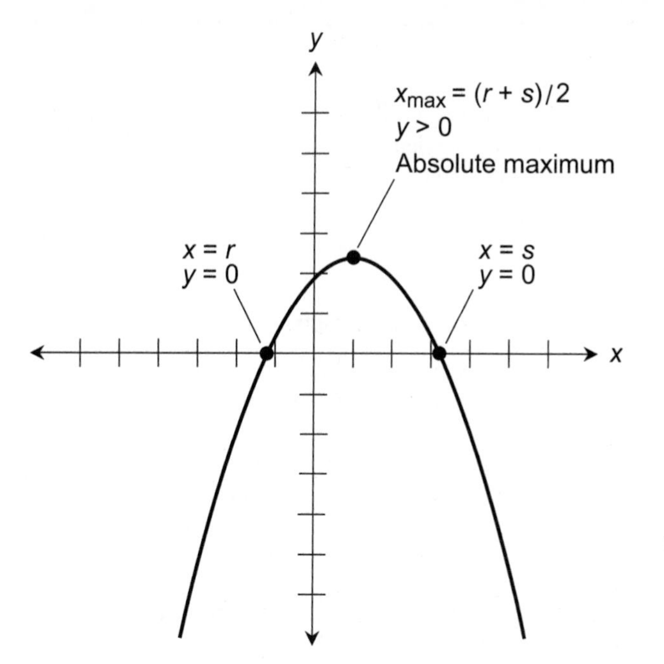

Figure 24-3 Graph of a quadratic function with two real zeros when the coefficient of x^2 is negative. The parabola opens downward, crosses the x axis twice, and has an absolute maximum with $y > 0$.

values when the parabola opens upward. When a parabola opens downward, it has a single point at which it "peaks." This point is called the *absolute maximum.*

Sometimes the term *extremum* is used in reference to an absolute maximum or an absolute minimum in the graph of a function. It means, as you might guess, "absolute extreme value." In a parabola, the extremum may also be called the *vertex.*

Alternative function notation

Sometimes a quadratic function is given a name such as f, and then its value is denoted by that name with the independent variable in parentheses afterward. For example, we might write

$$f(x) = x^2 + 2x + 1$$

instead of

$$y = x^2 + 2x + 1$$

Are you confused?

By now, you might ask, "How can we can tell whether the parabola for a particular quadratic function opens upward or downward?" This is easy to figure out. If the coefficient of x^2 in the polynomial is larger than 0 (that is, if $a > 0$ in the general form of the function), then the parabola opens upward. If $a < 0$, then the parabola opens downward. If $a = 0$, then we don't have a quadratic function at all, and its graph is not a parabola.

Finding the absolute minimum or maximum

When we know the x-intercepts of a quadratic function with two real zeros, it's easy to find the x-value of the absolute minimum or maximum of its graph. It's the arithmetic mean, or average, of the zeros r and s.

If $a > 0$ in the polynomial, then the parabola opens upward, and we have an absolute minimum somewhere. Let's call its x-value x_{min}. Then

$$x_{min} = (r + s)/2$$

If $a < 0$, then the parabola opens downward, and it has an absolute maximum. If we call its x-value x_{max}, then again

$$x_{max} = (r + s)/2$$

To find the y-value of the absolute minimum or maximum, we can plug the x-value into the function once we've found it. A little arithmetic will give us the y-value. Then we can plot the point on a coordinate grid.

When we have plotted the two x-intercepts along with the extremum, we can draw a fair approximation of the parabola representing the quadratic function.

Here's a challenge!

Consider the following quadratic function:

$$y = x^2 - 3x + 2$$

Determine whether the parabola for this function opens upward or downward. Then find the two real zeros, r and s. After that, find the x-value of the extremum. Then determine its y-value. Finally, plot the zeros and the absolute minimum or maximum, and draw an approximate curve through these three points that represents the graph of the function.

Solution

The coefficient of x^2 is positive, so we know that the parabola opens upward. We're lucky here because the polynomial equation factors into

$$(x - 1)(x - 2) = 0$$

The roots here are easily seen as $x = 1$ or $x = 2$. That means $r = 1$ and $s = 2$. Because the parabola opens upward, we should look for an absolute minimum. Its x-value is

$$\begin{aligned} x_{\min} &= (r + s)/2 \\ &= (1 + 2)/2 \\ &= 3/2 \end{aligned}$$

To find the y-value, we plug 3/2 into the quadratic function and grind out the arithmetic:

$$\begin{aligned} y &= (3/2)^2 - 3 \times 3/2 + 2 \\ &= 9/4 - 9/2 + 2 \\ &= 9/4 - 18/4 + 8/4 \\ &= (9 - 18 + 8) / 4 \\ &= -1/4 \end{aligned}$$

Now we know that the coordinates of the absolute minimum are (3/2, −1/4). We also know that the points (1, 0) and (2, 0) lie on the parabola. Figure 24-4 shows these points. They're close together, so it's difficult

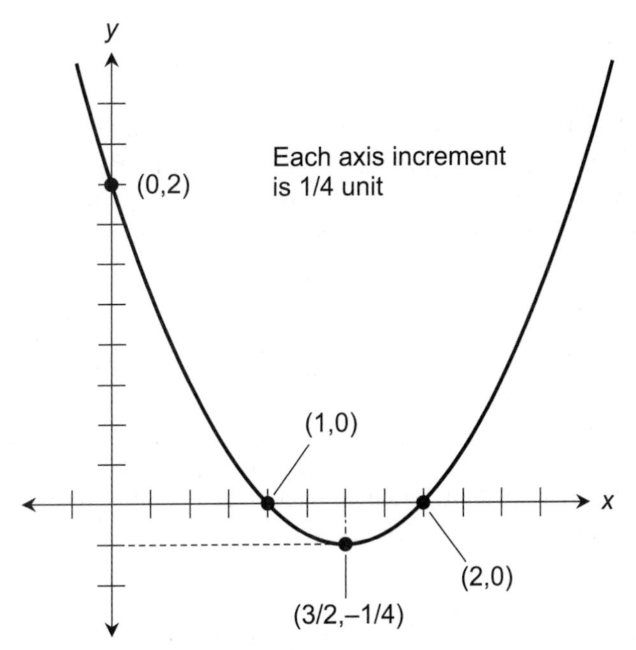

Figure 24-4 Approximate graph of $y = x^2 - 3x + 2$, where the independent variable is x and the dependent variable is y. On both axes, each increment represents 1/4 unit.

to get a clear picture of the parabola based on their locations. But we can find the y-intercept to help us draw the curve. When we plug the value 0 in for x, we get

$$\begin{aligned} y &= x^2 - 3x + 2 \\ &= 0^2 - 3 \times 0 + 2 \\ &= 0 - 0 + 2 \\ &= 2 \end{aligned}$$

This tells us that (0, 2) is on the curve. It is also shown in Fig. 24-4.

Here's a trick!

The graph of a quadratic function is *bilaterally symmetric* with respect to a vertical line passing through the vertex. That means the right-hand side of the curve is an exact "mirror image" of the left-hand side. In the challenge we just finished, this fact can be useful. Once we've drawn an approximation of the left-hand side of the curve using the points (3/2, −1/4), (1, 0), and (0, 2), we can fill out the right-hand side without having to find another distant point there.

One Real Zero

When a quadratic equation has one real root, its quadratic function has one real zero. When we graph such a function, it has a single point in common with the independent-variable axis.

Parabola opens upward

Figure 24-5 shows a generic graph of a quadratic function of x with one real zero r. At the point $(r, 0)$, the parabola is *tangent* to the x axis. Here, "tangent" means that the curve just brushes against the axis, touching it at a single point. If the quadratic function is

$$y = ax^2 + bx + c$$

then $a > 0$ because the parabola opens upward. The root of the equation

$$ax^2 + bx + c = 0$$

can be found using the quadratic formula. Because there is only one root, the discriminant in the quadratic formula must be 0. That means

$$b^2 - 4ac = 0$$

Therefore

$$\begin{aligned} r &= [-b \pm (b^2 - 4ac)^{1/2}] \,/\, (2a) \\ &= -b/(2a) \end{aligned}$$

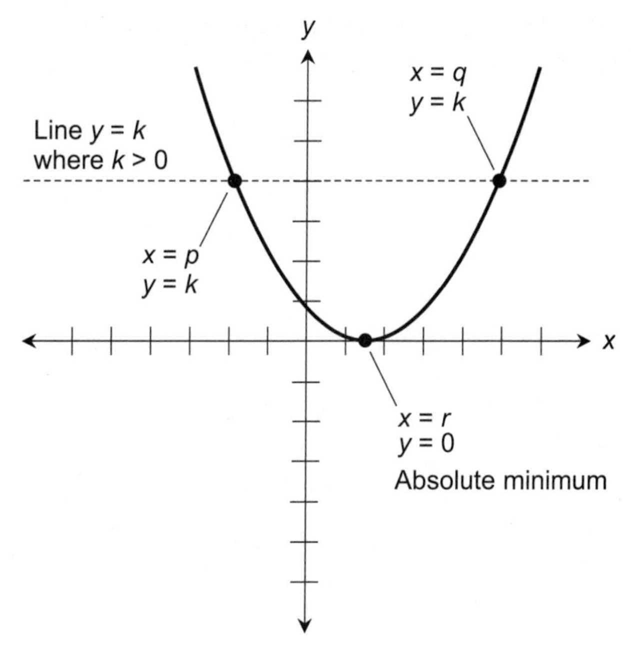

Figure 24-5 Graph of a quadratic function with one real zero when the coefficient of x^2 is positive. The parabola opens upward, is tangent to the x axis at a single point, and has an absolute minimum with $y = 0$. A line with the equation $y = k$, where $k > 0$, intersects the parabola twice.

The absolute minimum is at the point $(r, 0)$. The parabola seems to "sit upon" the x axis.

Parabola opens downward

Figure 24-6 illustrates the graph of another generic quadratic function with one real zero. Let's call it r, as before. Also as before, the point $(r, 0)$ on the parabola is tangent to the x axis. If the quadratic function is

$$y = ax^2 + bx + c$$

then $a < 0$ because the parabola opens downward. The root of the equation

$$ax^2 + bx + c = 0$$

can again be found using the quadratic formula. And just as before, the discriminant in the quadratic formula is equal to 0, so

$$r = -b/(2a)$$

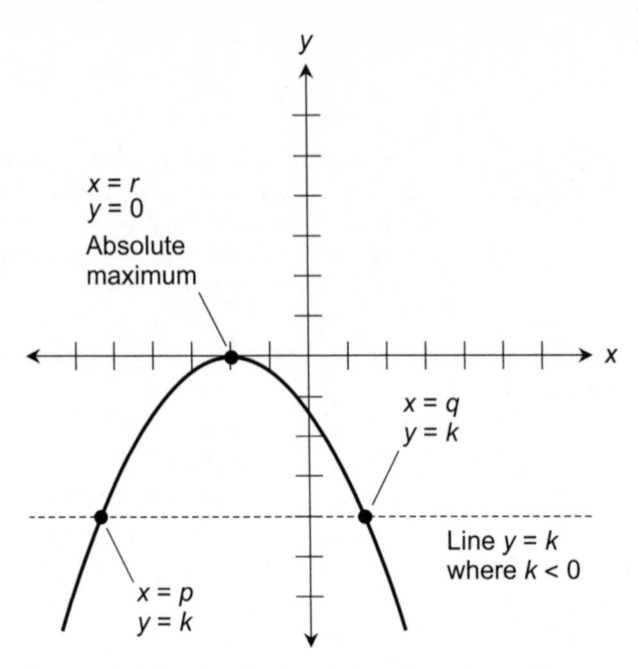

Figure 24-6 Graph of a quadratic function with one real zero when the coefficient of x^2 is negative. The parabola opens downward, is tangent to the x axis at a single point, and has an absolute maximum with $y = 0$. A line with the equation $y = k$, where $k < 0$, intersects the parabola twice.

The absolute maximum is at the point $(r, 0)$. The graph looks as if the parabola "hangs from" the x axis.

Are you confused?

In Figs. 24-5 and 24-6, you'll notice horizontal lines with the equation $y = k$. You might ask, "Why are the lines there? How do we find the points where the lines intersect the parabolas? Why do we need the points?" The answer: "Curve construction!" The lines allow us to find points that help us draw approximations of the graphs, once we've found the zero point $(r, 0)$ and have figured out whether the parabola opens upward or downward. If the parabola opens upward, we should choose a positive number for k. If the parabola opens downward, we should choose a negative number for k. Then we'll know that the line $y = k$ must intersect the parabola at two points. We can find the x-values of those points by letting $y = k$ in the quadratic function, and then "cooking up" a new quadratic equation out of that:

$$ax^2 + bx + c = k$$

This can be rearranged to get

$$ax^2 + bx + (c - k) = 0$$

which we can solve with the quadratic formula. The roots of this quadratic, which we'll call p and q, are the x-values of the two points where the parabola intersects the horizontal line $y = k$. By doing all this, we learn the coordinates of three points that lie on the parabola. Those points are (p, k), (q, k), and $(r, 0)$. If we've chosen the value of k wisely (made a lucky guess), we'll have three well-separated points, and we'll have an easy time drawing an approximation of the parabola.

Here's a challenge!

Look at the following quadratic function:

$$y = -4x^2 + 12x - 9$$

Find the zero or zeros. Determine whether the parabola for this function opens upward or downward. Find the x-value and the y-value of the extremum. Then find two more points so the curve can be drawn. Finally, draw an approximation of the curve.

Solution

We can find the zero or zeros of the function by applying the quadratic formula to the equation

$$-4x^2 + 12x - 9 = 0$$

In the general polynomial equation

$$ax^2 + bx + c = 0$$

we have $a = -4$, $b = 12$, and $c = -9$. When we plug these into the quadratic formula, we get

$$\begin{aligned} x &= [-b \pm (b^2 - 4ac)^{1/2}] / (2a) \\ &= \{-12 \pm [12^2 - 4 \times (-4) \times (-9)]^{1/2}\} / [2 \times (-4)] \\ &= [-12 \pm (144 - 144)^{1/2}] / (-8) \\ &= (-12 \pm 0^2) / (-8) \\ &= -12/(-8) \\ &= 3/2 \end{aligned}$$

The equation has only one root, so the function has only one zero. If we call it r, then we have $r = 3/2$, and we know that the point $(3/2, 0)$ is on the parabola. We also know that this point represents the extremum. The value of a, the coefficient of x^2 in the polynomial, is negative. That means the parabola opens downward, so $(3/2, 0)$ is the absolute maximum, and the parabola "hangs from the x axis." We can choose a negative number k and set it equal to y, getting a horizontal line that crosses the parabola twice. Let's try -5 for k, as shown in Fig. 24-7. The equation of the line is $y = -5$.

To find the x-values of the points where the line crosses the parabola, we have a new quadratic equation to solve:

$$-4x^2 + 12x - 9 = -5$$

We can put this into polynomial standard form by adding 5 to each side, getting

$$-4x^2 + 12x - 4 = 0$$

The quadratic formula tells us that the roots of this equation are

$$\begin{aligned} x &= [-b \pm (b^2 - 4ac)^{1/2}] / (2a) \\ &= \{-12 \pm [12^2 - 4 \times (-4) \times (-4)]^{1/2}\} / [2 \times (-4)] \\ &= [-12 \pm (144 - 64)^{1/2}] / (-8) \\ &= [-12 \pm 80^{1/2}] / (-8) \end{aligned}$$

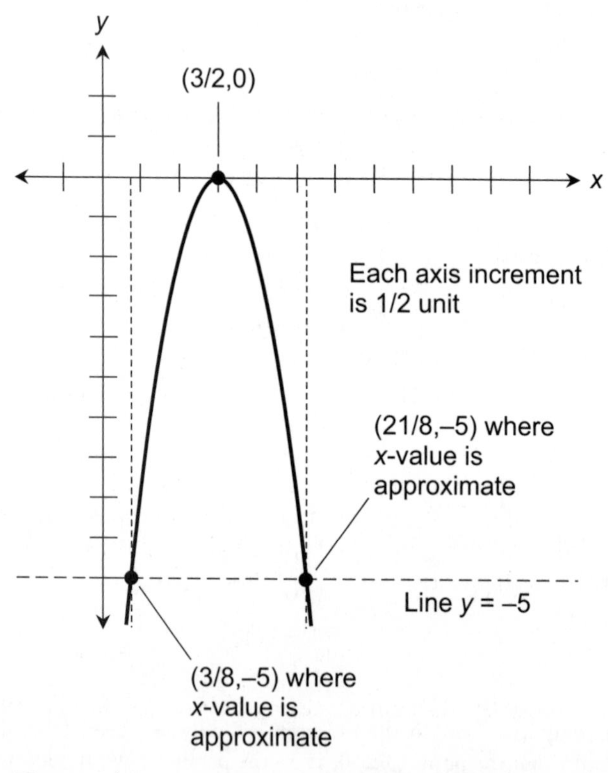

Figure 24-7 Approximate graph of $y = -4x^2 + 12x - 9$, where the independent variable is x and the dependent variable is y. On both axes, each increment represents 1/2 unit.

Our goal is to find the coordinates of points so we can draw an *approximation* of the parabola. We can't get perfection with pencil and paper, so we don't need exact values for the coordinates. If we can get within 1/10 of a unit, that ought to be good enough. When we use a calculator to find $80^{1/2}$, we get roughly 8.944. That's so close to 9 that it won't make a difference, when we draw the curve on paper, if we call it 9. So let's continue with our arithmetic, substituting 9 in for $80^{1/2}$, as follows:

$$x = (-12 \pm 9) / (-8)$$
$$= -3/(-8) \text{ or } -21/(-8)$$
$$= 3/8 \text{ or } 21/8$$

These represent the approximate x-values of the points where the horizontal line intersects the parabola in Fig. 24-7. The y-values are exactly equal to −5 for both points. We can plot the points as (3/8, −5) and (21/8, −5). We've already determined that (3/2, 0) is on the curve, and that it represents the absolute maximum. With these three points put down on our graph paper, we can fill in an approximation of the parabola.

No Real Zeros

When a quadratic equation has no real roots and its associated quadratic function has no real zeros, its graph is a parabola, but the curve lies entirely on one side of the independent-variable axis.

Parabola opens upward

Figure 24-8 is a generic graph of a quadratic function of x with no real zeros. The parabola does not cross the x axis anywhere. The curve has an absolute minimum that lies in the first or second quadrant. If the quadratic function is

$$y = ax^2 + bx + c$$

then $a > 0$ in this example, because the parabola opens upward.

It's reasonable to wonder, "Where do we start locating points on the curve in situations like this?" The task doesn't look easy at first, but we can find the x-value of the absolute minimum point. If we call that value x_{min}, then

$$x_{min} = -b/(2a)$$

The y-value of the absolute minimum point, y_{min}, is whatever we get when we plug x_{min} into the function:

$$y_{min} = ax_{min}^2 + bx_{min} + c$$

Once we've found the coordinates of the absolute minimum, we can find the coordinates of two other points. We can pick a number p somewhat smaller than x_{min}, and we can pick

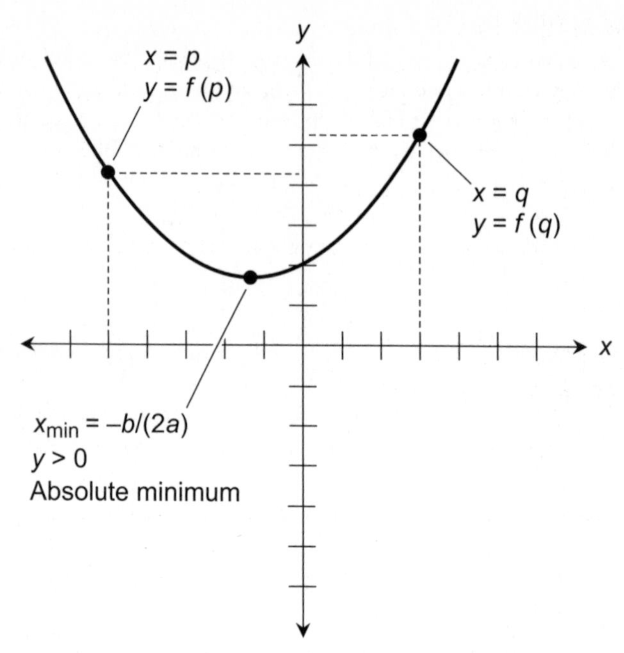

Figure 24-8 Graph of a quadratic function with no real zeros when the coefficient of x^2 is positive. The parabola opens upward, does not cross the x axis, has an absolute minimum with an x-value equal to $-b/(2a)$, and has a positive y-value.

another number q somewhat larger than x_{min}. We can choose integers for these numbers to make the arithmetic easy. The numbers p and q don't have to be equally smaller and larger than x_{min}, although we should try to get close to that ideal. Once we've chosen the numbers p and q, we can plug them into the function for x and find two more points on the curve. Then the parabola is easy to draw.

Are you astute?

Do you wonder why we didn't use the "x-value first" point-finding strategy earlier in this chapter for quadratic functions with one or two real zeros? "Isn't it easier," you might ask, "to choose a clean integer x-value for a point, and figure the y-value by plugging into the function? Isn't that better than picking a y-value and then grinding through the quadratic formula to get two x-values that are likely to come out irrational?" You tell me! By choosing x-values first, we have to go through the arithmetic twice, but it's usually simple. By choosing the y-value first, the arithmetic can be a little rough, but we only have to do it once. It's your choice. Either method will work fine.

Parabola opens downward

Figure 24-9 is another generic graph of a quadratic function of x with no real zeros. Again, the parabola does not cross the x axis. The curve has an absolute maximum in the third or fourth quadrant. If the function is

$$y = ax^2 + bx + c$$

then $a < 0$ because the parabola opens downward. The x-value of the absolute maximum point, x_{max}, is

$$x_{max} = -b/(2a)$$

The y-value of the absolute maximum point, y_{max}, is

$$y_{max} = ax_{max}^{\ 2} + bx_{max} + c$$

Once we've found the coordinates of the vertex, we can find the coordinates of two other points. We can pick a number p smaller than x_{max}, and we can pick another number q larger than x_{max}. We can then plug p and q into the function for x to find two more points on the curve. Then the parabola is easy to draw.

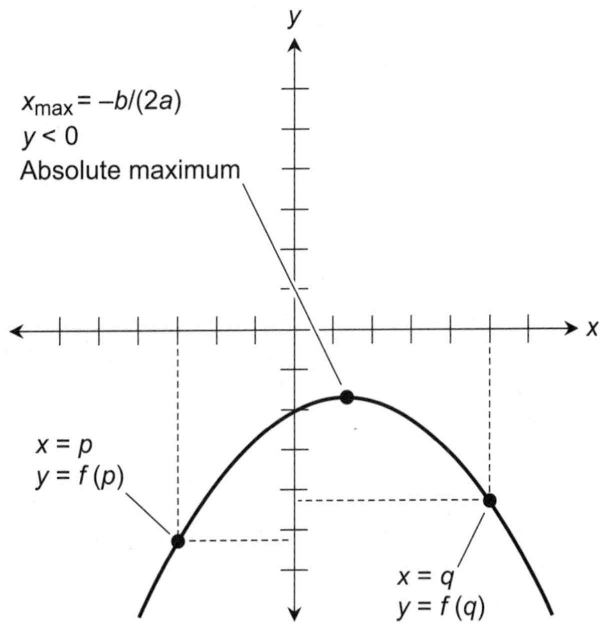

Figure 24-9 Graph of a quadratic function with no real zeros when the coefficient of x^2 is negative. The parabola opens downward, does not cross the x axis, has an absolute maximum with an x-value equal to $-b/(2a)$, and has a negative y-value.

Are you confused?

We've made a substantial claim: The x-value of the extremum for a quadratic function

$$y = ax^2 + bx + c$$

is equal to $-b/(2a)$ when there are no real zeros. This holds true whether the parabola opens upward or downward. It's also true for quadratic functions that have one or two real zeros. Do you wonder how we know this? I haven't proven it or even derived it in a general way.

This fact comes out of a maneuver that requires *differential calculus*. We won't get into calculus in this book, but the approach can be explained qualitatively. It involves finding the point on the parabola where a line tangent to the curve has a slope of 0. That's the point where the curve is "locally horizontal," and it's always the vertex in a quadratic function with real coefficients and a real constant. If you look at all the graphs in this chapter, you'll see that a straight line drawn tangent to the curve, and passing through the vertex, is always horizontal. Differential calculus gives us a formula for the slope m of a quadratic function at any point based on the x-value at that point:

$$m = 2ax + b$$

If we set $m = 0$ to make the slope horizontal, we get

$$0 = 2ax + b$$

That's a first-degree equation in x. When we solve it, we get

$$x = -b/(2a)$$

Here's a challenge!

Consider the following quadratic function with no real zeros:

$$y = 2x^2 + 4x + 3$$

Find out whether the parabola for this function opens upward or downward. After that, find the x-value and the y-value of the extremum. Then locate two other points on the curve, and draw an approximation of the parabola representing the function.

Solution

This parabola opens upward, because the coefficient of x^2 is positive. That means it has an absolute minimum. Remember again the general polynomial form for a quadratic function:

$$y = ax^2 + bx + c$$

The x-value at the absolute minimum point is

$$\begin{aligned} x_{min} &= -b/(2a) \\ &= -4 / (2 \times 2) \\ &= (-4)/4 \\ &= -1 \end{aligned}$$

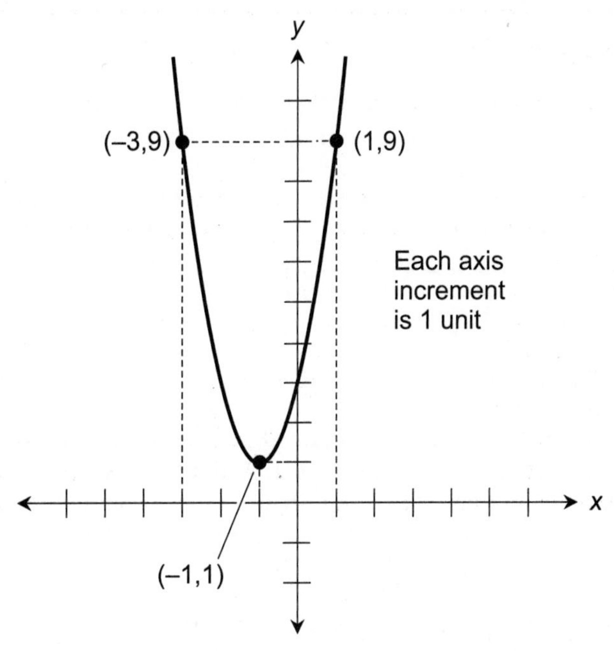

Figure 24-10 Approximate graph of $y = 2x^2 + 4x + 3$. On both axes, each increment represents 1 unit.

The y-value at the absolute minimum point is

$$\begin{aligned} y_{min} &= 2x_{min}^2 + 4x_{min} + 3 \\ &= 2 \times (-1)^2 + 4 \times (-1) + 3 \\ &= 2 - 4 + 3 \\ &= -2 + 3 \\ &= 1 \end{aligned}$$

From this, we know that the coordinates of the vertex are (−1,1). As the basis for our next point, let's choose $x = -3$. That's 2 units smaller than the x-value of the absolute minimum. We can plug that into the function to get

$$\begin{aligned} y &= 2x^2 + 4x + 3 \\ &= 2 \times (-3)^2 + 4 \times (-3) + 3 \\ &= 18 - 12 + 3 \\ &= 6 + 3 \\ &= 9 \end{aligned}$$

This gives us (−3, 9) as the coordinates of a second point on the curve. Finally, let's pick an x-value that's 2 units larger than x_{min}; that would be $x = 1$. Plugging it in, we obtain

$$\begin{aligned} y &= 2x^2 + 4x + 3 \\ &= 2 \times 1^2 + 4 \times 1 + 3 \\ &= 2 + 4 + 3 \\ &= 9 \end{aligned}$$

The third point is therefore (1, 9). We now have three points on the curve: (−3, 9), (−1, 1), and (1, 9). Figure 24-10 shows these points along with an approximation of the parabola passing through them.

Practice Exercises

This is an open-book quiz. You may (and should) refer to the text as you solve these problems. Don't hurry! You'll find worked-out answers in App. C. The solutions in the appendix may not represent the only way a problem can be figured out. If you think you can solve a particular problem in a quicker or better way than you see there, by all means try it!

1. Examine this quadratic function:

$$y = (x - 3)(4x - 1)$$

Does the parabola representing the graph of this function open upward or downward?

2. What are the real zeros, if any, of the function stated in Prob. 1?
3. What are the coordinates of the point representing the extremum of the function stated in Prob. 1?
4. Draw an approximate graph of the function stated in Prob. 1.
5. The graph of the following quadratic function lies entirely above the x axis. How can we know this without plotting any points?

$$y = 7x^2 + 5x + 2$$

6. What are the coordinates of the vertex point on the parabola representing the function stated in Prob. 5?
7. Consider the following quadratic function:

$$y = -2x^2 + 2x - 5$$

Does the parabola representing the graph of this function open upward or downward?

8. What are the real zeros, if any, of the function stated in Prob. 7?
9. What are the coordinates of the point representing the vertex of the parabola for the function stated in Prob. 7?
10. Draw an approximate graph of the function stated in Prob. 7.

CHAPTER

25

Cubic Equations in Real Numbers

Let's move into *single-variable cubic equations,* also called *cubics* or *third-degree equations.* This type of equation always has a term in which the variable is cubed. There may also be a term with the variable squared, a term with the variable itself (to the first power), and a stand-alone constant. We'll be concerned only with the real-number roots of single-variable cubics having real coefficients and a real constant.

Cube of Binomial

Some cubics can be expressed as the cube of a binomial with a real coefficient and a real constant. Cubics in this form are easy to solve. They have one real root, which can be derived from the binomial by setting it equal to zero.

Binomial-cubed form

Suppose x is a variable, a is the nonzero real-number coefficient of the variable, and b is a real-number constant. Consider the expression

$$(ax + b)^3$$

If we set this equal to 0, we get

$$(ax + b)^3 = 0$$

This is a cubic equation in *binomial-cubed form.* Here are three examples:

$$x^3 = 0$$
$$(x + 3)^3 = 0$$
$$(2x - 3)^3 = 0$$

In the first case, the stand-alone constant is 0. If we want to be formal, we can write

$$(x + 0)^3 = 0$$

In the second case, the coefficient of x in the binomial is 1 and the stand-alone constant is 3. In the third case, the coefficient of x in the binomial is 2, and the constant is -3.

Multiplying out

Now look at the last equation shown in the above set of three. We can multiply it out so the left side becomes a polynomial. First, we can rewrite it as

$$(2x - 3)(2x - 3)(2x - 3) = 0$$

Then we can multiply the second two factors using the product of sums rule. (Remember that subtraction is the same as the addition of a negative.) The result is

$$(2x - 3)(4x^2 - 12x + 9) = 0$$

If we multiply again using the expanded product of sums rule, consolidate the terms for x^2 and x, and pay close attention to the signs, we get

$$8x^3 - 36x^2 + 54x - 27 = 0$$

What's the real root?

A single-variable equation in binomial-cubed form has one real root, assuming the coefficient and the constant are both real numbers. That root can be found by taking away the exponent from the binomial, and then solving the first-degree equation that remains. In general, if we have the third-degree equation

$$(ax + b)^3 = 0$$

then the real root is the solution to

$$ax + b = 0$$

That solution is obtained by subtracting b from both sides, and then dividing through by a. We can get away with division by a, because $a \neq 0$. Therefore, the single real root to the cubic is

$$x = -b/a$$

Are you confused?

Simple cubics can sometimes look complicated, and complicated cubics can sometimes look simple. We often cannot know by glancing at a third-degree equation whether solving it will be easy, challenging, or difficult. The following example can illustrate.

Here's a challenge!

Solve the following cubic equation, and discover that the real-number root is an integer:

$$(3^{1/2}x - 12^{1/2})^3 = 0$$

Solution

Remember that the 1/2 power of a number is the positive square root. The above equation is therefore not ambiguous. It's a legitimate cubic equation in x, because the 1/2 powers involve only the coefficient and the constant, not the variable x itself. Both the coefficient and the stand-alone constant are irrational, so this equation looks difficult! But for the moment, let's forget about that. We can solve it using the above general formula. We have $a = 3^{1/2}$ and $b = -(12^{1/2})$. Therefore

$$\begin{aligned} x &= -b/a \\ &= -[-(12^{1/2})] / 3^{1/2} \\ &= 12^{1/2} / 3^{1/2} \end{aligned}$$

Using the power of quotient rule from Chap. 9, we can simplify this root and then reduce it to an integer:

$$\begin{aligned} x &= 12^{1/2} / 3^{1/2} \\ &= (12/3)^{1/2} \\ &= 4^{1/2} \\ &= 2 \end{aligned}$$

We should check to be sure that this root really works in the original equation! Let's plug it in and find out:

$$(3^{1/2}x - 12^{1/2})^3 = 0$$
$$(3^{1/2} \times 2 - 12^{1/2})^3 = 0$$
$$(3^{1/2} \times 4^{1/2} - 12^{1/2})^3 = 0$$

Using the power of product rule from Chap. 9, we can simplify this to

$$[(3 \times 4)^{1/2} - 12^{1/2}]^3 = 0$$
$$(12^{1/2} - 12^{1/2})^3 = 0$$
$$0^3 = 0$$
$$0 = 0$$

Three Binomial Factors

When a cubic equation can be expressed as three binomial factors, those factors are rarely all the same. Two of them might be identical, but often all three are different. Cubics of this sort are in *binomial-factor form.*

Binomial-factor form

Here's the general form of a cubic broken down into three binomial factors, assuming that the equation can be expressed that way with real coefficients and real constants:

$$(a_1x + b_1)(a_2x + b_2)(a_3x + b_3) = 0$$

The three coefficients are a_1, a_2, and a_3. They must all be nonzero, so the multiplied-out equation contains a nonzero multiple of x^3. The three stand-alone constants are b_1, b_2, and b_3. Here are some examples of binomial-factor cubics:

$$(x - 1)(x + 2)(x - 3) = 0$$
$$(3x + 2)(5x + 6)(-7x - 1) = 0$$
$$x(x + 6)(x + 8) = 0$$
$$x^2(-4x - 1) = 0$$

The third and fourth of these equations have one and two stand-alone constants equal to 0, respectively. In "unabridged" binomial-factor form, these equations are

$$(x + 0)(x + 6)(x + 8) = 0$$

and

$$(x + 0)(x + 0)(-4x - 1) = 0$$

Multiplying out

We're fortunate if we can reduce a cubic to three binomial factors. Consider the equation

$$x^3 - 2x^2 - 5x + 6 = 0$$

This doesn't advertise that it can be factored into a product of binomials! But try multiplying this out:

$$(x - 1)(x + 2)(x - 3) = 0$$

Let's work an example backward. Start with this:

$$x(x + 6)(x + 8) = 0$$

If we multiply the first factor by the second, we obtain

$$(x^2 + 6x)(x + 8) = 0$$

Multiplying these two factors out gives us

$$x^3 + 14x^2 + 48x = 0$$

Imagine that we were presented with the above equation for the first time, and we had never seen it in factored form. We could factor x out, getting

$$x(x^2 + 14x + 48) = 0$$

We could try to break up the trinomial part of this equation into binomial factors. With a few trials and errors, we'd find them. We would notice that the stand-alone constants must add up to 14 and multiply to 48; it wouldn't take us long to guess that they are 6 and 8. Once we had the three binomial factors of the cubic, finding the real roots would be easy, as the next example will show. Even if we couldn't factor the trinomial, we could attack it with the quadratic formula. Later, we'll see an example of that tactic.

What are the real roots?

To see how we can solve a binomial-factor cubic with real coefficients and a real constant, let's find the real roots of

$$(3x + 2)(5x + 6)(-7x - 1) = 0$$

To find the first real root, we solve the equation created by setting the first binomial factor equal to 0. That equation is

$$3x + 2 = 0$$

Subtracting 2 from each side and then dividing through by 3, we get $x = -2/3$. To find the second real root, we do the same thing with the second binomial factor. We have

$$5x + 6 = 0$$

Subtracting 6 from each side and then dividing through by 5, we get $x = -6/5$. To find the third real root, we solve the equation

$$-7x - 1 = 0$$

Adding 1 to each side and then dividing through by −7, we get $x = 1/(-7) = -1/7$. The real roots of the original cubic are therefore

$$x = -2/3 \text{ or } x = -6/5 \text{ or } x = -1/7$$

and the solution set X is $\{-2/3, -6/5, -1/7\}$. In Practice Exercises 3 and 4 at the end of this chapter, you'll multiply out the factors of the original cubic, and then test these roots in the resulting equation.

Are you confused?

You might wonder, "Why not divide the following equation through by x in an attempt to simplify and solve it?"

$$x^3 + 14x^2 + 48x = 0$$

That's a reasonable question. Dividing through by x can "knock the equation down" from third degree to second degree—or so we might be tempted to believe. Let's try it and see what happens! We get

$$(x^3 + 14x^2 + 48x) / x = 0/x$$

which apparently simplifies to

$$x^3/x + 14x^2/x + 48x/x = 0$$

and further to

$$x^2 + 14x + 48 = 0$$

This is a quadratic that can be factored into

$$(x + 6)(x + 8) = 0$$

The roots $x = -6$ or $x = -8$ come out of this process. We conclude that the solution set X for the original cubic must be $\{-6, -8\}$. That's easy, isn't it? Not so fast! It's also wrong, because it's incomplete. Let's find out how to get it right.

Here's a challenge!

Solve the following cubic equation in a way that works properly:

$$x^3 + 14x^2 + 48x = 0$$

Solution

We've seen this equation in its binomial factor form. Suppose that we could look at the above equation and see the binomial factor version immediately:

$$x(x + 6)(x + 8) = 0$$

To find the first root, we can take literally the simple first-degree equation

$$x = 0$$

To find the second root, we take the first-degree equation

$$x + 6 = 0$$

which resolves to $x = -6$. To find the third root, we take the first-degree equation

$$x + 8 = 0$$

which resolves to $x = -8$. The real roots of the cubic are therefore

$$x = 0 \text{ or } x = -6 \text{ or } x = -8$$

and the solution set X is $\{0, -6, -8\}$. A new root shows up this time: $x = 0$! The fact that one of the roots is 0 caused us to inadvertently divide by 0 when we divided the equation through by x. This "blinded" us to the existence of that root.

Binomial Times Trinomial

When a cubic can be expressed as a binomial multiplied by a trinomial, the equation is in *binomial-trinomial form.* (Actually, I've never seen that expression used in other texts. But it's easy to remember, don't you think?) A cubic in this form is not particularly difficult to solve for real roots. The technique shown in this section will also reveal the complex-number roots of a cubic equation, if any such roots exist.

Binomial-trinomial form

Suppose that a_1 and a_2 are nonzero real numbers. Also suppose that b_1, b_2, and c are real numbers, any or all of which can equal 0. The binomial-trinomial form of a cubic equation in the variable x can be written as follows:

$$(a_1x + b_1)(a_2x^2 + b_2x + c) = 0$$

Here are some examples of cubics in the binomial-trinomial form:

$$(-4x - 3)(-7x^2 + 6x - 13) = 0$$
$$(3x + 5)(16x^2 - 56x + 49) = 0$$
$$(3x)(4x^2 - 7x - 10) = 0$$
$$(-21x + 2)(3x^2 - 14) = 0$$

In the third case above, the stand-alone constant is 0 in the binomial. In the fourth case, the coefficient of x is 0 in the trinomial.

Multiplying out

Let's take a specific example of a cubic in binomial-trinomial form and multiply it out. Here's a good one, with plenty of sign changes to make it interesting:

$$(-4x - 3)(-7x^2 + 6x - 13) = 0$$

Using the expanded product of sums rule, we obtain

$$28x^3 - 24x^2 + 52x + 21x^2 - 18x + 39 = 0$$

Consolidating the terms for x^2 and x, we get

$$28x^3 - 3x^2 + 34x + 39 = 0$$

What are the real roots?

The process of finding the real roots of a cubic in the binomial-trinomial form is straightforward, as long as all the coefficients and constants are real numbers. First, we "manufacture" a first-degree equation from the binomial, setting it equal to 0. In the general form above, that would be

$$a_1x + b_1 = 0$$

which solves to

$$x = -b_1 / a_1$$

This will always give us one real root for the cubic. After that, we set the trinomial equal to 0, obtaining a quadratic equation. In the general form shown above, we get

$$a_2x^2 + b_2x + c = 0$$

We can find the real roots of this equation, if any exist, using techniques we've already learned for solving quadratics. (I like to use the quadratic formula, because it always works! Also, if the root or roots are complex but not real, the quadratic formula will produce them.) Expressed for the above general equation, the quadratic formula is

$$x = [-b_2 \pm (b_2^2 - 4a_2c)^{1/2}] / (2a_2)$$

Are you confused?

You've learned that a quadratic equation can have two different real roots, or only one real root, or none at all. How about cubics? You've already seen an example of a cubic with three real roots. You're about to see that a cubic equation in the binomial-trinomial form with real coefficients and a real constant can have two real roots. Then you'll discover that a cubic in the binomial-trinomial form with real coefficients and a real constant can have only one real root, along with two others that are complex.

"Okay," you say. Then you ask, "How about *no* real roots?" The answer: *Any* cubic in the binomial-trinomial form with real coefficients and a real constant always has at least one real root. That's because a first-degree equation can always be created from the binomial factor, and that equation always has a real solution. We can take this statement further: A cubic equation, *no matter what the form,* has at least one real root if all the coefficients and constants are real numbers.

Here's a challenge!

Find the real roots of the following cubic equation using the method described in this section, and state the real solution set X.

$$(3x + 5)(16x^2 - 56x + 49) = 0$$

Solution

First, we can construct a first-degree equation by setting the binomial factor equal to 0. That gives us

$$3x + 5 = 0$$

Subtracting 5 from each side and then dividing through by 3, we get $x = -5/3$. Next, we apply the quadratic formula to the trinomial factor. If we let $a_2 = 16$, $b_2 = -56$, and $c = 49$, then

$$\begin{aligned} x &= [-b_2 \pm (b_2{}^2 - 4a_2c)^{1/2}] / (2a_2) \\ &= \{56 \pm [(-56)^2 - 4 \times 16 \times 49]^{1/2}\} / (2 \times 16) \\ &= [56 \pm (3{,}136 - 3{,}136)^{1/2}] / 32 \\ &= 56/32 \\ &= 7/4 \end{aligned}$$

In this case, 7/4 is the only real root of the quadratic we obtain by setting the trinomial factor equal to 0. The original cubic therefore has two real roots:

$$x = -5/3 \text{ or } x = 7/4$$

and the real solution set X is $\{-5/3, 7/4\}$. The root $x = 7/4$ has multiplicity 2. In Practice Exercise 7 at the end of this chapter, you'll check these roots.

Here's another challenge!

Find the real roots of the following cubic equation using the approach described in this section, and state the real solution set X.

$$(-2x + 11)(2x^2 - 2x + 5) = 0$$

Solution

Our first step, as in the previous "challenge," is to create a first-degree equation by setting the binomial factor equal to 0. That gives us

$$-2x + 11 = 0$$

When we subtract 11 from each side and then divide the entire equation through by -2, we get the solution $x = -11/(-2)$, which is equal to 11/2. Then we apply the quadratic formula to the trinomial factor. If we let $a_2 = 2$, $b_2 = -2$, and $c = 5$, we have

$$\begin{aligned} x &= [-b_2 \pm (b_2{}^2 - 4a_2c)^{1/2}] / (2a_2) \\ &= \{2 \pm [(-2)^2 - 4 \times 2 \times 5]^{1/2}\} / (2 \times 2) \\ &= [2 \pm (4 - 40)^{1/2}] / 4 \\ &= [2 \pm (-36)^{1/2}] / 4 \end{aligned}$$

We can stop right here, because the discriminant is negative. The quadratic we obtain by setting the trinomial factor equal to 0 has no real roots (although it has two complex roots). The original cubic therefore has only one real root, $x = 11/2$. The real solution set X contains only that single root, so $X = \{11/2\}$.

Polynomial Equation of Third Degree

Any single-variable cubic equation can be written so it appears in *polynomial standard form.* Often, this is the form you'll first see.

Polynomial standard form

When a cubic equation is in polynomial standard form, the left side of the equals sign contains a third-degree polynomial, and the right side contains 0 alone. Here's the general form in the realm of the real numbers:

$$ax^3 + bx^2 + cx + d = 0$$

where a, b, c, and d are real numbers, $a \neq 0$, and x is the variable. All of the following equations are cubics in this form:

$$x^3 + 3x^2 + 2x + 5 = 0$$
$$-2x^3 + 3x^2 - 4x = 0$$
$$5x^3 - 7x^2 - 5 = 0$$
$$-7x^3 - 4x^2 = 0$$
$$4x^3 + 3x = 0$$
$$-7x^3 - 6 = 0$$

If you set $a = 0$ in the polynomial standard form of a single-variable cubic equation, you end up with the polynomial standard form for a single-variable quadratic, or even a first-degree equation (depending on the other coefficients). However, the coefficients of x^2 or x, as well as the stand-alone constant, can equal 0 in a cubic polynomial.

What can we do?

When we see a cubic equation in polynomial standard form, the roots may be apparent right away, but often they are not. If we come across the equation

$$x^3 - 8 = 0$$

we can see, perhaps without having to do any manipulations, that there's one real root, $x = 2$. But if we encounter

$$x^3 + 3x^2 + 2x + 5 = 0$$

the situation is more challenging. When we see an equation like this, we can try to factor it into a product of binomials, or into binomial-trinomial form. If we can manage to do that, then we can find the roots as described in the previous sections. But factoring a cubic polynomial is not always easy. If the coefficients and the constant of the polynomial equation are all real numbers, a binomial-trinomial expression for the equation *must exist,* but the coefficients and constants might not be integers. They might even turn out to be irrational numbers.

If we see a cubic in polynomial standard form and we find ourselves staring at it, paralyzed with uncertainty, there's some good news, some fair news, and some bad news. First, the good news: We can try to factor a binomial out of a polynomial using a process called *synthetic division.* That often works, but we'll probably have to go through the process several times before we find the right binomial. Now for the fair news: A general formula for solving a cubic equation exists. Finally, the bad news: That formula is so complicated that most mathematicians would rather try every other option first.

Before we see how synthetic division works, we should formalize an important principle.

The binomial factor rule

A little while ago, I made a strong claim: Any cubic equation with real coefficients and a real constant has at least one real root. Suppose we call that real number k. If we plug in k for the variable in the equation—no matter what form that equation happens to be in—and work out the arithmetic, the result will be 0.

Now imagine that we are faced with a cubic equation in polynomial standard form, and we can't figure out how it can be factored. It seems reasonable to suppose that if we can find the real root k, which *must* exist, then the binomial $(x - k)$ can be factored out of the cubic. That's because if we set x equal to k, then the value of $(x - k)$ becomes 0; and with 0 as one of the factors, the whole expression attains the value 0. We can package all this reasoning up into a formal statement called the *binomial factor rule* or the *factor theorem:*

- A real number k is a root of a cubic equation in the variable x if and only if $(x - k)$ is a factor of the cubic polynomial.

Remember what "if and only if" means in logical terms: the "if-then" reasoning works both ways. We can therefore rewrite the above rule as two separate statements:

- If a real number k is a root of a cubic equation in the variable x, then $(x - k)$ is a factor of the cubic polynomial.
- If k is a real number and $(x - k)$ is a factor of a cubic polynomial in the variable x, then k is a real root of the cubic equation.

This rule, which has been proven as a *theorem* by mathematicians, can be generalized to polynomial equations in the fourth degree (called *quartic equations*), the fifth degree (called *quintic equations*), and, in fact, any positive-integer degree (called *nth-degree equations*).

We try, we fail

Now that we're armed with plenty of theoretical facts, it's time to take aim at a real polynomial cubic equation using synthetic division. Let's try this:

$$x^3 - 7x - 6 = 0$$

Remember the general form:

$$ax^3 + bx^2 + cx + d = 0$$

To set up the synthetic division "grid," we write down the coefficients and the stand-alone constant in an array along with "wildcard" entries symbolized by pound signs (#), like this:

#	a	b	c	d
		#	#	#
	#	#	#	#

In our example, we have $a = 1$, $b = 0$, $c = -7$, and $d = -6$. We put those numbers into the top row from left to right, starting immediately to the right of the pound sign:

#	1	0	−7	−6
		#	#	#
	#	#	#	#

Now the guessing game begins! At first, we have no clue as to what the real root might be. Let's try 2 as a "test root." We put that number in the top left slot:

2	1	0	−7	−6
		#	#	#
	#	#	#	#

Next, we copy the second number in the top row to the "wildcard" slot directly beneath it in the bottom row:

2	1	0	−7	−6
		#	#	#
	1	#	#	#

We multiply 2 by 1 and place the product in place of the first wildcard in the second row:

2	1	0	−7	−6
		2	#	#
	1	#	#	#

We add the numbers in the third column, writing the sum in the bottom row:

2	1	0	−7	−6
		2	#	#
	1	2	#	#

We multiply this sum by the "test root" and place the product beneath the entry –7:

2	1	0	–7	–6
		2	4	#
	1	2	#	#

We add the numbers in the fourth column, writing the sum in the bottom row:

2	1	0	–7	–6
		2	4	#
	1	2	–3	#

Do you see the pattern? Next, we multiply the "test root" by –3, and write the product under the entry –6:

2	1	0	–7	–6
		2	4	–6
	1	2	–3	#

Adding the numbers in the last column yields the final result of our test. Let's write this last number, in the bottom row:

2	1	0	–7	–6
		2	4	–6
	1	2	–3	–12

This final number is called the *remainder*. What does this particular result mean? Well, the news is not good. We want the remainder to be 0! That is always the ultimate goal of synthetic division when we're looking for a binomial factor of a cubic equation. We must try another "test root" and go through this ritual again.

We try, we fail, we learn

Let's try 4 as our "test root" this time. We put a bold numeral 4 in the top left slot. Then the process goes along with the new numbers, in the same way as before. Here's the sequence of steps:

4	1	0	–7	–6
		#	#	#
	#	#	#	#

4	1	0	−7	−6
		#	#	#
	1	#	#	#

4	1	0	−7	−6
		4	#	#
	1	#	#	#

4	1	0	−7	−6
		4	#	#
	1	4	#	#

4	1	0	−7	−6
		4	16	#
	1	4	#	#

4	1	0	−7	−6
		4	16	#
	1	4	9	#

4	1	0	−7	−6
		4	16	36
	1	4	9	#

4	1	0	−7	−6
		4	16	36
	1	4	9	30

We've failed again! But it's not a total loss. We've learned something. The first "test root" produced a negative remainder. The second "test root" produced a positive remainder. This tells us that the number we're seeking lies between 2 and 4.

We try, we succeed

Let's see what happens with a "test root" of 3. By now you must be thinking, "A computer would be a big help with all this." That's true, but things are going to work out neatly this time. We won't need a computer. Let's go ahead:

3	1	0	−7	−6
		#	#	#
	#	#	#	#

3	1	0	−7	−6
		#	#	#
	1	#	#	#

3	1	0	−7	−6
		3	#	#
	1	#	#	#

3	1	0	−7	−6
		3	#	#
	1	3	#	#

3	1	0	−7	−6
		3	9	#
	1	3	#	#

3	1	0	−7	−6
		3	9	#
	1	3	2	#

3	1	0	−7	−6
		3	9	6
	1	3	2	#

3	1	0	−7	−6
		3	9	6
	1	3	2	0

Mission accomplished! We have produced a remainder of 0. Now we know that 3 is a real root of the cubic equation

$$x^3 - 7x - 6 = 0$$

and also that $(x - 3)$ is a factor of the polynomial.

What are the other roots?

We can write the binomial-trinomial form of a cubic equation straightaway, once we have performed synthetic division on the original cubic and managed to come up with a remainder of 0. Take a close look at the numbers in the bottom line of the last step in the synthetic division process for the "test root" of 3. Those numbers are 1, 3, 2, and 0. The first three of these, in the order shown, are the coefficients and the stand-alone constant in the trinomial factor of the cubic equation.

We know, in the above example, that $(x - 3)$ is the binomial factor. The factor theorem tells us so. So the trinomial factor is $(x^2 + 3x + 2)$. We can therefore write the binomial-trinomial version of the original cubic as

$$(x - 3)(x^2 + 3x + 2) = 0$$

To be certain of this, and to get a little complementary credit, you can multiply the left side of the equation out and see that it produces the original polynomial.

The other two real roots, if they exist, can be found by solving the quadratic equation

$$x^2 + 3x + 2 = 0$$

The roots of this equation turn out to be $x = -1$ and $x = -2$. You can verify this fact as another complementary-credit exercise. That gives us three real roots for the original cubic:

$$x = 3 \text{ or } x = -1 \text{ or } x = -2$$

and a real-number solution set $X = \{3,-1,-2\}$. For some more complementary credit, you can substitute each of these roots for x in the original cubic to demonstrate that they are correct.

Are you confused?

Synthetic division sometimes works out nicely, but in some cases it does not. What if the real roots of a cubic are all complicated fractions—or worse, all irrational numbers? What if all the methods to "crack a cubic" that we've seen in this chapter fail us? There are other schemes available, some of which will be covered in the next chapter.

Here's a challenge!

Consider the following cubic in polynomial standard form:

$$6x^3 + 13x^2 + 8x + 3$$

Show by synthetic division that $-3/2$ is a real root of this equation. Then, from the results of that process, show that $-3/2$ is the only real root.

Solution

First, let's set up the synthetic division array with the "test root," $x = -3/2$, and the coefficients in the top row.

−3/2	6	13	8	3
		#	#	#
	#	#	#	#

Subsequent steps proceed as follows:

−3/2	6	13	8	3
		#	#	#
	6	#	#	#

−3/2	6	13	8	3
		−9	#	#
	6	#	#	#

−3/2	6	13	8	3
		−9	#	#
	6	4	#	#

−3/2	6	13	8	3
		−9	−6	#
	6	4	#	#

−3/2	6	13	8	3
		−9	−6	#
	6	4	2	#

−3/2	6	13	8	3
		−9	−6	−3
	6	4	2	#

−3/2	6	13	8	3
		−9	−6	−3
	6	4	2	0

We get a remainder of 0. This tells us that $x = -3/2$ is a real root of the original cubic equation. We can now write it in binomial-trinomial form. The binomial factor is what we get when we subtract $-3/2$ from x.

That's $(x + 3/2)$. In the trinomial factor, the coefficient of x^2 is 6, the coefficient of x is 4, and the standalone constant is 2. We know this because 6, 4, and 2 appear, in that order, in the bottom row before the remainder 0. Here's the binomial-trinomial cubic:

$$(x + 3/2)(6x^2 + 4x + 2) = 0$$

Let's examine the trinomial factor. If we let $a_2 = 6$, $b_2 = 4$, and $c = 2$, we can find the discriminant, d, as follows:

$$\begin{aligned} d &= b_2^2 - 4a_2c \\ &= 4^2 - 4 \times 6 \times 2 \\ &= 16 - 48 \\ &= -32 \end{aligned}$$

If we set the trinomial factor equal to 0 to get a quadratic equation, then that quadratic has no real roots because $d < 0$. The only way the original cubic polynomial can attain the value 0 is if either the binomial factor is 0, the trinomial factor is 0, or both. The binomial becomes 0 if and only if $x = -3/2$. We've just discovered that no real number x can make the trinomial become 0. It follows that $x = -3/2$ is the only real root of the original cubic equation.

Practice Exercises

This is an open-book quiz. You may (and should) refer to the text as you solve these problems. Don't hurry! You'll find worked-out answers in App. C. The solutions in the appendix may not represent the only way a problem can be figured out. If you think you can solve a particular problem in a quicker or better way than you see there, by all means try it!

1. Multiply out the following equation to obtain a polynomial cubic:

   $$(ax + b)^3 = 0$$

   where x is the variable, and a and b are real numbers with $a \neq 0$.

2. Multiply out the following equation to get a polynomial cubic:

   $$(3^{1/2}x - 12^{1/2})^3 = 0$$

3. In the chapter text, we solved this cubic in binomial factor form:

   $$(3x + 2)(5x + 6)(-7x - 1) = 0$$

   We found that the real roots are

   $$x = -2/3 \text{ or } x = -6/5 \text{ or } x = -1/7$$

   Multiply this equation out to get it into the polynomial standard form.

4. Substitute the real roots that we found for the equation stated in Prob. 3, one by one, in place of x in the polynomial equation derived in the solution to Prob. 3. Verify that these three roots work in that equation.

5. Multiply out the general binomial-trinomial equation to get an equation in polynomial standard form for a cubic:

$$(a_1x + b_1)(a_2x^2 + b_2x + c) = 0$$

6. How can we tell from the coefficients and the constants alone how many real roots the equation stated in Prob. 5 has (before multiplying it out)?

7. One of the "challenges" in the text required that we find the roots of the following cubic equation in binomial-trinomial form:

$$(3x + 5)(16x^2 - 56x + 49) = 0$$

We found that the real roots are

$$x = -5/3 \text{ or } x = 7/4$$

Substitute these roots into the original equation, and go through the arithmetic to verify that they're accurate. Consider the following cubic equation in polynomial standard form:

$$-9x^3 + 21x^2 + 104x + 80 = 0$$

8. By means of synthetic division, verify that $x = 5$ is a real root of this equation.

9. Using the quadratic formula on the results of the synthetic division performed in the solution to Prob. 8, find the other root or roots of the original cubic, if any exist. Then state the solution set.

10. The final "challenge" in the text revealed that $x = -3/2$ is the only real root of the following cubic as expressed in the binomial-trinomial form:

$$(x + 3/2)(6x^2 + 4x + 2) = 0$$

How many real roots will the cubic have if the coefficient of x in the trinomial is changed from 4 to –4? What will the new real roots be, if there are any?

CHAPTER

26

Polynomial Equations in Real Numbers

In this chapter, we'll examine some ways to look for real roots of *higher-degree equations* with real coefficients and real constants. In this context, the term *higher-degree* applies to any equation of degree 4 or more. The tactics described in the last part of this chapter can be helpful in finding the real roots of stubborn third-degree (cubic) equations as well.

Binomial to the *n*th Power

Sometimes, a higher-degree equation can be written as a positive-integer power of a binomial with a real coefficient and a real constant. Such an equation always has exactly one real root, which can be found by setting the binomial equal to 0 to get a first-degree equation. The root has multiplicity equal to the value of the power to which the binomial is raised. That's the same as the degree of the equation.

The general form

Suppose x is a variable, a is the nonzero real-number coefficient of x, and b is a real-number constant. Consider the expression

$$(ax + b)^n$$

where n is a positive integer. If we set this equal to 0, we obtain

$$(ax + b)^n = 0$$

which is an equation in *binomial to the* n*th form.* (We don't have to include the word "power" because it's understood.) Here are some examples of binomial to the *n*th equations, where $n > 3$:

$$(2x - 3)^4 = 0$$
$$(6x + 1)^5 = 0$$
$$(23x + 77)^6 = 0$$

$$(-7x-12)^7=0$$
$$(-118x+59)^{13}=0$$
$$(-3x)^{17}=0$$

In the last case, the stand-alone constant is 0.

Multiplying out

You can multiply out an equation in binomial to the *n*th form to get a polynomial equation, but the result usually looks more complicated. In some instances, the multiplied-out equation becomes messy indeed, as would happen in the fifth example above. But the reverse process, if you can carry it out, is useful. Once in awhile, a formidable polynomial equation can be reduced to binomial to the *n*th form. Then it's easy to solve!

What's the real root?

Whenever we see an equation in binomial to the *n*th form, we can find the real root by considering the binomial as a first-degree equation. When we do that with the above higher-degree equations, we get

$$2x-3=0$$
$$6x+1=0$$
$$23x+77=0$$
$$-7x-12=0$$
$$-118x+59=0$$
$$-3x=0$$

These all resolve easily. The original equations, which are of degree 4, 5, 6, 7, 13, and 17, have one real root apiece, of multiplicity 4, 5, 6, 7, 13, and 17 respectively.

Are you confused?

Perhaps you still wonder, "What's all this fuss about root multiplicity? If an equation has one real root, isn't that all there is to be said about it?" The answer is, "Not exactly." Let's look at three equations that have identical solution sets. The first-degree equation

$$2x-3=0$$

has one real solution, $x = 3/2$. The fourth-degree equation

$$(2x-3)^4=0$$

has a single real root, $x = 3/2$. But there's something about the fourth-degree equation that makes it conceptually different than the first-degree equation. We can rewrite the fourth-degree equation as

$$(2x-3)(2x-3)(2x-3)(2x-3)=0$$

If we substitute 3/2 for x here, we get

$$(2 \times 3/2 - 3)(2 \times 3/2 - 3)(2 \times 3/2 - 3)(2 \times 3/2 - 3) = 0$$

which reduces to

$$(3 - 3)(3 - 3)(3 - 3)(3 - 3) = 0$$

and further to

$$0 \times 0 \times 0 \times 0 = 0$$

To carry out the substitution and simplification process completely, we must repeat it for each of the four binomials. The root $x = 3/2$ exists, in effect, "four times over." Now look at this:

$$(2x - 3)^{345} = 0$$

Here, the single real root $x = 3/2$ exists "345 times over."

The solution sets are identical for the first-degree, the fourth-degree, and the 345th-degree equations in this example. But the equations themselves are vastly different!

Here's a challenge!

Consider the following equation, which is in the form of a trinomial squared. Find all the real roots, and state the multiplicity of each.

$$(x^2 + 2x + 1)^2 = 0$$

Solution

Look closely at the trinomial. Suppose we set it equal to 0, so it becomes the quadratic

$$x^2 + 2x + 1 = 0$$

We can factor this to get

$$(x + 1)^2 = 0$$

If we substitute $(x + 1)^2$ for the trinomial in the original equation, we have

$$[(x + 1)^2]^2 = 0$$

The product of powers rule from Chap. 9 tells us that this is the same as

$$(x + 1)^{(2 \times 2)} = 0$$

which can be simplified to

$$(x + 1)^4 = 0$$

We now have a binomial to the nth equation with $n = 4$. We can find the solitary real root by setting the binomial equal to 0:

$$x + 1 = 0$$

This first-degree equation resolves to $x = -1$. The real solution set of the original trinomial-squared equation is $X = \{-1\}$. The single root has multiplicity 4.

Binomial Factors

A higher-degree equation can appear as a product of binomials. Such an equation has one real root for each factor. You can find each root by setting every binomial equal to 0 and then creating first-degree equations from them. Each root has multiplicity equal to the number of times its binomial appears in the product. If a binomial is raised to a power, then its root has multiplicity equal to that power.

The general form

Let x be a real-number variable. Suppose that $a_1, a_2, a_3, \ldots a_n$ are nonzero real coefficients of x, and $b_1, b_2, b_3, \ldots b_n$ are real stand-alone constants. Consider the equation

$$(a_1x + b_1)(a_2x + b_2)(a_3x + b_3) \cdots (a_nx + b_n) = 0$$

This is the binomial factor form for an equation of degree n. Here are some examples of equations in this form where $n > 3$:

$$(x + 1)(x - 2)(x + 3)(x - 4) = 0$$
$$(x + 1)(x - 2)^2(x + 3)^3(x - 4)^4 = 0$$
$$(3x + 7)^6(4x - 5)^2 = 0$$
$$(-x + 1)(-7x + 2)^7 = 0$$
$$(-3x + 21)^2(-8x + 5)^3 = 0$$
$$(x + 6)(2x - 5)^2(7x)(-3x) = 0$$

In all but the first of these equations, some of the binomials are raised to powers. That's the equivalent of repeating those binomials in the products by the number of times the power indicates. In the last equation, two of the stand-alone constants are equal to 0.

Multiplying out

You can multiply out an equation that appears in the binomial factor form to get a polynomial equation. If you want to do this for each of the six equations in the previous paragraph, go ahead! You'll end up with equations that are difficult to solve for anyone who comes across them for the first time.

What are the real roots?

As we search for real roots to a higher-degree equation in binomial factor form, we know we've "made a hit" when any of the factors becomes 0. That means we can set the binomial factors equal to 0, one by one, and then solve each of them to get all the real roots.

Consider the first of the six binomial factor equations listed above. Let's find all the real roots of

$$(x+1)(x-2)(x+3)(x-4)=0$$

We take each binomial individually, set it equal to 0, and then solve the resulting first-degree equations:

$$x+1=0$$
$$x-2=0$$
$$x+3=0$$
$$x-4=0$$

Any value of x that satisfies one of these is a root of the higher-degree equation. There are four such values:

$$x=-1 \text{ or } x=2 \text{ or } x=-3 \text{ or } x=4$$

so the real solution set of the higher-degree equation is $X=\{-1, 2, -3, 4\}$.

Now consider another higher-degree equation. This one is a little tricky, because three of the four binomials have exponents attached:

$$(x+1)(x-2)^2(x+3)^3(x-4)^4=0$$

If we remove the exponents from the binomials and set each equal to 0, we get the same four first-degree equations as before. That means the real roots of the higher-degree equation are the same, too. But three of the four roots have multiplicity greater than 1. The root $x=2$ has multiplicity 2, the root $x=-3$ has multiplicity 3, and the root $x=4$ has multiplicity 4.

Are you confused?

Does the concept of multiplicity still seem esoteric? In the example we just finished, it can help if we write out every binomial factor individually so none is raised to a power (other than the first power). Grouped according to their constants, those factors are

$$(x+1)$$
$$(x-2)(x-2)$$
$$(x+3)(x+3)(x+3)$$
$$(x-4)(x-4)(x-4)(x-4)$$

We have a root in every single case where one of these factors becomes equal to 0. There are not four such instances here, but 10! We "hit the target" once when $x = -1$, twice when $x = 2$, three times when $x = -3$, and four times when $x = 4$. This is true even though the real solution set has only four elements:

$$X = \{-1, 2, -3, 4\}$$

Incidentally, the degree of the original equation is equal to the sum of the exponents attached to the binomial factors. That's $1 + 2 + 3 + 4 = 10$. On that basis, we can immediately see that

$$(x + 1)(x - 2)^2(x + 3)^3(x - 4)^4 = 0$$

is a 10th-degree equation in the variable x.

Here's a challenge!

State the real roots of the following equation. Also state the real solution set X and the multiplicity of each root. What is the degree of the equation?

$$(x + 6)(2x - 5)^2(7x)(-3x) = 0$$

Solution

We take each binomial individually, set it equal to 0, and then solve the resulting first-degree equations:

$$x + 6 = 0$$
$$2x - 5 = 0$$
$$7x = 0$$
$$-3x = 0$$

The real roots are

$$x = -6 \text{ or } x = 5/2 \text{ or } x = 0$$

and the real solution set is $X = \{-6, 5/2, 0\}$. The root -6 has multiplicity 1. The root $5/2$ has multiplicity 2. The root 0 has multiplicity 2. These facts can be clarified by stating the original equation as

$$(x + 6)(2x - 5)(2x - 5)(7x)(-3x) = 0$$

The degree of the original equation is the sum of the exponents attached to the factors. In this example, whether we write the equation in the original form or in the fully expanded binomial factor form, that sum is 5, indicating that it's a fifth-degree equation.

Polynomial Standard Form

A higher-degree equation in polynomial standard form contains a sum of multiples of the variable raised to powers in descending order on the left side of the equals sign. The right side of the equals sign has 0 all by itself.

General polynomial equation

The polynomial standard form of a higher-degree equation can be written as

$$a_n x^n + a_{n-1} x^{n-1} + a_{n-2} x^{n-2} + \cdots + a_1 x + b = 0$$

where a_1, a_2, a_3, ... a_n are coefficients, b is the stand-alone constant, and n is a positive integer greater than 3. Here are some examples:

$$6x^4 - 3x^3 + 3x^2 + 2x + 5 = 0$$
$$3x^5 - 4x^3 = 0$$
$$-7x^7 - 5x^4 + 3x^3 - x^2 - 29 = 0$$
$$-4x^{11} = 0$$

In all but the first of these equations, some of the coefficients are equal to 0. The coefficient a_n, by which x^n is multiplied, can never be 0 in an nth-degree polynomial equation. If you set $a_n = 0$, you end up with

$$0x^n + a_{n-1} x^{n-1} + a_{n-2} x^{n-2} + \cdots + a_1 x + b = 0$$

That's the polynomial standard form for a single-variable equation of degree $n - 1$:

$$a_{n-1} x^{n-1} + a_{n-2} x^{n-2} + \cdots + a_1 x + b = 0$$

Mutants in the *n*th degree

As you can imagine, many single-variable equations can be morphed into the polynomial standard form. Here are some examples:

$$x^2 = 2x + 3x^7 - 4x^{15}$$
$$x = 4x^{21} - 7x^{17} + 2x^{11} + 2x^7$$
$$x^8 + 4x^6 + 7x^4 - x^2 + 3 = x + x^3 + x^5 + x^7$$

The only requirements for membership in the "single-variable nth-degree equation club" are that the equation be convertible into polynomial standard form, and that n be a positive integer. If $n = 1$, we have a first-degree equation; if $n = 2$, we have a quadratic; if $n = 3$, we have a cubic; if $n > 3$, we have a higher-degree equation.

Digging for Real Roots

Now that we've learned how to recognize a polynomial equation, it's time to think about finding the real roots of such an equation, if any exist. The next few sections offer some ways to look for the roots of higher-degree equations. But there are no guarantees. Anyone but the purest mathematician would likely concede that these types of situations lend themselves to computer programming.

A prefabricated problem

To illustrate how the real roots can be sought when we're confronted with a polynomial equation, let's look for the real roots of

$$x^4 - 2x^3 - 13x^2 + 14x + 24 = 0$$

This equation was built up from factors. Here's what it looks like in binomial factor form:

$$(x + 1)(x - 2)(x + 3)(x - 4) = 0$$

We solved this awhile ago. The roots are

$$x = -1 \text{ or } x = 2 \text{ or } x = -3 \text{ or } x = 4$$

Now imagine that we're looking at the polynomial version of this equation, and we've never seen it in any other form. We've been told to find the real roots.

How many roots?

When you embark on a quest to find the real roots of a polynomial equation, you might wonder how long you should keep trying before you give up (or let your computer take over). In part, it depends how much time and patience you have. Here's an important principle to keep in mind:

- A polynomial equation can never have more roots than its degree. That includes not only the real roots, but all of the complex roots.

If you're working on a polynomial equation of degree n and you've found n roots, you can terminate your quest. You've resolved the mystery. There is nothing more to find.

Bounds for real roots

The real roots of a polynomial equation always lie between two extremes. We can identify an interval that contains all the elements in the real solution set if we can discover an *upper bound* that's big enough, and if we can discover a *lower bound* that's small enough. In this context, we consider only *non-inclusive bounds.* That means an upper bound must be greater than the largest real root of the equation, and a lower bound must be less than the smallest real root of the equation.

Finding an upper bound

How can we find an upper bound for the roots of the polynomial equation under investigation? Here's the equation again:

$$x^4 - 2x^3 - 13x^2 + 14x + 24 = 0$$

The coefficients and constant are 1, −2, −13, 14, and 24 in order of descending powers of x. Let's set up a synthetic division array for this equation:

#	1	−2	−13	14	24
		#	#	#	#
	#	#	#	#	#

If we plug in a positive real number as a "test root," we can tell if it's an upper bound for the real solution set by looking at the values we get in the last row. If we get a nonzero remainder and none of the numbers in the last row are negative, then our "test root" is an upper bound. Let's take absolute values of the coefficients and constant in the equation and pick out the largest. In this case, that's 24. It seems reasonable that this might be larger than or equal to all the real roots. Let's input 24 and see what happens:

24	1	−2	−13	14	24
		#	#	#	#
	#	#	#	#	#

24	1	−2	−13	14	24
		#	#	#	#
	1	#	#	#	#

24	1	−2	−13	14	24
		24	#	#	#
	1	#	#	#	#

24	1	−2	−13	14	24
		24	#	#	#
	1	22	#	#	#

24	1	−2	−13	14	24
		24	528	#	#
	1	22	#	#	#

24	1	−2	−13	14	24
		24	528	#	#
	1	22	515	#	#

24	1	−2	−13	14	24
		24	528	12,360	#
	1	22	515	#	#

24	1	−2	−13	14	24
		24	528	12,360	#
	1	22	515	12,374	#

24	1	−2	−13	14	24
		24	528	12,360	296,976
	1	22	515	12,374	#

24	1	−2	−13	14	24
		24	528	12,360	296,976
	1	22	515	12,374	297,000

None of the numbers in the bottom row are negative. Therefore, 24 is an upper bound for the real solution set. The fact that the numbers increase so fast (we might even say that they "blow up") suggests that 24 is a much larger than the largest root of the equation. We can try something smaller, but still positive, and do the synthetic division again. As long as we input positive "test roots" and never see a negative number in the last row, we know that we're inputting upper bounds.

Finding a lower bound

If we plug in a negative number as a "test root," grind out the synthetic division process, get a nonzero remainder, and discover that the numbers in the last row alternate between positive and negative, it tells us that we've found a lower bound for the real solution set. Let's take the absolute values of the coefficients and constant, and pick out the *negative* of the largest result. That's −24. It's a good bet that this is smaller than all the real roots. Let's input it to the array and find out:

−24	1	−2	−13	14	24
		#	#	#	#
	#	#	#	#	#

−24	1	−2	−13	14	24
		#	#	#	#
	1	#	#	#	#

−24	1	−2	−13	14	24
		−24	#	#	#
	1	#	#	#	#

−24	1	−2	−13	14	24
		−24	#	#	#
	1	−26	#	#	#

−24	1	−2	−13	14	24
		−24	624	#	#
	1	−26	#	#	#

−24	1	−2	−13	14	24
		−24	624	#	#
	1	−26	611	#	#

−24	1	−2	−13	14	24
		−24	624	−14,664	#
	1	−26	611	#	#

−24	1	−2	−13	14	24
		−24	624	−14,664	#
	1	−26	611	−14,650	#

−24	1	−2	−13	14	24
		−24	624	−14,664	351,600
	1	−26	611	−14,650	#

−24	1	−2	−13	14	24
		−24	624	−14,664	351,600
	1	−26	611	−14,650	351,624

The numbers in the last row alternate in sign, telling us that −24 is a lower bound for the real solution set. The fact that the absolute values diverge rapidly from 0 suggests that our input number is far smaller than necessary. We can try something larger, but still negative, and do the synthetic division again. As long as we plug in negative "test roots" and get numbers in the last line that alternate in sign, we know that we're inputting lower bounds.

Narrowing the interval

We can reduce the size of the interval containing the real solutions by repeatedly testing positive numbers as upper bounds, and by repeatedly testing negative numbers as lower bounds. Once we get a positive "test root" that produces a negative number anywhere in the bottom line, or a negative "test root" that fails to cause the numbers in the bottom line to alternate in sign, we know that we are in the interval containing the real roots.

Suppose we gradually reduce the positive "test root" and gradually increase the negative "test root" in the synthetic division array

#	1	−2	−13	14	24
		#	#	#	#
	#	#	#	#	#

If we use integers for simplicity, we'll eventually get down to a *smallest upper bound*, and we'll also get up to a *greatest lower bound.* At that point, we can test numbers between those bounds to look for real roots. In this particular example, we'll get a smallest upper bound of 5 and a greatest lower bound of −4.

Rational roots

There's a lengthy but straightforward process you can use to find all the rational roots of a polynomial equation in the standard form

$$a_n x^n + a_{n-1} x^{n-1} + a_{n-2} x^{n-2} + \cdot \cdot + a_1 x + b = 0$$

where a_1, a_2, a_3, ... a_n are nonzero rational-number coefficients of the variable x, b is a nonzero rational constant, and n is a positive integer greater than 3. If $b = 0$, then you can't use the process, but you will at least know that 0 is a root. If $b \neq 0$, then you can go through the following steps.

- Make certain that all the numbers a_1, a_2, a_3, ... a_n, and b are integers. If that is not the case, multiply the equation through by the smallest constant that will turn all the numbers a_1, a_2, a_3, ... a_n, and b into integers.
- Find all the positive and negative integer factors of b, the stand-alone constant. Call these by the general name m.
- Find all the positive and negative integer factors of a_n, the coefficient of x^n (sometimes called the *leading coefficient*). Call these by the general name n.
- Write down all the possible ratios m/n. Call them by the general name r.

- With synthetic division, check every r to see if it is a root of the original polynomial equation. If you find a root, you'll get a remainder of 0 at the end of the synthetic division process.
- If none of the numbers r produces a remainder of 0, then the original polynomial equation has no rational roots.
- If one or more of the ratios r produces a remainder of 0, then every one of those numbers is a rational root of the equation.
- List all of the rational roots found after carrying out the preceding steps. Call them r_1, r_2, r_3, and so on.
- Create binomials of the form $(x - r_1)$, $(x - r_2)$, $(x - r_3)$, and so on. Each of these binomials is a factor of the original equation.
- If you're lucky, you'll end up with an equation in binomial to the nth form, or an equation in binomial factor form.
- If you're less lucky, you'll end up with one or more binomial factors and a quadratic factor. That factor can be set equal to 0, and then the quadratic formula can be used to find its roots. Neither of those roots will be rational. They might even be complex.
- If you're unlucky, you'll be stuck with one or more binomial factors and a cubic or higher-degree factor. If you set that factor equal to 0 to form a polynomial equation, you'll know that none of the roots of that equation are rational. Some might even be complex. You can try to solve it, but you should not expect the task to be easy.

Are you confused?

At this point, you must wonder, "Suppose we're left with a cubic or higher-degree polynomial as one of the factors, and its associated equation has some irrational roots. What can we do to find those roots?" The best answer is, "We can use a computer program to generate an approximate graph of the function produced by the polynomial equation, see how many times that graph crosses the x axis, and then use the computer to approximate the zeros of that function." This method will not allow us to find non-real roots.

Here's a challenge!

Use the above-described procedure to find all the rational roots of the polynomial equation we contrived earlier in this chapter, and for which we know the smallest upper bound is 5 and the greatest lower bound is –4. Once again, that equation is

$$x^4 - 2x^3 - 13x^2 + 14x + 24 = 0$$

Solution

Here is an outline of the process. You might want to work out the arithmetic, particularly the synthetic division problems, to verify.

- All the coefficients, as well as the stand-alone constant, are integers, so we don't have to multiply the equation through by anything.
- The positive and negative integer factors of the stand-alone constant, 24, are all the integers m that divide 24 without remainders. These numbers are 24, 12, 8, 6, 4, 3, 2, and 1, along with all their negatives.

- There are only two integer factors n of the leading coefficient: 1 and -1.
- All the possible ratios m/n are the same as the integers m: 24, 12, 8, 6, 4, 3, 2, and 1, along with all their negatives.
- Now let's cheat a little. Imagine that we've narrowed down the interval by doing synthetic division repeatedly, finding a smallest upper bound of 5 and a greatest lower bound of -4. That leaves us with rational numbers r of 4, 3, 2, 1, -1, -2, and -3 to check as possible roots.
- We input 4, 3, 2, 1, -1, -2, and -3 to synthetic division arrays, and see if we get a remainder of 0 for any of them.
- We get a remainder of 0 when $r = 4$, $r = 2$, $r = -1$, or $r = -3$. Now we know that every one of those numbers is a rational root of the equation.
- We have found four rational roots for a fourth-degree equation. There are no more rational roots to find! In fact, these are all the roots of any sort. Remember: A polynomial equation can never have more roots than its degree. That includes not only the rational roots, but the irrational and complex roots.

Practice Exercises

This is an open-book quiz. You may (and should) refer to the text as you solve these problems. Don't hurry! You'll find worked-out answers in App. C. The solutions in the appendix may not represent the only way a problem can be figured out. If you think you can solve a particular problem in a quicker or better way than you see there, by all means try it!

1. Rewrite each of the following equations in binomial to the nth form.

   (a) $(x^2 + 6x + 9)^2 = 0$

   (b) $(x^2 - 4x + 4)^3 = 0$

   (c) $(16x^2 - 24x + 9)^4 = 0$

2. What are the real roots for each of the equations stated in Prob. 1? What is the multiplicity in each case?

3. Rewrite each of the following equations in binomial factor form.

   (a) $(x^2 - 3x + 2)^2 = 0$

   (b) $(-3x^2 - 5x + 2)^5 = 0$

   (c) $(4x^2 - 9)^3 = 0$

4. What are the real roots for each of the equations stated in Prob. 3? What is the multiplicity in each case?

5. State the real roots of the following equation. Also state the real solution set X and the multiplicity of each root. What is the degree of the equation?

$$(x - 3/2)^2(2x - 7)^2(7x)^3(-3x + 5)^5 = 0$$

6. State the real roots of the following equation. Also state the real solution set X and the multiplicity of each root. What is the degree of the equation?

$$(x + 4)(2x - 8)^2(x/3 + 12)^3 = 0$$

7. Using synthetic division, find a lower bound and an upper bound for the interval containing all the real roots of the equation

$$2x^5 - 3x^3 - 2x + 2 = 0$$

8. Determine all the rational roots of the equation stated in Prob. 7.

9. Using synthetic division, find a lower bound and an upper bound for the interval containing all the real roots of the equation

$$3x^5 - 3x^2 + 2x - 2 = 0$$

10. Determine all the rational roots of the equation stated in Prob. 9.

CHAPTER

27

More Two-by-Two Systems

We've seen how pairs of linear equations can be solved as two-by-two systems. What if the equations are more complicated? In this chapter, we'll solve some two-by-two systems in which one or both of the equations are quadratic or cubic.

Linear and Quadratic

When we want to solve a pair of two-variable equations together as a system, we can go through these steps in order.

- Decide which variable to call independent, and which variable to call dependent.
- Morph both equations so they express the dependent variable in terms of the independent variable.
- Mix the independent-variable parts of the equations to get an equation in a single variable.
- Find the root or roots of that equation.
- Plug the root or roots into one of the morphed original equations, and calculate the corresponding value or values of the dependent variable.
- Express the solutions as ordered pairs with the independent variable listed first.

First, we morph

Let's try an example. Consider these two equations, the first of which is linear and the second of which is quadratic:

$$2x - y + 1 = 0$$

and

$$y = x^2 + x - 5$$

If we call x the independent variable, then the quadratic equation is already expressed as a function of x, so we don't have to manipulate it. The linear equation can be rearranged so it appears as a function of x by adding y to each side and then transposing the left and right sides, getting

$$y = 2x + 1$$

We now have two functions in which x is the independent variable and y is the dependent variable.

Next, we mix

When we mix the independent-variable parts of the above functions, we have

$$x^2 + x - 5 = 2x + 1$$

We can subtract 1 from both sides to obtain

$$x^2 + x - 6 = 2x$$

Then we can subtract $2x$ from both sides to get

$$x^2 - x - 6 = 0$$

Next, we solve

We've derived a quadratic equation that can be solved using any of the methods from Chaps. 22 and 23. We're lucky here, because this equation can be factored into

$$(x + 2)(x - 3) = 0$$

The roots are found by solving the two first-degree equations

$$x + 2 = 0$$

and

$$x - 3 = 0$$

giving us $x = -2$ or $x = 3$. We can substitute these two values for x into either of the morphed original equations to obtain corresponding values for y. The simpler of the two is

$$y = 2x + 1$$

For $x = -2$, we have

$$\begin{aligned} y &= 2 \times (-2) + 1 \\ &= -4 + 1 \\ &= -3 \end{aligned}$$

For $x = 3$, we have

$$\begin{aligned} y &= 2 \times 3 + 1 \\ &= 6 + 1 \\ &= 7 \end{aligned}$$

The two solutions of the system are therefore $(x, y) = (-2, -3)$ and $(x, y) = (3, 7)$.

Finally, we check

Let's check both solutions in the original two-variable equations to be certain that we did our algebra and arithmetic right. This exercise can also help us to see how the system "plays out." First, let's check $(-2, -3)$ in the original two-variable linear equation:

$$\begin{aligned} 2x - y + 1 &= 0 \\ 2 \times (-2) - (-3) + 1 &= 0 \\ -4 + 3 + 1 &= 0 \\ -1 + 1 &= 0 \\ 0 &= 0 \end{aligned}$$

Next, we check $(3, 7)$ in that same equation:

$$\begin{aligned} 2x - y + 1 &= 0 \\ 2 \times 3 - 7 + 1 &= 0 \\ 6 - 7 + 1 &= 0 \\ -1 + 1 &= 0 \\ 0 &= 0 \end{aligned}$$

Next, we check $(-2, -3)$ in the original two-variable quadratic:

$$\begin{aligned} y &= x^2 + x - 5 \\ -3 &= (-2)^2 + (-2) - 5 \\ -3 &= 4 + (-2) - 5 \\ -3 &= 2 - 5 \\ -3 &= -3 \end{aligned}$$

To finish up, we check $(3, 7)$ in the original quadratic:

$$\begin{aligned} y &= x^2 + x - 5 \\ 7 &= 3^2 + 3 - 5 \\ 7 &= 9 + 3 - 5 \\ 7 &= 12 - 5 \\ 7 &= 7 \end{aligned}$$

Are you confused?

We've found two ordered pairs (x, y) that solve the above pair of equations as a two-by-two system. How do we know that these are the only two solutions for this system? We can demonstrate this visually for the real-number solutions by graphing both equations together on the coordinate plane. The linear equation shows up as a straight line, and the quadratic shows up as a parabola. In the next chapter, we'll plot these graphs. You'll see that they intersect at the points corresponding to the ordered pairs $(-2, -3)$ and $(3, 7)$, but nowhere else.

Here's a challenge!

Suppose that a_1, a_2, b_1, b_2, and c are real numbers, and neither a_1 nor a_2 are equal to 0. Consider these two functions of x:

$$y = a_1x + b_1$$

and

$$y = a_2x^2 + b_2x + c$$

Derive a general formula for solving this linear-quadratic system.

Solution

We can follow the same procedure as we did for the example we just solved, but we must use letter constants instead of specific numbers, and we can't simplify the expressions as much. These two equations are ready to mix; we have no morphing to do. When we set the right sides equal to each other, we get

$$a_1x + b_1 = a_2x^2 + b_2x + c$$

We can subtract b_1 from each side to get

$$a_1x = a_2x^2 + b_2x + c - b_1$$

We can then subtract a_1x from each side, obtaining

$$0 = a_2x^2 + b_2x + c - b_1 - a_1x$$

The commutative, distributive, and grouping principles, followed by left-to-right transposition, allow us to rearrange this equation to get

$$a_2x^2 + (b_2 - a_1)x + (c - b_1) = 0$$

If you're confused by this rearrangement, or if you aren't convinced that it's correct, then check it out for yourself. Change the subtractions to negative additions, move things around, and then change them back again.

Now we have a quadratic equation in standard form. That means we can solve it directly with the quadratic formula. Let's state the quadratic formula once again as we originally learned it. For the general quadratic

$$ax^2 + bx + c = 0$$

the roots are given by

$$x = [-b \pm (b^2 - 4ac)^{1/2}] / (2a)$$

To make the quadratic formula work in the current equation, let's make these substitutions in the classical version:

- Write a_2 in place of a
- Write $(b_2 - a_1)$ in place of b
- Write $(c - b_1)$ in place of c

When we do that, we get

$$x = \{-(b_2 - a_1) \pm [(b_2 - a_1)^2 - 4a_2(c - b_1)]^{1/2}\} / (2a_2)$$

The roots defined by this formula give us the x-values for the solution of the original linear-quadratic system. We can simplify this slightly by getting rid of the minus sign in the first expression on the right side of the equals sign, and then reversing the positions of a_1 and b_2 inside the parentheses, getting

$$x = \{(a_1 - b_2) \pm [(b_2 - a_1)^2 - 4a_2(c - b_1)]^{1/2}\} / (2a_2)$$

Any attempt to further simplify this will eliminate some grouping symbols, but will not make the formula easier to use.

When we've found the x-values of the solutions, we can plug them into either of the original functions to obtain the y-values. In this case, the linear function is less messy than the quadratic. If we call the x-values x_1 and x_2, then

$$y_1 = a_1x_1 + b_1$$

and

$$y_2 = a_1x_2 + b_1$$

The solutions of the whole system can be written as the ordered pairs (x_1, y_1) and (x_2, y_2).

Two Quadratics

Now let's solve a two-by-two system consisting of these quadratic equations in the two variables x and y:

$$4x^2 + 6x + 2y + 8 = 0$$

and

$$3x^2 + y + 5x - 11 = 0$$

First, we morph

In both of these equations, a multiple of y can be separated out and placed alone on the left side of the equals sign, producing quadratic functions of x. In the first equation, we can subtract $2y$ from each side and then transpose the sides to get

$$-2y = 4x^2 + 6x + 8$$

Dividing through by −2, we get the function

$$y = -2x^2 - 3x - 4$$

In the second original equation, we can subtract y from each side and then transpose the sides, getting

$$-y = 3x^2 + 5x - 11$$

Multiplying through by −1, we get

$$y = -3x^2 - 5x + 11$$

Next, we mix

When we directly mix the right sides of the above two quadratic functions, we get a single equation in one variable:

$$-2x^2 - 3x - 4 = -3x^2 - 5x + 11$$

We can add the quantity $(3x^2 + 5x - 11)$ to each side, obtaining

$$x^2 + 2x - 15 = 0$$

which is a quadratic equation in polynomial standard form.

Next, we solve

We now have an equation that can be easily factored. It does not take long to figure out that the above quadratic is equivalent to

$$(x + 5)(x - 3) = 0$$

The roots are found by solving the equations

$$x + 5 = 0$$

and

$$x - 3 = 0$$

giving us $x = -5$ or $x = 3$. We can substitute these two values for x into either of the original functions to get the y-values. Let's use the first one:

$$y = -2x^2 - 3x - 4$$

For $x = -5$, we have

$$\begin{aligned} y &= -2 \times (-5)^2 - 3 \times (-5) - 4 \\ &= -2 \times 25 - (-15) - 4 \\ &= -50 + 15 - 4 \\ &= -35 - 4 \\ &= -39 \end{aligned}$$

For $x = 3$, we have

$$\begin{aligned} y &= -2 \times 3^2 - 3 \times 3 - 4 \\ &= -2 \times 9 - 9 - 4 \\ &= -18 - 9 - 4 \\ &= -27 - 4 \\ &= -31 \end{aligned}$$

The two solutions of the system are therefore $(x, y) = (-5, -39)$ and $(x, y) = (3, -31)$.

Finally, we check

Even though it may seem redundant, we should check both solutions in the original equations to remove all doubt about their correctness. Let's begin by plugging $(-5, -39)$ into the first equation, paying careful attention to the signs:

$$\begin{aligned} 4x^2 + 6x + 2y + 8 &= 0 \\ 4 \times (-5)^2 + 6 \times (-5) + 2 \times (-39) + 8 &= 0 \\ 4 \times 25 + (-30) + (-78) + 8 &= 0 \\ 100 + (-30) + (-78) + 8 &= 0 \\ 70 + (-78) + 8 &= 0 \\ -8 + 8 &= 0 \\ 0 &= 0 \end{aligned}$$

Next, we check $(3, -31)$ in that equation:

$$\begin{aligned} 4x^2 + 6x + 2y + 8 &= 0 \\ 4 \times 3^2 + 6 \times 3 + 2 \times (-31) + 8 &= 0 \end{aligned}$$

$$4 \times 9 + 18 + (-62) + 8 = 0$$
$$36 + 18 + (-62) + 8 = 0$$
$$54 + (-62) + 8 = 0$$
$$-8 + 8 = 0$$
$$0 = 0$$

Next, we check (−5, −39) in the second original equation:

$$3x^2 + y + 5x - 11 = 0$$
$$3 \times (-5)^2 + (-39) + 5 \times (-5) - 11 = 0$$
$$3 \times 25 + (-39) + (-25) - 11 = 0$$
$$75 + (-39) + (-25) - 11 = 0$$
$$36 + (-25) - 11 = 0$$
$$11 - 11 = 0$$
$$0 = 0$$

Completing the job, we check (3, −31) in that equation:

$$3x^2 + y + 5x - 11 = 0$$
$$3 \times 3^2 + (-31) + 5 \times 3 - 11 = 0$$
$$3 \times 9 + (-31) + 15 - 11 = 0$$
$$27 + (-31) + 15 - 11 = 0$$
$$-4 + 15 - 11 = 0$$
$$11 - 11 = 0$$
$$0 = 0$$

Are you confused?

You might ask, "Is it possible for a two-by-two system, in which one or both of the equations is quadratic, to have imaginary or complex solutions?" The answer is "Yes. In that case, when we use the quadratic formula, those solutions will appear." You'll see this happen in later in this chapter.

Here's a challenge!

Suppose that a_1, a_2, b_1, b_2, c_1, and c_2 are real numbers, and neither a_1 nor a_2 is equal to 0. Consider these two functions of x:

$$y = a_1x^2 + b_1x + c_1$$

and

$$y = a_2x^2 + b_2x + c_2$$

Derive a general formula for solving this system.

Solution

Both of these equations are presented as functions of x, so we can mix the right sides directly without having to manipulate. When we do that, we get

$$a_1x^2 + b_1x + c_1 = a_2x^2 + b_2x + c_2$$

The next several steps are tricky. It's easy to make mistakes with the signs and the grouping. To minimize the risk of error, let's do negative additions, rearrange and regroup the symbols, and change the negative additions to subtractions after we've put everything in the proper order. We can add the negatives of a_2x^2, b_2x, and c_2 to each side, getting

$$a_1x^2 + b_1x + c_1 + (-a_2x^2) + (-b_2x) + (-c_2) = 0$$

Using the commutative law for addition, we can rewrite this as

$$a_1x^2 + (-a_2x^2) + b_1x + (-b_2x) + c_1 + (-c_2) = 0$$

The distributive law of multiplication over addition allows us to group the coefficients and the constants to get

$$[a_1 + (-a_2)]x^2 + [b_1 + (-b_2)]x + [c_1 + (-c_2)] = 0$$

Changing the negative additions back to subtractions, we obtain

$$(a_1 - a_2)x^2 + (b_1 - b_2)x + (c_1 - c_2) = 0$$

We can solve this equation with the quadratic formula. For reference, here it is, yet one more time, in its classical form. (Have you memorized it yet?) When we see the quadratic equation

$$ax^2 + bx + c = 0$$

the roots are given by

$$x = [-b \pm (b^2 - 4ac)^{1/2}] / (2a)$$

To solve the equation at hand, we can make these substitutions in the classical version of the quadratic formula:

- Write $(a_1 - a_2)$ in place of a
- Write $(b_1 - b_2)$ in place of b
- Write $(c_1 - c_2)$ in place of c

When we make these changes in "copy-and-paste" fashion, we get

$$x = \{-(b_1 - b_2) \pm [(b_1 - b_2)^2 - 4(a_1 - a_2)(c_1 - c_2)]^{1/2}\} / [2(a_1 - a_2)]$$

We can simplify this slightly by getting rid of the minus sign in the first expression on the right side of the equals sign, and then reversing the positions of b_1 and b_2 inside the parentheses. We can also multiply out the denominator. But there's no good reason to expand the products of the binomials. That would produce a formula with fewer grouping symbols, but the arithmetic would be more cumbersome in practical use. Let's express the formula as

$$x = \{(b_2 - b_1) \pm [(b_1 - b_2)^2 - 4(a_1 - a_2)(c_1 - c_2)]^{1/2}\} / (2a_1 - 2a_2)$$

After we've found the x-values of the solutions by plugging in the coefficients, plugging in the constants, and going through the arithmetic, we can put those x-values into either of the original two functions and calculate the y-values. If we call the x-values x_1 and x_2, and if we use the first of the original functions, we have

$$y_1 = a_1 x_1^2 + b_1 x_1 + c_1$$

and

$$y_2 = a_1 x_2^2 + b_1 x_2 + c_1$$

The solutions of the whole system can be written as ordered pairs in the form (x_1, y_1) and (x_2, y_2).

Enter the Cubic

Let's solve a two-by-two system in which one equation is linear and the other is cubic. Consider these:

$$x^3 + 6x^2 + 14x - y = -7$$

and

$$-6x + 2y = 2$$

First, we morph

Again, it appears as if we ought to let x be the independent variable, and then derive two functions of that variable. In the first equation, we can add 7 to each side, getting

$$x^3 + 6x^2 + 14x - y + 7 = 0$$

Then we can add y to each side and transpose the equation left-to-right, obtaining y as a function of x:

$$y = x^3 + 6x^2 + 14x + 7$$

In the second equation, we can divide through by -2 to obtain

$$3x - y = -1$$

Adding 1 to each side gives us

$$3x - y + 1 = 0$$

We can add y to each side and transpose the equation left-to-right, getting the function

$$y = 3x + 1$$

Next, we mix

When we mix the independent-variable parts of the above functions, we obtain one equation in one variable:

$$x^3 + 6x^2 + 14x + 7 = 3x + 1$$

If we subtract the quantity $(3x + 1)$ from both sides, we get

$$x^3 + 6x^2 + 11x + 6 = 0$$

This is a straightforward cubic equation in polynomial standard form. The roots aren't obvious from casual inspection, but we can use the techniques from Chap. 25 to solve it.

Next, we solve

Now that we have derived a cubic equation in one variable, our mission is to find its roots. We can use synthetic division several times to obtain factors. Ultimately, we find that the cubic factors into

$$(x + 1)(x + 2)(x + 3) = 0$$

The roots can be found by solving the three equations we get when we set each binomial equal to 0. Those roots are $x = -1$, $x = -2$, and $x = -3$. The y-values can be found by plugging these roots into either of the original functions. Let's use the linear one; it's the less messy of the two! For $x = -1$, we have

$$\begin{aligned} y &= 3x + 1 \\ &= 3 \times (-1) + 1 \\ &= -3 + 1 \\ &= -2 \end{aligned}$$

Now we know that our first solution is $(x, y) = (-1, -2)$. When $x = -2$, we have

$$\begin{aligned} y &= 3x + 1 \\ &= 3 \times (-2) + 1 \\ &= -6 + 1 \\ &= -5 \end{aligned}$$

Our second solution is $(x, y) = (-2, -5)$. Plugging in $x = -3$, we have

$$\begin{aligned} y &= 3x + 1 \\ &= 3 \times (-3) + 1 \\ &= -9 + 1 \\ &= -8 \end{aligned}$$

Our third solution is $(x, y) = (-3, -8)$.

Finally, we check

There are six arithmetic exercises to do! It's tedious, but if we want to be sure our solutions are right, it's mandatory. We'd better be careful with the signs, using negative additions rather than subtractions as much as possible! We check $(-1, -2)$ in the first original equation:

$$\begin{aligned} x^3 + 6x^2 + 14x - y &= -7 \\ (-1)^3 + 6 \times (-1)^2 + 14 \times (-1) - (-2) &= -7 \\ -1 + 6 \times 1 + (-14) + 2 &= -7 \\ -1 + 6 + (-14) + 2 &= -7 \\ -7 &= -7 \end{aligned}$$

Next, we check $(-2, -5)$ in the first original equation:

$$\begin{aligned} x^3 + 6x^2 + 14x - y &= -7 \\ (-2)^3 + 6 \times (-2)^2 + 14 \times (-2) - (-5) &= -7 \\ -8 + 6 \times 4 + (-28) + 5 &= -7 \\ -8 + 24 + (-28) + 5 &= -7 \\ -7 &= -7 \end{aligned}$$

Next, we check $(-3, -8)$ in the first original equation:

$$\begin{aligned} x^3 + 6x^2 + 14x - y &= -7 \\ (-3)^3 + 6 \times (-3)^2 + 14 \times (-3) - (-8) &= -7 \\ -27 + 6 \times 9 + (-42) + 8 &= -7 \\ -27 + 54 + (-42) + 8 &= -7 \\ -7 &= -7 \end{aligned}$$

That completes the check for the original cubic. Now we plug $(-1, -2)$ into the second original equation:

$$\begin{aligned} -6x + 2y &= 2 \\ -6 \times (-1) + 2 \times (-2) &= 2 \\ 6 + (-4) &= 2 \\ 2 &= 2 \end{aligned}$$

Next, we check (−2, −5) in the second original equation:

$$-6x + 2y = 2$$
$$-6 \times (-2) + 2 \times (-5) = 2$$
$$12 + (-10) = 2$$
$$2 = 2$$

Finishing up, we check (−3, −8) in the second original equation:

$$-6x + 2y = 2$$
$$-6 \times (-3) + 2 \times (-8) = 2$$
$$18 + (-16) = 2$$
$$2 = 2$$

Are you confused (or bemused)?

Think back to the notion of multiplicity for the roots of certain quadratic, cubic, and higher-degree equations. Do you wonder if the same concept applies to the solutions of two-by-two systems when at least one of the equations is of degree 2 or more? Well, it does! If the equation we create by mixing has a root with multiplicity of 2 or more, the corresponding solution of the whole system has the same multiplicity. You'll see this happen as you work out the last two practice exercises at the end of this chapter.

Here's a challenge!

Solve these cubic equations as a two-by-two system:

$$y = 5x^3 + 3x^2 + 5x + 7$$

and

$$y = 2x^3 + x^2 + 2x + 5$$

Solution

Solving this problem requires some keen intuition, a lot of trial and error, or both. We are lucky in one respect, at least: These equations are already functions of x, so we have no morphing to do. We can directly mix the right sides to get

$$5x^3 + 3x^2 + 5x + 7 = 2x^3 + x^2 + 2x + 5$$

Let's subtract the entire right side of this equation, as a single quantity, from both sides. That changes each term on the left and sets the right side equal to 0, giving us

$$3x^3 + 2x^2 + 3x + 2 = 0$$

Now here's the trick! If we spend enough time playing around with this, we'll find that it can be factored into

$$(3x + 2)(x^2 + 1) = 0$$

When we set the first binomial equal to 0, we get

$$3x + 2 = 0$$

which resolves to $x = -2/3$. When we set the second binomial equal to 0, we get

$$x^2 + 1 = 0$$

Subtracting 1 from both sides gives us

$$x^2 = -1$$

This resolves to $x = j$ or $x = -j$. We have found three roots to the cubic we got by mixing:

$$x = -2/3 \text{ or } x = j \text{ or } x = -j$$

We can substitute these roots into either of the original cubics. Let's use the second one. For $x = -2/3$, we have

$$\begin{aligned}
y &= 2x^3 + x^2 + 2x + 5 \\
&= 2 \times (-2/3)^3 + (-2/3)^2 + 2 \times (-2/3) + 5 \\
&= 2 \times (-8/27) + 4/9 + (-4/3) + 5 \\
&= -16/27 + 12/27 + (-36/27) + 135/27 \\
&= 95/27
\end{aligned}$$

Our first solution is $(x, y) = (-2/3, 95/27)$. For $x = j$, we have

$$\begin{aligned}
y &= 2x^3 + x^2 + 2x + 5 \\
&= 2j^3 + j^2 + 2j + 5 \\
&= 2(-j) + (-1) + j2 + 5 \\
&= -j2 + (-1) + j2 + 5 \\
&= -j2 + j2 + (-1) + 5 \\
&= 0 + 4 \\
&= 4
\end{aligned}$$

Our second solution is $(x, y) = (j, 4)$. For $x = -j$, we have

$$\begin{aligned}
y &= 2x^3 + x^2 + 2x + 5 \\
&= 2(-j)^3 + (-j)^2 + 2(-j) + 5 \\
&= 2j + (-1) + (-j2) + 5
\end{aligned}$$

$$= j2 + (-1) + (-j2) + 5$$
$$= j2 + -j2 + (-1) + 5$$
$$= 0 + 4$$
$$= 4$$

Our third solution is $(x, y) = (-j, 4)$.

And finally ...

Now it's time to check these solutions to be sure that they work. I can't blame you if you're weary of doing arithmetic, and you'd prefer to take these solutions on faith. That's all right for now. Tomorrow, when your mind is rested, come back and check them for complementary credit.

Practice Exercises

This is an open-book quiz. You may (and should) refer to the text as you solve these problems. Don't hurry! You'll find worked-out answers in App. C. The solutions in the appendix may not represent the only way a problem can be figured out. If you think you can solve a particular problem in a quicker or better way than you see there, by all means try it!

1. Solve the following pair of equations as a two-by-two system, including the complex-number solutions, if any. Let x be the independent variable, and then define y as a function of x in each equation:

$$3x + y - 1 = 0$$

and

$$2x^2 - y + 1 = 0$$

2. Check the solution(s) to Prob. 1 in the original equations for correctness.
3. Solve the following pair of equations as a two-by-two system, including the complex-number solutions, if any. Let x be the independent variable, and then define y as a function of x in each equation:

$$3x + y - 1 = 0$$

and

$$2x^2 - 3x - y + 3 = 0$$

4. Check the solution(s) to Prob. 3 in the original equations for correctness.

5. Solve the following pair of equations as a two-by-two system, including the complex-number solutions, if any. Let x be the independent variable, and then define y as a function of x in each equation:

$$x^2 + x - y = -1$$

and

$$x^2 - 2x - y = 2$$

6. Check the solution(s) to Prob. 5 in the original equations for correctness.
7. Solve the following pair of equations as a two-by-two system, including the complex-number solutions, if any. Let x be the independent variable, and then define y as a function of x in each equation:

$$x^2 + y = 0$$

and

$$2x^3 - y = 0$$

Explain why one of the solutions has multiplicity 2.

8. Check the solution(s) to Prob. 7 in the original equations for correctness.
9. Solve the following pair of equations as a two-by-two system, including the complex-number solutions, if any. Let x be the independent variable, and then define y as a function of x in each equation:

$$4x^3 + 2x^2 + 2x - 2y - 8 = 0$$

and

$$3x^3 - 2x^2 + 4x - y - 5 = 0$$

What is the multiplicity of each solution?

10. Check the solution(s) to Prob. 9 in the original equations for correctness.

CHAPTER

28

More Two-by-Two Graphs

In this chapter, we'll graph the systems we solved in Chap. 27 to see how they work in the realm of real numbers. Real solutions always appear as intersection points between the graphs of functions. However, when a two-by-two system has non-real complex solutions, those solutions don't show up as intersections between real-number graphs.

Linear and Quadratic

The first example we worked out in Chap. 27 involved a system of two equations, one linear and the other quadratic:

$$2x - y + 1 = 0$$

and

$$y = x^2 + x - 5$$

We called x the independent variable, so we rewrote the first equation as a function of x. The two-function system then became

$$y = 2x + 1$$

and

$$y = x^2 + x - 5$$

First, we tabulate some points

The graphs we're about to create aren't meant to be precise. We aren't trying to pinpoint the real solutions using the graphs; we've already calculated them! The graphs serve only to geometrically illustrate the functions and their solutions.

Table 28-1. Selected values for graphing the functions $y = 2x + 1$ and $y = x^2 + x - 5$. Bold entries indicate real solutions.

x	$2x + 1$	$x^2 + x - 5$
−4	−7	7
−3	−5	1
−2	**−3**	**−3**
0	1	−5
1	3	−3
3	**7**	**7**
4	9	15
5	11	25

Table 28-1 compares some values of x, some values of the first function, and some values of the second function. The left-hand column contains selected values of x. The middle column contains values of the linear function that we get when we input the chosen values of x. The right-hand column contains values of the quadratic function that we get for the indicated values of x.

Are you confused?

Do you wonder why we chose the values in Table 28-1 as we did? The two solutions can be tabulated easily, so it makes sense to include them:

$$(x, y) = (-2, -3)$$

and

$$(x, y) = (3, 7)$$

These solutions are written down in boldface. When we graph the functions, the points corresponding to these solutions will be the points where the graphs intersect.

The other points are chosen strategically. We want to find ordered pairs that lie in the vicinity of the solutions. That means we should choose values of x that are somewhat less than −2, somewhere between −2 and 3, and somewhat larger than 3. Then, when we plot the graphs, we can expect to get a good view with the solutions near the middle. We calculate the values of the functions in the middle and right-hand columns by plugging in the values of x and going through the arithmetic.

Next, we plot the solution(s)

To plot the solutions of this system, we could use a strict Cartesian plane with each division on both axes equal to 1 unit. But we must cover a span of −4 to 5 for the independent variable, and a span of −7 to 25 for the dependent variable, based on the function values we have found in Table 28-1. A true Cartesian graph having that span would be as big as a road map!

We could shrink it by making each division represent, say, 5 units on both axes. But then the solutions wouldn't show up well; they'd be too close together.

In situations like this, the best approach is to use rectangular coordinates, but not strict Cartesian coordinates. Let's make each increment on the x axis represent 1 unit, and each increment on the y axis represent 5 units. When we make the axis increments different in size, we distort the slopes and contours of the lines and curves, but that's not important here. Our goal is only to get a good fit for the ranges of the values we determined when we created Table 28-1, and a clear picture of how the graphs intersect. To begin, we plot the two solution points.

Finally, we plot the rest

Once we've plotted the solution points as dots on the coordinate grid, we can locate and plot the rest of the points in the table. If we do this on paper, we can use pencil and draw the points lightly, so we can erase them later.

As we draw the points, we must remember the increments we've chosen for each axis. They aren't the same, and it's easy to get them confused. It's also important to keep track of which graph goes with which function!

We can draw the points for one graph with a pencil, fill in its line or curve with a pen, draw the points for the second graph with a pencil, fill in its line or curve with a pen, and then erase all the penciled-in points when the ink is dry. If we've chosen the axis increments wisely, the solutions will be near the center of the coordinate area, and the other points will be scattered fairly well over the rest of it. Fig. 28-1 shows the final result.

Here's a challenge!

By examining Fig. 28-1, describe how the linear function

$$y = 2x + 1$$

(shown by the solid line) can be modified to produce a system with no real solutions, assuming that the quadratic function (shown by the dashed curve) stays the same, and assuming the slope of the line stays the same.

Solution

If we change the linear function so its graph doesn't intersect the parabola, then the resulting two-by-two system will have no real solutions. Imagine moving the straight line downward in Fig. 28-1. As we do this, the solution points get closer together, eventually merging. At the moment the two points become one, the system has a single real solution with multiplicity 2.

The linear function is expressed in slope-intercept form, so we can see that the line has a slope of 2 and a y-intercept of 1. Each division on the y axis is 5 units, so we can see that the quadratic function has an absolute minimum of approximately −5. Suppose we change the linear function so the y-intercept is −10. That will put the line completely below the parabola, and will give us the two-by-two system

$$y = 2x - 10$$

and

$$y = x^2 + x - 5$$

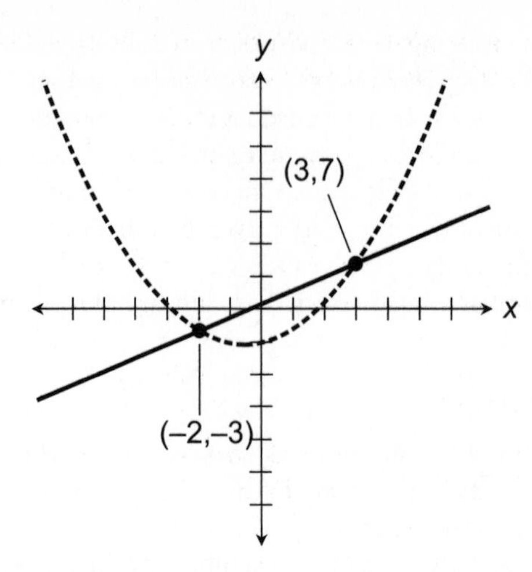

Figure 28-1 Graphs of $y = 2x + 1$ and $y = x^2 + x - 5$. The first function is graphed as a solid line; the second function is graphed as a dashed curve. Real-number solutions appear as points where the line and the curve intersect. On the x axis, each increment is 1 unit. On the y axis, each increment is 5 units.

If we solve this system, we'll get two results, but they'll both be non-real complex numbers. For extra credit, you can solve the "revised" system and see for yourself.

Are you confused?

Does the slope of the line in Fig. 28-1 seem smaller than 2? In a way, it is! In the *algebraic* sense the slope is 2, but in the *geometric* sense it's only 2/5. The increments on the vertical axis are five times as large as the ones on the horizontal axis. That distorts the slopes and contours of the graphs, expanding everything horizontally (or compressing everything vertically) by a factor of 5. If we had used a true Cartesian plane, the line would have the steepness we should expect for a slope of 2 when drawn. The parabola would also look different; it would seem "five times sharper."

Two Quadratics

In the second example in Chap. 27, we solved this system of quadratic equations in two variables:

$$4x^2 + 6x + 2y + 8 = 0$$

and

$$3x^2 + y + 5x - 11 = 0$$

Again, we let x be the independent variable, and then we manipulated the equations to obtain y as a function of x in both cases. That gave us

$$y = -2x^2 - 3x - 4$$

and

$$y = -3x^2 - 5x + 11$$

First, we tabulate some points

Table 28-2 shows some values of x, along with the results of plugging those values into the above functions and churning out the arithmetic. We can start building the table by entering the two solutions, which we determined in Chap. 27. They are

$$(x, y) = (-5, -39)$$

and

$$(x, y) = (3, -31)$$

These solutions are written as bold numerals. The other values are chosen to produce graph points in the vicinity of the solutions. We can choose a couple of x-values less than -5, two more between -5 and 3, and two more larger than 3. Then we can calculate the values of the functions, and write them in the middle and right-hand columns of the table.

Table 28-2. Selected values for graphing the functions $y = -2x^2 - 3x - 4$ and $y = -3x^2 - 5x + 11$. Bold entries indicate real solutions.

x	$-2x^2 - 3x - 4$	$-3x^2 - 5x + 11$
−10	−174	−239
−7	−81	−101
−5	**−39**	**−39**
−2	−6	9
0	−4	11
3	**−31**	**−31**
6	−94	−127
9	−193	−277

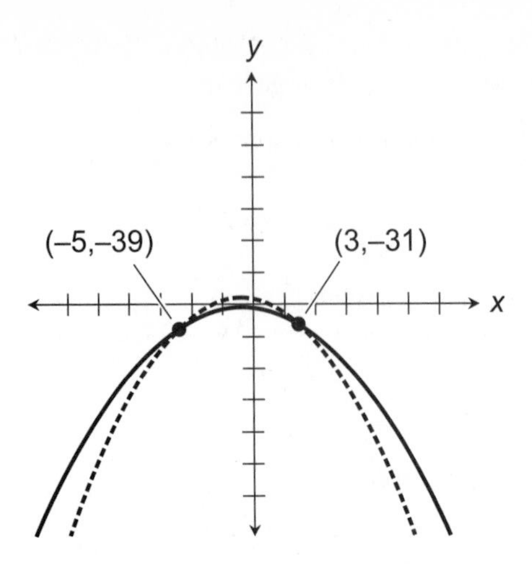

Figure 28-2 Graphs of $y = -2x^2 - 3x - 4$ and $y = -3x^2 - 5x + 11$. The first function is graphed as a solid curve; the second function is graphed as a dashed curve. Real-number solutions appear as points where the curves intersect. On the x axis, each increment is 2 units. On the y axis, each increment is 50 units.

Next, we plot the solution(s)

Once again, let's use rectangular coordinates. In this case, the span of absolute values for the input (the independent variable x) is from 0 to 10, while the absolute values of the functions go as high as 277. Let's make each increment on the x axis represent 2 units, and each increment on the y axis represent 50 units. With six "hash marks" going out from 0 in each of the four directions along the axes, that gives us absolute-value spans from 0 to 12 for x and 0 to 300 for y, as shown in Fig. 28-2. We then plot the two solution points.

Finally, we plot the rest

We can fill in the graphs by plotting the remaining points indicated in Table 28-2. As before, it's a good idea to draw the points for one graph with a pencil, fill in its curve with a pen, draw the points for the other graph with a pencil, fill in its curve with a pen, wait for the ink to dry, and finally run an eraser over the curves to get rid of the pencil marks. In Fig. 28-2, the approximate graph for

$$y = -2x^2 - 3x - 4$$

is a solid parabola, and the approximate curve for

$$y = -3x^2 - 5x + 11$$

is a dashed parabola.

Here's a challenge!

By examining Fig. 28-2, describe how the quadratic function

$$y = -2x^2 - 3x - 4$$

(shown by the solid curve) can be modified to produce a system with no real solutions, assuming that the other quadratic function (shown by the dashed curve) stays the same, and also assuming that the contour of the graph for the modified function stays the same.

Solution: Phase 1

We can move the solid parabola straight upward. As we do that, the two intersection points get closer together. When the solid parabola reaches a certain "critical altitude," the intersection points merge, indicating that the system has a single real solution with multiplicity 2. If we move the solid parabola upward beyond the "critical altitude," the two parabolas no longer intersect. We can intuitively see this by comparing the "sharpness" of the two curves. The solid parabola is not as "sharp" as the dashed one, so the two curves diverge once we have raised the solid parabola high enough to completely clear the dashed one.

Are you confused?

"How," you ask, "can we change a quadratic function to move its parabola straight upward?" The answer is simple. We can increase the value of the stand-alone constant, leaving the rest of the equation unchanged. That increases the y-values of all the points without changing their x-values. Every point on the parabola is displaced straight upward by the same amount as every other point.

Solution: Phase 2

"All right," you say. "How much must we increase the constant in the first function to be sure that the solid parabola clears the dashed parabola?" We can find out using some creative algebra. Look again at the process we used in Chap. 27 to derive an equation for the x-value of the solution to this system. We started with the two quadratic functions

$$y = -2x^2 - 3x - 4$$

and

$$y = -3x^2 - 5x + 11$$

When we mixed the right sides, we got

$$-2x^2 - 3x - 4 = -3x^2 - 5x + 11$$

which simplified to

$$x^2 + 2x - 15 = 0$$

Examine the discriminant d for this equation:

$$\begin{aligned} d &= 2^2 - 4 \times 1 \times (-15) \\ &= 4 - (-60) \\ &= 4 + 60 \\ &= 64 \end{aligned}$$

How can we change the stand-alone constant, which is −15 right now, to bring the discriminant down to 0? We can find out by substituting 0 for d, and by substituting a letter constant for −15 in the above equation. If use k as the letter constant, we get

$$0 = 2^2 - 4 \times 1 \times k$$

which simplifies to

$$0 = 4 - 4k$$

and further to

$$4k = 4$$

This resolves to $k = 1$. In the quadratic we got by mixing, let's change the stand-alone constant k from −15 to 1. That gives us

$$x^2 + 2x + 1 = 0$$

If we increase the constant any more, then d becomes negative, and the system has no real solutions.

Now we know that if we increase the stand-alone constant by 16 in the quadratic represented by the solid parabola in Fig. 28-2, the two solution points will merge in the graph. So let's increase that constant by 17! When we do that, the function becomes

$$y = -2x^2 - 3x + 13$$

The graph of this function is a parabola with the same contour as the original one, but displaced by 17 units upward in the coordinate plane. For extra credit, try solving the system

$$y = -2x^2 - 3x + 13$$

and

$$y = -3x^2 - 5x + 11$$

and see for yourself that it has no real solutions.

Here's another challenge!

By examining Fig. 28-2, describe how the quadratic function

$$= -3x^2 - 5x + 11$$

(shown by the dashed curve) can be modified to produce a system with no real solutions, assuming that the other quadratic function (shown by the solid curve) stays the same, and also assuming that the contour of the graph for the modified function stays the same.

Solution

To cause the two solution points to merge, we must move the dashed parabola downward to the same extent we moved the solid parabola upward in the previous example. That means we must reduce the constant in the function for the dashed parabola by 16, giving us

$$y = -3x^2 - 5x - 5$$

If we reduce the constant to anything smaller than −5, the real solutions of the two-by-two system vanish, and the parabolas no longer intersect.

Enter the Cubic

The third time around in Chap. 27, we solved this two-by-two system:

$$x^3 + 6x^2 + 14x - y = -7$$

and

$$-6x + 2y = 2$$

As with all the other examples, we let x be the independent variable, and then we morphed to get y as a function of x in both cases. In this situation we got

$$y = x^3 + 6x^2 + 14x + 7$$

and

$$y = 3x + 1$$

We found these solutions:

$$(x, y) = (-1, -2)$$
$$(x, y) = (-2, -5)$$
$$(x, y) = (-3, -8)$$

First, we tabulate some points

Table 28-3 shows several different values of x, along with the results of plugging those values into the functions and calculating. All three solutions are included, and are written in bold. We also include two x- values less than −3, and two larger than −1. Because the x-values for the solutions are consecutive integers, it makes sense to choose consecutive integers on either side of them.

Table 28-3. Selected values for graphing the functions $y = x^3 + 6x^2 + 14x + 7$ and $y = 3x + 1$. Bold entries indicate real solutions.

x	$x^3 + 6x^2 + 14x + 7$	$3x + 1$
−5	−38	−14
−4	−17	−11
−3	**−8**	**−8**
−2	**−5**	**−5**
−1	**−2**	**−2**
0	7	1
1	28	4

Next, we plot the solution(s)

By examining Table 28-3, we can see that the span of absolute values for the input is from 0 to 5, while the absolute values of the functions go up to 38. This time, let's make each increment on the x axis represent 1 unit, and each increment on the y axis represent 5 units. With six divisions going out from 0 to the left and six to the right, that gives us an absolute-value span from 0 to 6 for x, and that's enough. For y, we have six divisions going up and eight going down, and that's sufficient to include all the function values in Table 28-3. Now that we've decided on the dimensions of the coordinate grid, we can plot the three solution points as shown in Fig. 28-3.

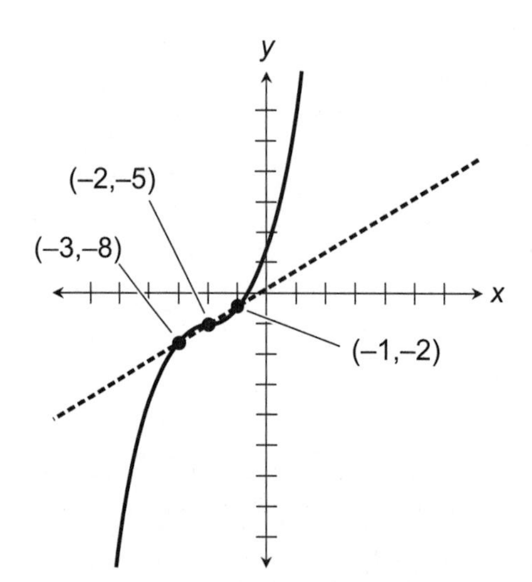

Figure 28-3 Graphs of $y = x^3 + 6x^2 + 14x + 7$ and $y = 3x + 1$. The first function is graphed as a solid curve; the second function is graphed as a dashed line. Real-number solutions appear as points where the curve and the line intersect. On the x axis, each increment is 1 unit. On the y axis, each increment is 5 units.

Finally, we plot the rest

We can fill in the graphs by plotting the remaining points in the table. In Fig. 28-3, the approximate graph for

$$y = x^3 + 6x^2 + 14x + 7$$

is the solid curve, and the approximate graph for

$$y = 3x + 1$$

is the dashed line.

Are you confused?

Figure 28-3 doesn't show the relationship between the curve and the line very well in the vicinity of the solution points. If you want to get a "finer" graph in that region, you can plot points at intervals of 1/2 unit, 1/5 unit, or even 1/10 unit for x-values between −4 and 0 or between −5 and 1. You can also include more points "farther out," say for x-values of −7, −10, and −15 on the negative side and 5, 10, and 15 on the positive side. A programmable calculator, or a personal computer with calculating software installed, makes an excellent assistant for this process, and can save you from having to do a lot of tedious arithmetic. You might also find a site on the Internet that can calculate values of a linear, quadratic, cubic, or higher-degree function based on coefficients, the constant, and input values you choose.

Here's a challenge!

In the "challenge" at the end of Chap. 27, we solved the following two cubic functions as a two-by-two system:

$$y = 5x^3 + 3x^2 + 5x + 7$$

and

$$y = 2x^3 + x^2 + 2x + 5$$

We got one real solution, $(x, y) = (-2/3, 95/27)$, and two complex-conjugate solutions. Draw a graph showing these two functions, along with the real solution point.

Solution

Table 28-4 shows several values of x, along with the resulting function values. The solution is in the middle, written in bold. The span of values for the input is from −3 to 2, while the span of values of the functions is from −116 to 69. Let's make each increment on the x axis represent 1/2 unit, and each increment on the y axis represent 10 units. With six divisions going out from 0 to the left and six to the right, that gives us a span from −3 to 3 for x. For y, we have eight divisions going up and 12 divisions going down, and that's a span

Table 28-4. Selected values for graphing the functions $y = 5x^3 + 3x^2 + 5x + 7$ and $y = 2x^3 + x^2 + 2x + 5$. The bold entry indicates the real solution.

x	$5x^3 + 3x^2 + 5x + 7$	$2x^3 + x^2 + 2x + 5$
−3	−116	−46
−2	−31	−11
−1	0	2
−2/3	**95/27**	**95/27**
Approx. −0.67	**Approx. 3.52**	**Approx. 3.52**
0	7	5
1	20	10
2	69	29

of −120 to 80, more than enough to include all the function values in Table 28-4. To plot the solution point, we can convert the values to decimal form and go to a couple of decimal places. Then we get

$$(x, y) = (-0.67, 3.52)$$

This point is shown as a solid dot in Fig. 28-4. Once we've plotted it, we fill in the graphs of the functions. The approximate graph for

$$y = 5x^3 + 3x^2 + 5x + 7$$

is the solid curve, and the approximate graph for

$$y = 2x^3 + x^2 + 2x + 5$$

is the dashed curve.

Are you still confused?

Do you wonder about the "cubic curves" in Figs. 28-3 and 28-4? They're a lot different from the graphs of quadratics! The graph of a cubic function always has one of six characteristic shapes, as shown in Fig. 28-5. They all look rather like distorted images of the letter "S" tipped on its side, perhaps flipped over backward, and then extended forever upward and down.

Unlike a quadratic function, which has a limited range with an absolute maximum or an absolute minimum, a cubic function always has a range that spans the entire set of real numbers, although it can have a *local maximum* and a *local minimum.* The graph of a cubic function also has something else that you'll never see in the graph of a quadratic: an *inflection point,* where the curvature reverses direction. The contour of the graph depends on the signs and values of the function's coefficients and constant.

If you want to get familiar with how the graphs of various cubic functions look, you can conjure up a few cubic functions with assorted coefficients and constants. Then plot a couple of dozen points for each function, and connect the points with smooth curves. But don't spend too much time at this. A book devoted to the art of graphing cubic and higher-degree functions could consume thousands of pages! You'll learn more about graphing functions when you take a course in calculus.

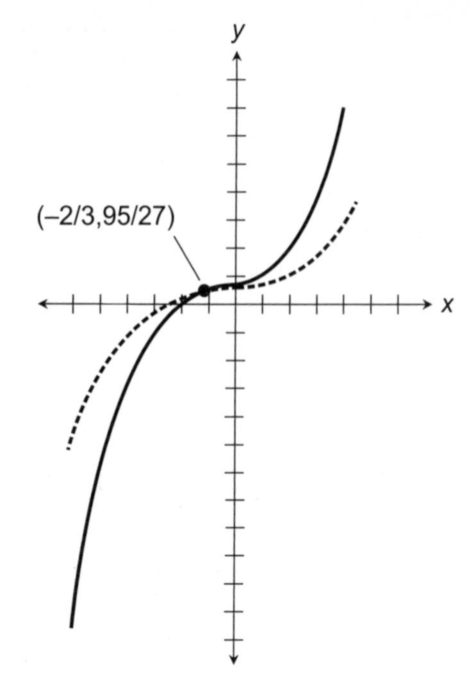

Figure 28-4 Graphs of $y = 5x^3 + 3x^2 + 5x + 7$ and $y = 2x^3 + x^2 + 2x + 5$. The first function is graphed as a solid curve; the second function is graphed as a dashed curve. The real-number solution appears as the point where the curves intersect. On the x axis, each increment is 1/2 unit. On the y axis, each increment is 10 units.

Practice Exercises

This is an open-book quiz. You may (and should) refer to the text as you solve these problems. Don't hurry! You'll find worked-out answers in App. C. The solutions in the appendix may not represent the only way a problem can be figured out. If you think you can solve a particular problem in a quicker or better way than you see there, by all means try it!

1. Look again at Practice Exercise 1 and its solution from Chap. 27. Create a table for both functions based on x-values of -3, -2, $-3/2$, -1, $-1/2$, 0, 1, and 2. Here are the functions that came from the original equations:

$$y = -3x + 1$$

and

$$y = 2x^2 + 1$$

Use bold numerals to indicate the real solutions, if any exist.

2. Plot an approximate graph showing the curves based on the table you created when you worked out Prob. 1. On the x axis, let each increment represent 1/2 unit. On the y axis, let each increment represent 3 units. Draw the first function's graph as a solid line or curve. Draw the second function's graph as a dashed line or curve. Plot and label all real solution points, if any exist.

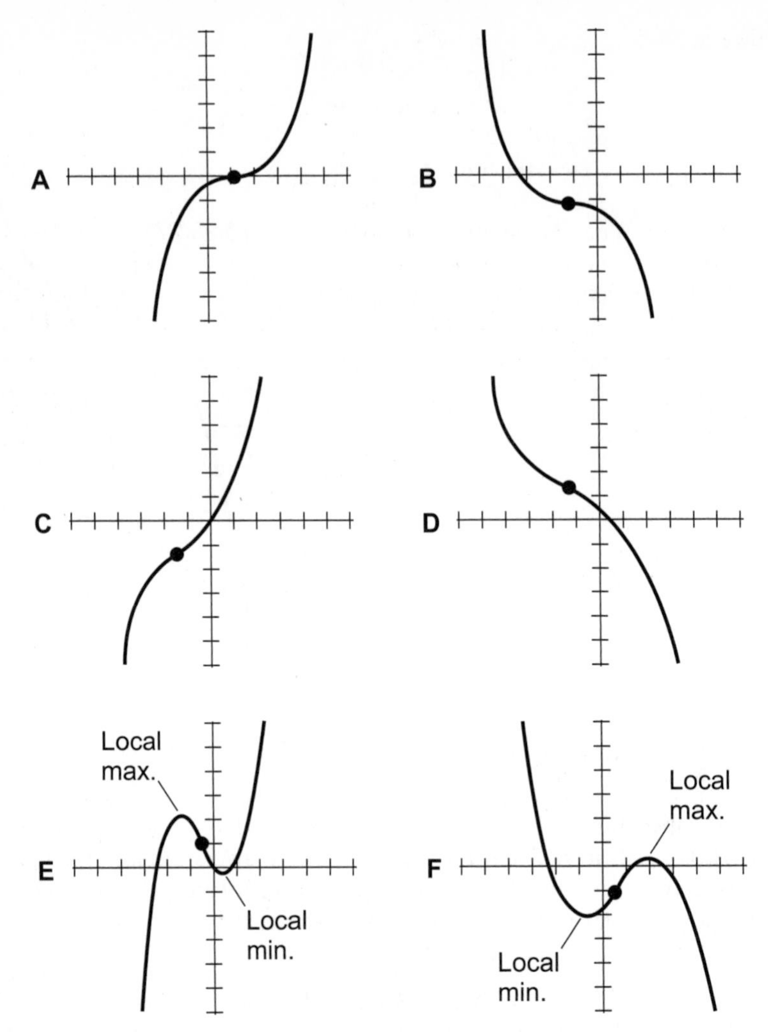

Figure 28-5 Characteristic shapes for graphs of cubic functions. At A and B, the curves maintain upward or downward trends, although they level off at the inflection points (solid dots). At C and D, the curves maintain upward or downward trends, and they don't level off at the inflection points. At E and F, the curves trend mainly upward or downward but "back up" between local maxima and local minima, with inflection points in the "backward zones."

3. Look again at Practice Exercise 3 and its solution from Chap. 27. Create a table of values for both functions, based on *x*-values of −3, −2, −1, 0, 1, 2, 3, and 4. Here are the functions that came from the original equations:

$$y = -3x + 1$$

and

$$y = 2x^2 - 3x + 3$$

Use bold numerals to indicate the real solutions, if any exist.

4. Plot an approximate graph showing the curves based on the table you created when you worked out Prob. 3. On the x axis, let each increment represent 1 unit. On the y axis, let each increment represent 5 units. Draw the first function's graph as a solid line or curve. Draw the second function's graph as a dashed line or curve. Plot and label all real solution points, if any exist.

5. Look again at Practice Exercise 5 and its solution from Chap. 27. Create a table of values for both functions, based on x-values of -4, -3, -2, -1, 0, 1, 2, 3, and 4. Here are the functions that came from the original equations:

$$y = x^2 + x + 1$$

and

$$y = x^2 - 2x - 2$$

Use bold numerals to indicate the real solutions, if any exist.

6. Plot an approximate graph showing the curves based on the table you created when you worked out Prob. 5. On the x axis, let each increment represent 1 unit. On the y axis, let each increment represent 4 units. Draw the first function's graph as a solid line or curve. Draw the second function's graph as a dashed line or curve. Plot and label all real solution points, if any exist.

7. Look again at Practice Exercise 7 and its solution from Chap. 27. Create a table of values for both functions, based on x-values of -3, -2, -1, 0, 1, 2, and 3. Here are the functions that came from the original equations:

$$y = -x^2$$

and

$$y = 2x^3$$

Use bold numerals to indicate the real solutions, if any exist.

8. Plot an approximate graph showing the curves based on the table you created when you worked out Prob. 7. On the x axis, let each increment represent 1/2 unit. On the y axis, let each increment represent 10 units. Draw the first function's graph as a solid line or curve. Draw the second function's graph as a dashed line or curve. Plot and label all real solution points, if any exist.

9. Look again at Practice Exercise 9 and its solution from Chap. 27. Create a table of values for both functions, based on *x*-values of −3, −2, −1, 0, 1, 2, 3, and 4. Here are the functions that came from the original equations:

$$y = 2x^3 + x^2 + x - 4$$

and

$$y = 3x^3 - 2x^2 + 4x - 5$$

Use bold numerals to indicate the real solutions, if any exist.

10. Plot an approximate graph showing the curves based on the table you created when you worked out Prob. 9. On the *x* axis, let each increment represent 1 unit. On the *y* axis, let each increment represent 20 units. Draw the first function's graph as a solid line or curve. Draw the second function's graph as a dashed line or curve. Plot and label all real solution points, if any exist.

CHAPTER

29

Logarithms and Exponentials

Logarithms and exponentials show up in many branches of mathematics, sometimes unexpectedly. If you're getting ready for more advanced subjects such as analysis, calculus, or engineering mathematics, you should know something about logarithms and exponentials.

What Is a Logarithm?

A *logarithm* (sometimes called a *log*) of a quantity is a power to which a positive real constant is raised to get that quantity. The constant is called the *base*, which is usually 10 (the base familiar to most of us) or *e*, an irrational number called *Euler's constant.*

Finding e

If your calculator lacks a key that displays *e* directly, here's a trick that will put it there for you. Enter the number 1, then hit the inverse-function key or put a check in the appropriate box if you're using a computer (it might be labeled "Inv" or "Rev"). Then hit the "ln" or "$\log_e$" key. You should see 2.71828 ... and then some more digits.

The log base

Imagine a positive real constant b. We raise b to some real variable power y, getting another real number x as the result. We can write this as

$$b^y = x$$

where $b > 0$. In this equation, the exponent y is the *base-b logarithm* of x. When we say it that way, we write it as

$$y = \log_b x$$

A logarithm is an exponent in a situation where it is also the dependent variable in a function. The above equation tells us that y is a *base-b logarithmic function* of x.

We can raise negative numbers to real-number powers, but this is rarely done with logarithmic functions. We aren't likely to encounter any "base-(−10)" or "base-(−*e*)" logarithms, for example, although technically there is no reason why such things can't exist.

Common logs

Base-10 logarithms are also known as *common logarithms* or *common logs*. In equations, common logs are denoted by writing "log" with a subscript 10, and then the number, called the *argument*, for which you want to find the logarithm. Here are a few examples that you can verify with your calculator:

$$\log_{10} 100 = 2$$
$$\log_{10} 45 \approx 1.653$$
$$\log_{10} 10 = 1$$
$$\log_{10} 6 \approx 0.7782$$
$$\log_{10} 1 = 0$$
$$\log_{10} 0.5 \approx -0.3010$$
$$\log_{10} 0.1 = -1$$
$$\log_{10} 0.07 \approx -1.155$$
$$\log_{10} 0.01 = -2$$

The squiggly equals sign means "is approximately equal to." You'll often see it in scientific and engineering papers, articles, and books. The first equation above is another way of writing

$$10^2 = 100$$

You could also say "The common log of 100 is equal to 2." The second equation is another way of writing

$$10^{1.653} \approx 45$$

You could also say "The common log of 45 is approximately equal to 1.653."

Natural logs

Base-*e* logarithms are also called *natural logs* or *Napierian logs*. In equations, the natural-log function is usually denoted by writing "ln" or "$\log_e$" followed by the argument. Here are some equations using the natural-log function, which you can check out with your calculator:

$$\ln 100 \approx 4.605$$
$$\ln 45 \approx 3.807$$
$$\ln 10 \approx 2.303$$
$$\ln 6 \approx 1.792$$
$$\ln e = 1$$
$$\ln 1 = 0$$

$$\ln(1/e) = -1$$
$$\ln 0.5 \approx -0.6931$$
$$\ln 0.1 \approx -2.303$$
$$\ln 0.07 \approx -2.659$$
$$\ln 0.01 \approx -4.605$$

The first equation above is another way of writing

$$e^{4.605} \approx 100$$

You would say it as "The natural log of 100 is equal to approximately 4.605." The second equation is an alternative way writing

$$e^{3.807} \approx 45$$

You could also say "The natural log of 45 is approximately equal to 3.807."

The arguments in common or natural log functions don't have to be whole numbers, fractions, or terminating decimals. You can have logs of arguments that are irrational, such as π, e, the positive square root of 2, or the cube root of 100.

Don't let them confuse you!

Authors don't all agree on what the notation "log" means. In some texts, natural logs (that is, base-e logs) are denoted by writing "log" without a subscript, followed by the argument. But in other texts and in most calculators, "log" means the common (base-10) log.

To avoid confusion, always include the base as a subscript whenever you write "log" followed by an argument. For example, write "$\log_{10}$" or "$\log_e$" instead of "log" all by itself. You don't have to use a subscript when you write "ln" for the natural log.

If you aren't sure what the "log" key on a calculator does, do a trial calculation to find out. If the "log" of 10 equals 1, then it's the common log. If the "log" of 10 equals an irrational number slightly larger than 2.3, then it's the natural log.

Here's a challenge!

Compare the common logarithms of 0.01, 0.1, 1, 10, and 100. Then compare the natural logarithms of those same arguments.

Solution

Note that $0.01 = 10^{-2}$, $0.1 = 10^{-1}$, $1 = 10^0$, $10 = 10^1$, and $100 = 10^2$. Therefore:

$$\log_{10} 0.01 = -2$$
$$\log_{10} 0.1 = -1$$
$$\log_{10} 1 = 0$$
$$\log_{10} 10 = 1$$
$$\log_{10} 100 = 2$$

These values are all exact! But now suppose you want to compare the natural logs of these same five arguments. The base-e logarithm of a number is the power of e that produces that number. You must use a calculator to find natural logs. The results are as follows, rounded off to four decimal places in each case (except the natural log of 1, which is an exact value):

$$\ln 0.01 \approx -4.6052$$
$$\ln 0.1 \approx -2.3026$$
$$\ln 1 = 0$$
$$\ln 10 \approx 2.3026$$
$$\ln 100 \approx 4.6052$$

The log of 1 is always equal to 0, no matter what the base, because any positive real number raised to the zeroth power is equal to 1.

How Logarithms Work

With logarithms, you can convert products into sums, ratios into differences, powers into products, and roots into ratios.

Changing a product to a sum

Imagine two positive real number variables x and y. The logarithm of their product, no matter what the base b happens to be (as long as $b > 0$), is always equal to the sum of the logarithms of the individual numbers. You can write this as

$$\log_b xy = \log_b x + \log_b y$$

Let's look at a numerical example. Consider the arguments exact. Use your calculator to follow along:

$$\log_{10} (3 \times 4) = \log_{10} 3 + \log_{10} 4$$

Working out both sides and approximating the results to four decimal places, you should get

$$\log_{10} 12 \approx 0.4771 + 0.6021$$
$$\approx 1.0792$$

You shouldn't expect to get perfect answers every time you use logarithms, because the results are almost always irrational numbers. That means they are endless, non-repeating decimals. Approximation is the best you can do.

Changing a ratio to a difference

Again, suppose that x and y are positive real numbers. Then the logarithm of their ratio, regardless of the base b, is equal to the difference between the logarithms of the individual numbers:

$$\log_b (x/y) = \log_b x - \log_b y$$

You can work out an example using the same numerical arguments as before. Again, follow along with your calculator:

$$\log_{10} (3/4) = \log_{10} 3 - \log_{10} 4$$

Working out both sides and approximating the results to four decimal places:

$$\begin{aligned} \log_{10} 0.7500 &\approx 0.4771 - 0.6021 \\ &\approx -0.1250 \end{aligned}$$

Are you confused?

If you take the base-10 log of 0.7500 directly with your calculator and round it off to four decimal places, you'll see −0.1249, not −0.1250. What's going on? Why is there a discrepancy between these two methods of determining the base-10 log of 3/4? It's not a flaw in the calculator, and it's not your imagination. This is an example of a phenomenon called *rounding error*, which often occurs when repeated calculations are done using approximate values. This isn't the last time you're going to see it!

Changing a power to a product

Logarithms simplify the raising of a number to a power. This tactic is useful when the argument does not have two whole numbers, and it's especially handy when irrationals enter the scene. Let x be a positive real number, and let y be any real number. The base-b logarithm of x raised to the power y can be rearranged as a product:

$$\log_b x^y = y \log_b x$$

Again, this works for any positive base b. Here's an example using the same arguments as before, carried out to four decimal places:

$$\begin{aligned} \log_{10} (3^4) &= 4 \log_{10} 3 \\ \log_{10} 81 &\approx 4.0000 \times 0.4771 \\ &\approx 1.9084 \end{aligned}$$

Now let's try an example in which both of the numbers in the input argument are decimals. We'll go to three places:

$$\begin{aligned} \log_{10} (2.635^{1.078}) &= 1.078 \log_{10} 2.635 \\ &\approx 1.078 \times 0.421 \\ &\approx 0.454 \end{aligned}$$

We'll return to this result later.

Changing a reciprocal to a negative

The logarithm (to any base b) of the reciprocal of a number is equal to the negative of the logarithm of that number. Stated mathematically, if x is a positive real number, then

$$\log_b (1/x) = -(\log_b x)$$

This is a special case of division simplifying to subtraction. Now let's look at a numerical example. Suppose $x = 3$ (exactly) and we use natural logs, as follows:

$$\ln (1/3) = -(\ln 3)$$

Using a calculator, we can evaluate both expressions. This time, let's go to nine decimal places.

$$\ln 0.333333333 \approx -(\ln 3.000000000)$$
$$-1.098612290 \approx -1.098612289$$

We have some rounding error here, but it's only at the ninth decimal place, equivalent to one part in 1,000,000,000!

Reciprocal within an exponent

What happens when we have a reciprocal in an exponent? Suppose x is a positive real number, and y is any real number except zero. Then the logarithm (to any positive base b) of the yth root of x (also denoted as x to the $1/y$ power) is equal to the log of x, divided by y:

$$\log_b (x^{1/y}) = (\log_b x) / y$$

Let's try this with $x = 8$ and $y = 1/3$, considering both values exact, and using natural logs evaluated to five decimal places. We get

$$\ln (8^{1/3}) = (\ln 8) / 3$$

We know that the 1/3 power (or cube root) of 8 is equal to 2. Therefore

$$\ln 2 = (\ln 8) / 3$$
$$0.69315 \approx 2.07944 / 3$$
$$0.69315 = 0.69315$$

The results agree to all five decimal places here, so we can write a plain equals sign instead of a squiggly one.

Log conversions

Here are a couple of useful rules for converting natural logs to common logs and vice-versa. You'll sometimes have to do this, especially if you get into physics or engineering.

Suppose that x is a positive real number. The common logarithm of x can be expressed in terms of the natural logarithms of x and 10. If we go to six decimal places to approximate the natural log of 10, we have

$$\begin{aligned}\log_{10} x &= (\ln x) / (\ln 10)\\ &\approx (\ln x) / 2.302585\\ &\approx 0.434294 \ln x\end{aligned}$$

Let's try a numerical example. Let $x = 3.537$. Working to three decimal places, we can use a calculator to find

$$\ln 3.537 \approx 1.263$$

We multiply by 0.434294 and round the answer off to three decimal places, getting

$$0.434294 \times 1.263 \approx 0.549$$

We can compare this with the common log of 3.537 as a calculator determines it, again rounding off to three decimal places:

$$\log_{10} 3.537 \approx 0.549$$

Now let's go the other way. Suppose x is a positive real number. The natural logarithm of x can be expressed in terms of the common logarithms of x and e. If we go to six decimal places to approximate the common log of e, we have

$$\begin{aligned}\ln x &= (\log_{10} x) / (\log_{10} e)\\ &\approx (\log_{10} x) / 0.434294\\ &\approx 2.302588 \log_{10} x\end{aligned}$$

If you're astute, you'll notice that this value differs slightly from the constant 2.302585 we obtained above. Again, this is an example of rounding error. To demonstrate with a calculator, the reciprocal of 0.434294 comes out as 2.302588 when you round it off to six decimal places. But if you take the constant derived from ln 10 earlier, which is 2.302525 when rounded to six decimal places, its reciprocal comes out as 0.434294.

Are you confused?

"Is there any way," you ask, "to eliminate rounding errors in calculations?" The answer is, "Yes, sometimes; but it can be a little tricky." When working with irrational numbers, or any other numbers where the values can be approximated but never exactly written down, you can carry out your calculations to many more decimal places than necessary until the very end, and then—but only then—round the values off to the number of places you want.

If you have a high-end calculator that goes to two or three dozen digits such as the sort found in some computers, you can leave all the displayed digits in the figures as you go through the arithmetic. Using the memory store and recall functions, those extra digits can be kept in the process, and that will keep the rounding error extremely small. When you get the final answer and round it off to, say, six decimal places instead of the two or three dozen, the error won't show up. You can play around with a good calculator to see how this works.

Are you still confused?

You might also ask, "What about the logarithm of 0, or of any negative number?" The answer is, "These are not defined in the set of real numbers, no matter what the base." To understand why, check out what happens if you try to calculate the common log of −2. Suppose that

$$\log_{10} -2 = x$$

This can be rewritten as

$$10^x = -2$$

What's the value of x? That's hard to say, but it's not a real number! No matter what real number you choose for x, the value of 10^x is positive. If you change −2 to any other negative number, or to 0, you run into the same problem. It's impossible to find any real number x, such that 10^x is not a positive real.

Here's a challenge!

Find the natural log of 238.967 from its common log using the above conversion formula. Express your answer to three decimal places. Then take the natural log of 238.967 directly with your calculator, round it off to three decimal places, and compare that result with the first one.

Solution

Working with a calculator, you should get

$$\log_{10} 238.967 \approx 2.378$$

Now remember the conversion formula:

$$\ln x \approx 2.302588 \log_{10} x$$

In this example, you should multiply 2.302588 by 2.378. When you round the answer off to three decimal places, you should get

$$2.302588 \times 2.378 \approx 5.476$$

When you take the natural log of 238.967 directly with a calculator and then round off to three decimal places, you'll see that

$$\ln 238.967 \approx 5.476$$

What Is an Exponential?

The *exponential* of a quantity is what you get when you raise a certain positive real number, called the base, to a power equal to that quantity. As is the case with logarithms, the two bases you'll most often see are 10 and *e*. When Euler's constant, *e*, is raised to a variable power, it's often called the *exponential constant.*

The exponential base

An exponential function has a base and an exponent, just as a logarithm has. In fact, an exponential function is an "inside-out" way of looking at a logarithmic function! Suppose you have three real numbers b, x, and y, where $b > 0$, and you raise b to the xth power to get y, like this:

$$y = b^x$$

Then y is the *base b exponential* of x. In the expression $10^2 = 100$, 10 is the base and 100 is the exponential. In the expression $e^3 \approx 20.0855$, e is the base and 20.0855 is the exponential.

Here's a twist!

It's possible to raise imaginary numbers to real-number powers and get negative numbers; you've already learned about this. You could talk about base-j exponentials, for example, and make mathematical sense. Consider this:

$$j^2 = -1$$

You won't be likely to hear anybody state this fact as "Minus 1 is the base-j exponential of 2." But theoretically, it's a valid statement.

Common exponentials

Base-10 exponentials are also known as *common exponentials.* Here are a few examples that you can verify with your calculator, rounding to three decimal places except in those cases when the resultants are exact:

$$10^2 = 100$$
$$10^{1.478} \approx 30.061$$
$$10^1 = 10$$
$$10^{0.8347} \approx 6.834$$
$$10^0 = 1$$
$$10^{-0.5} \approx 0.316$$
$$10^{-1} = 0.1$$
$$10^{-1.7} \approx 0.020$$
$$10^{-2} = 0.01$$

In the first equation, 2 is the argument (the value on which the function depends), and 100 is the resultant. You would say is as "The common exponential of 2 is equal to 100." In the second equation, you would say "The common exponential of 1.478 is approximately 30.061." As with logarithms, the arguments in an exponential function need not be whole numbers. They can even be irrational; you could speak of 10^{π}, for example. (It's approximately equal to 1,385.)

Natural exponentials

Base-e exponentials are also called *natural exponentials.* Here are some examples using the same arguments as above, rounding to three decimal places except for e^0, which is exactly 1:

$$
\begin{aligned}
e^2 &\approx 7.389 \\
e^{1.478} &\approx 4.384 \\
e^1 &\approx 2.718 \\
e^{0.8347} &\approx 2.304 \\
e^0 &= 1 \\
e^{-0.5} &\approx 0.607 \\
e^{-1} &\approx 0.368 \\
e^{-1.7} &\approx 0.183 \\
e^{-2} &\approx 0.135
\end{aligned}
$$

Logarithms vs. exponentials

The exponential function is the inverse of the logarithm function, and vice-versa. When two functions are inverses, they "undo" each other, as long as both functions are defined for the all the arguments of interest. A logarithm can be "undone" by the exponential function of the same base. The reverse of this is also true: an exponential can be "undone" by the log function of the same base.

Sometimes, the common exponential of a quantity is called the *common antilogarithm* (antilog_{10}) or the *common inverse logarithm* ($\log^{-1}$) of that number. The natural exponential of a quantity is sometimes called the *natural antilogarithm* (antiln) or the *natural inverse logarithm* ($\ln^{-1}$) of that number.

We can illustrate the relationship between a common log and a common exponential with two equations. If we let the abbreviation "log" represent the base-10 logarithm, then

$$\log(10^x) = x$$

for any real number x, and

$$10^{(\log y)} = y$$

for any positive real number y. A similar pair of equations holds for the natural logarithms. We can replace 10 with e, and replace "log" with "ln" to get

$$\ln(e^x) = x$$

for any real number x, and

$$e^{(\ln y)} = y$$

for any positive real number y.

An example

Now let's see how a common antilog can be used to find the value of a non-whole number raised to the power of another non-whole number. Recall the example set out earlier in this chapter:

$$\begin{aligned} \log_{10} (2.635^{1.078}) &= 1.078 \log_{10} 2.635 \\ &\approx 1.078 \times 0.421 \\ &\approx 0.454 \end{aligned}$$

If we get rid of all the intermediate expressions, we have

$$\log_{10} (2.635^{1.078}) \approx 0.454$$

Taking the common antilog of both sides, we get

$$\text{antilog}_{10} [\log_{10} (2.635^{1.078})] \approx \text{antilog}_{10}\, 0.454$$

We can use a calculator to find the common antilog of 0.454, and simplify the left side of the equation based on the fact that the antilog function "undoes" the log function. When we do that, we get

$$2.635^{1.078} \approx 2.844$$

We can use the "x^y" key in a calculator to verify the above result. When I enter the original numbers into my calculator and use that key, I get

$$2.635^{1.078} \approx 2.842$$

This answer disagrees from the previous answer by 0.002 (or 2 parts in 1,000) because we rounded off the calculations at every step, introducing a rounding error.

The log-antilog scheme is the way most calculators work out powers when the input values are not whole numbers. Before the invention of logs and antilogs, expressions such as $2.635^{1.078}$ were mysterious, indeed. We can use logs and antilogs of any base to evaluate any number raised to the power of any other number, as long as the log and the antilog are both defined for all the arguments.

Another example

What do you get if you raise e to the power of π? You should remember from basic geometry what π (the lowercase Greek letter called "pi") means. It's the ratio of any circle's circumference to its diameter, and is an irrational number equal to approximately 3.14159.

Suppose you have a calculator that can't directly raise one number to the power of another, but it does have a common log key. You can calculate e^{π} using the rule for converting a power to a product:

$$\log_{10} e^{\pi} = \pi \log_{10} e$$

Take the value 2.71828 as an approximation of e, and 3.14159 as an approximation of π. Then, going to five decimal places in each step, the above equation becomes

$$\begin{aligned}\log_{10} e^{\pi} &\approx 3.14159 \log_{10} 2.71828 \\ &\approx 3.14159 \times 0.43429 \\ &\approx 1.36436\end{aligned}$$

When you take the common antilog of this, you'll get the value of e^{π}, because

$$\text{antilog}_{10} (\log_{10} e^{\pi}) = e^{\pi}$$

Calculating, you should get

$$\text{antilog}_{10} 1.36436 \approx 23.13982$$

This makes intuitive sense. Think of it like this: Because π is a little more than 3, and because e is a little less than 3, the value of e^{π} should be somewhere near 3^3, which is 27. It's not terribly close, but it's "in the ball park"!

If your calculator has a key for raising one number to the power of another (in general), you can check this out. When I input the numbers into my calculator, I get

$$2.71828^{3.14159} \approx 23.14058$$

As usual, rounding error has crept into this process. If you like, you can repeat this exercise, letting your calculator keep all the extra digits it can along the way, and rounding off to five decimal places when you get to the very end.

Here's a challenge!

Using a calculator, find π^{e}. Use the same process as you did to find e^{π} in the example you just finished. Round off the values to five decimal places. Verify this result by using the "x^y" key if your calculator has one.

Solution

Once again, you can take advantage of the rule for converting a power to a product. This time, you have

$$\log_{10} \pi^{e} = e \log_{10} \pi$$

Consider π equal to 3.14159, and consider e equal to 2.71828. Then

$$\begin{aligned}\log_{10} \pi^e &\approx 2.71828 \log_{10} 3.14159 \\ &\approx 2.71828 \times 0.49715 \\ &\approx 1.35139\end{aligned}$$

When you take the common antilog of this, you'll get π^e because

$$\text{antilog}_{10} (\log_{10} \pi^e) = \pi^e$$

Calculating, you should get

$$\text{antilog}_{10}\ 1.35139 \approx 22.45898$$

When I input the numbers into my calculator and use the "x^y" key, I get

$$3.14159^{2.71828} \approx 22.45906$$

Rounding error strikes again! And as before, if you wish, you can see for yourself how this error shrinks to the vanishing point if you let your calculator keep all of its extra digits until the final step.

How Exponentials Work

In most real-life applications of exponentials, the base b is either 10 or e. However, once in a while you'll come across a situation where b is some other positive real number. This section describes the basic properties that hold for exponentials in general.

Reciprocal vs. negative exponent

Suppose that x is some real number. The reciprocal of the exponential of x is equal to the exponential of the negative of x, as follows:

$$1 / (b^x) = b^{-x}$$

when $b > 0$. You should recognize this from your work with powers and roots. Here's a familiar example. You know that 1/8 is equal to $1 / (2^3)$. This is the same as saying that 1/8 is equal to 2^{-3}. You also know that 1/100 equals $1 / (10^2)$, which is the same as saying that 1/100 equals 10^{-2}. Now consider this, rounded to four decimal places:

$$\begin{aligned}1 / (e^3) &\approx 1 / (2.718^3) \\ &\approx 1 / 20.079 \\ &\approx 0.0498\end{aligned}$$

Compare the above with the result of entering −3 into a scientific calculator, then hitting "Inv," then hitting "ln," and finally rounding to four decimal places:

$$e^{-3} \approx 0.0498$$

Rounding error stayed out of this little exercise!

Product vs. sum

Exponential functions can express the relationship between sums and products, just as logarithms do. Suppose that x and y are real numbers. Then

$$b^x b^y = b^{(x+y)}$$

when $b > 0$. To demonstrate, let $b = 10$, $x = 4$, and $y = -6$. We can plug in the numbers for the product of exponentials and get and get

$$\begin{aligned} 10^4 \times 10^{-6} &= 10{,}000 \times 0.000001 \\ &= 0.01 \end{aligned}$$

When we evaluate the right side, we get

$$\begin{aligned} 10^{[4+(-6)]} &= 10^{(4-6)} \\ &= 10^{-2} \\ &= 0.01 \end{aligned}$$

The results agree. You'll find that this is always true, no matter what base and arguments you use, as long as the base is positive. Of course, if you get nonterminating decimals for any of the values in the calculation, you should expect some rounding error.

Ratio vs. difference

Again, suppose that x and y are real numbers. Then

$$b^x / b^y = b^{(x-y)}$$

when $b > 0$. Using the same numerical values as before, we can demonstrate this. We plug in the numbers on the left side of the equation and get

$$\begin{aligned} 10^4 / 10^{-6} &= 10{,}000 / 0.000001 \\ &= 10{,}000 \times 1{,}000{,}000 \\ &= 10{,}000{,}000{,}000 \\ &= 10^{10} \end{aligned}$$

Then we can evaluate the right side to see that

$$\begin{aligned} 10^{[4-(-6)]} &= 10^{(4+6)} \\ &= 10^{10} \end{aligned}$$

Ratio in an exponent

Here's a more complicated property of exponentials. Let x and y be real numbers, with the restriction that y cannot be equal to 0. Then

$$b^{(x/y)} = (b^x)^{(1/y)}$$

when $b > 0$. Let's try an example where the base b is 10, with exponents $x = 4$ and $y = 7$. Evaluating the left side first, letting $4/7 \approx 0.5714$ and using the "x^y" or "x^y" function key on a calculator, we get

$$\begin{aligned} 10^{(4/7)} &\approx 10^{0.5714} \\ &\approx 3.727 \end{aligned}$$

Alternatively, we can enter 0.5714, hit the "Inv" key, and then hit "log" to find 10 to the power of 0.5714. Now when we plug the numbers into the right side of the general equation and work it out, we obtain

$$(10^4)^{(1/7)} = 10{,}000^{(1/7)}$$

To work this out on a calculator, we must first figure 1/7 to four decimal places. That gives us 0.1429. Then, we enter 10,000, hit the "x^y" or "x^y" key, and enter 0.1429. The result is 3.729. There's a discrepancy, because we've taken a rounding error to the seventh power!

Power of a power vs. product

Exponentials can show the relationship between a "power of a power" and a product. Suppose that x and y are real numbers. Then

$$(b^x)^y = b^{(xy)}$$

when $b > 0$. To demonstrate this, let $b = e$, $x = 2$, and $y = 3$. Let's evaluate the left side first, using 2.718 as the value of e and going to three decimal places during the calculation process:

$$\begin{aligned} (e^2)^3 &\approx (2.718^2)^3 \\ &\approx 7.388^3 \\ &\approx 403.256 \end{aligned}$$

Now the right side:

$$\begin{aligned} e^{(2\times 3)} &= e^6 \\ &\approx 2.718^6 \\ &\approx 403.178 \end{aligned}$$

If we let the calculator keep all its extra digits during the calculation process, we don't get the rounding error when we go back to three decimal places at the end. (Try it and see.) Now let's use $b = 10$ instead of $b = e$. In that case, the left side becomes

$$\begin{aligned}(10^2)^3 &= 100^3 \\ &= 1{,}000{,}000 \\ &= 10^6\end{aligned}$$

and the right side becomes

$$10^{(2\times 3)} = 10^6$$

There's no rounding error here, because the values are exact throughout!

Mixing exponentials in a product

We can express the product of a common exponential and a natural exponential a "mutant exponential" whose base is 10 times e. If x is the argument of both exponentials, then

$$(10^x)(e^x) = (10e)^x$$

This is an adaptation of the power of product rule from Chap. 9. Let's try a numerical example. Let $x = 4$. If we express the value of e to five decimal places, the left side of the above equation works out as

$$\begin{aligned}(10^4)(e^4) &\approx 10{,}000 \times 2.71828^4 \\ &\approx 10{,}000 \times 54.59800 \\ &\approx 545{,}980\end{aligned}$$

and the right side becomes

$$\begin{aligned}(10e)^4 &\approx (10 \times 2.71828)^4 \\ &\approx 27.1828^4 \\ &\approx 545{,}980\end{aligned}$$

Mixing exponentials in a ratio

How about ratios of mixed common and natural exponentials? If x is a real number, then

$$10^x / e^x = (10/e)^x$$

This is an adaptation of the power of quotient rule from Chap. 9. We can work this out using $x = 4$, as in the previous example. Expressing the value of e to five decimal places but rounding off our final answer to only two decimal places, the left side of the above equation becomes

$$\begin{aligned}10^4 / e^4 &\approx 10{,}000 / 2.71828^4 \\ &\approx 10{,}000 / 54.59800 \\ &\approx 183.16\end{aligned}$$

and the right side becomes

$$(10/e)^4 \approx (10 / 2.71828)^4$$
$$\approx 3.67880^4$$
$$\approx 183.16$$

Now let's invert this ratio. Suppose x is a real number. Then we can write

$$e^x / 10^x = (e/10)^x$$

Again, let's use $x = 4$ and go through this with a calculator. Expressing the value of e to five decimal places and leaving our final answer at five decimal places too, the left side of the above equation works out as

$$e^4 / 10^4 \approx 2.71828^4 / 10{,}000$$
$$\approx 54.59800 / 10{,}000$$
$$\approx 0.00546$$

and the right side works out as

$$(e/10)^4 \approx (2.71828 / 10)^4$$
$$\approx 0.271828^4$$
$$\approx 0.00546$$

Are you confused?

To get a "snapshot" of how exponentials of different bases work out, here's a trick to get the general idea. First, we determine the values of 1 to the powers of −4, −3, −2, −1, 0, 1, 2, 3, and 4. (That's right, the base is 1!) Then we do the same thing with the bases 2, e, and 10. Finally, we compare the values as shown in Table 29-1. For the bases 1, 2, and 10, all the results are exact. For the base e, we approximate everything to four decimal places except e^0, which is exactly 1.

Here's a challenge!

Find the number whose natural exponential function value is exactly 1,000,000, and the number whose natural exponential function value is exactly 0.0001.

Solution

To solve this problem, we must be sure that we know what we're trying to get! Suppose we call the solution x. In the first case, we can solve the following equation for x:

$$e^x = 1{,}000{,}000$$

Table 29-1. Comparison of exponentials for bases 1, 2, e, and 10. Values for base e are approximate except for e^0, which is exactly 1.

x	1^x	2^x	e^x	10^x
−4	1	0.0625	0.0183	0.0001
−3	1	0.125	0.0498	0.001
−2	1	0.25	0.1353	0.01
−1	1	0.5	0.3679	0.1
0	1	1	1	1
1	1	2	2.7183	10
2	1	4	7.3891	100
3	1	8	20.0855	1,000
4	1	16	54.5982	10,000

Taking the natural logarithm of each side, we get

$$\ln (e^x) = \ln 1{,}000{,}000$$
$$x = \ln 1{,}000{,}000$$

This simplifies to a matter of finding a natural logarithm with a calculator. Rounded off to two decimal places, we have

$$\ln 1{,}000{,}000 \approx 13.82$$

In the second case, we must solve the following equation for x:

$$e^x = 0.0001$$

Taking the natural logarithm of each side, we get

$$\ln (e^x) = \ln 0.0001$$
$$x = \ln 0.0001$$

This simplifies, as in the first case, to a matter of finding a natural logarithm with a calculator. When we do that, and then round off to two decimal places, we get

$$\ln 0.0001 \approx -9.21$$

Practice Exercises

This is an open-book quiz. You may (and should) refer to the text as you solve these problems. Don't hurry! You'll find worked-out answers in App. C. The solutions in the appendix may not represent the only way a problem can be figured out. If you think you can solve a particular problem in a quicker or better way than you see there, by all means try it!

1. Let $x = 2.3713018568$ and $y = 0.902780337$. Find xy to three decimal places using common logarithms.
2. Approximate the product xy from Prob. 1 using natural logarithms. Show that the result is the same as that obtained with common logs when rounded off to three decimal places.
3. The *power gain* of an electronic circuit, in units called *decibels* (abbreviated dB), can be calculated according to this formula:

$$G = 10 \log (P_{out}/P_{in})$$

   where G is the gain, P_{out} is the output signal power, and P_{in} is the input signal power, both specified in watts. Suppose the audio input to the left channel of high-fidelity amplifier is 0.535 watts, and the output is 23.7 watts. What is the power gain of this circuit in decibels? Round off the answer to the nearest tenth of a decibel.
4. Suppose the audio output signal in the scenario of Prob. 3 is run through a long length of speaker wire, so instead of the 23.7 watts that appears at the left-channel amplifier output, the speaker only gets 19.3 watts. What is the power gain of the length of speaker wire, in decibels? Round off the answer to three decimal places.
5. If a positive real number increases by a factor of exactly 10, how does its common (base-10) logarithm change?
6. Show that the solution to Prob. 5 is valid for all positive real numbers.
7. If a positive real number decreases by a factor of exactly 100 (becomes 1/100 as great), how does its common logarithm change?
8. Show that the solution to Prob. 7 is valid for all positive real numbers.
9. If a positive real number is divided by a factor of 357, how does its natural (base-e) logarithm change? Express the answer to two decimal places.
10. Show that the solution to Prob. 9 is valid for all positive real numbers.

CHAPTER

30

Review Questions and Answers

Part Three

This is not a test! It's a review of important general concepts you learned in the previous nine chapters. Read it though slowly and let it "sink in." If you're confused about anything here, or about anything in the section you've just finished, go back and study that material some more.

Chapter 21

Question 21-1

There are two distinct numbers that, when squared, produce −1. What are those two numbers?

Answer 21-1

One of them is the unit imaginary number, which we call j. That's the engineer's and applied mathematician's notation. Many mathematics texts use the letter i to represent it. The other is the negative of the unit imaginary number, $-j$.

Question 21-2

How can we show that squaring $-j$ produces the same result as squaring j?

Answer 21-2

By definition, we know that

$$j^2 = -1$$

We can multiply the left side of this equation by $(-1)^2$ without having any effect on its value, because $(-1)^2 = 1$. When we do that, we get

$$(-1)^2 j^2 = -1$$

The power of product rule allows us to rewrite the left side of this equation, so it becomes

$$[(-1)j]^2 = -1$$

But $(-1)j$ is the same thing as $-j$, so we can simplify further to get

$$(-j)^2 = -1$$

This is the same result as we obtain by squaring j.

Question 21-3

How is an imaginary number "put together"?

Answer 21-3

An imaginary number is the product of j and a real number. Suppose we call the real number b. If b is positive, then we write the product of j and b as jb. If b is negative, we write the product as $-jb$. (We always put the minus sign first.) If $b = 0$, then we can write the product as $j0$. But because $j0 = 0$, we would more likely write $j0$ simply as 0.

Question 21-4

How can we show that $j3 + j7$ is the same as $j7 + j3$, assuming that the familiar distributive law of arithmetic works with the unit imaginary number?

Answer 21-4

When we apply the distributive law for multiplication over addition "backward" to the first expression, we get

$$j3 + j7 = j(3 + 7)$$

The commutative law for addition tells us that $3 + 7 = 7 + 3$. By substitution on the right side, we get

$$j3 + j7 = j(7 + 3)$$

Applying the distributive law "forward" on the right side, we get

$$j3 + j7 = j7 + j3$$

Question 21-5

How can we show, again assuming that the distributive law works with the unit imaginary number, that if a and b are any two real numbers, then $ja + jb$ is the same as $jb + ja$?

Answer 21-5

We can use the same proof procedure as we did in Answer 21-4, using letter constants instead of numbers! The distributive law tells us that

$$ja + jb = j(a + b)$$

The commutative law allows us to rewrite this as

$$ja + jb = j(b + a)$$

and the distributive law, applied again, gives us

$$ja + jb = jb + ja$$

Question 21-6

What is the absolute value of $j25$? What is the absolute value of $-j25$?

Answer 21-6

Remember that in general, if b is any nonnegative real number, then

$$|jb| = b$$

and

$$|-jb| = b$$

The absolute value of $j25$ and the absolute value of $-j25$ are therefore both equal to 25.

Question 21-7

How is a complex number "put together"?

Answer 21-7

A complex number is the sum of a real number and an imaginary number. If we have two real numbers a and b, then $a + jb$ is a complex number. Conversely, any complex number can be written in the form $a + jb$, where a and b are real numbers. There are no restrictions on the values of a or b. They can be negative, positive, or 0.

Question 21-8

What is the sum of the complex numbers $a_1 + jb_1$ and $a_2 + jb_2$, where a_1, a_2, b_1, and b_2 are real numbers?

Answer 21-8

When we want to add two complex numbers, we add the real parts and the complex parts separately. Therefore,

$$\begin{aligned}(a_1 + jb_1) + (a_2 + jb_2) &= (a_1 + a_2) + (jb_1 + jb_2)\\ &= (a_1 + a_2) + j(b_1 + b_2)\end{aligned}$$

Question 21-9

What is the product of the complex numbers $a_1 + jb_1$ and $a_2 + jb_2$, where a_1, a_2, b_1, and b_2 are real numbers?

Answer 21-9

When we want to multiply one complex number by another, we treat both factors as binomials, keeping in mind the fact that $j^2 = -1$. Therefore,

$$\begin{aligned}(a_1 + jb_1)(a_2 + jb_2) &= a_1 a_2 + ja_1 b_2 + jb_1 a_2 + j^2 b_1 b_2 \\ &= a_1 a_2 + ja_1 b_2 + jb_1 a_2 - b_1 b_2 \\ &= a_1 a_2 - b_1 b_2 + ja_1 b_2 + jb_1 a_2 \\ &= (a_1 a_2 - b_1 b_2) + j(a_1 b_2 + b_1 a_2)\end{aligned}$$

Question 21-10

What is the conjugate of a complex number $a + jb$, where a and b are real numbers? What happens when we add a complex number to its conjugate? What happens when we multiply a complex number by its conjugate?

Answer 21-10

We can get the conjugate of any complex number $a + jb$ by reversing the sign of the imaginary part. Therefore, $a + jb$ and $a - jb$ are conjugates of each other. When we add a complex number to its conjugate using the rule from Answer 21-8, we get

$$\begin{aligned}(a + jb) + (a - jb) &= (a + a) + (jb - jb) \\ &= 2a + j0 \\ &= 2a\end{aligned}$$

When we multiply a complex number by its conjugate using the rule from Answer 21-9, we get

$$\begin{aligned}(a + jb)(a - jb) &= a^2 - jab + jba - j^2 b^2 \\ &= a^2 - jab + jab + b^2 \\ &= a^2 + b^2\end{aligned}$$

Chapter 22

Question 22-1

What is the polynomial standard form for a quadratic equation in the variable x?

Answer 22-1

When a quadratic equation in x is written in polynomial standard form, it's formatted like this:

$$ax^2 + bx + c = 0$$

where a and b are coefficients of the variable x, and c is a constant. For the equation to be a quadratic, the coefficient of x^2 (in this case a) must not be equal to 0.

Question 22-2

How can we write the following equation as a quadratic in polynomial standard form when a_1, a_2, b_1, and b_2 are real numbers, and x is the variable?

$$(a_1x + b_1)(a_2x + b_2) = 0$$

Answer 22-2

We can start by multiplying the two binomials on the left side together, using the product of sums rule from Chap. 9. When we do that, we get

$$a_1a_2x^2 + a_1b_2x + b_1a_2x + b_1b_2 = 0$$

which can be rearranged to obtain

$$(a_1a_2)x^2 + (a_1b_2 + b_1a_2)x + b_1b_2 = 0$$

This equation is in polynomial standard form, provided $a_1 \neq 0$ and $a_2 \neq 0$. The coefficient a from the "template" (Answer 22-1) is the quantity a_1a_2. The coefficient b from the "template" is the quantity $(a_1b_2 + b_1a_2)$. The constant c from the "template" is the quantity b_1b_2.

Question 22-3

Which of the following equations are quadratics in one variable? Which are not?

$$x^2 = 8x + 3x^3$$
$$-3x = 7x^2 - 12$$
$$x^2 + 2x = 7 - x$$
$$x^4 - 2 = -8x^2 - 7x^3$$
$$13 + 3x = 12x^2$$

Answer 22-3

A quadratic equation in one variable always contains a nonzero multiple of the variable squared, and no higher powers of the variable. There may also be terms containing the variable itself (to the first power) along with terms that are simple constants. On that basis, the second, third, and fifth equations are quadratics in one variable. The first and fourth equations are not.

Question 22-4

What are the roots of the following quadratic equation, where a_1, a_2, b_1, and b_2 are real numbers, and x is the variable? Assume that neither a_1 nor b_1 is equal to 0:

$$(a_1x + b_1)(a_2x + b_2) = 0$$

Answer 22-4

This equation is a product of binomials. The roots are the values of x that make either factor equal to 0. We can find those roots by setting each binomial equal to 0, creating two separate first-degree equations. Then we can morph the equations to get the variable all alone on the left side of the equals sign, and some combination of the coefficient and constant on the right side. For the first term, the process goes like this:

$$a_1x + b_1 = 0$$
$$a_1x = -b_1$$
$$x = -b_1/a_1$$

For the second term, it's the same, but with subscripts of 2 instead of 1:

$$a_2x + b_2 = 0$$
$$a_2x = -b_2$$
$$x = -b_2/a_2$$

Question 22-5

How can we solve the following quadratic equation, where a and b are real numbers, and x is the variable? Assume that $a \neq 0$:

$$(ax + b)^2 = 0$$

Answer 22-5

We can look at this equation as a product of two identical binomials:

$$(ax + b)(ax + b) = 0$$

We can solve this quadratic in the same way as we solved the equation in Answer 22-4. But this time, we get only one root:

$$ax + b = 0$$
$$ax = -b$$
$$x = -b/a$$

Question 22-6

What's special about the root of the equation we solved in Answer 22-5?

Answer 22-6

This root, $-b/a$, occurs "twice over." In technical terms, the root has multiplicity 2.

Question 22-7

Suppose we see a quadratic equation in polynomial standard form. Again, here's the general form:

$$ax^2 + bx + c = 0$$

where a and b are coefficients of the variable x, $a \neq 0$, and c is a constant. What is the general formula for solving this quadratic?

Answer 22-7

The formula, known as the quadratic formula, is

$$x = [-b \pm (b^2 - 4ac)^{1/2}] / (2a)$$

This is worth memorizing!

Question 22-8

What is the discriminant in the quadratic formula? Why is it significant?

Answer 22-8

The discriminant, sometimes symbolized as d, is the quantity $(b^2 - 4ac)$. It tells us whether or not a quadratic equation with real-number coefficients and a real-number constant has any real roots. If $d > 0$, then the equation has two different real roots. If $d = 0$, then the equation has one real root with multiplicity 2. If $d < 0$, then the equation has no real roots.

Question 22-9

How can we use the quadratic formula to find the roots of the following equation?

$$9x^2 - 42x = -49$$

Answer 22-9

Before applying the formula, we must get the equation into polynomial standard form. We can do that by adding 49 to each side, obtaining

$$9x^2 - 42x + 49 = 0$$

In the general polynomial standard equation

$$ax^2 + bx + c = 0$$

we have $a = 9$, $b = -42$, and $c = 49$. Plugging these into the quadratic formula, we get

$$\begin{aligned} x &= [-b \pm (b^2 - 4ac)^{1/2}] / (2a) \\ &= [42 \pm (42^2 - 4 \times 9 \times 49)^{1/2}] / (2 \times 9) \\ &= [42 \pm (1{,}764 - 1{,}764)^{1/2}] / 18 \\ &= 42/18 \\ &= 7/3 \end{aligned}$$

This equation has the single root $x = 7/3$ with multiplicity 2.

Question 22-10

How can we use the quadratic formula to find the roots of the following equation?

$$-x = 3x^2 - 4$$

Answer 22-10

First, let's get the equation into polynomial standard form. We can do that by adding x to both sides and then switching the right and left sides. That gives us

$$3x^2 + x - 4 = 0$$

In the general polynomial standard equation

$$ax^2 + bx + c = 0$$

we have $a = 3$, $b = 1$, and $c = -4$. Plugging these into the quadratic formula, we get

$$\begin{aligned} x &= [-b \pm (b^2 - 4ac)^{1/2}] \,/\, (2a) \\ &= \{-1 \pm [1^2 - 4 \times 3 \times (-4)]^{1/2}\} \,/\, (2 \times 3) \\ &= [-1 \pm (1 + 48)^{1/2}] \,/\, 6 \\ &= (-1 \pm 49^{1/2}) \,/\, 6 \\ &= (-1 \pm 7) \,/\, 6 \\ &= 6/6 \text{ or } -8/6 \\ &= 1 \text{ or } -4/3 \end{aligned}$$

The roots of the quadratic equation are $x = 1$ or $x = -4/3$.

Chapter 23

Question 23-1

Does a quadratic equation with real coefficients and a real constant, but with a negative discriminant, have any roots at all? If so, what are they like?

Answer 23-1

When a quadratic equation with real coefficients and a real constant has a negative discriminant, the equation has two roots, both of which are non-real complex numbers.

Question 23-2

How can we use the quadratic formula to find the roots of the following equation?

$$-x = 3x^2 + 4$$

Answer 23-2

First, we must get the equation into polynomial standard form. We can do that by adding x to both sides and then switching the right and left sides, getting

$$3x^2 + x + 4 = 0$$

In the general polynomial standard equation

$$ax^2 + bx + c = 0$$

we have $a = 3$, $b = 1$, and $c = 4$. Plugging these into the quadratic formula, we get

$$\begin{aligned} x &= [-b \pm (b^2 - 4ac)^{1/2}] / (2a) \\ &= [-1 \pm (1^2 - 4 \times 3 \times 4)^{1/2}] / (2 \times 3) \\ &= [-1 \pm (1 - 48)^{1/2}] / 6 \\ &= [-1 \pm (-47)^{1/2}] / 6 \end{aligned}$$

The quantity $(-47)^{1/2}$ is the imaginary number $j(47^{1/2})$. Therefore

$$x = [-1 \pm j(47^{1/2})] / 6$$

If we want to express these roots individually, we can write

$$x = [-1 + j(47^{1/2})] / 6 \text{ or } x = [-1 - j(47^{1/2})] / 6$$

We can reduce these to standard complex-number form using the right-hand distributive law for division over addition or subtraction, getting

$$x = -1/6 + j(47^{1/2}/6) \text{ or } x = -1/6 - j(47^{1/2}/6)$$

Question 23-3

Suppose we examine the discriminant of a quadratic equation, and we discover that the equation has two complex roots. How can we tell whether or not these are pure imaginary numbers?

Answer 23-3

If the discriminant is a negative real and the coefficient of x is 0, then the roots are pure imaginary. If the discriminant is a negative real and the coefficient of x is a nonzero real, then the roots are complex but not pure imaginary.

Question 23-4

In a quadratic equation with real coefficients but pure imaginary roots, how are those roots related? In a quadratic equation with real coefficients but complex roots that aren't pure imaginary, how are those roots related?

Answer 23-4

If the roots are pure imaginary, then they are additive inverses. That is, one is the negative of the other. If the roots are complex but not pure imaginary, then they are conjugates.

Question 23-5

Consider the following quadratic equation in polynomial standard form:

$$4x^2 + 64 = 0$$

How can this equation be rewritten in binomial factor form?

Answer 23-5

If we subtract 64 from both sides of the equation and then divide through by 4, we get

$$x^2 = -16$$

This tells us that the roots are $x = j4$ or $x = -j4$. We can write the negatives of these roots as constants in a pair of binomials and then set their product equal to 0, obtaining

$$(x - j4)(x + j4) = 0$$

That's the binomial factor form of the original equation.

Question 23-6

Consider the general quadratic equation

$$px^2 + q = 0$$

where p and q are both positive real numbers. What are the roots of this equation?

Answer 23-6

Subtracting q from each side, we get

$$px^2 = -q$$

Dividing through by p, which we know is not 0 because we've been told that it's positive, we obtain

$$x^2 = -q/p$$

which can be rewritten as

$$x^2 = -1(q/p)$$

Because p and q are both positive, we know that the ratio q/p is positive. Its positive and negative square roots are therefore both real numbers. We can take the square root of both sides of the above equation, getting

$$\begin{aligned} x &= \pm[-1(q/p)]^{1/2} \\ &= \pm(-1)^{1/2}\ [\pm(q/p)^{1/2}] \\ &= \pm j[(q/p)^{1/2}] \end{aligned}$$

The roots are therefore $x = j[(q/p)^{1/2}]$ or $x = -j[(q/p)^{1/2}]$.

Question 23-7

Is it possible for the roots of a quadratic equation to be pure imaginary but not additive inverses? If so, provide an example of such an equation. If not, explain why not.

Answer 23-7

A quadratic equation can have roots that are pure imaginary but not additive inverses. Here is an example of such an equation in binomial factor form:

$$(x + j2)(x + j3) = 0$$

The roots of this equation are $x = -j2$ or $x = -j3$, as we can verify by plugging them in. They are not additive inverses.

Question 23-8

We have learned in the last several chapters (but not explicitly stated in full, until now!), that if the polynomial standard form of a quadratic equation has real coefficients and a real constant, then one of these things must be true:

- There are two different real roots
- There is a single real root with multiplicity 2
- There are two different pure imaginary roots, and they are additive inverses
- There are two different complex roots, and they are conjugates

In Answer 23-7, we found a quadratic equation that has two pure imaginary roots that are not additive inverses. How is this possible?

Answer 23-8

The coefficients and constant in the polynomial standard form of this equation are not all real numbers. To see that, we can multiply the product of binomials out:

$$\begin{aligned} (x + j2)(x + j3) &= x^2 + j3x + j2x + (j2j3) \\ &= x^2 + j5x - 6 \end{aligned}$$

In this case, the coefficient of x is imaginary. The coefficient of x^2, as well as the stand-alone constant, are real. The complete polynomial quadratic equation is

$$x^2 + j5x - 6 = 0$$

Question 23-9

Is it possible for one root of a quadratic equation to be pure real and the other pure imaginary? If so, provide an example of such an equation in polynomial standard form. If not, explain why not.

Answer 23-9

Yes, this is possible. Here is an example of such an equation in binomial factor form:

$$(x + 2)(x + j3) = 0$$

The roots of this equation are $x = -2$ or $x = -j3$, as we can verify by plugging them in. To convert this to polynomial standard form, we multiply the product of binomials out:

$$\begin{aligned}(x + 2)(x + j3) &= x^2 + j3x + 2x + j6 \\ &= x^2 + (2 + j3)x + j6\end{aligned}$$

Here, the coefficient of x is complex but not pure imaginary, and the stand-alone constant is pure imaginary. The coefficient of x^2 is real. The complete polynomial quadratic equation is

$$x^2 + (2 + j3)x + j6 = 0$$

Question 23-10

Is it possible for a quadratic equation to have two nonconjugate complex roots, neither or which is pure imaginary? If so, provide an example of such an equation in polynomial standard form. If this sort of situation is impossible, explain why.

Answer 23-10

This, too, is possible! Suppose the roots are $1 + j$ and $2 + j$. These are non-conjugate complex numbers, and neither of them is pure imaginary. We can construct a binomial factor quadratic with these numbers as roots by subtracting the roots from x, like this:

$$[x - (1 + j)][x - (2 + j)] = 0$$

When we multiply the left side of this equation out to obtain a polynomial, taking extra precautions to be sure that we don't mess up with the signs, we obtain

$$\begin{aligned}[x - (1 + j)][x - (2 + j)] &= [x + (-1) + (-j)][x + (-2) + (-j)] \\ &= x^2 + (-2x) + (-jx) + (-x) + 2 + j + (-jx) + j2 + (-1) \\ &= x^2 + (-3x) + (-j2x) + j3 + 1 \\ &= x^2 + (-3 - j2)x + (1 + j3)\end{aligned}$$

Here, the coefficient of x^2 is real, the coefficient of x is complex, and the stand-alone constant is complex. The complete polynomial quadratic is

$$x^2 + (-3 - j2)x + (1 + j3) = 0$$

Chapter 24

Question 24-1

Consider the general form of a quadratic function where x is the independent variable, y is the dependent variable, and a, b, and c are real numbers with $a \neq 0$:

$$y = ax^2 + bx + c$$

The graph of this function in Cartesian coordinates is always a parabola that opens either straight upward or straight downward. How can we tell which way the parabola opens by simply looking at a specific function of this type?

Answer 24-1

The parabola opens straight upward if and only if $a > 0$. The parabola opens straight downward if and only if $a < 0$.

Question 24-2

Suppose we see a quadratic function written as shown in Question 24-1, with specific numbers in place of a, b, and c. We plot several points (x, y) on the Cartesian plane by plugging in various values of x and calculating the results for y. How can we determine how many real zeros the function has, assuming we plot enough points to get a "clear picture" of the parabola?

Answer 24-2

The quadratic function has two different real zeros if and only if the parabola crosses the x axis twice. The function has one real zero with multiplicity 2 if and only if the parabola is tangent to ("brushes up against") the x axis at the absolute maximum point or the absolute minimum point. The function has no real zeros if and only if the parabola doesn't intersect the x axis at all.

Question 24-3

Parabolas that open upward always have an absolute minimum. Parabolas that open downward always have an absolute maximum. Imagine a quadratic function in which x is the independent variable and y is the dependent variable. Its graph is a parabola. If the function has two real zeros where $x = p$ and $x = q$, what is the x-value of the absolute maximum or minimum (that is, the vertex point) of the parabola? Let's call it x_v in this example.

Answer 24-3

The value x_v is the average of the two zeros. That's also known as the arithmetic mean, and is equal to the sum of the values divided by 2:

$$x_v = (p + q) / 2$$

Question 24-4

Imagine another quadratic function in which x is the independent variable and y is the dependent variable. If this function has a single real zero with multiplicity 2 where $x = p$, what is x_v, the x-value of the vertex point on its graph?

Answer 24-4

When a quadratic function has only one real zero, the parabola is tangent to the x axis at the vertex point. That's also the x-value of the real zero. Therefore, $x_v = p$.

Question 24-5

Suppose we come across the following quadratic function in binomial factor form, where x is the independent variable and y is the dependent variable:

$$y = (x + 2)(x - 4)$$

Does the parabola representing this function in Cartesian coordinates open upward or downward?

Answer 24-5

To determine this, we must get the right side of the equation in polynomial standard form by multiplying the binomials. When we do that, we get

$$y = x^2 - 2x - 8$$

Because the coefficient of x^2 is positive, the parabola opens upward.

Question 24-6

What are the real zeros of the function stated in Question 24-5? What are the coordinates (x_v, y_v) of the vertex point in its graph? Is the vertex an absolute maximum or an absolute minimum?

Answer 24-6

The zeros can be seen by looking at the original form of the function. The right side of that equation is a product of binomials. If we set it equal to 0, getting a quadratic equation in x, we have

$$(x + 2)(x - 4) = 0$$

The zeros of the function are the same as the roots of this quadratic. Without doing any algebra or arithmetic, we can see that these roots are $x = -2$ or $x = 4$.

To find the vertex point, let's remember the general polynomial standard form for a quadratic function:

$$y = ax^2 + bx + c$$

The x-coordinate of the vertex point, x_v, can be found by the formula

$$x_v = -b/2a$$

In this quadratic, $a = 1$ and $b = -2$. Therefore

$$\begin{aligned} x_v &= -(-2) / (2 \times 1) \\ &= 2/2 \\ &= 1 \end{aligned}$$

Plugging this in and working out the arithmetic using the product of binomials, we get

$$\begin{aligned} y_v &= (x_v + 2)(x_v - 4) \\ &= (1 + 2)(1 - 4) \\ &= 3 \times (-3) \\ &= -9 \end{aligned}$$

Therefore, $(x_v, y_v) = (1, -9)$. This vertex is an absolute minimum because, as we found in Answer 24-5, the parabola opens upward.

Question 24-7

Based on the information in Answers 24-5 and 24-6, how can we sketch an approximate graph of the quadratic function stated in Question 24-5?

Answer 24-7

We have found that the zeros of the function are $x = -2$ and $x = 4$. The points representing them are at $(-2, 0)$ and $(4, 0)$. The vertex point is at $(1, -9)$, and the graph is a parabola that opens upward. Knowing all this, we can sketch the graph as shown in Fig. 30-1.

Question 24-8

Suppose we come across the following quadratic function in binomial factor form, where x is the independent variable and y is the dependent variable:

$$y = -3x^2 + 7x - 11$$

Does the parabola representing this function in Cartesian or rectangular coordinates open upward or downward? How many real zeros does the function have?

Answer 24-8

The parabola opens downward because the coefficient of x^2 is negative. To determine how many real zeros the function has, we can evaluate the discriminant d. Once again, recall the general polynomial standard form for a quadratic function:

$$y = ax^2 + bx + c$$

Here, we have $a = -3$, $b = 7$, and $c = -11$. Therefore

$$\begin{aligned} d &= b^2 - 4ac \\ &= 7^2 - 4 \times (-3) \times (-11) \\ &= 49 - 132 \\ &= -83 \end{aligned}$$

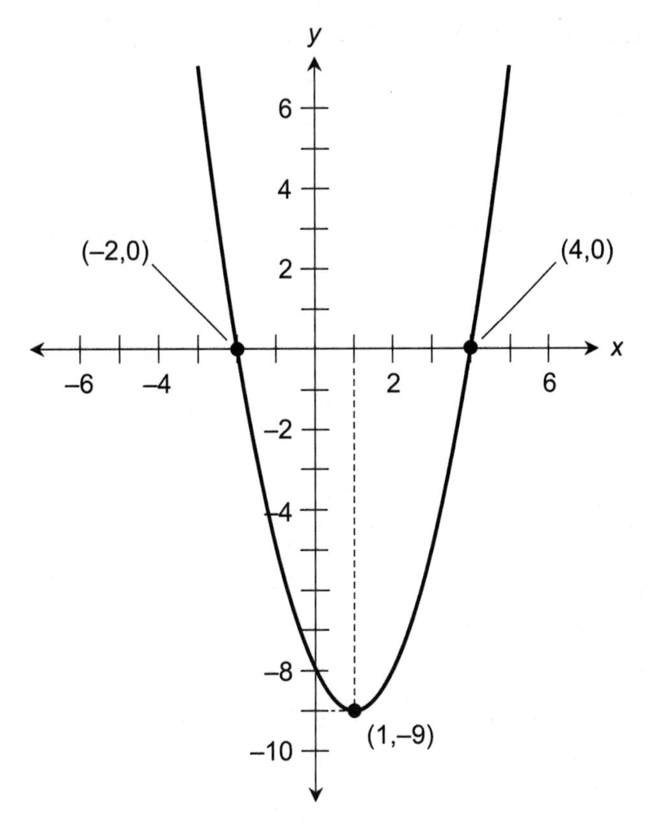

Figure 30-1 Illustration for Answer 24-7. This is the graph of $y = x^2 - 2x - 8$.

The fact that $d < 0$ tells us that the function has no real zeros, just as it tells us that the quadratic equation

$$-3x^2 + 7x - 11 = 0$$

has no real roots.

Question 24-9

What are the coordinates (x_v, y_v) of the vertex point in the graph of the function stated in Question 24-8? Is the vertex an absolute maximum or an absolute minimum?

Answer 24-9

The x-coordinate of the vertex point, x_v, can be found by the formula

$$x_v = -b/2a$$

In this function, $a = -3$ and $b = 7$. Therefore

$$\begin{aligned} x_v &= -7 / [2 \times (-3)] \\ &= -7/(-6) \\ &= 7/6 \end{aligned}$$

Plugging this in and working out the arithmetic using the function, we get

$$\begin{aligned} y_v &= -3x_v^2 + 7x_v - 11 \\ &= -3 \times (7/6)^2 + 7 \times (7/6) - 11 \\ &= -49/12 + 49/6 - 11 \\ &= -49/12 + 98/12 - 132/12 \\ &= (-49 + 98 - 132) / 12 \\ &= -83/12 \end{aligned}$$

Therefore, $(x_v, y_v) = (7/6, -83/12)$. This vertex is an absolute maximum, because the parabola opens downward.

Question 24-10

Locate two points on the graph of the function stated in Question 24-8, other than the vertex. How can we sketch an approximate graph of the function?

Answer 24-10

We can locate two points by plugging in a value of x somewhat smaller than x_v, and another value of x somewhat larger than x_v. Let's try $x_1 = 0$ and $x_2 = 2$. In the first case, we have

$$\begin{aligned} y_1 &= -3x_1^2 + 7x_1 - 11 \\ &= -3 \times 0^2 + 7 \times 0 - 11 \\ &= 0 + 0 - 11 \\ &= -11 \end{aligned}$$

The first non-vertex point is $(x_1, y_1) = (0, -11)$. In the second case,

$$\begin{aligned} y_2 &= -3x_2^2 + 7x_2 - 11 \\ &= -3 \times 2^2 + 7 \times 2 - 11 \\ &= -12 + 14 - 11 \\ &= 2 - 11 \\ &= -9 \end{aligned}$$

The second non-vertex point is $(x_2, y_2) = (2, -9)$. Now we know these things:

- The vertex point is $(7/6, -83/12)$
- The parabola contains two other points $(0, -11)$ and $(2, -9)$
- The parabola opens downward

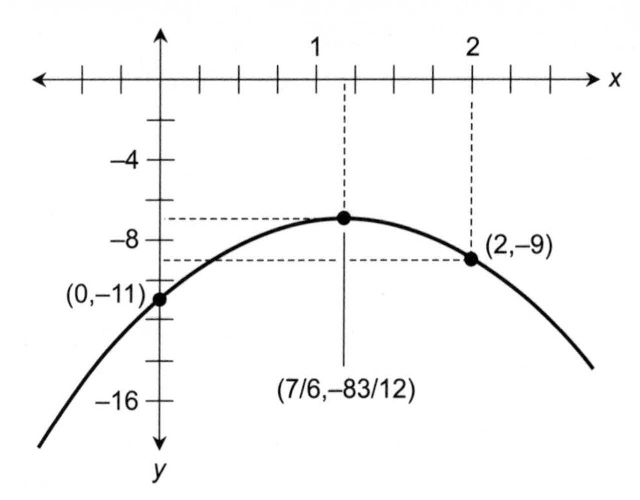

Figure 30-2 Illustration for Answer 24-10. This is the graph of $y = -3x^2 + 7x - 11$. On the x axis, each increment is 1/4 unit. On the y axis, each increment is 2 units.

Armed with this information, we can sketch the graph. To fit it into a neat space, let's use rectangular coordinates where each horizontal division is 1/4 unit and each vertical division is 2 units, with the origin all the way up at the top. We can use a calculator as a "point-plotting aid" and approximate the vertex (7/6, −83/12) as (1.17, −6.92). The result is shown in Fig. 30-2.

Chapter 25

Question 25-1

What is the polynomial standard form of a cubic equation in the variable x?

Answer 25-1

When a cubic equation is in polynomial standard form and the variable is x, the equation can be written like this:

$$ax^3 + bx^2 + cx + d = 0$$

where a, b, c, and d are real numbers, and $a \neq 0$.

Question 25-2

What is the binomial-cubed form of a cubic equation in the variable x?

Answer 25-2

An equation in binomial-cubed form can always be written like this when x is the variable:

$$(ax + b)^3 = 0$$

where a is a nonzero real number, and b is a real number.

Question 25-3

What does the binomial-cubed equation from Answer 25-2 look like when multiplied out into polynomial standard form?

Answer 25-3

Let's begin by separating the left side of the equation into a product of three identical binomials:

$$(ax + b)(ax + b)(ax + b) = 0$$

When we multiply the second two binomials together and then simplify the result into a trinomial, we get the equation

$$(ax + b)(a^2x^2 + 2abx + b^2) = 0$$

Multiplying the binomial by the trinomial and simplifying gives us

$$a^3x^3 + 3a^2bx^2 + 3ab^2x + b^3 = 0$$

Question 25-4

How many real roots does the equation stated in Answer 25-2 have? What is that root, or what are they? What is the real solution set X?

Answer 25-4

There is one real root with multiplicity 3. It is the solution to the equation we obtain when we set the binomial equal to 0:

$$ax + b = 0$$

When we subtract b from both sides and then divide through by a (which is okay because we know that $a \neq 0$), we get

$$x = -b / a$$

The real solution set is therefore

$$X = \{-b / a\}$$

Question 25-5

What is the binomial-factor form of a cubic equation in the variable x?

Answer 25-5

An equation in binomial-factor form can always be written like this when x is the variable:

$$(a_1x + b_1)(a_2x + b_2)(a_3x + b_3) = 0$$

where a_1, a_2, and a_3 are nonzero real numbers, and b_1, b_2, and b_3 are real numbers.

Question 25-6

What does the binomial-factor equation from Answer 25-5 look like when multiplied out into polynomial standard form?

Answer 25-6

Let's start by multiplying the second and third binomials together. When we do that, we obtain

$$(a_1x + b_1)(a_2a_3x^2 + a_2b_3x + b_2a_3x + b_2b_3) = 0$$

Now let's multiply the binomial by the polynomial on the left side. That gives us

$$a_1a_2a_3x^3 + a_1a_2b_3x^2 + a_1b_2a_3x^2 + a_1b_2b_3x + b_1a_2a_3x^2 + b_1a_2b_3x + b_1b_2a_3x + b_1b_2b_3 = 0$$

When we group the terms for x^2 and x together and apply the distributive law for multiplication over addition, we get the polynomial standard form

$$a_1a_2a_3x^3 + (a_1a_2b_3 + a_1b_2a_3 + b_1a_2a_3)x^2 + (a_1b_2b_3 + b_1a_2b_3 + b_1b_2a_3)x + b_1b_2b_3 = 0$$

Question 25-7

How many real roots does the equation stated in Answer 25-5 have? What is that root, or what are they? What is the real solution set X?

Answer 25-7

There are three real roots. They are the solutions to the equations we obtain when we set the binomials equal to 0:

$$a_1x + b_1 = 0$$
$$a_2x + b_2 = 0$$
$$a_3x + b_3 = 0$$

When we subtract the "*b*-constant" from both sides in each of these equations and then divide through by the "*a*-coefficient" (which is okay because we know that a_1, a_2, and a_3 are all nonzero), we get these roots:

$$x = -b_1 / a_1 \text{ or } x = -b_2 / a_2 \text{ or } x = -b_3 / a_3$$

The real solution set is therefore

$$X = \{-b_1 / a_1, -b_2 / a_2, -b_3 / a_3\}$$

Question 25-8

It is possible for two, or even all three, of the roots in Answer 25-7 to be the same? If so, give examples. If not, explain why not.

Answer 25-8

Yes, this can happen. If two of the roots are identical, then the cubic equation has a total of two real roots, one of them of multiplicity 2. Here's an example:

$$(x - 3)(x + 2)(-2x - 4) = 0$$

In this equation, the root for each binomial is the value of x that makes the binomial equal to 0. In order from left to right, those values are

$$x = 3 \text{ or } x - 2 \text{ or } x = -2$$

If all three of the roots are identical, then the cubic equation has one real root with multiplicity 3. Consider this:

$$(x - 5)(2x - 10)(-3x - 15) = 0$$

Here, as before, the root for each binomial is the *x*-value that makes it 0. The roots for each binomial, in order from left to right, are

$$x = 5 \text{ or } x = 5 \text{ or } x = 5$$

Question 25-9

What is the binomial factor rule?

Answer 25-9

Imagine that we come across a cubic equation and we put it into the polynomial standard form, like this:

$$ax^3 + bx^2 + cx + d = 0$$

where a, b, c, and d are real numbers, and $a \neq 0$. The binomial factor rule tells us that a real number k is a root of this equation if and only if $(x - k)$ is a factor of the polynomial.

Question 25-10

What is the smallest number of real roots that a single-variable cubic equation with real-number coefficients and a real-number constant can have? What is the largest number of real roots that such an equation can have?

Answer 25-10

A cubic equation in one variable, with real coefficients and a real constant, can have one real root, two real roots, or three real roots. There must always be at least one, but there can never be more than three.

Chapter 26

Question 26-1

What is the polynomial standard form of an *n*th-degree equation in the variable x, where n is a natural number larger than 3?

Answer 26-1

The polynomial standard form of such an equation is

$$a_n x^n + a_{n-1} x^{n-1} + a_{n-2} x^{n-2} + \cdots + a_1 x + b = 0$$

where a_1, a_2, a_3, ... a_n, and b are real numbers. In addition, the leading coefficient, a_n, must not be equal to 0.

Question 26-2

In an equation of the form shown in Answer 26-1, what would happen if the coefficient a_n, by which x^n is multiplied, were equal to 0?

Answer 26-2

If $a_n = 0$, we get the equation

$$0x^n + a_{n-1} x^{n-1} + a_{n-2} x^{n-2} + \cdots + a_1 x + b = 0$$

The term for x^n has vanished, leaving us with the polynomial standard form for a single-variable equation of degree $n - 1$:

$$a_{n-1} x^{n-1} + a_{n-2} x^{n-2} + \cdots + a_1 x + b = 0$$

Question 26-3

What is the binomial-to-the-*n*th form of an *n*th-degree equation in the variable x?

Answer 26-3

An *n*th-degree equation in binomial-to-the-*n*th form can always be written like this when x is the variable:

$$(ax + b)^n = 0$$

where a is a nonzero real number, b is a real number, and n is a positive integer. Theoretically, n can be any positive integer. If $n = 1$, the equation is of the first degree; if $n = 2$, the equation is quadratic; if $n = 3$, the equation is cubic. If $n > 3$, the equation is of higher degree.

Question 26-4

How many real roots does the equation stated in Answer 26-3 have? What is that root, or what are they? What is the real solution set X?

Answer 26-4

There is one real root with multiplicity n. It is the solution to the equation we obtain when we set the binomial equal to 0:

$$ax + b = 0$$

When we subtract b from both sides and then divide through by a (which is okay because we know that $a \neq 0$), we get

$$x = -b/a$$

The real solution set is therefore

$$X = \{-b/a\}$$

Question 26-5

Find all the real roots of the following equation, state the multiplicity of each, and state the real solution set X.

$$(x^2 - 6x + 9)^2 = 0$$

Answer 26-5

Let's set the quantity $(x^2 - 6x + 9)$ equal to 0, so it becomes the quadratic equation

$$x^2 - 6x + 9 = 0$$

We can factor this into

$$(x - 3)^2 = 0$$

If we substitute the quantity $(x-3)^2$ for the trinomial in the original equation, we get

$$[(x-3)^2]^2 = 0$$

which can be simplified to

$$(x-3)^4 = 0$$

We can solve by setting the binomial equal to 0:

$$x - 3 = 0$$

This first-degree equation resolves to $x = 3$. This is the only real root of the original higher-degree equation, and it has multiplicity 4. The real solution set is $X = \{3\}$.

Question 26-6

What is the binomial factor form of an *n*th-degree equation in the variable *x*?

Answer 26-6

Suppose that $a_1, a_2, a_3, \ldots, a_n$ are nonzero real numbers, and $b_1, b_2, b_3, \ldots, b_n$ are real stand-alone constants. Let *x* be the variable in the following equation:

$$(a_1x + b_1)(a_2x + b_2)(a_3x + b_3) \cdots (a_nx + b_n) = 0$$

This is the binomial factor form for an *n*th-degree equation in the variable *x*.

Question 26-7

How many real roots does the equation stated in Answer 25-5 have? What is that root, or what are they? What is the real solution set *X*?

Answer 26-7

There are *n* real roots. They are the solutions to the equations we obtain when we set the binomials equal to 0:

$$a_1x + b_1 = 0$$
$$a_2x + b_2 = 0$$
$$a_3x + b_3 = 0$$
$$\downarrow$$
$$a_nx + b_n = 0$$

When we subtract the "*b*-constant" from both sides in each of these equations and then divide through by the "*a*-coefficient" (which is okay because we know that $a_1, a_2, a_3, \ldots, a_n$ are all nonzero), we get these roots:

$$x = -b_1/a_1 \text{ or } x = -b_2/a_2 \text{ or } x = -b_3/a_3$$
$$\ldots \text{ or } x = -b_n/a_n$$

The real solution set is therefore

$$X = \{-b_1/a_1, -b_2/a_2, -b_3/a_3, \cdots -b_n/a_n\}$$

Question 26-8

What are the real roots of the following equation? What is the multiplicity of each root? What is the real solution set *X*? What is the degree of the equation?

$$(x+4)(2x-8)^2(3x)^5 = 0$$

Answer 26-8

We take each binomial individually, set it equal to 0, and then solve the resulting first-degree equations:

$$x + 4 = 0$$
$$2x - 8 = 0$$
$$3x = 0$$

These equations resolve to $x = -4$, $x = 4$, and $x = 0$ respectively. Therefore, the real roots of the original equation are

$$x = -4 \text{ or } x = 4 \text{ or } x = 0$$

and the real solution set is $X = \{-4, 4, 0\}$. The root -4 has multiplicity 1. The root 4 has multiplicity 2. The root 0 has multiplicity 5. The degree of the equation is the sum of the exponents attached to the factors, in this case $1 + 2 + 5$, or 8.

Question 26-9

What is the largest number of real roots that a single-variable equation of the *n*th degree can have? What is the largest number of real or complex roots that such an equation can have?

Answer 26-9

A single-variable equation of the *n*th degree can have, at most, *n* roots in total, considering the real-number roots and the complex-number roots combined.

Question 26-10

There's a way to find all the *rational-number* roots of an *n*th-degree equation in the single variable *x* when it appears in the polynomial standard form

$$a_n x^n + a_{n-1}x^{n-1} + a_{n-2}x^{n-2} + \cdots + a_1 x + b = 0$$

where a_1, a_2, a_3, ... a_n, and *b* are nonzero rationals, and *n* is a positive integer greater than 3. What is that process?

Answer 26-10

We must do the following things, in the order shown below. The process can be tedious, but it's often more likely to produce useful results than tackling the equation "head-on" or hoping for an intuitive breakthrough.

- Make sure that a_1, a_2, a_3, ... a_n, and b are all integers. If they aren't, multiply the equation through by the smallest constant that will turn them all into integers.
- Find all the positive and negative integer factors of b. Call those factors m.
- Find all the positive and negative integer factors of a_n. Call those factors n.
- Write down all the possible ratios m / n. Call those ratios r.
- With synthetic division, check every r, one at a time, to see if we get a remainder of 0.
- If none of the numbers r produces a remainder of 0, then the original equation has no rational roots.
- If one or more of the ratios r produces a remainder of 0, then every one of those numbers is a rational root of the equation.
- List all of the rational roots. Call them r_1, r_2, r_3, and so on.
- Create binomials of the form $(x - r_1)$, $(x - r_2)$, $(x - r_3)$, and so on. Each of these binomials is a factor of the original equation.
- If we're lucky, this process will give us an equation in binomial to the nth form, or an equation in binomial factor form.
- If we're less lucky, we'll get one or more binomial factors and a quadratic factor. That factor can be set equal to 0, and then the quadratic formula can be used to find its roots. Neither of those roots will be rational. They might even be complex.
- If we're unlucky, we'll get one or more binomial factors and a cubic or higher-order polynomial factor. If we set the polynomial factor equal to 0, we can be sure that none of the roots associated with it are rational.

Chapter 27

Question 27-1

Suppose we're confronted with a pair of equations in two variables, and one or both of the equations is nonlinear. How can we solve these equations as a two-by-two system?

Answer 27-1

When we want to solve a general two-by-two system of equations, we can go through these steps in order.

- Decide which variable to call independent, and which one to call dependent.
- Morph both equations so they express the dependent variable in terms of the independent variable.
- Mix the independent-variable parts of the equations to get an equation in one variable.
- Find the root(s) of that equation.

- Plug the root(s) into one of the morphed original equations, and calculate the corresponding value(s) of the dependent variable.
- Express the solution(s) as one or more ordered pairs.

Question 27-2

Consider the following pair of quadratic equations as a two-by-two system:

$$y = x^2$$

and

$$y = -x^2$$

How can we find the real solutions of this system?

Answer 27-2

Let's call x the independent variable. Then we can mix the right sides of the equations to get

$$x^2 = -x^2$$

Adding x^2 to each side, we get

$$2x^2 = 0$$

Dividing through by 2 gives us

$$x^2 = 0$$

This equation has one real root, $x = 0$. When we plug this into either of the original equations, we get $y = 0$. Therefore, the single real solution to this system is (0,0).

Question 27-3

Consider the following pair of quadratic equations as a two-by-two system:

$$y = x^2 - 1$$

and

$$y = -x^2 + 1$$

How can we find the real solutions of this system?

Answer 27-3

Again, let x be the independent variable. When we mix the right sides, we get

$$x^2 - 1 = -x^2 + 1$$

Adding the quantity $(x^2 + 1)$ to each side, we get

$$2x^2 = 2$$

Dividing this equation through by 2, we obtain

$$x^2 = 1$$

It's apparent, without doing any algebra, that the roots of this are $x = 1$ or $x = -1$. Plugging $x = 1$ into the first original equation, we get $y = 0$. Plugging $x = -1$ into that same equation, we again get $y = 0$. This system therefore has two real solutions, (1,0) and (−1,0).

Question 27-4

Consider the following pair of quadratic equations as a two-by-two system:

$$y = a_1x^2 + b_1$$

and

$$y = a_2x^2 + b_2$$

where a_1 and a_2 are nonzero real numbers that are not equal to each other, and b_1 and b_2 are real numbers. How can we find the real solutions of this system?

Answer 27-4

Once again, let's call x the independent variable. When we mix the right sides, we get

$$a_1x^2 + b_1 = a_2x^2 + b_2$$

We can subtract a_2x^2 from each side and then apply the distributive law to get

$$(a_1 - a_2)x^2 + b_1 = b_2$$

Subtracting b_1 from each side produces

$$(a_1 - a_2)x^2 = b_2 - b_1$$

Dividing through by the quantity $(a_1 - a_2)$, which we know is okay because we've been told that $a_1 \neq a_2$, we get

$$x^2 = (b_2 - b_1) / (a_1 - a_2)$$

This means that

$$x = [(b_2 - b_1) / (a_1 - a_2)]^{1/2}$$

or

$$x = -[(b_2 - b_1) / (a_1 - a_2)]^{1/2}$$

When we plug either of these into the first original equation, we have to square it. We'll get the same thing in both cases, so we might as well substitute directly for x^2 into the first original equation to obtain

$$y = a_1(b_2 - b_1) / (a_1 - a_2) + b_1$$

Multiplying the right-hand side out, we get

$$y = (a_1 b_2 - a_1 b_1) / (a_1 - a_2) + b_1$$

Expressing b_1 as the ratio $b_1/1$, applying the general rule for adding ratios, and then canceling out identical terms that subtract from each other, we obtain

$$y = (a_1 b_2 - a_2 b_1) / a_1 a_2$$

Writing these solutions as ordered pairs is tricky and messy. We might as well state the x and y values separately and take advantage of the plus-or-minus sign. Then we can write

$$x = \pm[(b_2 - b_1) / (a_1 - a_2)]^{1/2} \text{ and } y = (a_1 b_2 - a_2 b_1) / a_1 a_2$$

Question 27-5

What happens when we plug the solution for x^2 into the second original equation in the above system (instead of the first one, which we already did) and solve for y?

Answer 27-5

When we substitute directly for x^2 into the second original equation, we get

$$y = a_2(b_2 - b_1) / (a_1 - a_2) + b_2$$

Multiplying the right-hand side out, we get

$$y = (a_2 b_2 - a_2 b_1) / (a_1 - a_2) + b_2$$

Expressing b_2 as the ratio $b_2/1$, applying the general rule for adding ratios, and then canceling out identical terms that subtract from each other, we obtain the same result as before:

$$y = (a_1 b_2 - a_2 b_1) / a_1 a_2$$

Question 27-6

Consider the following pair of quadratic equations as a two-by-two system:

$$y = (x + 1)^2$$

and

$$y = (x - 1)^2$$

How can we find the real solutions of this system?

Answer 27-6

The best approach here is to multiply out the right sides of both of the binomial-squared equations, obtaining the quadratics in polynomial standard form. When we do that, we get

$$y = x^2 + 2x + 1$$

and

$$y = x^2 - 2x + 1$$

Mixing the right sides of these quadratics gives us

$$x^2 + 2x + 1 = x^2 - 2x + 1$$

Subtracting the quantity $(x^2 + 1)$ from each side, we obtain

$$2x = -2x$$

We can add $2x$ to each side and then divide through by 4 to get $x = 0$ as the sole root of this "mixed quadratic." When we substitute this value for x into either of the original equations, we get $y = 1$. Therefore, the system has the single real solution $(0,1)$.

Question 27-7

Consider the following pair of quadratic equations as a two-by-two system:

$$y = (ax + b)^2$$

and

$$y = (ax - b)^2$$

Where a and b are real numbers, neither of which are equal to 0. How can we find the real solutions of this system?

Answer 27-7

Again, let's multiply out the right sides of the binomial-squared equations. When we do that, we obtain

$$y = a^2x^2 + 2abx + b^2$$

and

$$y = a^2x^2 - 2abx + b^2$$

Setting the right sides equal gives us the "mixed quadratic"

$$a^2x^2 + 2abx + b^2 = a^2x^2 - 2abx + b^2$$

When we subtract the quantity $(a^2x^2 + b^2)$ from each side, we get

$$2abx = -2abx$$

We can add $2abx$ to each side and then divide through by $4ab$ (which is "legal" because we've been told that $a \neq 0$ and $b \neq 0$, so we can be sure that $4ab \neq 0$). When we do that, we obtain $x = 0$ as the only root of the "mixed quadratic." We can substitute this value for x into the original equations, obtaining $y = b^2$ in both cases. Therefore, the system has the single solution $(0, b^2)$.

Question 27-8

Imagine that we're trying to solve a general two-by-two system of equations, and we mix them to get a single equation in one variable. When we solve that "mixed" equation, we get two different roots, one of which has multiplicity 1, and the other of which has multiplicity 2. What does this say about the solutions of the original system?

Answer 27-8

The solutions of the original two-by-two system have the same multiplicity pattern as the roots of the "mixed" equation. In this case, that means there is one solution with multiplicity 1, and another solution with multiplicity 2.

Question 27-9

Consider the following pair of equations:

$$y = (x + 1)^3$$

and

$$y = x^3 + 2x^2 + x$$

How can we find the real solutions to this two-by-two system?

Answer 27-9

Let's multiply out the first equation to get it into polynomial standard form. When we do that, the system becomes

$$y = x^3 + 3x^2 + 3x + 1$$

and

$$y = x^3 + 2x^2 + x$$

We can mix the right sides of these equations, getting

$$x^3 + 3x^2 + 3x + 1 = x^3 + 2x^2 + x$$

Now let's subtract the entire right side of this equation from the left side. When we do that, we obtain

$$x^2 + 2x + 1 = 0$$

which factors into

$$(x + 1)^2 = 0$$

This equation has the single real root $x = -1$, with multiplicity 2. When we plug that into the first original equation, we get

$$\begin{aligned} y &= (x + 1)^3 \\ &= (-1 + 1)^3 \\ &= 0^3 \\ &= 0 \end{aligned}$$

Therefore, the original system has the single real solution $(-1,0)$, with multiplicity 2.

Question 27-10

Consider the following pair of equations:

$$y = (x + 1)^3$$

and

$$y = (x + 2)^3$$

How can we find the real solutions of this two-by-two system, if any exist?

Answer 27-10

Let's multiply both of these equations out to get cubics in polynomial standard form. That gives us

$$y = x^3 + 3x^2 + 3x + 1$$

and

$$y = x^3 + 6x^2 + 12x + 8$$

When we mix the right sides of these equations, we get

$$x^3 + 3x^2 + 3x + 1 = x^3 + 6x^2 + 12x + 8$$

Subtracting the entire left side from the right side and then switching right-to-left, we obtain

$$3x^2 + 9x + 7 = 0$$

The discriminant of this quadratic is negative, telling us that it has no real roots. The x-values of any solutions we can derive for the original system will not be real numbers. Therefore, no real solutions exist.

Chapter 28

Question 28-1

How do we graph a general two-by-two system of equations when we want to see approximately where the curves intersect at the real solutions, but we don't need a lot of precision?

Answer 28-1

First, we can calculate several ordered pairs for both functions individually, including the real solutions, if any exist. It can be helpful to put the values in a table. Next, we figure out the scales we should have on our graph so as to provide a good "picture" of the situation. Then we plot the real solution point or points, if any exist, on the coordinate grid. After that, we plot the rest of the points based on the values in the table we've created. Finally, we fill in the lines or curves for both functions.

Question 28-2

Consider the system of equations we solved in Answer 27-2:

$$y = x^2$$

and

$$y = -x^2$$

How can we sketch an approximate graph of this system, showing the real solution?

Answer 28-2

We can tabulate and plot several points in both functions including the real solution, (0,0). Table 30-1 compares some values of x, some values of the first function, and some values of

Table 30-1. Selected values for graphing the functions $y = x^2$ and $y = -x^2$. The bold entry indicates the real solution.

x	x^2	$-x^2$
−2	4	−4
−1	1	−1
0	**0**	**0**
1	1	−1
2	4	−4

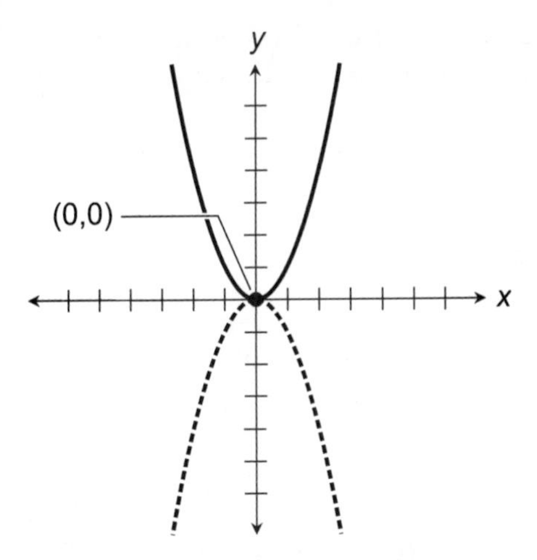

Figure 30-3 Illustration for Answer 28-2. The first function is graphed as a solid curve; the second function is graphed as a dashed curve. The real-number solution appears as a point where the curves intersect. On both axes, each increment is 1 unit.

the second function. Figure 30-3 shows the curves and the solution point. On both axes, each increment represents 1 unit.

Question 28-3

Consider the system of equations we solved in Answer 27-3:

$$y = x^2 - 1$$

and

$$y = -x^2 + 1$$

How can we sketch an approximate graph of this system, showing the two real solutions?

Answer 28-3

We can tabulate and plot several points in both functions including the real solutions, (1,0) and (−1,0). Table 30-2 compares some values of x, some values of the first function, and some values of the second function. Figure 30-4 shows the curves and the solution points. On both axes, each increment represents 1 unit.

Table 30-2. Selected values for graphing the functions $y = x^2 - 1$ and $y = -x^2 + 1$. Bold entries indicate real solutions.

x	$x^2 - 1$	$-x^2 + 1$
−2	3	−3
−1	**0**	**0**
0	−1	1
1	**0**	**0**
2	3	−3

Question 28-4

When we compare the systems stated in Questions 28-2 and 28-3 and graphed in Figs. 30-3 and 30-4, we can see that the curves have the same shapes in both situations. But in the second case, the upward-opening parabola has been moved vertically down by 1 unit, while the downward-opening parabola has been moved vertically up by 1 unit. This has caused the single intersection point (Fig. 30-3) to "break in two" (Fig. 30-4). What will happen if we move the upward-opening parabola, shown by the solid curve, further down, and move the downward-opening parabola, shown by the dashed curve, further up by the same distance? How will the equations in the system change if we do this?

Answer 28-4

If we move the parabolas this way, the intersection points will move farther from each other. The negative x-value of one real solution will become more negative, and the positive x-value

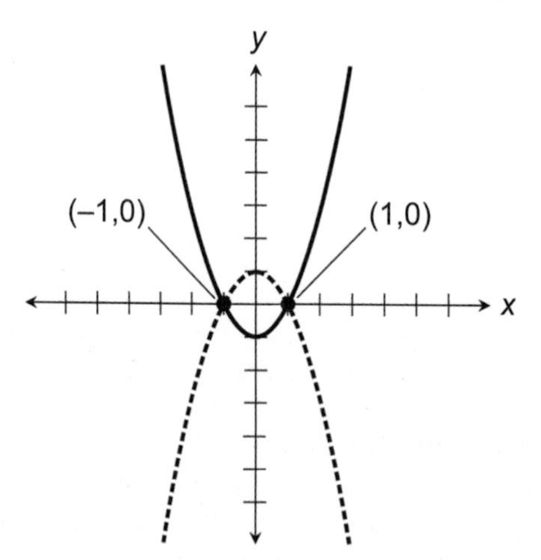

Figure 30-4 Illustration for Answer 28-3. The first function is graphed as a solid curve; the second function is graphed as a dashed curve. Real-number solutions appear as points where the curves intersect. On both axes, each increment is 1 unit.

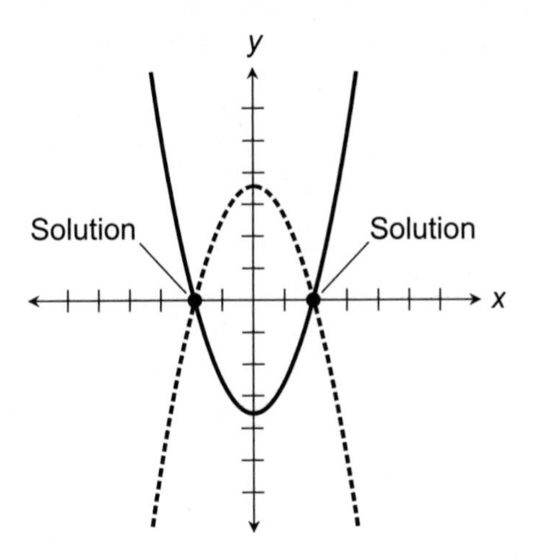

Figure 30-5 Illustration for Answer 28-4. The first function is graphed as a solid curve; the second function is graphed as a dashed curve. Real-number solutions appear as points where the curves intersect. On both axes, each increment is 1 unit.

of the other real solution will become more positive to the same extent. The y-values of both solutions will remain at 0; the points will stay on the x axis. Figure 30-5 shows an example. On both axes, each increment represents 1 unit. The stand-alone constants in the equations will change. The negative constant in the first equation will become more negative, and the positive constant in the second equation will become more positive to the same extent.

Question 28-5

Let's modify the system presented in Question 28-4 and graphed in Fig. 30-5. Suppose that we move the upward-opening parabola even further straight down, but leave the downward-opening parabola in the same place. What will happen to the solution points? How will the equations in the system change?

Answer 28-5

The intersection points will move even farther from each other. The negative x-value of one real solution will become more negative, and the positive x-value of the other real solution will become more positive to the same extent. The y-values of both solutions will become negative to an equal extent. The solution points will move off the x axis into the third and fourth quadrants of the coordinate plane. This assumes that we move the upward-opening parabola exactly in the negative-y direction. Figure 30-6 shows an example. On both axes, each increment represents 1 unit. The negative constant in the first equation will become more negative, and the positive constant in the second equation will stay the same.

Question 28-6

Consider the system of equations we solved in Answer 27-6:

$$y = (x + 1)^2$$

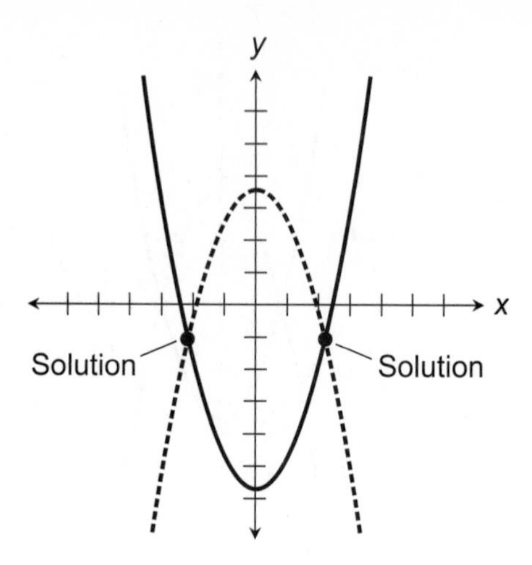

Figure 30-6 Illustration for Answer 28-5. The first function is graphed as a solid curve; the second function is graphed as a dashed curve. Real-number solutions appear as points where the curves intersect. On both axes, each increment is 1 unit.

and

$$y = (x - 1)^2$$

How can we sketch an approximate graph of this system, showing the real solution?

Answer 28-6

We can tabulate and plot several points in both functions including the real solution, (0,1). Table 30-3 compares some values of x, some values of the first function, and some values of the second function. Figure 30-7 shows the curves and the solution point. On the x axis, each increment is 1/2 unit. On the y axis, each increment is 2 units.

Table 30-3. Selected values for graphing the functions $y = (x + 1)^2$ and $y = (x - 1)^2$. The bold entry indicates the real solution.

x	$(x + 1)^2$	$(x - 1)^2$
−3	4	16
−2	1	9
−1	0	4
0	**1**	**1**
1	4	0
2	9	1
3	16	4

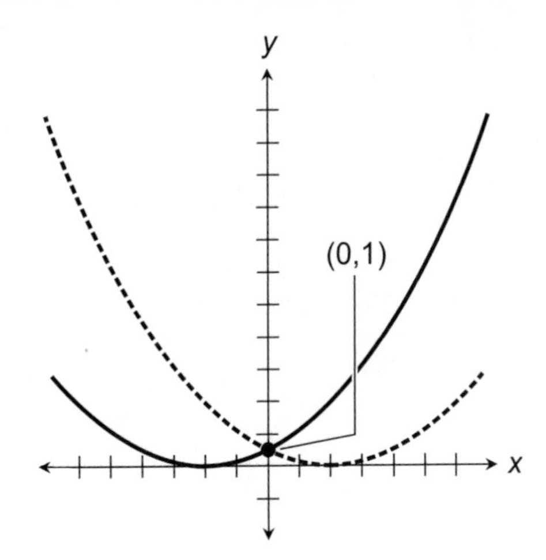

Figure 30-7 Illustration for Answer 28-6. The first function is graphed as a solid curve; the second function is graphed as a dashed curve. The real-number solution appears as a point where the curves intersect. On the x axis, each increment is 1/2 unit. On the y axis, each increment is 2 units.

Question 28-7

Let's modify the system of equations stated in Question 28-6. Suppose we multiply the right side of the second equation by −1, producing this two-by-two system:

$$y = (x + 1)^2$$

and

$$y = -(x - 1)^2$$

How will this affect the graph of the second equation, shown by the dashed curve? How will it affect the real solution set?

Answer 28-7

If we multiply the right side of this equation by −1, we multiply all values of the function by −1. This inverts the entire graph of the function with respect to the x axis. Figure 30-8 shows the result. On the x axis, each increment is 1/2 unit. On the y axis, each increment is 2 units. We can see that the system has no real solutions because the curves don't intersect. The real solution set is empty.

Question 28-8

Let's modify the system of equations stated in Question 28-6 in a different way. Suppose we subtract 12 from the right side of the second equation, producing this two-by-two system:

$$y = (x + 1)^2$$

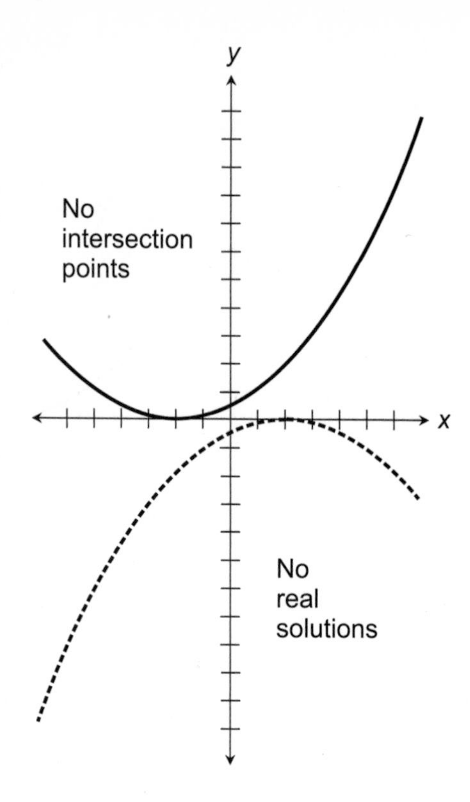

Figure 30-8 Illustration for Answer 28-7. The first function is graphed as a solid curve; the second function is graphed as a dashed curve. The curves do not intersect, indicating that the system has no real solutions. On the x axis, each increment is 1/2 unit. On the y axis, each increment is 2 units.

and

$$y = (x - 1)^2 - 12$$

How will this affect the graph of the second equation, shown by the dashed curve? How will it affect the real solution set?

Answer 28-8

If we subtract 12 from the right side of this equation, we reduce all values of the function by 12. This moves the entire graph of the function vertically down by 12 units. Figure 30-9 shows the result. On the x axis, each increment is 1/2 unit. On the y axis, each increment is 2 units. This graph suggests that the resulting system still has one real solution, but it has changed. If we want to find the solution, we must solve the new system, starting all over again from scratch. (For extra credit, you can do this.)

Question 28-9

Consider the system of equations we solved in Answer 27-9:

$$y = (x + 1)^3$$

and

$$y = x^3 + 2x^2 + x$$

How can we sketch an approximate graph of this system, showing the real solution?

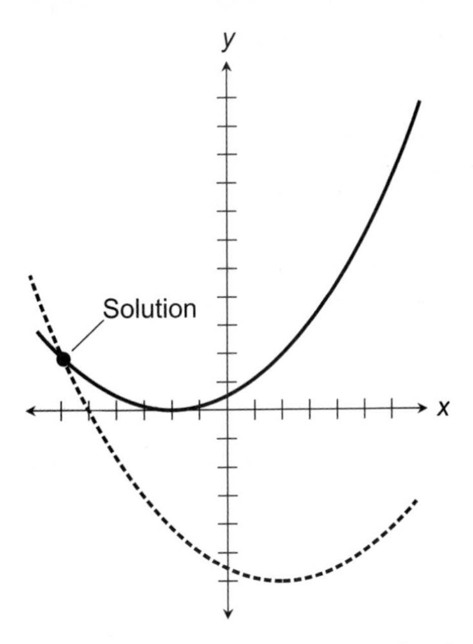

Figure 30-9 Illustration for Answer 28-8. The first function is graphed as a solid curve; the second function is graphed as a dashed curve. The solution point has moved. To find its coordinates, we must solve the new system algebraically. On the x axis, each increment is 1/2 unit. On the y axis, each increment is 2 units.

Answer 28-9

We can tabulate and plot several points in both functions including the real solution, (−1,0). Table 30-4 compares some values of x, some values of the first function, and some values of the second function. Figure 30-10 shows the curves and the solution point. On the x axis, each increment is 1/2 unit. On the y axis, each increment is 5 units.

Table 30-4. Selected values for graphing the functions $y = (x + 1)^3$ and $y = x^3 + 2x^2 + x$. The bold entry indicates the real solution.

x	$(x + 1)^3$	$x^3 + 2x^2 + x$
−3	−8	−12
−2	−1	−2
−1	**0**	**0**
0	1	0
1	8	4
2	27	18

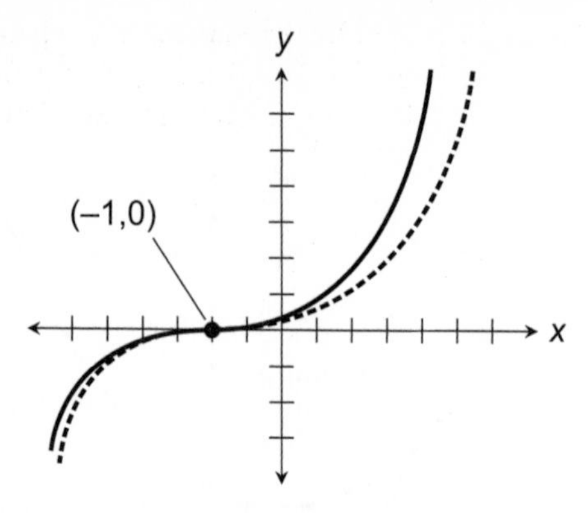

Figure 30-10 Illustration for Answer 28-9. The first function is graphed as a solid curve; the second function is graphed as a dashed curve. The real-number solution appears as a point where the curves intersect. On the x axis, each increment is 1/2 unit. On the y axis, each increment is 5 units.

Question 28-10

Consider the system of equations we evaluated in Answer 27-10, where we found that there are no real solutions:

$$y = x^3 + 3x^2 + 3x + 1$$

and

$$y = x^3 + 6x^2 + 12x + 8$$

How can we sketch an approximate graph of this system, showing that there are no real solutions?

Answer 28-10

We can tabulate and plot enough points in both functions to get clear images of their curves, and to show that they do not intersect (although they come close). Table 30-5 compares some values of x, some values of the first function, and some values of the second function. Figure 30-11 shows the curves. On the x axis, each increment is 1 unit. On the y axis, each increment is 10 units.

Chapter 29

Question 29-1

If we say that the common logarithm of a certain number p is equal to q, what do we mean?

Answer 29-1

The common logarithm (or common log) of a number is the power to which we must raise 10 to get that number. If we say that the common log of p is equal to q, we mean

$$p = 10^q$$

Table 30-5. Selected values for graphing the functions $y = x^3 + 3x^2 + 3x + 1$ and $y = x^3 + 6x^2 + 12x + 8$. This system has no real solutions.

x	$x^3 + 3x^2 + 3x + 1$	$x^3 + 6x^2 + 12x + 8$
−5	−64	−27
−4	−27	−8
−3	−8	−1
−2	−1	0
−1	0	1
0	1	8
1	8	27
2	27	64

The common log is sometimes called the base-10 log, because 10 is the base that we raise to various powers.

Question 29-2

According to the definition in Answer 29-1, what is the common log of 10? Of 100? Of 1,000? Of 10,000? What happens to the common log of a positive number, as that number grows larger and larger indefinitely?

Answer 29-2

The common log is related to a growing positive number like this:

- The common log of 10 is 1, because $10^1 = 10$.
- The common log of 100 is 2, because $10^2 = 100$.

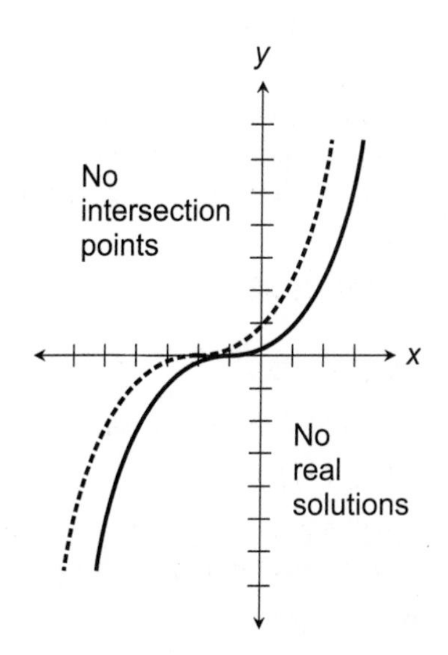

Figure 30-11 Illustration for Answer 28-10. The first function is graphed as a solid curve; the second function is graphed as a dashed curve. The curves do not intersect (although they come close!), so there are no real solutions. On the x axis, each increment is 1 unit. On the y axis, each increment is 10 units.

- The common log of 1,000 is 3, because $10^3 = 1{,}000$.
- The common log of 10,000 is 4, because $10^4 = 10{,}000$.

As a number gets larger without limit, so does its common log. The size of the logarithm grows much more slowly than the size of the number.

Question 29-3

According to the definition in Answer 29-1, What is the common log of 1? Of 1/10? Of 1/100? Of 1/1000? What happens to the common log of a positive real number whose absolute value keeps shrinking, that is, as the number approaches 0 from the positive direction? What happens to the common log of a shrinking positive real number when it actually becomes 0?

Answer 29-3

As the absolute value of a positive number keeps shrinking, its common log changes like this:

- The common log of 1 is 0, because $10^0 = 1$.
- The common log of 1/10 is -1, because $10^{-1} = 1/10$.
- The common log of 1/100 is -2, because $10^{-2} = 1/100$.
- The common log of 1/1,000 is -3, because $10^{-3} = 1/1{,}000$.

As a shrinking positive number approaches 0, its common log becomes more negative. As the shrinking positive number "closes in" on 0, the common log decreases—that is, it increases negatively—without limit. When the shrinking positive number actually reaches 0, its common log is no longer defined in the set of real numbers. (Perhaps it's non-real but complex, or maybe it's some other kind of number entirely. Evaluating it is beyond the scope of this book.)

Question 29-4

If we say that the natural logarithm of a certain number p is equal to q, what do we mean?

Answer 29-4

The natural logarithm (or natural log) of a number is the power to which we must raise Euler's constant, e, to get that number. If we say that the natural log of p is equal to q, we mean

$$p = e^q$$

The common log is sometimes called the base-e log, because e is the base that we raise to various powers. The value of e is approximately 2.71828. It's an irrational number, however, so it cannot be fully written out in decimal form.

Question 29-5

According to the definition in Answer 29-4, what is the natural log of e? Of e^2? Of e^3? Of e^4? What happens to the natural log of a number as that number grows larger without limit?

Answer 29-5

The natural log is related to a growing number like this:

- The natural log of e is 1, because $e^1 = e$.
- The natural log of e^2 is 2.
- The natural log of e^3 3.
- The natural log of e^4 is 4.

As a number gets larger without limit, so does its natural log, but the size of the log grows much more slowly than the size of the number.

Question 29-6

According to the definition in Answer 29-4, What is the natural log of 1? Of $1/e$? Of $1/e^2$? Of $1/e^3$? What happens to the natural log of a positive real number whose absolute value keeps shrinking? What happens to the natural log of a shrinking positive real number when it actually becomes 0?

Answer 29-6

As the absolute value of a positive number keeps shrinking, its natural log changes like this:

- The natural log of 1 is 0, because $e^0 = 1$.
- The natural log of $1/e$ is -1.
- The natural log of $1/e^2$ is -2.
- The natural log of $1/e^3$ is -3.

As a positive number approaches 0, its natural log becomes more negative. There is no limit to how large negatively the log can get. When the shrinking positive number actually reaches 0, its natural log is no longer defined in the set of real numbers. (Perhaps it's non-real but complex, or maybe it's some other kind of number entirely. Evaluating it is beyond the scope of this book.)

Question 29-7

How can logarithms be used to change products into sums, or ratios into differences? Do these properties of logs depend on the base?

Answer 29-7

The logarithm of the product of two numbers is equal to the sum of their logarithms. The logarithm of the ratio of two numbers is equal to the difference between their logarithms. These rules work for common logs as well as for natural logs. In fact, they work no matter what the base happens to be, as long as we don't change the base during the calculation!

Question 29-8

What is the common exponential of a number? What is the natural exponential of a number?

Answer 29-8

The common exponential of a number is what we get when we raise 10 to a power equal to that number. The natural exponential of a number is what we get when we raise e to a power equal to that number. When working with natural exponentials, it's customary to call e the exponential constant.

Question 29-9

How can we find the number x whose common exponential is 100,000? How can we find the number y whose natural exponential is 100,000?

Answer 29-9

To find the number x whose common exponential is 100,000, we must find the power of 10 that gives us 100,000. We want to solve the equation

$$10^x = 100{,}000$$

It's easy see that $x = 5$ in this case. But if we want to go through the motions of solving the above equation formally, we can take the common log (symbolized $\log_{10}$) of both sides, obtaining

$$\log_{10}(10^x) = \log_{10} 100{,}000$$

The common log function "undoes" the common exponential function, so we can simplify this equation to

$$x = \log_{10} 100{,}000$$

A calculator tells us that $\log_{10} 100{,}000$ is exactly equal to 5.

Finding the number y whose natural exponential is 100,000 is a little more involved. We want to find the power of e that gives us 100,000, so we must solve the equation

$$e^y = 100{,}000$$

If we take the natural log (symbolized ln) of both sides of this equation, we get

$$\ln (e^y) = \ln 100{,}000$$

The natural log function "undoes" the natural exponential function, so we have

$$y = \ln 100{,}000$$

A calculator tells us that $y = 11.513$, rounded off to three decimal places.

Question 29-10

How can we find the number x whose natural exponential is $1/e^5$? How can we find the number y whose common exponential is $1/e^5$?

Answer 29-10

To find the number x whose natural exponential is $1/e^5$, we must find the power of e that gives us $1/e^5$. We want to solve the equation

$$e^x = 1/e^5$$

This is almost trivial, because $1/e^5$ is just another way of writing e^{-5}. Now we have

$$e^x = e^{-5}$$

Obviously, this means $x = -5$. If, despite the simplicity of this, we insist on solving formally and including every step, we can take the natural log of both sides of the above equation, obtaining

$$\ln (e^x) = \ln (e^{-5})$$

We can simplify both sides to get

$$x = -5$$

Finding the number y whose common exponential is $1/e^5$ requires more work, but not much. We want to find the power of 10 that gives us e^{-5}, so we must solve the equation

$$10^y = e^{-5}$$

If we take the common log of both sides of this equation, we obtain

$$\log_{10} (10^y) = \log_{10} (e^{-5})$$

The common log function "undoes" the common exponential function, so we have

$$y = \log_{10} (e^{-5})$$

A calculator tells us that $e^{-5} = 0.006737947$, rounded off to nine decimal places. That ought to be plenty of digits to give us a good idea of the final answer, which is the common log of 0.006737947. Rounding off the end result to three decimal places, we get

$$\begin{aligned} y &= \log_{10} (0.006737947) \\ &= -2.171 \end{aligned}$$

Final Exam

Do not refer to the text when taking this test. A good score is at least 80 percent of the answers correct. Answers are in the back of the book. It's best to have a friend check your score the first time, so you won't memorize the answers if you want to take the test again. This test is long. Don't try to do it all 547in one sitting!

1. Which of the following procedures, if any, is *not* an acceptable way to modify an equation?
   (a) Add a positive integer to the part of the equation to the left of the equals sign, and add the same positive integer to the part of the equation to the right of the equals sign.
   (b) Subtract a positive integer from the part of the equation to the left of the equals sign, and subtract the same positive integer from the part of the equation to the right of the equals sign.
   (c) Multiply the part of the equation to the left of the equals sign by a positive integer, and multiply the part of the equation to the right of the equals sign by the same positive integer.
   (d) Divide the part of the equation to the left of the equals sign by a positive integer, and divide the part of the equation to the right of the equals sign by the same positive integer.
   (e) All of the above procedures are acceptable.
2. Consider the following pair of equations as a two-by-two linear system:

$$x = 4y - 3$$

and

$$x = 4y + 2$$

What is the solution to this system?

(a) $(x, y) = (0, 4)$.

(b) $(x, y) = (3/4, -1/2)$.

(c) $(x, y) = (-4/3, 2)$.

(d) There are infinitely many solutions.

(e) There is no solution.

3. If the discriminant in a quadratic equation is equal to 0, it means that
   (a) the equation has two different real roots.
   (b) the equation has one real root with multiplicity 2.
   (c) the equation has one imaginary root with multiplicity 2.
   (d) the equation has two different nonreal, complex roots.
   (e) the equation has no roots at all.
4. Which of the following is *not* a rational number?
   (a) 22/7
   (b) 25.25252525 ...
   (c) π
   (d) $64^{1/2}$
   (e) $(64/9)^{1/2}$
5. Suppose that m and n are integers, and p is a nonzero integer. All of the following equations *except one* are generally true. Which one is the exception?
   (a) $(m + n)/p = m/p + n/p$
   (b) $p/(m - n) = p/m - p/n$
   (c) $p(m + n) = pm + pn$
   (d) $(m + n)p = mp + np$
   (e) $(m - n)p = mp - np$
6. Suppose we are trying to solve a three-by-three linear system using matrices. We are sure we haven't made any mistakes along the way. We come up with this matrix:

2	2	2	8
3	3	3	−7
−1	−1	−1	5

We can conclude that the original three-by-three system has

(a) no solutions.

(b) one solution.

(c) two solutions.

(d) three solutions.

(e) infinitely many solutions.

7. Fill in the blank to make the following statement true. "Suppose that a is a nonzero number. Also suppose that m and n are rational numbers, with $n \neq 0$. If we raise a to the mth power and then take the nth root of that quantity, we get the same result as if we ________."
   (a) raise a to the power of $(m + n)$
   (b) raise a to the power of $(m - n)$
   (c) raise a to the power of mn
   (d) raise a to the power of m/n
   (e) raise a to the power of $1/(mn)$

8. When we want to add two complex numbers, we must
   (a) add the real parts and multiply the imaginary parts, getting a real number.
   (b) multiply the real parts and add the imaginary parts, getting a complex number.
   (c) multiply the real parts and the imaginary parts separately, getting a real number.
   (d) add the real parts and the imaginary parts separately, getting a complex number.
   (e) find their absolute values and add them, getting a real number.

9. In Fig. FE-1, which of the following statements is true, assuming A, B, C, and D are all non-empty sets?
   (a) Sets A and B are disjoint, and sets C and D are disjoint.
   (b) Sets A and C are disjoint, and sets B and D are disjoint.
   (c) Sets B and C are disjoint, and sets A and D are disjoint.
   (d) Sets A and C are disjoint, and sets A and D are disjoint.
   (e) None of the above.

10. In Fig. FE-1, which of the following statements is true, assuming A, B, C, and D are all non-empty sets?
    (a) $B \subseteq A$
    (b) $B \in A$
    (c) $A \cap B = A$
    (d) $A \cap B = \varnothing$
    (e) $A \cup B = B$

11. Suppose it's 12:00 noon on the twenty-fifth day of June. How many 24-hour days will pass between this moment and 12:00 noon on the fourth day of July, in the same year and in the same time zone? (June has 30 days.)
    (a) Eight days.
    (b) Nine days.
    (c) Ten days.
    (d) Eleven days.
    (e) Twelve days.

12. The numerical value of $13/(-3)$ is the same as the value of
    (a) $-4\text{-}1/3$.
    (b) $4\text{-}1/3$.

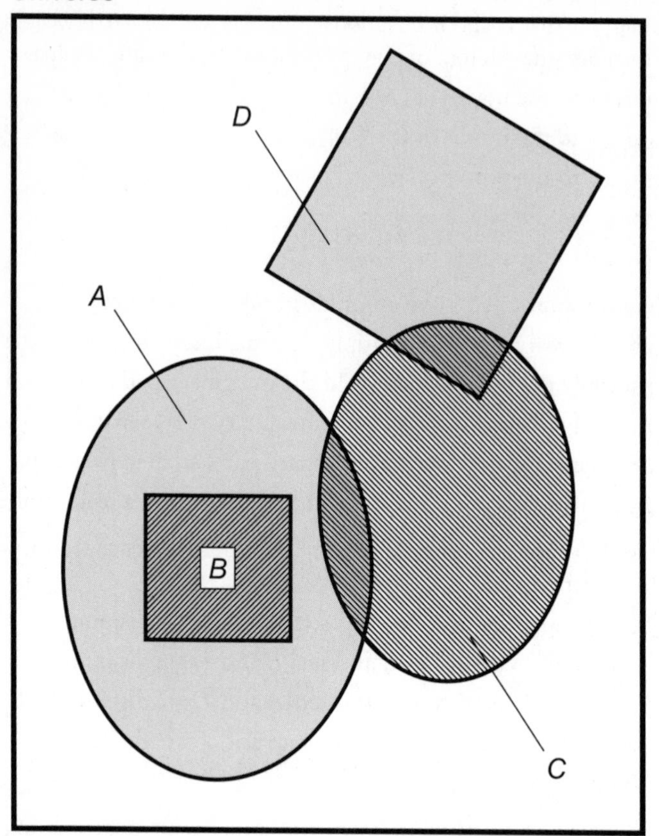

Figure FE-1 Illustration for Final Exam Questions 9 and 10.

(c) −3/13.
(d) 3/(−13).
(e) −3/(−13).

13. What is (−3/7) / (13/2) in lowest terms?
    (a) −13/4
    (b) −6/91
    (c) −39/14
    (d) −3/46
    (e) −5/26

14. Consider the following third-degree equation:

$$(x + 7)^3 = 0$$

How many real roots does this equation have?

(a) None.

(b) One.

(c) Two.

(d) Three.

(e) Infinitely many.

15. Which of the following is a 4th root of 1?

(a) 1

(b) −1

(c) j

(d) $-j$

(e) All of the above

16. When we take the 5th root of a number, it's the same thing as raising that number to the power of

(a) −5.

(b) 1/5.

(c) −1/5.

(d) 25.

(e) −25.

17. Suppose x and y are real numbers. The quantity $[x/(y+1)]^2$ makes sense

(a) for all possible values of x and y.

(b) for all possible values of x and y, as long as $y \neq 0$.

(c) for all possible values of x and y, as long as $x \neq 0$.

(d) for all possible values of x and y, as long as $y \neq -1$.

(e) for all possible values of x and y, as long as $y \neq 1$.

18. The decimal expansion of an irrational number between 0 and 1

(a) terminates after a finite number of digits.

(b) is endless and repeating.

(c) is endless and non-repeating.

(d) can be converted to a ratio of two integers.

(e) None of the above.

19. Consider the following equation:

$$(x - j5)^7 = 0$$

This equation has an imaginary root $j5$ with multiplicity

(a) 1.

(b) j.

(c) 5.
(d) $j5$.
(e) 7.

20. What's the binary equivalent of the decimal 127?
    (a) 1001000
    (b) 1111111
    (c) 1000001
    (d) 10000000
    (e) 11100111

21. In Fig. FE-2, what is being done to the original number $16n$?
    (a) It is being repeatedly multiplied by 2.
    (b) It is being repeatedly multiplied by −2.

Figure FE-2 Illustration for Final Exam Questions 21 and 22.

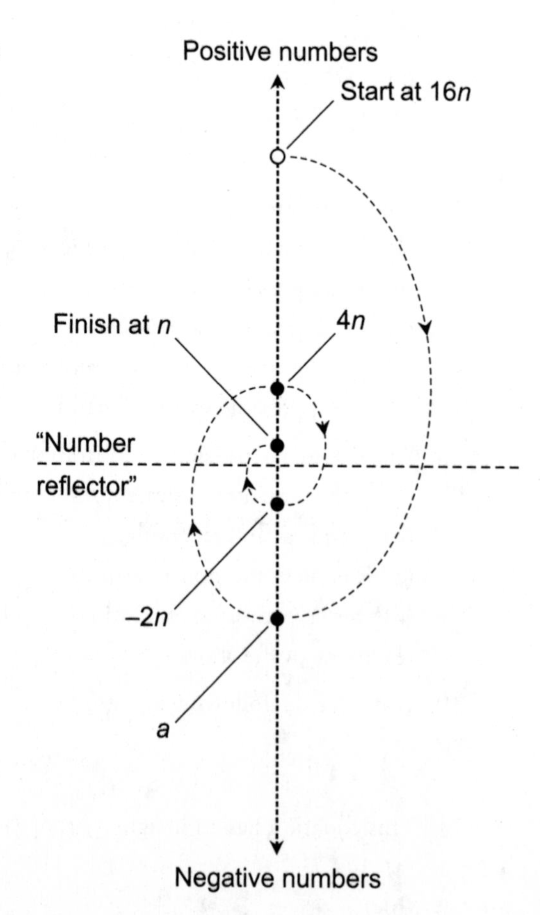

(c) It is being repeatedly divided by 2.
(d) It is being repeatedly divided by −2.
(e) It is being repeatedly squared.

22. In Fig. FE-2, the value of a is
   (a) impossible to determine without more information.
   (b) $-6n$.
   (c) $-8n$.
   (d) $-10n$.
   (e) $-12n$.

23. Imagine that we come across a cubic function and a linear function, where all the coefficients and constants are real numbers. We want to solve these functions as a two-by-two system. What is the *maximum* number of real solutions that such a system can have?
   (a) One.
   (b) Two.
   (c) Three.
   (d) Infinitely many.
   (e) We can't answer this without more information.

24. Which of the following statements is false?
   (a) All integers are rational.
   (b) All real numbers are irrational.
   (c) All negative rational numbers are real.
   (d) All natural numbers are integers.
   (e) All integers are real.

25. When we take a quantity to the power of $-1/5$, it's the same thing as
   (a) taking the 5th root of the quantity and then taking the reciprocal of the result.
   (b) taking the 5th power of the quantity and then taking the reciprocal of the result.
   (c) taking the −5th root of the quantity and then taking the reciprocal of the result.
   (d) taking the −5th power of the quantity and then taking the reciprocal of the result.
   (e) None of the above.

26. Consider the following third-degree equation:

$$(x - 1)(x + 7)^2 = 0$$

   How many real roots does this equation have?
   (a) None.
   (b) One.
   (c) Two.

(d) Three.
(e) Infinitely many.

27. Which of these integers is the smallest?
(a) −8
(b) −2
(c) 0
(d) 3
(e) 7

28. Which of these integers has the largest absolute value?
(a) −8
(b) −2
(c) 0
(d) 3
(e) 7

29. Imagine a two-by-two linear system in variables x and y. Suppose the graphs of the equations are parallel, but not coincident, lines in Cartesian coordinates, where y is the dependent variable. Such a system has
(a) solutions corresponding to the y-intercepts of the lines.
(b) solutions corresponding to the x-intercepts of the lines.
(c) solutions corresponding to the x-intercept of one line and the y-intercept of the other line.
(d) no solutions.
(e) infinitely many solutions.

30. Consider the following pair of equations as a two-by-two system:

$$y = a_1x + b_1$$

and

$$y = a_2x^2 + b_2x + c$$

where a_1, a_2, b_1, b_2, and c are real numbers, and neither a_1 nor a_2 are equal to 0. What are the smallest and largest numbers of elements that the solution set of such a system can have?
(a) None, and one.
(b) None, and two.
(c) None, and three.
(d) One, and three.
(e) None, and infinitely many.

31. Suppose we have a positive integer a. We subtract $-a$ from it. The result is
    (a) equal to 0.
    (b) equal to a.
    (c) equal to $a + a$.
    (d) equal to $-a - a$.
    (e) undefined.

32. Fill in the blank to make the following statement true. "Suppose that a is a nonzero number. Also suppose that m and n are rational numbers. If we raise a to the mth power and then raise that quantity to the nth power, we get the same result as if we ________."
    (a) raise a to the power of $(m + n)$
    (b) raise a to the power of $(m - n)$
    (c) raise a to the power of mn
    (d) raise a to the power of m/n
    (e) raise a to the power of $1/(mn)$

33. Suppose we see two cubic functions. All of the coefficients and constants are real numbers. As we solve these functions as a two-by-two system, we create a single-variable equation by mixing the independent-variable parts of the functions. When we factor that equation, we discover that it can be written in binomial-cubed form. Based on that knowledge, what can we say about the multiplicities of the real solutions of the original system?
    (a) Nothing, because there are no real solutions at all.
    (b) There are three different real solutions, each of which has multiplicity 1.
    (c) There is one real solution with multiplicity 1, and a second, different real solution with multiplicity 2.
    (d) There is one real solution with multiplicity 3.
    (e) There is one real solution with multiplicity 6.

34. Which of the following sets is nondenumerable?
    (a) The set of all natural numbers.
    (b) The set of all negative integers.
    (c) The set of all integers.
    (d) The set of all rational numbers.
    (e) The set of all irrational numbers.

35. The intersection of the null set with any other set is always equal to
    (a) that other set.
    (b) the null set.
    (c) the set containing 0.
    (d) the set containing the null set.
    (e) the universal set.

36. Consider the following quadratic equation:

$$2x^2 + 5x + 8 = 0$$

What is the discriminant in this equation?

(a) $39^{1/2}$
(b) $-39^{1/2}$
(c) 39
(d) −39
(e) $\pm j39$

37. The quadratic equation stated in Question 36 has
(a) two distinct real roots.
(b) one real root with multiplicity 2.
(c) one imaginary root with multiplicity 2.
(d) two imaginary roots that are additive inverses of each other.
(e) two complex roots that are conjugates of each other.

38. Suppose somebody tells us that there's a general law about how the terms in a subtraction problem can be grouped. According to that person, if m, n, and p are integers, then it is always true that

$$(m - n) - p = m - (n - p)$$

What can we say about this? Is the person right? Is this a legitimate law of mathematics? If so, what is it called?

(a) This isn't a legitimate law of mathematics.
(b) Yes. It is called the associative law.
(c) Yes. It is called the distributive law.
(d) Yes. It is called the commutative law.
(e) Yes. It is called the law of additive inverses.

39. Figure FE-3 represents all the rational numbers in power-of-10 form. Three points are shown on the lines: X, Q, and P. Suppose the numbers corresponding to these points are called x, q, and p respectively. Based on the information in the drawing,
(a) $|x|$ is one order of magnitude larger than $|q|$.
(b) $|x|$ is one order of magnitude smaller than $|q|$.
(c) $|x|$ is five orders of magnitude larger than $|q|$.
(d) $|x|$ and $|q|$ have the same order of magnitude.
(e) the order-of-magnitude relationship between $|x|$ and $|q|$ can't be defined.

40. Based on the information in Fig. FE-3,
(a) $|x|$ is six orders of magnitude larger than $|p|$.
(b) $|x|$ is six orders of magnitude smaller than $|p|$.

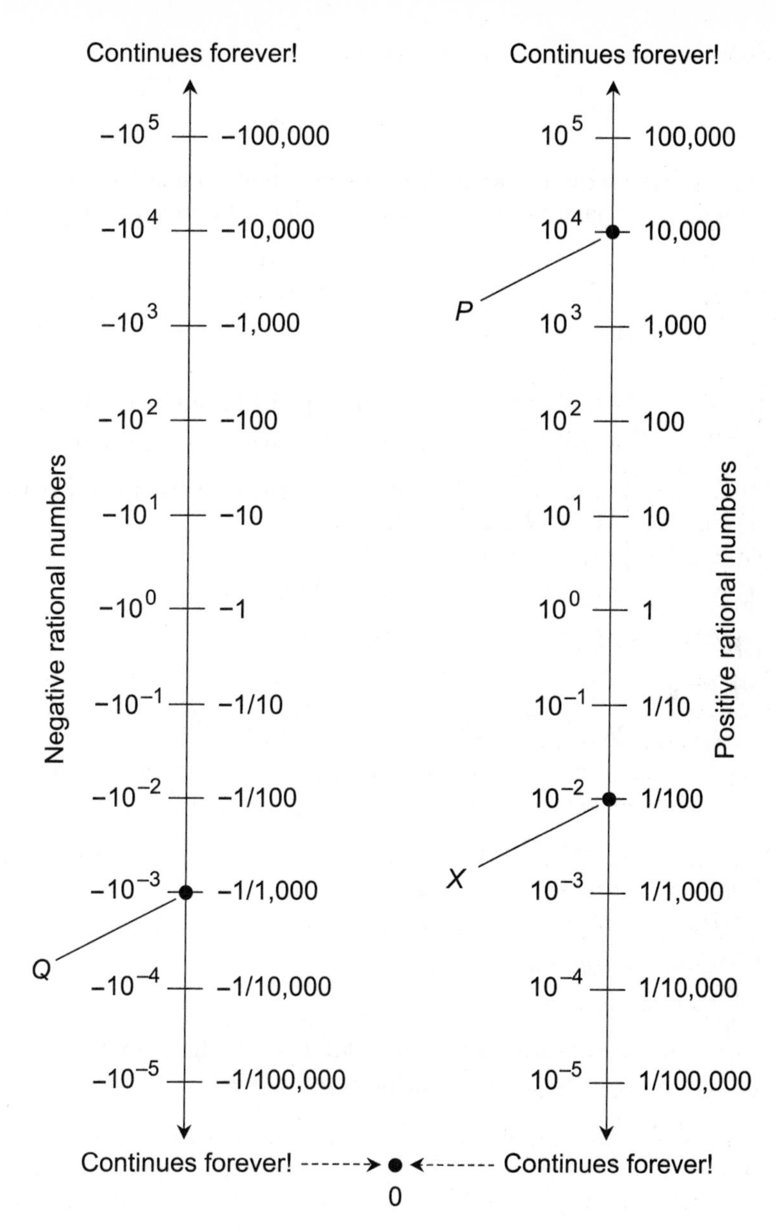

Figure FE-3 Illustration for Final Exam Questions 39 and 40.

(c) $|x|$ is two orders of magnitude larger than $|p|$.

(d) $|x|$ is two orders of magnitude smaller than $|p|$.

(e) $|x|$ is six times smaller than $|p|$.

41. Consider the following pair of equations as a two-by-two linear system:

$$2x + 7y = 8$$

and

$$-4x + y = -2$$

How can we add multiples of these two equations to make x vanish, leaving us with a solvable first-degree equation in y alone, with coefficients that are all integers?

(a) We can't.

(b) We can multiply the top equation through by 2, and then add it to the bottom equation.

(c) We can multiply the bottom equation through by $-1/2$, and then add it to the top equation.

(d) We can multiply both equations through by 2, and then add them.

(e) We can multiply both equations through by 0, and then add them.

42. All of the following equations *except one* are generally true for any two integers p and q. Which one is the exception?

(a) $p + q = q + p$

(b) $p + (-q) = (-q) + p$

(c) $(-p) + q = q + (-p)$

(d) $p - q = q - p$

(e) $(-p) + (-q) = (-q) + (-p)$

43. Suppose that m and n are integers, and p and q are nonzero integers. Which of the following statements is always true?

(a) If $m/p = n/q$, then $mq = np$.

(b) If $m/p = n/q$, then $mn = pq$.

(c) If $m/p = n/q$, then $mp = nq$.

(d) If $m/p = n/q$, then $m/q = n/p$.

(e) None of the above.

44. Suppose a, b, and c are single digits, all different. What is the fractional equivalent of the endless repeating decimal $0.abcaabcaabcaabca...$?

(a) $a,bca / 9{,}999$

(b) $c,aab / 9{,}999$

(c) $a,abc / 9{,}999$

(d) $b,caa / 9{,}999$

(e) None of the above

45. Fill in the blank to make the following statement true. "Suppose that a is a nonzero number. Consider a^m and a^n, where m and n are integers. If we multiply these two quantities, we get the same result as if we ________."

(a) divide a by $(m + n)$

(b) multiply a by $(m + n)$

(c) raise a to the power of $(m + n)$

(d) raise a to the power of mn

(e) raise a to the power of $1/(mn)$

46. When we add $a + jb$ to its conjugate, we get
    (a) $a^2 + b^2$
    (b) $a^2 - b^2$
    (c) $2a$
    (d) $2b$
    (e) $4ab$

47. Consider this general equation, where p, q, r, and s are all positive real numbers:

$$(px + q)(rx + s) = 0$$

What is the solution set X for this equation?

(a) $X = \{q/p, s/r\}$
(b) $X = \{-q/p, -s/r\}$
(c) $X = \{pq, rs\}$
(d) $X = \{-pq, -rs\}$
(e) We need more information to answer this

48. We can tell right away that a fraction is in lowest terms if
    (a) its numerator is "cleanly" divisible by its denominator.
    (b) its denominator is "cleanly" divisible by its numerator.
    (c) its numerator is a product of primes.
    (d) its denominator is a product of primes.
    (e) its numerator and denominator are both prime.

49. Imagine that we're working exclusively in the hexadecimal system. What numeral represents a quantity one less than 100?
    (a) 9F.
    (b) 7F.
    (c) AF.
    (d) FF.
    (e) There is no such numeral as 100 in the hexadecimal system, so this question is meaningless!

50. The commutative law for addition can be applied to finite sums of
    (a) natural numbers only.
    (b) integers only.
    (c) rational numbers only.
    (d) irrational numbers only.
    (e) real numbers.

51. Imagine that we come across two cubic functions, where all the coefficients and constants are real numbers. We want to solve these functions as a two-by-two system. What is the *minimum* number of real solutions that such a system can have?
    (a) None.
    (b) One.
    (c) Two.
    (d) Three.
    (e) We can't answer this without more information.

52. Figure FE-4 shows four mappings. In each case, the maximal domain and the co-domain are represented by the solid gray regions. Which of these drawings illustrates the concept of a surjection onto the co-domain?
    (a) Drawing A.
    (b) Drawing B.
    (c) Drawing C.
    (d) Drawing D.
    (e) None of them.

Figure FE-4 Illustration for Final Exam Question 52.

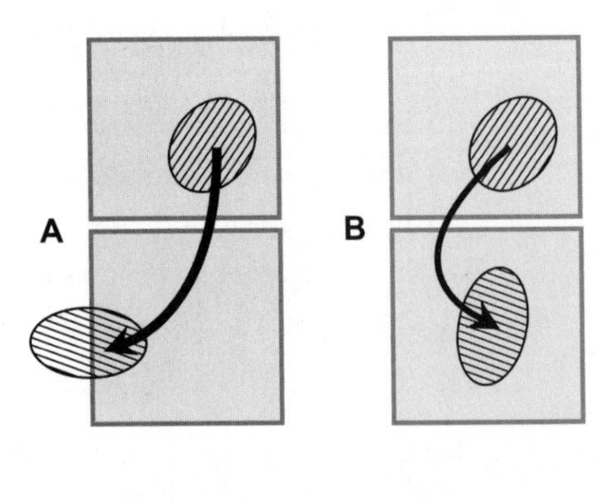

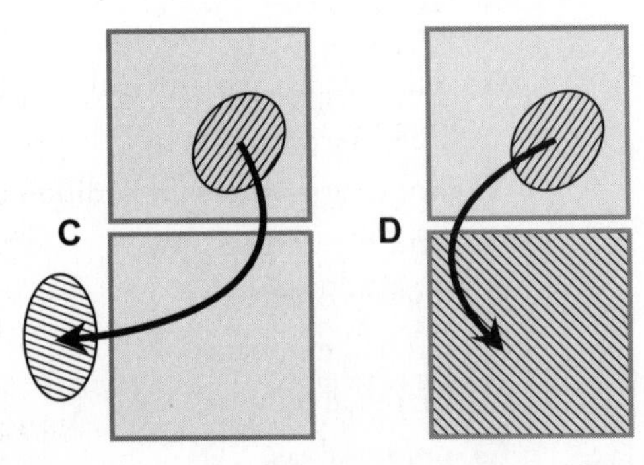

53. Consider the equation $4x + 2y = -7$, where x is the independent variable and y is the dependent variable. The slope of the graph of this equation is
    (a) 2.
    (b) −2.
    (c) −7/2.
    (d) 2/7.
    (e) undefined.

54. The y-intercept of the graph of the equation stated in Question 53 is
    (a) 2.
    (b) −2.
    (c) −7/2.
    (d) 2/7.
    (e) undefined.

55. The x-intercept of the graph of the equation stated in Question 53 is
    (a) 1.
    (b) −1.
    (c) 7/4.
    (d) −7/4.
    (e) impossible to determine without more information.

56. What is the solution to the equation $-x + a + 5 = 0$, where x is the unknown and a is a constant?
    (a) $x = 5 + a$
    (b) $x = 5 - a$
    (c) $x = a - 5$
    (d) $x = -a - 5$
    (e) It can't be determined without more information

57. When graphing a two-by-two system to illustrate the real solutions, a rectangular coordinate system often works better than a strict Cartesian coordinate system because
    (a) the rectangular system shows the true slopes of the lines or curves, but the Cartesian system does not.
    (b) the rectangular system shows negative as well as positive solutions, but the Cartesian system shows only positive solutions.
    (c) the rectangular system arranges the four quadrants in a more sensible way than the Cartesian system.
    (d) the rectangular system can provide exact values for the solutions merely by observation, but the Cartesian system can provide only approximate values.
    (e) the rectangular system can often provide a better pictorial fit than the Cartesian system for the range of values we want to show.

58. In Fig. FE-5, a scheme is shown for graphing the inverse of a relation. For any point that's part of the graph of the original relation, we can find its counterpart in the graph of the inverse relation by going to the opposite side of the "point reflector," exactly the same distance away. What is the equation of this "point reflector" line?
    (a) $y = 1$
    (b) $x + y = 0$
    (c) $x - y = 0$
    (d) $x = 1$
    (e) None of the above
59. In Fig. FE-5, one of the points is labeled *P*. In which quadrant is this point?
    (a) The first quadrant.
    (b) The second quadrant.
    (c) The third quadrant.
    (d) The fourth quadrant.
    (e) It is not in any quadrant.
60. Under what circumstances can we add the same variable to both sides of an equation and get another valid equation?
    (a) Never.
    (b) Only if the variable can never become equal to 0.

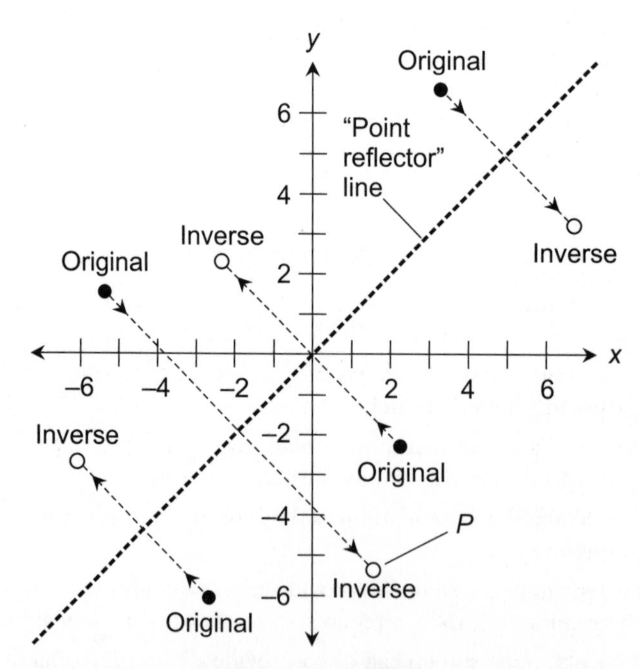

Figure FE-5 Illustration for Final Exam Questions 58 and 59.

(c) Only if the variable can never become negative.
(d) Only if the variable can never become irrational.
(e) Always.

61. Completing the square is a method of solving a quadratic equation for real roots by
    (a) squaring the left and right sides.
    (b) adding a constant to both sides in order to get the square root of a binomial on one side and a positive real number on the other side.
    (c) adding a constant to both sides in order to get the square of a binomial on one side and a positive real number on the other side.
    (d) taking the square root of both sides, discovering the imaginary roots if any exist.
    (e) converting it into a cubic equation that turns out to be easier to solve than the original quadratic.

62. Which of the following statements is false?
    (a) If a real number k is a root of a cubic equation in the variable x, then $(x + k)$ is a factor of the cubic polynomial.
    (b) If a real number k is a root of a cubic equation in the variable x, then $(x - k)$ is a factor of the cubic polynomial.
    (c) If k is a real number and $(x - k)$ is a factor of a cubic polynomial in the variable x, then k is a real root of the cubic equation.
    (d) All cubic equations have at least one real root, as long as the coefficients and the stand-alone constant are all real numbers.
    (e) A cubic equation can have one real root and two other roots that are complex conjugates of each other.

63. A mapping in which each element in the domain corresponds to one, but only one, element in the range is called
    (a) a rejection.
    (b) a bijection.
    (c) an injection.
    (d) a surjection.
    (e) onto.

64. Consider the following quadratic equation in binomial factor form:

$$(x - j3)(x + j3) = 0$$

What is the solution set X for this equation?
    (a) $X = \{3\}$
    (b) $X = \{j3\}$
    (c) $X = \{-3\}$
    (d) $X = \{-j3\}$
    (e) None of the above

65. When we say that a real number u is the natural logarithm of some other real number v, and that e is Euler's constant (an irrational number equal to about 2.71828), we are in effect saying that
    (a) v equals e to the uth power.
    (b) u equals e to the vth power.
    (c) v equals u to the eth power.
    (d) u equals v to the eth power.
    (e) v to the uth power equals e.

66. A three-by-three linear system is consistent and not redundant if and only if
    (a) it has a single, unique solution.
    (b) it has two distinct solutions.
    (c) it has three distinct solutions.
    (d) it has infinitely many solutions.
    (e) it has no solutions.

67. Under what circumstances can we divide both sides of an equation by the same variable and get another valid equation?
    (a) Never.
    (b) Only if the variable can never become equal to 0.
    (c) Only if the variable can never become negative.
    (d) Only if the variable can never become irrational.
    (e) Always.

68. The relation graphed in Fig. FE-6 is not a function of x if we think of it as a mapping from values of x to values of y. We can see this because
    (a) the curve is not a straight line.
    (b) the domain is not the entire set of real numbers.
    (c) the relation is not one-to-one.
    (d) there are values of x that map into more than one value of y.
    (e) there no values of y that map into more than one value of x.

69. How can we restrict the range of the relation graphed in Fig. FE-6 so that it becomes a function of x if we think of it as a mapping from values of x to values of y?
    (a) We can restrict the range to the set of non-negative reals.
    (b) We can restrict the range to the set of negative reals.
    (c) We can restrict the range to the set of reals larger than 1.
    (d) We can restrict the range to the set of reals smaller than −1.
    (e) Any of the above.

70. The relation graphed in Fig. FE-6 is a function of y if we think of it as a mapping from values of y to values of x. We can see this because
    (a) there are values of x that map into more than one value of y.
    (b) there no values of y that map into more than one value of x.

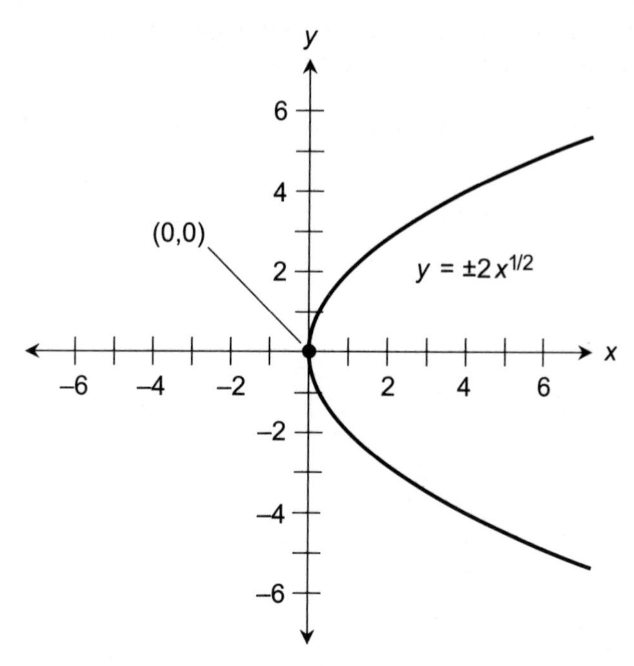

Figure FE-6 Illustration for Final Exam Questions 68, 69, and 70.

(c) the curve is symmetrical.
(d) the domain is the entire set of real numbers.
(e) the relation is a surjection.

71. Suppose that the coefficients and constant in a polynomial equation are placed in a synthetic division array. We try a negative real-number "test root" and get a nonzero remainder. The numbers in the last row alternate between positive and negative. This tells us that our "test root" is
(a) larger than all the real roots of the equation.
(b) equal to the largest real root of the equation.
(c) somewhere between the smallest and the largest real roots of the equation.
(d) equal to the smallest real root of the equation.
(e) smaller than all the real roots of the equation.

72. State the solution set X for the quadratic equation

$$x^2 + 100 = 0$$

(a) $X = \{j10, -j10\}$
(b) $X = \{10, -10\}$
(c) $X = \{(10 + j10), (10 - j10)\}$

(d) $X = \{j10\}$

(e) $X = \varnothing$

73. Which of the following equations is true for all positive real numbers u and v?

(a) $\ln u + \ln v = \ln (u + v)$

(b) $\ln u + \ln v = \ln (uv)$

(c) $\ln u + \ln v = \ln (u^v)$

(d) $\ln u \ln v = \ln (uv)$

(e) None of the above

74. Imagine that we come across two *different* functions, both of which are of degree larger than 2, and where all the coefficients and constants are real numbers. We want to solve these functions as a two-by-two system. What is the *maximum* number of real solutions that such a system can have?

(a) Three.

(b) Four.

(c) Five.

(d) Infinitely many.

(e) We must have more information before we can answer this.

75. When we say that the common logarithm of a real number w is equal to z, we are in effect saying that

(a) z equals 10 to the wth power.

(b) w equals 10 to the zth power.

(c) z equals w to the 10th power.

(d) w equals z to the 10th power.

(e) w to the z power equals 10.

76. The octal numeral 700 represents the same quantity as the base-10 numeral

(a) 448.

(b) 589.

(c) 600.

(d) 816.

(e) 1,023.

77. In Fig. FE-7, the straight line is the graph of a function where t is the independent variable and x is the dependent variable. What is the slope of this line? (Be careful! Note that the *horizontal* axis represents t, and the *vertical* axis represents x!)

(a) 1/2

(b) 1

(c) −1/2

(d) −1

(e) None of the above

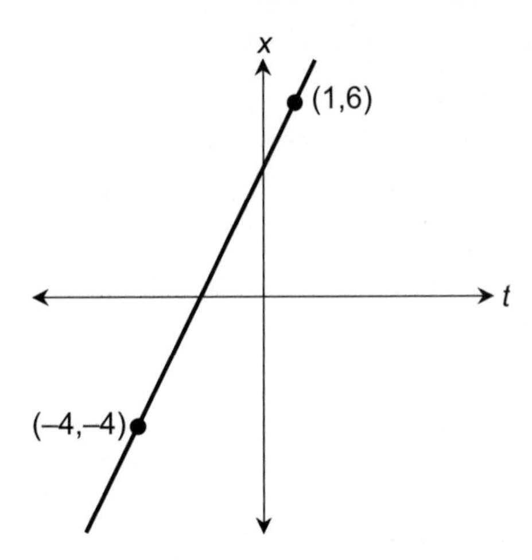

Figure FE-7 Illustration for Final Exam Questions 77, 78, and 79.

78. In the graph of Fig. FE-7, the *x*-intercept of the line is
    (a) 13/4.
    (b) 7/2.
    (c) 4.
    (d) 17/4.
    (e) impossible to determine without more information.
79. In the graph of Fig. FE-7, the *t*-intercept of the line is
    (a) −11/6.
    (b) −2.
    (c) −13/6.
    (d) −7/3.
    (e) impossible to determine without more information.
80. Suppose we see the following cubic equation and we want to find its real roots:

$$x^3 - 4x^2 + 7x = 0$$

We're tempted to divide this through by x, so we can reduce it to a quadratic that will be easier to solve. Is that a good idea?
    (a) No, because one of the real roots is $x = 0$.
    (b) No, because a cubic equation can never be divided through directly.
    (c) No, because we must always factor cubics into binomials to solve them.
    (d) Yes, it will work fine.
    (e) Yes, but only because the leading coefficient is equal to 1.

81. Within the set of real numbers, the base-(−1) logarithm of 99
    (a) is not defined.
    (b) is equal to −1 to the 99th power, or −1.
    (c) is equal to 99 to the −1st power, or 1/99.
    (d) is equal to the 99th root of −1, or −1.
    (e) is equal to (ln 99) to the −1st power, or approximately 0.22.

82. What is the sum of 3/5 and 7/12 in lowest terms?
    (a) 71/60
    (b) 10/17
    (c) 60/71
    (d) 17/35
    (e) 73/69

83. Imagine an integer n. We add n to 2/3 of itself, and then we subtract 3/4 of n from that sum. We end up with −11. What is the original integer n?
    (a) 11
    (b) −13
    (c) 29
    (d) −12
    (e) 0

84. For any three-by-three linear system to have a single, unique solution, the graphs of all the equations must
    (a) not intersect anywhere.
    (b) intersect at the origin of Cartesian three-space.
    (c) intersect in a single flat plane.
    (d) intersect in a single straight line.
    (e) intersect at a single point.

85. Which of the following equations is *not* a quadratic in one variable?
    (a) $x^2 = 2x + 3$
    (b) $1/x = x^2 - 7$
    (c) $x^2 + 4x = 27$
    (d) $x - 21 = -8x^2 - 22$
    (e) $6 + x = 2x^2$

86. Which of the following is a first-degree equation in one variable? Here, x is the variable, while a, b, c, and d are constants.
    (a) $3x^2 + 5x - 5 = 0$
    (b) $ax + bx^{1/2} + cx^{1/3} = d$

(c) $5x - a + b^2 = 8$

(d) $x^3 + x^2 + x = 1$

(e) $x - x^{1/2} - x^{1/3} = a + b$

87. When we morph a matrix representing a three-by-three linear system, we can perform any of the following operations except one. Which one is the "illegal move"?

(a) Interchange all the elements between two rows, while keeping the elements of both rows in the same order from left to right.

(b) Divide all the elements in a row by a nonzero constant, keeping the elements in the same order from left to right.

(c) Multiply all the elements in a row by a nonzero constant, keeping the elements in the same order from left to right.

(d) Add all the elements in any row to all the elements in another row, and then replace the elements in either row by the sum, taking care to keep the elements of both rows in the same order from left to right.

(e) Add a constant to all the elements in a row, keeping the elements in the same order from left to right.

88. A prime number is

(a) a natural number that can be factored into a product of composite numbers.

(b) a composite number that cannot be divided by any other natural number except 1 without a remainder.

(c) a natural number that is 2 or larger, and that can only be factored into a product of itself and 1.

(d) a natural number that is the square of some composite number.

(e) a natural number that is the cube of some composite number.

89. In a rectangular coordinate plane, both axes are linear. This is another way of saying that

(a) along either axis, the change in value is directly proportional to the distance we move along that axis.

(b) both axes have increments of the same size.

(c) the origin is at the point (0, 0).

(d) linear equations always have graphs that fall exactly on one of the axes.

(e) the axes intersect at a right angle.

90. Figure FE-8 is the graph of a two-by-two linear system. What is the slope-intercept equation of line L, considering x as the independent variable and y as the dependent variable?

(a) $y = (-2/3)x - 2$

(b) $y = (2/3)x + 3/2$

(c) $y = (3/2)x + 3$

(d) $y = (-3/2)x - 8/5$

(e) More information is necessary to answer this

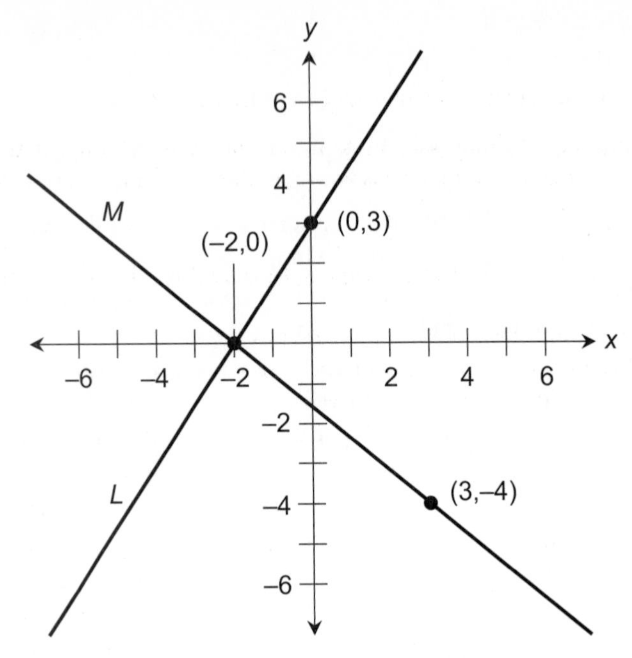

Figure FE-8 Illustration for Final Exam Questions 90 through 93.

91. In Fig. FE-8, what is the slope-intercept equation of line *M*, considering *x* as the independent variable and *y* as the dependent variable?
    (a) $y = (-5/4)x - 2$
    (b) $y = (5/4)x + 3/2$
    (c) $y = (4/5)x + 3$
    (d) $y = (-4/5)x - 8/5$
    (e) More information is necessary to answer this

92. In Fig. FE-8, what is the slope-intercept equation of line *L*, considering *y* as the independent variable and *x* as the dependent variable?
    (a) $x = (2/3)y - 2$
    (b) $x = (-2/3)y - 2$
    (c) $x = (3/2)y + 3$
    (d) $x = (-3/2)y - 3$
    (e) More information is necessary to answer this

93. In Fig. FE-8, what is the slope-intercept equation of line *M*, considering *y* as the independent variable and *x* as the dependent variable?
    (a) $x = (4/5)y - 2$
    (b) $x = (4/5)y + 2$

(c) $x = (-5/4)y - 2$
(d) $x = (-5/4)y + 2$
(e) More information is necessary to answer this

94. Here are several things we can do to "smaller than or equal" statements and still have valid statements. Something is wrong with one of these claims. Which claim is wrong, and how can it be corrected?

- We can reverse the left and right sides only if we change the inequality to "larger than or equal."
- We can add the same quantity to both sides.
- We can subtract the same quantity from both sides.
- We can add one statement to another.
- We can multiply both sides by the same quantity.

(a) The first statement is wrong. We can never reverse the left and right sides of any inequality.
(b) The second statement is wrong. We can only add the same quantity to both sides of a "smaller than or equal" statement if that quantity is positive.
(c) The third statement is wrong. We can only subtract the same quantity from both sides of a "smaller than or equal" statement if that quantity is negative.
(d) The fourth statement is wrong. We cannot, in general, add one "smaller than or equal" statement to another.
(e) The fifth statement is wrong. It works if the quantity is nonnegative; but if we multiply both sides by a negative quantity, we must change the relation to "larger than or equal."

95. In Cartesian three-space, the equation $2x + 4y - 6z = 7$ represents
(a) a straight line.
(b) a flat plane.
(c) a parabola.
(d) a circle.
(e) None of the above.

96. Consider the following first-degree equation in the variable x, where a, b, c, d, e, and f are constants:

$$-3a + x/(bcd) = 24ef$$

This equation has meaning only under certain conditions. Which of these statements *fully* states those conditions?
(a) We cannot let b, c, and d all equal 0 at the same time.
(b) We cannot allow b, c, or d to equal 0 at any time.
(c) We cannot allow a to equal 0 at any time.

(d) We cannot allow e to equal 0 at any time.

(e) We cannot allow f to equal 0 at any time.

97. Assuming the condition in Question 96 has been satisfied, what is the solution of this equation in terms of the constants?

(a) $x = 3abcd + 24bcdef$

(b) $x = abcd/3 + bcdef/24$

(c) $x = 3abcd - 24bcdef$

(d) $x = abcd/3 - bcdef/24$

(e) More information is needed to answer this

98. When a quantity is raised to the -1 power,

(a) we must be sure the quantity can never equal 0.

(b) we always get the additive inverse of the quantity.

(c) we always get a rational number.

(d) we always get a negative number.

(e) we always get 1, except when the quantity equals 0.

99. Refer to Fig. FE-9. Imagine two straight lines in this Cartesian plane, one passing through points P and Q, and the other passing through lines R and S. Now think about the two-by-two system of linear equations represented by these lines. What can we say about this system if we consider both of the equations as functions of x?

(a) The system has one solution.

(b) The system is inconsistent.

(c) Neither function has an inverse that is also a function.

(d) The system is redundant.

(e) One of the functions has an inverse that is also a function, but the other function does not.

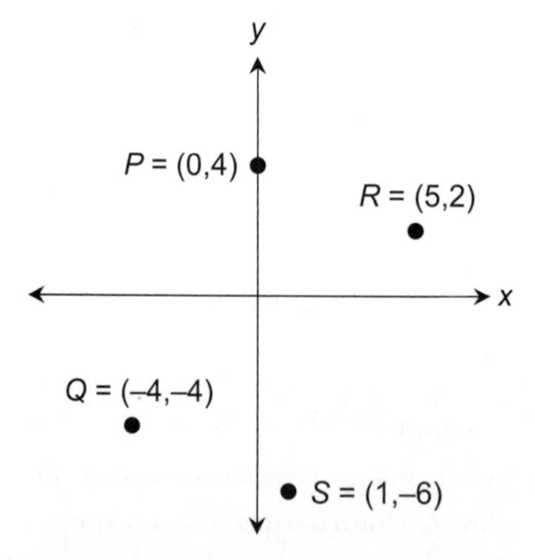

Figure FE-9 Illustration for Final Exam Questions 99 and 100.

100. Look again at Fig. FE-9. This time, imagine two straight lines, one passing through points P and S, and the other passing through lines Q and R. Consider the two-by-two system of linear equations represented by these lines. This system has
(a) one solution.
(b) two solutions.
(c) three solutions.
(d) four solutions.
(e) infinitely many solutions.

101. Imagine a three-by-three linear system of equations, each of which is in the following form:

$$ax + by + cz = d$$

where a, b, c, and d are constants, and x, y, and z are variables. Now suppose that the following matrix represents this system:

2	0	0	6
0	−3	0	12
0	0	5	0

This matrix is in
(a) linear form.
(b) dependent form.
(c) diagonal form.
(d) unit diagonal form.
(e) redundant form.

102. The matrix shown in Question 101 contains enough information so that we can infer the solution to the linear system it represents. How can that solution be expressed as an ordered triple of the form (x, y, z)?
(a) (3, −4, 0)
(b) (6, 12, 0)
(c) (2, −3, 5)
(d) (8, 9, 5)
(e) (12, −36, 0)

103. Suppose we are trying to solve a three-by-three linear system using matrices. We are sure we haven't made any mistakes along the way. We come up with this matrix:

7	7	7	28
1	1	1	4
−15	−15	−15	−60

We can conclude that the original three-by-three system has

(a) no solutions.

(b) one solution.

(c) two solutions.

(d) three solutions.

(e) infinitely many solutions.

104. Are the expressions $2 + j2$ and $2 + j^2$ different? If so, how?

(a) These two expressions look different, but they actually represent the same number.

(b) The first expression represents a complex (but not real) number. The second expression represents a real number.

(c) The first expression represents a real number. The second expression represents a complex (but not real) number.

(d) The first expression represents a pure imaginary number. The second expression represents a number that is neither pure real nor pure imaginary.

(e) The first expression represents a complex number that is neither pure real nor pure imaginary. The second expression represents a pure imaginary number.

105. In Fig. FE-10, what do all the numbers corresponding to points on the circle have in common?

(a) They are all pure real numbers.

(b) They are all pure imaginary numbers.

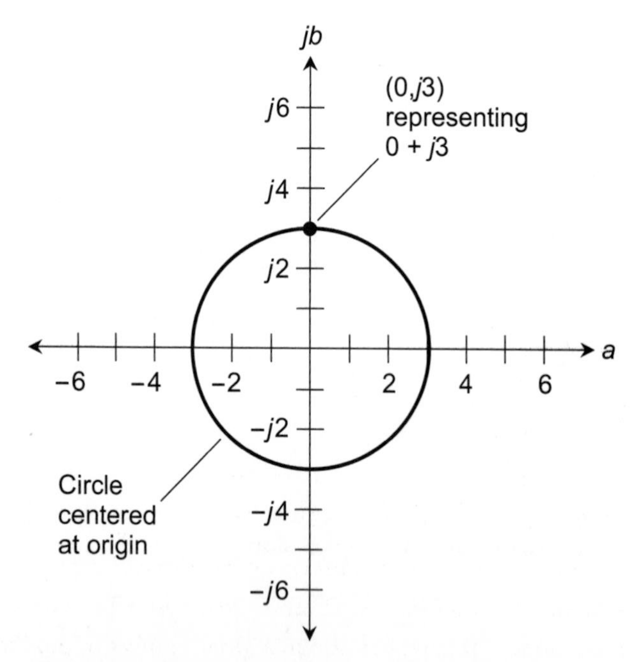

Figure FE-10 Illustration for Final Exam Question 105.

(c) They all have absolute values equal to 3.

(d) They are all complex conjugates.

(e) They are all equal to the square root of −9.

106. Imagine a two-by-two linear system in variables x and y. Suppose the graphs of the equations are straight lines that intersect at a point in the second quadrant of a Cartesian coordinate system where y is the dependent variable. From this information, we know that

(a) the solution values for x and y are both positive.

(b) the solution value for x is positive, and the solution value for y is negative.

(c) the solution value for x is negative, and the solution value for y is positive.

(d) the solution values for x and y are both negative.

(e) the solution value for x is the additive inverse of the solution value for y.

107. Which of the following statements is false?

(a) $j^7 = j^3$

(b) $j^8 = j^4$

(c) $j^{12} = j^8$

(d) $j^{16} = j^{12}$

(e) $j^{21} = j^{18}$

108. Here is a little mathematical verse:
For every x, y, and z:

If x is smaller than y

and

y is equal to z,

then

x is smaller than z.

How can we write this in mathematical symbols?

(a) $(\forall\ x, y, z) : [(x \leq y)\ \&\ (y \leq z)] \Rightarrow (x < z)$

(b) $(\forall\ x, y, z) : [(x < y)\ \&\ (y = z)] \Rightarrow (x < z)$

(c) $(\forall\ x, y, z) : [(x \geq y)\ \&\ (y \leq z)] \Rightarrow (x < z)$

(d) $(\forall\ x, y, z) : [(x < y)\ \&\ (y > z)] \Rightarrow (x < z)$

(e) $(\forall\ x, y, z) : [(x = y)\ \&\ (y < z)] \Rightarrow (x < z)$

109. What is meant by the term *additive identity element*?

(a) It's a number that we can add to a given number to produce a sum of 0.

(b) It's a number that we can add to a given number to produce a sum equal to the original number.

(c) It's a number that we can add to a given number to produce a sum equal to the negative of the original number.
(d) It's a number that we can add to a given number to produce a sum of 1.
(e) It's a number that we can add to a given number to produce a sum equal to twice the original number.

110. Consider the following pair of equations as a two-by-two linear system:

$$2r = 6s - 8$$

and

$$-7r = -21s + 28$$

What is the solution to this system?
(a) $(r, s) = (-2, 7)$.
(b) $(r, s) = (-8/7, 6/7)$.
(c) $(r, s) = (1/4, -3/4)$.
(d) There are infinitely many solutions.
(e) There is no solution.

111. The essential domain of a mapping is always a subset of the
(a) maximal domain.
(b) codomain.
(c) independent variable.
(d) dependent variable.
(e) range.

112. Figure FE-11 is graph of a quadratic function where x is the independent variable. Based on the information shown, what can we say about the coefficient of x^2 in the polynomial standard form of the function?
(a) It is an imaginary number.
(b) It is a complex (but not real) number.
(c) It is a positive real number.
(d) It is a negative real number.
(e) We need more information to answer this.

113. Based on the information shown in Fig. FE-11, how many real zeros does the quadratic function have?
(a) More than two.
(b) Two.
(c) One.
(d) None.
(e) We need more information to answer this.

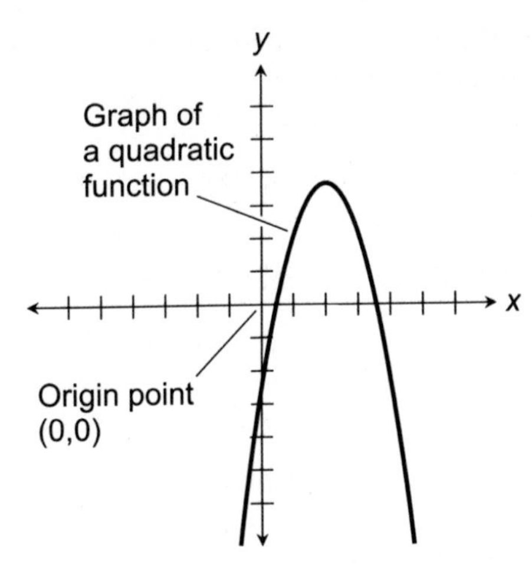

Figure FE-11 Illustration for Final Exam Questions 112 and 113.

114. What is the union of the sets $A = \{1, 2, 3\}$ and $B = \{3, 4, 5\}$?
    (a) The set {3}.
    (b) The set {1, 2, 3}.
    (c) The set {3, 4, 5}.
    (d) The set {1, 2, 3, 4, 5}.
    (e) The null set.

115. What is the intersection of the sets $A = \{1, 2, 3\}$ and $B = \{3, 4, 5\}$?
    (a) The set {3}.
    (b) The set {1, 2, 3}.
    (c) The set {3, 4, 5}.
    (d) The set {1, 2, 3, 4, 5}.
    (e) The null set.

116. The first few primes are 2, 3, 5, 7, 11, 13, 17, 19, and 23. Using this information along with a calculator, we can determine that one of the following numbers is prime. Which one?
    (a) 407
    (b) 423
    (c) 437
    (d) 457
    (e) 473

117. A rational number can always be expressed as
    (a) an endless, nonrepeating decimal.
    (b) a ratio of 1 to an integer.

(c) a ratio of an integer to a positive integer.
(d) a ratio of an integer to 1.
(e) a terminating decimal.

118. If the discriminant in a quadratic equation is a positive real number, it means that
(a) the equation has two different real roots.
(b) the equation has one real root with multiplicity 2.
(c) the equation has one imaginary root with multiplicity 2.
(d) the equation has two different imaginary roots.
(e) the equation has no roots at all, real or imaginary.

119. Consider the following cubic equation in binomial-trinomial form:

$$(x - 5)(x^2 + 2x + 1) = 0$$

One of the real roots of this equation has multiplicity 2. Which one?
(a) The root $x = 5$.
(b) The root $x = -5$.
(c) The root $x = 0$.
(d) The root $x = 1$.
(e) The root $x = -1$.

120. Consider the following quadratic equation, which is expressed as a product of trinomials:

$$(x + 2 - j5)(x + 2 + j5) = 0$$

What is the solution set X for this equation?
(a) $X = \{(-2 - j5), (-2 + j5)\}$
(b) $X = \{(2 - j5), (2 + j5)\}$
(c) $X = \{2, -5\}$
(d) $X = \{j2, -j5\}$
(e) None of the above

121. Suppose that a_1, a_2, b_1, b_2, c_1, and c_2 are real numbers, and neither a_1 nor a_2 are equal to 0. Consider

$$y = a_1x^2 + b_1x + c_1$$

and

$$y = a_2x^2 + b_2x + c_2$$

What is the largest number of elements that the solution set of such a system can have? Assume that the two equations are not identical, and are not some constant multiple of each other.

(a) One.

(b) Two.

(c) Three.

(d) Four.

(e) Infinitely many.

122. Suppose someone gives us a list of the first few elements of an infinite set S, and assures us that the numbers keep doubling as we move down the list:

$$S = \{2, 4, 8, 16, 32, 64, ...\}$$

Which of the following is *not* an element of S?

(a) 4,096

(b) 8,192

(c) 16,384

(d) 1,048,576

(e) 4,194,305

123. Consider an integer expressed as a digit sequence without any other symbols (except a minus sign if the integer is negative). We can make the absolute value of this integer 10,000 times as large by

(a) adding 10,000.

(b) adding four ciphers to the left-hand end of the digit sequence.

(c) adding four ciphers to the right-hand end of the digit sequence.

(d) increasing the left-most digit by 4.

(e) inserting a decimal point four digits in from the right-hand end of the digit sequence.

124. Figure FE-12 is graph of a quadratic function where x is the independent variable. Based on the information shown, what can we say about the coefficient of x^2 in the polynomial standard form of the function?

(a) It is an imaginary number.

(b) It is a complex (but not real) number.

(c) It is a positive real number.

(d) It is a negative real number.

(e) We need more information to answer this.

125. Based on the information shown in Fig. FE-12, how many real zeros does the quadratic function have?

(a) More than two.

(b) Two.

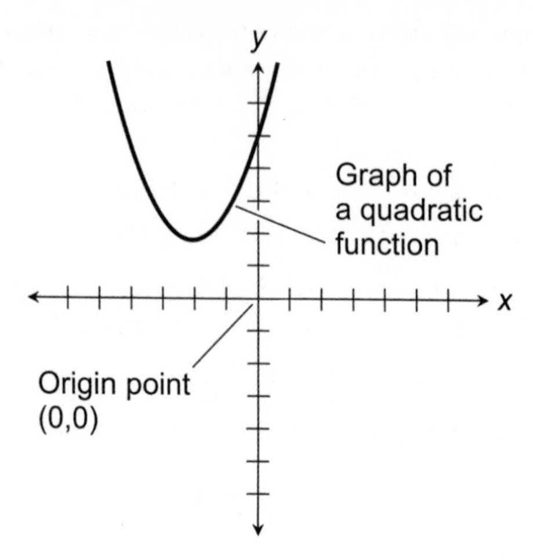

Figure FE-12 Illustration for Final Exam Questions 124 and 125.

(c) One.
(d) None.
(e) We need more information to answer this.

126. Imagine four nonzero real numbers w, x, y, and z such that $wx = yz$. From this, we can conclude that
(a) $w/y = x/z$.
(b) $w/z = x/y$.
(c) $w/x = y/z$.
(d) $z/x = w/y$.
(e) None of the above.

127. When using synthetic division in an attempt to factor a binomial out of a polynomial, our goal is to find a "test root" that produces
(a) a remainder of 0 at the end of the process.
(b) all 0s in the last line.
(c) all positive reals in the last line.
(d) alternating positive and negative reals in the last line.
(e) all negative reals in the last line.

128. According to the set-based number-building scheme, the natural number 24 is
(a) {1, 2, 3, ..., 22, 23, 24}.
(b) {1, 2, 3, ..., 21, 22, 23}.
(c) {0, 1, 2, ..., 22, 23, 24}.
(d) {0, 1, 2, ..., 21, 22, 23}.
(e) {24}.

129. Suppose that the coefficients and constant in a polynomial equation are placed in a synthetic division array. We try a positive real-number "test root" and get a nonzero remainder. None of the numbers in the last row are negative. This tells us that our "test root" is
   (a) larger than all the real roots of the equation.
   (b) equal to the largest real root of the equation.
   (c) somewhere between the smallest and the largest real roots of the equation.
   (d) equal to the smallest real root of the equation.
   (e) smaller than all the real roots of the equation.

130. Which of the following fractions is in lowest terms?
   (a) 7/91
   (b) 57/3
   (c) −23/115
   (d) −29/17
   (e) None of the above

131. Figure FE-13 shows the graphs of two quadratic functions in a real-number rectangular coordinate plane. The origin (0, 0) is where the *x* and *y* axes intersect. The curves do not intersect anywhere. Suppose we consider these two functions as a system, and we find two solutions. Based on the information shown,
   (a) both solutions are ordered pairs of positive real numbers.
   (b) both solutions are ordered pairs of negative real numbers.
   (c) one solution is an ordered pair of positive reals, and the other solution is an ordered pair of negative reals.

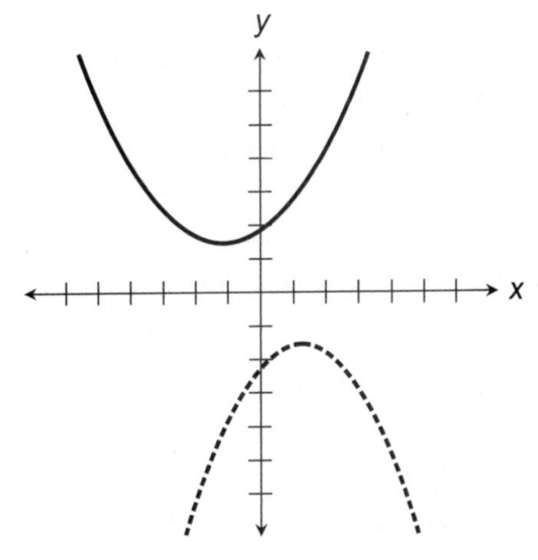

Figure FE-13 Illustration for Final Exam Question 131.

(d) one solution is an ordered pair of reals, and the other solution is an ordered pair of pure imaginary numbers.

(e) both solutions are ordered pairs of complex (but not real) numbers.

132. If a number decreases by a factor of 100, then its natural logarithm
    (a) decreases by a factor of ln 100, or approximately 4.6.
    (b) decreases by a factor of exactly 2.
    (c) increases by a factor of exactly 2.
    (d) increases by a factor of ln 100, or approximately 4.6.
    (e) increases by a factor of the square root of 100, or exactly 10.

133. Which of the following statements concerning natural numbers is false?
    (a) A number must be either even or odd, but can't be both.
    (b) An even number plus 1 is always odd.
    (c) An even number divided by 2 is always a natural number.
    (d) An odd number times 2 is always a natural number.
    (e) An odd number divided by 2 is always a natural number.

134. Suppose we want to solve a higher-degree polynomial equation, and we're told that all the roots are irrational numbers. What can we do to find, or at least approximate, those roots?
    (a) We can factor the equation into the nth power of a binomial (where n is the degree of the equation), make that binomial into a first-degree equation, and then solve that equation.
    (b) We can factor the equation into the nth power of a trinomial (where n is the degree of the equation), make that trinomial into a quadratic, and then solve that equation with the quadratic formula.
    (c) We can factor the equation into binomials with integer coefficients and integer constants, and then derive the roots from those binomials.
    (d) We can use synthetic division to find the upper and lower bounds of the real roots, and then find those roots using the rational roots theorem.
    (e) We can use a computer program to graph the function produced by the polynomial, and then use the computer to approximate the zeros of that function.

135. If a number increases by a factor of 10,000, then its common logarithm
    (a) decreases by a factor of ln 10,000, or approximately 9.2.
    (b) decreases by a factor of exactly 4.
    (c) increases by a factor of exactly 4.
    (d) increases by a factor of ln 10,000, or approximately 9.2.
    (e) increases by a factor of the square root of 10,000, or exactly 100.

136. Figure FE-14 shows the graphs of two quadratic functions in a real-number rectangular coordinate plane. The origin (0, 0) is where the x and y axes intersect. The

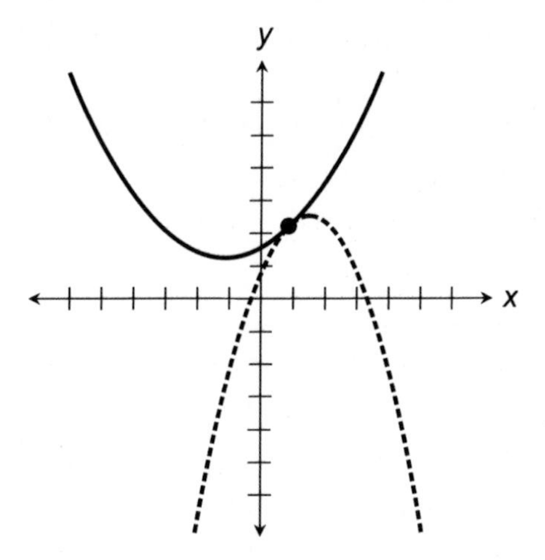

Figure FE-14 Illustration for Final Exam Question 136.

graphs intersect at a single point. Suppose we consider these two functions as a system. Based on the information shown,

(a) there is a single solution with multiplicity 2; it is an ordered pair of real numbers.
(b) there is a single solution with multiplicity 2; it is an ordered pair of pure imaginary numbers.
(c) there is a single solution with multiplicity 2; it is an ordered pair of complex (but not real) numbers.
(d) there are two solutions; one is an ordered pair of reals, and the other is an ordered pair of pure imaginary numbers.
(e) there are two solutions; one is an ordered pair of reals, and the other is an ordered pair of complex (but not real) numbers.

137. Consider the following equation, which represents a function of the variable x:

$$y = x^2 + 3x + 1$$

We can write down ordered pairs such as (0,1) or (1,5) to show specific examples of this function. In any such ordered pair, the second number represents a value of the

(a) codomain.
(b) dependent variable.
(c) bijection.
(d) inverse.
(e) essential domain.

138. When we multiply $a + jb$ by its conjugate, we get

(a) $a^2 + b^2$
(b) $a^2 - b^2$

(c) $2a$
(d) $2b$
(e) $4ab$

139. Which of the following is an equivalence relation?
(a) The "if and only if" relation.
(b) The "if/then" relation.
(c) The "strictly larger than" relation.
(d) The "larger than or equal" relation.
(e) None of the above.

140. Consider the following equation:

$$(x+1)(x-3)^3(x+5)^5(x-7)^7 = 0$$

What is the degree of this equation?
(a) 4
(b) 15
(c) 16
(d) 56
(e) We need more information to answer this

141. What is the solution to the equation $ax / (b+1) = 3$, where x is the unknown, a and b are constants, $a \neq 0$, and $b \neq -1$?
(a) $x = 3ab + 3a$
(b) $x = 3a/b + a/3$
(c) $x = 3b/a + 3ab$
(d) $x = 3b/a + 3/a$
(e) More information is needed to solve this equation

142. What is 25/45 – 2/9 in lowest terms?
(a) 29/45
(b) 23/45
(c) 1/3
(d) 4/9
(e) 5/11

143. The decimal-numeral equivalent of the Roman numeral CDXXXII is
(a) 932.
(b) 832.
(c) 732.
(d) 632.
(e) 432.

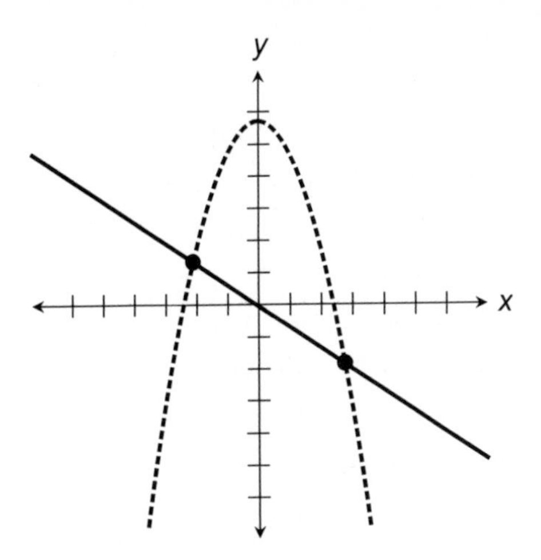

Figure FE-15 Illustration for Final Exam Question 144.

144. Figure FE-15 shows the graphs of a linear function and a quadratic function in real-number rectangular coordinates. The origin (0, 0) is where the *x* and *y* axes intersect. The graphs intersect at two points. Suppose we consider these two functions as a system. Based on the information shown,
   (a) there are two solutions; both are ordered pairs of real numbers.
   (b) there is a single solution with multiplicity 2; it is an ordered pair of reals.
   (c) there is a single solution with multiplicity 2; it is an ordered pair of complex numbers.
   (d) there are two solutions; one is an ordered pair of reals, and the other is an ordered pair of pure imaginary numbers.
   (e) there are two solutions; one is an ordered pair of reals, and the other is an ordered pair of complex (but not real) numbers.

145. Figure FE-16 shows the graphs of two quadratic functions in real-number rectangular coordinates. The origin (0, 0) is where the *x* and *y* axes intersect. The functions are

$$f(x) = a_1x^2 + b_1x + c_1$$

and

$$g(x) = a_2x^2 + b_2x + c_2$$

where all the coefficients and constants are real numbers, and a_1 and a_2 are both nonzero. The graphs intersect at a single point. How can we change this system into one that we can be certain has no real solutions?
   (a) We need more information to answer this.
   (b) We can decrease the value of c_1, leaving everything else the same.

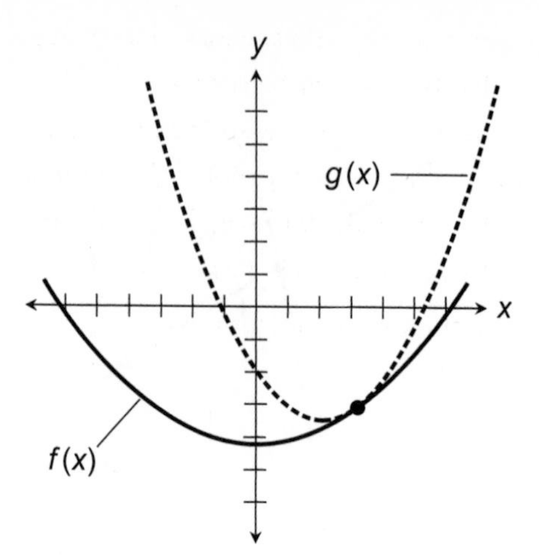

Figure FE-16 Illustration for Final Exam Question 145.

(c) We can increase the value of c_1, leaving everything else the same.

(d) We can decrease the value of c_2, leaving everything else the same.

(e) We can reverse the sign of a_2, leaving everything else the same.

146. Suppose someone claims that she has come up with a way to solve a three-by-three linear system in variables x, y, and z. She tells us to:

- Take the first equation and solve it for x in terms of y and z.
- Take the second equation and do the same thing.
- Make a two-by-two system in y and z from the first two steps.
- Solve that system.
- Substitute the solutions for y and z back into the first equation and solve it for x.

This process will not, in general, work to solve a three-by-three system. Which of the following constitutes a fatal flaw in our friend's scheme?

(a) The process can work only if the system is consistent.

(b) The process completely ignores the third equation.

(c) The process does not use the addition method at any point.

(d) The process does not employ matrix morphing.

(e) The process can produce a solution only if the system is inconsistent.

147. There's something else wrong with the process outlined in Question 146. What's that?

(a) The third "bulleted" step will not actually give us a two-by-two system.

(b) The two-by-two system derived in the third step will be inconsistent.

(c) There is no way to derive an expression for x in terms of y and z.

(d) It will cause us to unwittingly divide by 0.

(e) It will cause us to unwittingly subtract an equation from itself.

148. If the discriminant in a quadratic equation is a negative real number, it means that
   (a) the equation has two different real roots.
   (b) the equation has one real root with multiplicity 2.
   (c) the equation has one imaginary root with multiplicity 2.
   (d) the equation has two different imaginary or complex roots.
   (e) the equation has no roots at all, real or complex.

149. A mapping is a bijection if and only if it is
   (a) reflexive.
   (b) a function.
   (c) one-to-one and onto.
   (d) a subset of its co-domain.
   (e) transitive.

150. Consider the following pair of equations as a two-by-two system:

$$y = a_1 x + b_1$$

and

$$y = a_2 x + b_2$$

where a_1, a_2, b_1, and b_2 are real numbers, and neither a_1 nor a_2 are equal to 0. What are the smallest and largest numbers of elements that the solution set of such a system can have? Assume that the two equations are not identical, and are not some constant multiple of each other.
   (a) None, and one.
   (b) None, and two.
   (c) None, and infinitely many.
   (d) One, and infinitely many.
   (e) Two, and infinitely many.

APPENDIX

A

Worked-Out Solutions to Exercises: Chapters 1 to 9

These worked-out solutions do not necessarily represent the only way a problem can be figured out. If you think you can solve a particular problem in a quicker or better way than you see here, by all means go ahead! But always check your work to be sure your "alternative" answer is correct.

Chapter 1

1. There are several ways to figure this out, but they're all a little "messy." You can count the days out loud and write "hash marks" on a sheet of paper, or you can use a calendar and write numerals on it day by day, starting with 0 on June 24 and working up. If you want to avoid all that counting, you can figure out the number of days of interest in June, July, August, September, and October, and then add them up. Remember that in any year, June has 30 days, July has 31 days, August has 31 days, and September has 30 days. In June, you have 30 − 24 = 6 days of interest; in July you have all 31 days; in August you have all 31 days; in September you have all 30 days; in October you have 2 days of interest. The total is therefore 6 + 31 + 31 + 30 + 2 = 100 days.

2. Remember that 1 = I, 5 = V, 10 = X, 50 = L, and 100 = C. The answers are as follows, broken down into sums for clarification.
   (a) 200 = 100 + 100 = C + C = CC
   (b) 201 = 100 + 100 + 1 = C + C + I = CCI
   (c) 209 = 100 + 100 + 9 = C + C + IX = CCIX
   (d) 210 = 100 + 100 + 10 = C + C + X = CCX

3. Remember that M = 1,000, C = 100, X = 10, V = 5, and I = 1. The answers are as follows, broken down into sums for clarification.
   (a) MMXX = 1,000 + 1,000 + 10 + 10 = 2,020
   (b) MMXIX = 1,000 + 1,000 + 10 + 9 = 2,019

(c) MMIX = 1,000 + 1,000 + 9 = 2,009

(d) MMVI = 1,000 + 1,000 + 6 = 2,006

4. The number three hundred two trillion, seventy billion, one hundred forty-nine million, six thousand, one hundred ten looks like this as a decimal numeral:

302,070,149,006,110

Note the placement of the commas and the ciphers. Also note that we use the American billion, which is equivalent to a thousand million.

5. This is a slightly whimsical problem. You can add as many ciphers as you want to the left of the digit 3 in the answer to the previous problem, and it does not change the value of the number it represents. A mathematician might write it as

...,000,000,...,302,070,149,006,110

You can keep "attaching ciphers" forever in the left-hand direction!

6. You can make any number in the decimal system ten times as large by adding one cipher to its right and then repositioning the commas. Based on the numeral shown for the answer to Prob. 4 above, you would get

3,020,701,490,061,100

If you want to make a number a hundred times as large, add two ciphers to its right and then reposition the commas. Starting with the numeral shown for the answer to Prob. 4, that gives you

30,207,014,900,611,000

If you want to make a number a thousand times as large, add three ciphers to its right and then reposition the commas. Starting with the numeral shown for the answer to Prob. 4, that gives you

302,070,149,006,110,000

7. The number in the final answer to Prob. 6 would be written out in words as three hundred two quadrillion, seventy trillion, one hundred forty-nine billion, six million, one hundred ten thousand. That's based on the United States terminology where a billion is a thousand million, and a trillion is a million million.

8. To solve this problem, tally the values of each digit in the decimal system and then add them up, as follows.

- One times one gives you 1
- One times two gives you 2
- Zero times four gives you 0

- One times eight gives you 8
- Zero times sixteen gives you 0
- One times thirty-two gives you 32

The decimal-numeral equivalent is therefore $1 + 2 + 0 + 8 + 0 + 32 = 43$. This is the quantity you commonly imagine as forty-three.

9. There are at least three ways to solve this problem. You can "cheat" and use Table 1-2. To do it manually, note that forty-three, the decimal equivalent as derived in the solution to Prob. 8, is equal to five times eight, plus three more. That means you can define quantity forty-three as the octal numeral 53. You can also add up the values in *octal arithmetic*. That's tricky and hasn't been covered in this chapter, so we won't show it here.

10. As with the decimal-to-octal conversion, there are at least three ways to solve this problem. You can "cheat" and use Table 1-2. To do it manually, note that forty-three is equal to two times sixteen, plus eleven more. In hexadecimal notation, eleven is represented by B. That means you can define the quantity forty-three in hexadecimal form as the numeral 2B. The last method is to add up the values in *hexadecimal arithmetic*. Again, that's tricky and hasn't been discussed, so we won't do it here.

Chapter 2

1. The null set is a subset of any set. The null set lacks elements, like an empty bank account lacks money. You can say that it contains nothing as an element! If you have a certain set A with known elements, you can add nothing, and you always end up with the same set A. The null set is a subset of itself, although not a *proper* subset of itself. Look at an example. If you let the written word "nothing" actually stand for nothing, then

$$\varnothing = \{\ \} = \{\text{nothing}\}$$

and

$$\{\text{nothing}\} \subseteq \{\text{nothing}, 1, 2, 3\}$$

so therefore

$$\varnothing \subseteq \{1, 2, 3\}$$

Keep in mind that a *subset* is not the same thing as a set *element*. The null set *contains* nothing, but the null set is not *itself* nothing. An empty bank account is a perfectly valid account unless somebody closes it.

2. You can build up an infinite number of sets if you start out with nothing. First, take nothing and make it an element of a set. That gives you the null set. Then consider the set containing the null set, that is, $\{\varnothing\}$. This is a legitimate mathematical thing, but

it's not the same thing as the null set. It can be the sole element of another set, namely $\{\{\varnothing\}\}$. By now, you should be able to sense where this is leading:

$$\varnothing$$
$$\{\varnothing\}$$
$$\{\{\varnothing\}\}$$
$$\{\{\{\varnothing\}\}\}$$
$$\{\{\{\{\varnothing\}\}\}\}$$
$$\{\{\{\{\{\varnothing\}\}\}\}\}$$
$$\downarrow$$

and so on, forever

You started with nothing, and you have turned it into an infinite number of mathematical objects! A variation of this idea is handy for "building" numbers, as you'll see in Chap. 3.

3. In Fig. 2-6, set P, represented by the small dark-shaded triangle, is common to (that is, shared by) sets A and C. Therefore, region P represents the intersection of sets A and C. You can write this as

$$P = A \cap C$$

Set Q, represented by the small, dark-shaded, irregular four-sided figure, is fully shared by sets B and D. Therefore, region Q represents the intersection of sets B and D, which you can write as

$$Q = B \cap D$$

4. Whenever two regions are entirely separate, then the sets they represent are disjoint, and the intersection of those sets is the null set. You can see from Fig. 2-6 that the only pairs of regions that don't overlap are A and D, B and C, and C and D. Therefore, the only null-set intersection pairs are

$$A \cap D = \varnothing$$
$$B \cap C = \varnothing$$
$$C \cap D = \varnothing$$

5. The universal set (call it U) is a subset of itself. That's trivial, because any set is a subset of itself. But U is not a proper subset of itself. Remember, U is the set of all entities, real or imaginary. If U were a proper subset of itself, then there would be some entity that did not belong to U. That's impossible; it contradicts the very definition of U! This little argument is an example of a tactic called *reductio ad absurdum* (Latin for "reduction to absurdity") that mathematicians have used for thousands of years to prove or disprove "slippery statements." It can work well in a courtroom, too.

6. There are plenty of examples that will work here. The set of all even whole numbers, W_{even}, is a proper subset of the set of all whole numbers, W. Both of these sets have

infinitely many elements. Amazingly enough, you can pair the elements of both sets off one-to-one! The mechanics of this goes a little beyond the scope of this chapter, but you can get an idea of how it works if you divide every element of W_{even} by 2, one at a time, and then write down the first few elements of the resulting set. When you do that, you get

$$\{0/2, 2/2, 4/2, 6/2, 8/2, 10/2, \ldots\}$$

But that's exactly the same as W, because when you perform the divisions, you get

$$\{0, 1, 2, 3, 4, 5, \ldots\}$$

This is one of the strange things "infinity" can do. You can take away every other element of a set that has infinitely many elements and that can be written out as an "implied list," and the resulting set is exactly the same "size" as the original set.

7. Let's write out the sets again here as "implied lists":

$$A = \{1, 1/2, 1/3, 1/4, 1/5, 1/6, \ldots\}$$
$$G = \{1, 1/2, 1/4, 1/8, 1/16, 1/32, \ldots\}$$

It's not hard to see that set G contains all those elements, but only those elements, that belong to both sets. Therefore

$$A \cap G = G$$

If you start with set A and then toss in all the elements of G, you get the same set A (with certain elements listed twice, but they can count only once). That means set A contains precisely those elements that belong to one set or the other, or both, so

$$A \cup G = A$$

8. First, isolate all the individual elements of the set {1, 2, 3}. They are 1, 2, and 3. Then start the list of subsets by putting down the null set, which is a subset of any set. Then assemble all the sets you possibly can, using one or more of the elements 1, 2, 3, and list them. You should get

$$\varnothing$$
$$\{1\}$$
$$\{2\}$$
$$\{3\}$$
$$\{1, 2\}$$
$$\{1, 3\}$$
$$\{2, 3\}$$
$$\{1, 2, 3\}$$

9. The set {1, {2, 3}} has only two elements: the number 1 and the set {2, 3}. You can't break {2, 3} down and have it remain an element of the original set {1, {2, 3}}. Therefore, the set of all possible subsets of {1, {2, 3}} is

$$\varnothing$$
$$\{1\}$$
$$\{\{2, 3\}\}$$
$$\{1, \{2, 3\}\}$$

10. The set {1, {2, {3}}} has two elements: the number 1 and the set {2, {3}}. You cannot break {2, {3}} down and have it remain an element of the original set {1, {2, {3}}}. Therefore, the set of all possible subsets of {1, {2, {3}}} is

$$\varnothing$$
$$\{1\}$$
$$\{\{2, \{3\}\}\}$$
$$\{1, \{2, \{3\}\}\}$$

When writing down complicated sets like these, always be sure that the number of opening braces is the same as the number of closing braces. If they're different, you've made a mistake somewhere.

Chapter 3

1. This is the ultimate trivial question. In the number-building systems described in this chapter, nothing doesn't represent any number.
2. No. The number 6 is divisible by 3 without a remainder:

$$6/3 = 2$$

Of course, the quotient here, 2, is even. There are plenty of other even numbers that can be divided by 3 leaving no remainder.

3. When you multiply an odd number by 3, you always get an odd number as the product. The reason for this is similar to the reason why any even number times 7 gives you another even number. (Proving that was one of the "challenge" problems in this chapter.) For the first few odd numbers, multiplication by 3 always produces an odd number:

$$3 \times 1 = 3$$
$$3 \times 3 = 9$$
$$3 \times 5 = 15$$
$$3 \times 7 = 21$$
$$3 \times 9 = 27$$

You can prove that multiplying any odd number by 3 always gives you an odd number if you realize that the *last digit* of an odd number is always odd. Think of an odd number. However large it is, it looks like one of the following:

______1

______3

______5

______7

______9

where the long underscore represents any string of digits you want to put there. Now think of "long multiplication" by 3. Remember how you arrange the numerals on the paper and then do the calculations. You always start out by multiplying the last digits of the two numbers together, getting the last digit of the product. The odd number on top, which you are multiplying by the number on the bottom, must end in 1, 3, 5, 7, or 9. If the number on the bottom is 3, then the last digit in the product must be 3, 9, 5 (the second digit in 15), 1 (the second digit in 21), or 7 (the second digit in 27) respectively. Therefore, any odd number times 3 is always odd.

4. To figure out whether or not 901 is prime, try to break it down into a product of prime factors. If you succeed in doing that, then 901 is not prime. If you fail, then 901 is prime. First take its square root using a calculator. You'll get 30 with a decimal point and some digits. Round this up to the next whole number, which is 31. Using Table 3-1, you can list the set all the primes up to and including 31. That set is

$$\{2, 3, 5, 7, 11, 13, 17, 19, 23, 29, 31\}$$

Now, divide 901 by each of these numbers. As things turn out, 901 is divisible by 17 without a remainder. Therefore, 901 is not prime.

5. To find the prime factors of 1,081, start by taking its square root using a calculator. You should get 32 with a decimal point and some digits. Round this up to 33. Using Table 3-1, list the set all the primes less than or equal to 33. That set is

$$\{2, 3, 5, 7, 11, 13, 17, 19, 23, 29, 31\}$$

Divide 1,081 by each of these numbers. You will see that 1,081 is divisible by 23 without a remainder. You get

$$1{,}081 = 23 \times 47$$

Both of these factors are prime, as you can see by looking at Table 3-1. The number 1,081 therefore has prime factors of 23 and 47.

6. When you break the number 841 into a product of primes using the same process as you did to solve Prob. 5, you'll see that

$$841 = 29 \times 29$$

It is a perfect square of a prime.

7. When you factor 2,197 into a product of primes (again using the process you did to solve Probs. 5 and 6), you'll see that

$$2{,}197 = 13 \times 13 \times 13$$

It has three prime factors, all equal to 13. When a natural number is multiplied by itself and then the result is multiplied by the original number again, the product is a *perfect cube.* The number 2,197 is a perfect cube of a prime.

8. Any composite number can be factored into a product of primes. In the traditional sense, all the primes are positive because they are all natural numbers larger than 1. Remember from basic arithmetic that whenever you multiply a positive number times another positive number, the result is always positive. That means no negative number can be composite if we stick to the traditional definition of a prime number.

9. Suppose 0 were defined as prime. Then 0 would also be composite, because you can multiply 0 by any prime you want, and you always get 0. An even more serious problem occurs if we let 1 be called prime. If that were true, then 1 times any other prime would be equal to that same prime, making every prime number composite! That's why mathematicians generally refuse to call 0 or 1 prime numbers. But when we don't allow them to be prime, they can't be composite either, because they can't be broken down into factors from the traditional set of primes {2, 3, 5, 7, 11, 13, 17, 19, ...}.

10. In Fig. 3-5, you start with 0 and proceed through the positive and negative integers alternately. You can create an "implied one-ended list" of the entire set Z of integers starting with 0, one after the other, this way:

$$Z = \{0, 1, -1, 2, -2, 3, -3, \ldots\}$$

If you pick any integer, no matter how large positively or negatively, this "implied one-ended list" will eventually get to it. You can pair the set of all natural numbers one-to-one with the set of all integers like this, with natural numbers on the left and integers on the right:

$$0 \leftrightarrow 0$$
$$1 \leftrightarrow 1$$
$$2 \leftrightarrow -1$$
$$3 \leftrightarrow 2$$
$$4 \leftrightarrow -2$$
$$5 \leftrightarrow 3$$
$$6 \leftrightarrow -3$$
$$\downarrow$$

and so on, forever

Chapter 4

1. The first sum works out like this:

$$
\begin{aligned}
a &= |-3 + 4 + (-5) + 6| \\
&= |2| \\
&= 2
\end{aligned}
$$

   The second sum works out like this:

$$
\begin{aligned}
b &= |-3| + |4| + |-5| + |6| \\
&= 3 + 4 + 5 + 6 \\
&= 18
\end{aligned}
$$

   From these facts, you can conclude that for integers, the absolute value of a sum is not necessarily equal to the sum of the absolute values.

2. The first expression simplifies like this, step-by-step:

$$
\begin{aligned}
&(3 + 5) - (7 + 9) - (11 + 13) - 15 \\
&= 8 - 16 - 24 - 15 \\
&= -47
\end{aligned}
$$

   The second expression simplifies like this, step-by-step:

$$
\begin{aligned}
&3 + (5 - 7) + (9 - 11) + (13 - 15) \\
&= 3 + (-2) + (-2) + (-2) \\
&= -3
\end{aligned}
$$

3. When you see no parentheses in a string of sums and differences, all the numbers or variables are in effect "outside the parentheses," so you should perform the operations straightaway from left to right. Here is the original expression:

$$3 + 5 - 7 + 9 - 11 + 13 - 15$$

   In order, the sums and differences evolve as follows:

$$
\begin{aligned}
3 + 5 &= 8 \\
8 - 7 &= 1 \\
1 + 9 &= 10 \\
10 - 11 &= -1 \\
-1 + 13 &= 12 \\
12 - 15 &= -3
\end{aligned}
$$

4. It makes sense to consider warming as a positive temperature change, and cooling as a negative temperature change (although there is no technical reason why you couldn't do it the other way around). Here are the year-by-year records once again for reference:

- January 2005 averaged 5 degrees cooler than January 2004.
- January 2004 averaged 2 degrees warmer than January 2003.
- January 2003 averaged 1 degree cooler than January 2002.
- January 2002 averaged 7 degrees warmer than January 2001.
- January 2001 averaged the same temperature as January 2000.
- January 2000 averaged 6 degrees cooler than January 1999.
- January 1999 averaged 3 degrees warmer than January 1998.

If positive change represents warming, then going backward in time you have year-to-year changes of −5, 2, −1, 7, 0, −6, and 3 degrees. You add these integers to discover the change in average temperature between January 1998 and January 2005:

$$(-5) + 2 + (-1) + 7 + 0 + (-6) + 3 = 0$$

In Hoodopolis, there was no difference in the average temperature for January 2005 as compared with January 1998.

5. This works whenever $b = a$. You can then substitute a for b and get

$$a - a = a - a$$

You could also substitute b for a, getting

$$b - b = b - b$$

Obviously, either of these equations will hold true for all possible integers.

6. This always works if $c = 0$. With that restriction, a and b can be any integers you want. In order to see why, you can start with the original equation:

$$(a - b) + c = a - (b + c)$$

"Plug in" 0 for c. Then you get

$$(a - b) + 0 = a - (b + 0)$$

which simplifies to

$$a - b = a - b$$

That equation holds true for any a and b you could possibly choose. The values of a and b don't have to be the same.

Here are a couple of extra-credit exercises if you want a further challenge. Does the original rule work when $a = 0$, but you let b and c be any integers? Does it work when $b = 0$, but you let a and c be any integers? You're on your own!

7. This always works if $c = 0$. Then a and b can be any integers you want. Here's the original equation:

$$(a - b) - c = a - (b - c)$$

"Plug in" 0 for c. Then you get

$$(a - b) - 0 = a - (b - 0)$$

which simplifies to

$$a - b = a - b$$

Here are a couple more extra-credit exercises. Does the original rule work when $a = 0$, but you let b and c be any integers? Does it work when $b = 0$, but you let a and c be any integers? Have fun!

8. See Table A-1. Follow each statement and reason closely so you're sure how it follows from the statements before. This proves that for any four integers a, b, c, and d,

$$a + b + c + d = d + c + b + a$$

9. Let's start with the original expression, and then morph it into the second one. We begin with this:

$$(a + b + c) + d$$

Table A-1. Solution to Prob. 8 in Chap. 4. This S/R proof shows how you can reverse the order in which four integers a, b, c, and d are added, and get the same sum. As you read down the left-hand column, each statement is equal to every statement that came above it.

Statements	Reasons
$a + b + c + d$	Begin here
$a + (b + c + d)$	Group the last three integers
$a + (d + c + b)$	Result of the "challenge" where it was proved that you can reverse the order of a sum of three integers
$(d + c + b) + a$	Commutative law for the sum of a and $(d + c + b)$
$d + c + b + a$	Ungroup the first three integers
Q.E.D.	Mission accomplished

We can "zip up" the sum $b + c$, and call the "package" e. Now our expression looks like this:

$$(a + e) + d$$

The associative law for addition tells us that we can rewrite this as

$$a + (e + d)$$

Now let's "unzip" e so it turns back into the sum $b + c$ (which it always has been, after all), to get

$$a + (b + c + d)$$

That's the result we want, proving that for any four integers a, b, c, and d,

$$(a + b + c) + d = a + (b + c + d)$$

If this seems like a trivial exercise, keep in mind that you're getting your brain into shape to do more complicated tricks some day.

10. The first expression can be simplified as shown in Table A-2, which breaks the process down into statements and reasons. The second expression can be simplified as shown in Table A-3. These are called S/R derivations. They're not intended as proofs, but to show how an expression can be morphed into another expression. In this particular situation, the two original expressions turn out to be equivalent for all integers a, b, and c.

Table A-2. Solution to the first part of Prob. 10 in Chap. 4. As you read down the left-hand column, each statement is equal to every statement that came above it.

Statements	Reasons
$(a + b - c) + (a - b + c)$	Begin here
$[a + b + (-c)] + [a + (-b) + c]$	Change subtractions into negative additions; replace original parentheses with brackets for temporary clarification
$a + b + (-c) + a + (-b) + c$	Get rid of outer sets of brackets; they are not necessary in straight sums
$a + a + b + (-b) + c + (-c)$	Commutative law for addition, generalized
$a + a$	$b + (-b) = 0$ and $c + (-c) = 0$ so the b's and c's "cancel out"

Table A-3. Solution to the second part of Prob. 10 in Chap. 4. As you read down the left-hand column, each statement is equal to every statement above it.

Statements	Reasons
$a + (b - c) + (a - b) + c$	Begin here
$a + [b + (-c)] + [a + (-b)] + c$	Change subtractions into negative additions; replace original parentheses with brackets for temporary clarification
$a + b + (-c) + a + (-b) + c$	Get rid of brackets; they served a good purpose but are not necessary in straight sums
$a + a + b + (-b) + c + (-c)$	Commutative law for addition, generalized
$a + a$	$b + (-b) = 0$ and $c + (-c) = 0$ so the b's and c's "cancel out"

Chapter 5

1. If you start with a positive integer p and multiply by −3, you get a negative integer whose absolute value is $3\,|p|$. If you start with a negative integer n and multiply by −3, you get a positive integer whose absolute value is $3\,|n|$.

2. If you start with any integer and keep multiplying by −3 over and over, the *polarity* (that is, the "positivity" or "negativity") of the end product keeps alternating, and the absolute value increases by a factor of 3 every time. So with any integer q, whether positive or negative, you get

$$|-3q| = 3\,|q|$$
$$|-3 \times (-3q)| = 3 \times 3\,|q|$$
$$|-3 \times (-3) \times (-3q)| = 3 \times 3 \times 3\,|q|$$
$$\downarrow$$

and so on, forever

As you can see, the absolute value grows rapidly as you keep on multiplying by −3, unless the starting integer q happens to be 0.

3. Evaluating this requires close attention and patience. Here is the initial expression for reference:

$$4 + 32 / 8 \times (-2) + 20 / 5 / 2 - 8$$

Table A-4 is an S/R breakdown of the evaluation process.

Table A-4. Solution to Prob. 3 in Chap. 5. As you read down the left-hand column, each statement is equal to every statement above it.

Statements	Reasons
$4 + 32 / 8 \times (-2) + 20 / 5 / 2 - 8$	Begin here
$4 + 32 / [8 \times (-2)] + 20 / 5 / 2 - 8$	Group the multiplication
$4 + 32 / (-16) + 20 / 5 / 2 - 8$	Do the multiplication
$4 + [32/(-16)] + [(20/5)/2)] - 8$	Group the divisions
$4 + (-2) + [(20/5)/2] - 8$	Do the division $32/(-16) = -2$
$4 + (-2) + (4/2) - 8$	Do the division $20/5 = 4$
$4 + (-2) + 2 - 8$	Do the division $4/2 = 2$
$4 + (-2) + 2 + (-8)$	Convert the subtraction to negative addition
-4	Do the additions from left to right

Table A-5. Solution to Prob. 4 in Chap. 5. As you read down the left-hand column, each statement is equal to every statement above it.

Statements	Reasons
$abcd$	Begin here
$a(bcd)$	Group the last three integers
$a(dcb)$	Result of the "challenge" where it was proved that you can reverse the order of a product of three integers
$(dcb)a$	Commutative law for the product of a and (dcb)
$dcba$	Ungroup the first three integers
Q.E.D.	Mission accomplished

4. See Table A-5. Follow each statement and reason closely so you're sure how it follows from the statements before. This proves that for any four integers a, b, c, and d,

$$abcd = dcba$$

5. Here are the steps in the calculation process, using parentheses where needed:

$$-15 \times (-45) = 675$$
$$675 / (-25) = -27$$
$$-27 \times (-9) = 243$$
$$243 / (-81) = -3$$
$$-3 \times (-5) = 15$$

6. This always works if $c = 1$. Then a can be any integer, and b can be any integer except 0. Here's the original equation:

$$(a/b)/c = a/(b/c)$$

"Plug in" 1 for c. Then you get

$$(a / b) / 1 = a / (b / 1)$$

which simplifies to

$$a / b = a / b$$

Here is an opportunity get some extra credit. Does the original rule work when $a = 1$, but you let b and c be any integers except 0? Does it work when $b = 1$, but you let a be any integer and c be any integer except 0?

7. This, again, always works if $c = 1$. Then a and b can be any integers. Here's the original equation:

$$(ab) / c = a(b / c)$$

"Plug in" 1 for c. Then you get

$$(ab) / 1 = a(b / 1)$$

which simplifies to

$$ab = ab$$

Now, can you guess what's coming? Another extra-credit workout! Does the original rule hold true when $a = 1$, but you let b be any integer and c be any integer except 0? Does it work when $b = 1$, but you let a be any integer and c be any integer except 0?

8. Let's start with the distributive law of multiplication over addition in its left-hand form. You'll notice that we're using different letters of the alphabet. That will help keep you from sinking into an "*abc* rut" with the naming of variables. Otherwise, the law is stated in exactly the same way. We can write the original form of the law like this:

$$p(m + n) = pm + pn$$

where n, m, and p are integers. Now, the solution is only a matter of applying the commutative law for multiplication three times, once for each of the three products above:

$$(m + n)p = mp + np$$

Q.E.D. That's all there is to it!

9. The variables have unfamiliar names for the same reason as in Prob. 8. See Table A-6. This proves that for any two integers d and g,

$$-(d + g) = -d - g$$

10. Again, unfamiliar variable names can keep your attention on the way things work, without getting stuck in the "*abc* routine" of rote memorization. See Table A-7. This proves that for any two integers h and k,

$$-(h - k) = k - h$$

Table A-6. Solution to Prob. 9 in Chap. 5. As you read down the left-hand column, each statement is equal to every statement above it.

Statements	Reasons
$-(d+g)$	Begin here
$(d+g)(-1)$	Principle of the sign-changing element
$(-1)(d+g)$	Commutative law for multiplication
$(-1)d+(-1)g$	Distributive law for multiplication over addition
$-d+(-g)$	Principle of the sign-changing element (the other way around)
$-d-g$	Addition of a negative is the same thing as subtraction
Q.E.D.	Mission accomplished

Table A-7. Solution to Prob. 10 in Chap. 5. As you read down the left-hand column, each statement is equal to every statement above it.

Statements	Reasons
$-(h-k)$	Begin here
$(h-k)(-1)$	Principle of the sign-changing element
$h(-1)-k(-1)$	Right-hand distributive law for multiplication over subtraction
$-h-(-k)$	Principle of the sign-changing element (the other way around)
$-h+k$	Subtraction of a negative is the same thing as addition of a positive
$k+(-h)$	Commutative law for addition
$k-h$	Addition of a negative is the same thing as subtraction of a positive
Q.E.D.	Mission accomplished

Chapter 6

1. The ratio of the top wind speed in a "category 1" hurricane to the top wind speed in a "category 4" hurricane is 95:155. It's easy to see that both of these integers are divisible by 5, because they both end in 5. When you divide both the numerator and the denominator by 5, you get 19:31. These integers are both prime, so this expression of the ratio is in lowest terms. The ratio of the top wind speed in a "category 4" hurricane to the top wind speed in a "category 1" hurricane is found by switching the numerator and denominator, getting 31:19.

2. The ratio of the absolute temperature of the boiling point to the absolute temperature of the freezing point, expressed in kelvins, is 373:273. To get this in lowest terms, you must find an integer that "divides out" from both of these numbers until the denominator becomes prime. That's not as obvious here as in Prob. 1! Start by trying to factor 373 into a product of primes. The square root of 373 is equal to 19 followed by

a decimal point and some digits. Round this up to 20. The primes less than 20 are 2, 3, 5, 7, 11, 13, 17, and 19. As you divide 373 by each of these primes, you'll see that none of them gives you a quotient without a remainder. That means 373 is itself a prime number, so the ratio 373:273 is in lowest terms.

3. You can use the "brute-force" method and factor both the numerator and the denominator of 231/230 into primes, and you'll discover that the prime factors that make up the numerator are entirely different from the prime factors that make up the denominator. Therefore, the fraction 231/230 has to be in lowest terms. But there's an easier way to see this, and you don't have to do any work to figure it out. Note that the absolute values of the numerator and the denominator differ by only 1, and the denominator is positive. Now think: what will happen if you divide both the numerator and the denominator by any positive integer other than 1, in an attempt to reduce the fraction? The resulting numerator and denominator will always have absolute values that differ by less than 1, so they can't both be integers. But in order to be a "legitimate fraction," both the numerator and the denominator must be integers.

4. You want to see if −154/165 is in lowest terms, and if it is not, to reduce it. First, convert both the numerator and the denominator into products of primes and then attach the extra "factor" of −1 to the numerator, like this:

$$-154 = -1 \times 2 \times 7 \times 11$$

and

$$165 = 3 \times 5 \times 11$$

Next, use these products to build a fraction in which both the numerator and the denominator consist of prime factors, and the numerator has the extra "factor" −1:

$$(-1 \times 2 \times 7 \times 11) \; / \; (3 \times 5 \times 11)$$

The common prime factor is 11. Remove it from both the numerator and the denominator, getting

$$(-1 \times 2 \times 7) \; / \; (3 \times 5)$$

That's −14/15. You can sense immediately that this is in lowest terms because the numerator and the denominator have absolute values that differ by only 1, and the denominator is positive.

5. When you have two fractions in lowest terms and multiply them, the product is sometimes in lowest terms, but not always. First, consider this:

$$\begin{aligned} 3/5 \times 7/11 &= (3 \times 7) \; / \; (5 \times 11) \\ &= 21/55 \end{aligned}$$

This product is in lowest terms. You know this because the numerator is the product of the primes 3 and 7, and the denominator is the product of the primes 5 and 11. When the numerator and denominator of a fraction are both factored into primes, the denominator is positive, and no prime in the numerator is the same as any prime in the denominator, then that fraction is in lowest terms.

With this knowledge, it's not hard to think of a couple of fractions that are in lowest terms individually, but that produce a fraction that is not in lowest terms when they are multiplied by each other. Take a look at this:

$$\begin{aligned} 3/5 \times 5/7 &= (3 \times 5) / (5 \times 7) \\ &= 15/35 \end{aligned}$$

Both of the original fractions are in lowest terms here, but the product is not. You can factor 5 out of both the numerator and the denominator of 15/35, reducing it to 3/7.

6. When you have two fractions in lowest terms and divide one by the other, the quotient is sometimes in lowest terms, but not always. This can be seen by "mutating" the solution to Prob. 5. First, consider this:

$$\begin{aligned} (3/5) / (11/7) &= 3/5 \times 7/11 \\ &= (3 \times 7) / (5 \times 11) \\ &= 21/55 \end{aligned}$$

This is in lowest terms, as you saw in the solution to Prob. 5. Now try this:

$$\begin{aligned} (3/5) / (7/5) &= 3/5 \times 5/7 \\ &= (3 \times 5) / (5 \times 7) \\ &= 15/35 \end{aligned}$$

Again as you saw in the solution to Prob. 5, this quotient is not in lowest terms, even though the dividend and divisor fractions are.

7. Table A-8 is an S/R derivation proving that

$$(a/b) / [(c/d) / (e/f)] = ade / bcf$$

Table A-8. Solution to Prob. 7 in Chap. 6. The result is a formula for repeated division of fractions, where the second pair of fractions is divided first. As you read down the left-hand column, each statement is equal to all the statements above it.

Statements	Reasons
$(a/b) / [(c/d) / (e/f)]$	Begin here
$(a/b) / (cf / de)$	Apply the formula for division of c/d by e/f
$(a/b) / (g/h)$	Temporarily let $cf = g$ and $de = h$, and substitute the new names in the previous expression
ah / bg	Apply the formula for division of a/b by g/h
ade / bcf	Substitute cf for g and de for h in the previous expression

Table A-9. Solution to Prob. 8 in Chap. 6. This shows that the commutative law holds for the multiplication of fractions. As you read down the left-hand column, each statement is equal to all the statements above it.

Statements	Reasons
$(a/b)(c/d)$	Begin here
ac / bd	Formula for multiplication of fractions
ca / db	Commutative law for multiplication of integers applied to numerator and denominator
$(c/d)(a/b)$	Formula for multiplication of fractions applied "backwards"
Q.E.D.	Mission accomplished

Compare this with the result of the "challenge" problem, where you found out that

$$[(a/b) / (c/d)] / (e/f) = adf / bce$$

8. Table A-9 is an S/R proof that if a, b, c, and d are nonzero integers, then

$$(a/b)(c/d) = (c/d)(a/b)$$

Therefore, the commutative property holds for the multiplication of fractions, and indeed for the multiplication of any two rational numbers. Note the fourth step of this proof, in which the rule for multiplication of fractions is applied "backwards." We can get away with this because equality works in both directions! This fact, which may seem trivial to you but is really quite significant, is one of three aspects of equality known as the *reflexive*, *symmetric*, and *transitive* properties. You should know what these terms mean. The reflexive property tells us that for any quantity a,

$$a = a$$

The symmetric property tells us that for any two quantities a and b,

$$\text{If } a = b, \text{ then } b = a$$

The transitive property tells us that for any three quantities a, b, and c,

$$\text{If } a = b \text{ and } b = c, \text{ then } a = c$$

Whenever any means of comparing things has all three of these properties, then it's called an *equivalence relation*. Equality is the most common example of an equivalence relation. But there are others, such as the logical connector "if and only if" or "iff," symbolized by a double-shafted, double-headed arrow, often with a little extra space on either side ($\Leftrightarrow$).

Table A-10. Solution to Prob. 9 in Chap. 6. This shows that the associative law holds for the multiplication of fractions. As you read down the left-hand column, each statement is equal to all the statements above it.

Statements	Reasons
$[(a/b)(c/d)](e/f)$	Begin here
$(ac / bd)(e/f)$	Formula for multiplication of a/b by c/d
$(ac)e / (bd)f$	Formula for multiplication of (ac / bd) by e/f
$a(ce) / b(df)$	Associative law for multiplication of integers applied to numerator and denominator
$(a/b)(ce / df)$	Formula for multiplication of fractions applied "backward" to turn quotient of products into product of quotients
$(a/b)[(c/d)(e/f)]$	Formula for multiplication of fractions applied "backward" (again!) to turn quotient of products into product of quotients
Q.E.D.	Mission accomplished

9. Table A-10 shows that the associative property of multiplication holds for any three fractions. If we have fraction of the form a/b, another of the form c/d, and a third of the form e/f, where a, c, and e are integers, and b, d, and f are positive integers, then

$$[(a/b)(c/d)](e/f) = (a/b)[(c/d)(e/f)]$$

Do you notice that this proof has a certain symmetry? First, we mold the expression, in two "steps up," into a form where we can apply the associative law for multiplying integers. Then we invoke that law. Finally we mold the expression, in two "steps back down," into the form we want.

10. Let's explore this situation and see what sorts of "clues" we can find. We've been told that there are four integers a, b, c, and d, such that

$$(a/b) / (c/d) = (c/d) / (a/b)$$

If we use the rule for division of one fraction by another on both sides of this equation, we get

$$ad / bc = cb / da$$

Invoking the commutative law for multiplication in the numerator and denominator on the right-hand side of the equals sign, we see that

$$ad / bc = bc / ad$$

We aren't allowed to get away with trivial solutions, such as letting all the integers be equal to 1 or letting them all be equal to −1. But suppose that $a = 7$, $b = 5$, $c = 14$, and $d = 10$. (This isn't the only example we can use, but it should give you the general idea.) Then

$$\begin{aligned} ad / bc &= (7 \times 10) / (5 \times 14) \\ &= 70/70 \\ &= 1 \end{aligned}$$

and

$$\begin{aligned} bc / ad &= (5 \times 14) / (7 \times 10) \\ &= 70/70 \\ &= 1 \end{aligned}$$

Let's "plug in" the values $a = 7$, $b = 5$, $c = 14$, and $d = 10$ to the original equation and see what we get:

$$(a/b) / (c/d) = (c/d) / (a/b)$$

therefore

$$(7/5) / (14/10) = (14/10) / (7/5)$$

Note that 7/5 and 14/10 actually represent the same rational number. All we've really done here is show that 1 = 1, in a roundabout way.

Chapter 7

1. Figure A-1 shows a number line that covers the range of positive rational numbers from 10 up to 100,000. To find the number of orders of magnitude, subtract the powers of 10.

Figure A-1 Illustration for the solution to Prob. 1 in Chap. 7.

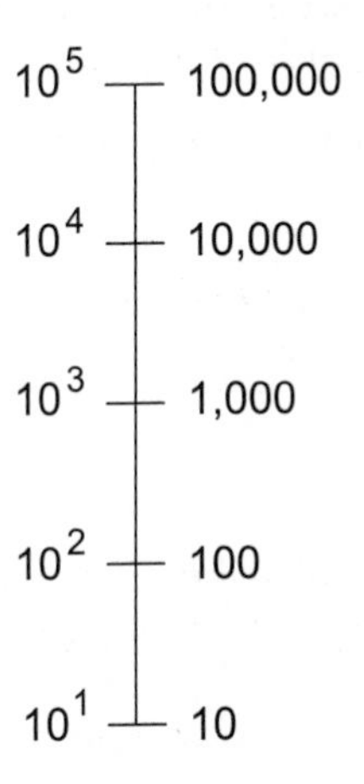

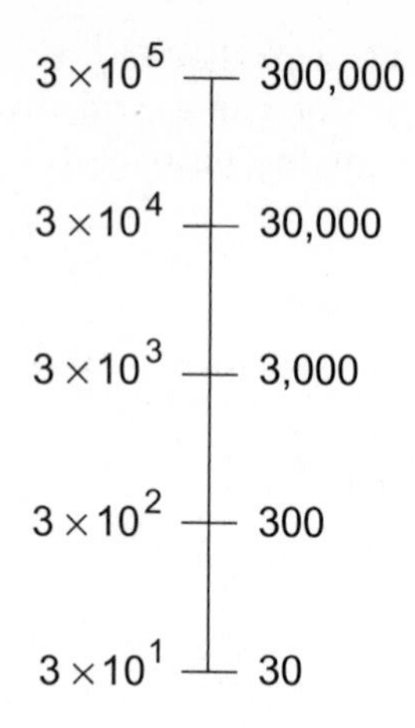

Figure A-2 Illustration for the solution to Prob. 2 in Chap. 7.

Here, $100{,}000 = 10^5$ and $10 = 10^1$. We have $5 - 1$, or 4, so there are four orders of magnitude in this span. Another way to see this is to count the number of intervals between "hash marks" on the number line.

2. Figure A-2 shows a number line that covers the range of positive rational numbers from 30 up to 300,000. In this case, we can divide the larger number by the smaller, and then count the number of ciphers in the quotient. We get

$$300{,}000 / 30 = 10{,}000$$

These two numbers differ by four orders of magnitude, the same extent as the two numbers differ in Prob. 1. We can also count the number of intervals between hash marks, just as in Prob. 1.

3. In this situation, the larger number is 75,000,000 and the smaller number is 330. If we divide the larger by the smaller, we get 227,272 and a fraction. That's between a factor of 100,000 (or 10^5) and 1,000,000 (or 10^6). We can therefore say that 75,000,000 is between five and six orders of magnitude larger than 330. Right now, that's the best we can do. (It's possible to come up with a more precise value, but that involves *logarithms*, which we have not yet studied.)

4. The answers, along with explanations, are as follows:
   (a) The number 4.7 is equivalent to 4-7/10. The fractional part here is in lowest terms, because the numerator, 7, is prime.
   (b) The number −8.35 is equivalent to −8-35/100. The fractional part can be reduced to 7/20, so the entire expression becomes −8-7/20. Note the difference in appearance, as well as the difference in purpose, between the minus sign and the dash!
   (c) The number 0.02 has no integer portion. The fractional part is 2/100, which reduces to 1/50. Therefore, the entire expression is 1/50.
   (d) The number −0.29 has no integer portion. The fractional part is −29/100, which is already in lowest terms because the absolute value of the numerator, 29, is prime and is not a prime factor of the denominator. Therefore, the entire expression is −29/100.

5. The answers, along with explanations, are as follows:
   (a) We have the number 4-7/10. We can convert the integer part, which is 4, into 10ths by multiplying it by 10 and then dividing the result by 10, getting

$$(4 \times 10) / 10 = 40/10$$

   The entire number is therefore

$$\begin{aligned} 40/10 + 7/10 &= (40 + 7) / 10 \\ &= 47/10 \end{aligned}$$

   (b) We start with the number −8-7/20. We convert the integer part, −8, into 20ths by multiplying and then dividing by 20, getting

$$(-8 \times 20) / 20 = -160/20$$

   The entire number is therefore

$$\begin{aligned} -160/20 + (-7/20) &= (-160 - 7) / 20 \\ &= -167/20 \end{aligned}$$

   (c) The number 1/50 is a fraction already, and has been reduced to lowest terms. There's nothing further for us to do here!
   (d) This situation is the same as in part (c). We already have the final expression in the form of the starting number, −29/100.

6. The easiest way to work out these problems is to input the numerator into a calculator, and then divide by the denominator. When we do that, we get the following results.
   (a) 44/16 = 2.75
   (b) −81/27 = −3
   (c) 51/13 = 3.923076923076923076...
   (d) −45/800 = −0.05625

   In case (c), you'll need a calculator that can display a lot of digits if you want to be certain of the repeating pattern of digits to the right of the decimal point. (It's 923076.) If you don't have such a calculator, you can perform old-fashioned, manual long division to discover the pattern.

7. This problem can be solved in two steps. First, we use a calculator or long division to determine the decimal equivalent of 1/17. We get this endless repeating decimal:

$$0.05882352941176470588235294117647...$$

   The repeating sequence of digits is 0588235294117647. The initial cipher is important here! Next, we count up the number of digits in this sequence, including the cipher at the beginning. There are 16 digits in the repeating string. We construct a fraction with the repeating sequence in the numerator and a string of 16 digits, all 9s, in the denominator, inserting commas to make the large numbers more readable. That gives us

0,588,235,294,117,647 / 9,999,999,999,999,999

The initial cipher can now be removed. It was only necessary to be sure we put the correct number of 9s in the denominator. The final answer is:

588,235,294,117,647 / 9,999,999,999,999,999

If you want, you can check this by dividing it out using a calculator with a large display, such as the one in a computer. You should get the same result as we got when we divided 1 by 17. Another way to check this is to divide the denominator of the above fraction by the numerator (that is, take the quotient "upside-down"). You should get exactly 17.

8. The number we are given to start with is 2.892892892.... To convert this to a ratio of integers, we first write down the part of the expression to the right of the decimal point, like this:

   . 892 892 892 ...

   From this, we know that fractional part of the expression is 892/999. We put back the whole-number part, getting

   2-892/999

   Now we must convert 2 to a fraction with a denominator 999. We multiply 2 by 999, getting 1,998. This goes into the numerator. Now we have two fractions that we can easily add to produce the final ratio:

   $$892 / 999 + 1{,}998 / 999 = (892 + 1{,}998) / 999$$
   $$= 2{,}890 / 999$$

   It is always a good idea to check the results of calculations like this by dividing out on a calculator. In this case, the quotient is 2.892892892..., the original number in decimal form.

9. We can be certain that this decimal expansion, which we have been told is an endless string of digits, has a repeating pattern. The original quotient is a rational number by definition. Remember, *any* rational number can be expressed as either a terminating decimal or an endless repeating decimal. The repeating pattern of digits in the decimal expansion might be incredibly long, but it is finite.

10. This is one of the most baffling problems in mathematics. The trouble comes up because we're trying to compare hard reality with pure theory. Even the most powerful supercomputer can be confused by a string-of-digits problem if the repeating pattern is complicated enough. But the fact that a pattern can't be discovered in a human lifetime does not prove conclusively that there is not a pattern! It works the other way, too. If we see a long string of digits repeating many times, we can't be sure it will repeat endlessly, unless we know that there's a ratio of integers with the same value.

Chapter 8

1. When a negative number is raised to an even positive integer power, the result is always a positive number. When a negative number is raised to an odd positive integer power, the result is always a negative number.

2. The answers, along with explanations, are as follows.

   (a) If we raise a base of −2 to increasing integer powers starting with 1, we get this sequence:

   $$(-2)^1, (-2)^2, (-2)^3, (-2)^4, (-2)^5, \ldots$$

   When we multiply these out, we get

   $$-2, 4, -8, 16, -32, \ldots$$

   The numbers alternate between negative and positive, and their absolute values double with each repetition. This sequence "runs away" toward both "positive infinity" and "negative infinity"!

   (b) If we do the same thing with a base of −1, we get

   $$(-1)^1, (-1)^2, (-1)^3, (-1)^4, (-1)^5, \ldots$$

   Multiplying these out gives us

   $$-1, 1, -1, 1, -1, \ldots$$

   The numbers simply alternate between −1 and 1.

   (c) If we carry out the same process with a base of −1/2, we get

   $$(-1/2)^1, (-1/2)^2, (-1/2)^3, (-1/2)^4, (-1/2)^5, \ldots$$

   Multiplying these out produces

   $$-1/2, 1/4, -1/8, 1/16, -1/32, \ldots$$

   The numbers again alternate between negative and positive, and their absolute values get half as large with each repetition. This sequence converges toward 0 "from both sides."

3. Here are the answers. Note how they "mirror" the results of Prob. 2.

   (a) If we raise a base of −2 to smaller and smaller negative integer powers starting with −1, we get this sequence:

   $$(-2)^{-1}, (-2)^{-2}, (-2)^{-3}, (-2)^{-4}, (-2)^{-5}, \ldots$$

   This is the same as

   $$1/(-2)^1, 1/(-2)^2, 1/(-2)^3, 1/(-2)^4, 1/(-2)^5, \ldots$$

When we multiply these out, we get

$$1/(-2),\ 1/4,\ 1/(-8),\ 1/16,\ 1/(-32),\ \ldots$$

which is the same as

$$-1/2,\ 1/4,\ -1/8,\ 1/16,\ -1/32,\ \ldots$$

The numbers alternate between negative and positive, and their absolute values get half as large with each repetition. This sequence is identical to the solution for Problem 2(c).

(b) If we do the same thing with a base of −1, we get

$$(-1)^{-1},\ (-1)^{-2},\ (-1)^{-3},\ (-1)^{-4},\ (-1)^{-5},\ \ldots$$

This is the same as

$$1/(-1)^1,\ 1/(-1)^2,\ 1/(-1)^3,\ 1/(-1)^4,\ 1/(-1)^5,\ \ldots$$

Multiplying these out gives us

$$1/(-1),\ 1/1,\ 1/(-1),\ 1/1,\ 1/(-1),\ \ldots$$

which is the same as

$$-1,\ 1,\ -1,\ 1,\ -1,\ \ldots$$

The numbers simply alternate between −1 and 1. This result is identical with the solution for Prob. 2(b).

(c) Finally, let's do the process with a base of −1/2. We get the sequence

$$(-1/2)^{-1},\ (-1/2)^{-2},\ (-1/2)^{-3},\ (-1/2)^{-4},\ (-1/2)^{-5},\ \ldots$$

This is the same as

$$1/(-1/2)^1,\ 1/(-1/2)^2,\ 1/(-1/2)^3,\ 1/(-1/2)^4,\ 1/(-1/2)^5,\ \ldots$$

which is the same as

$$1/(-1/2),\ 1/(1/4),\ 1/(-1/8),\ 1/(1/16),\ 1/(-1/32),\ \ldots$$

which can be simplified to

$$-2,\ 4,\ -8,\ 16,\ -32,\ \ldots$$

That's the same thing we got when we solved Problem 2(a).

4. The expression can be simplified to a sum of individual terms when we remember that squaring any quantity (that is, taking it to the second power) is the same thing as multiplying it by itself. Then we can use the results of the final "Challenge" in Chap. 5. Step-by-step, we get:

$$\begin{aligned}(y+1)^2 &= (y+1)(y+1)\\ &= yy + y1 + 1y + (1 \times 1)\\ &= y^2 + y + y + 1\\ &= y^2 + 2y + 1\end{aligned}$$

5. This problem can be solved just like Prob. 4, but we have to pay careful attention because of the minus sign:

$$\begin{aligned}(y-1)^2 &= (y-1)(y-1)\\ &= yy + y(-1) + (-1y) + [(-1) \times (-1)]\\ &= y^2 + (-y) + (-y) + 1\\ &= y^2 - 2y + 1\end{aligned}$$

6. Let's start with the generalized addition-of-exponents (GAOE) rule as it is stated in the chapter text. Here it is again, with the exponent names changed for variety! For any number a except 0, and for any rational numbers p and q,

$$a^p a^q = a^{(p+q)}$$

Suppose r is another rational number. Let's multiply both sides of the above equation by a^r. This gives us

$$(a^p a^q)a^r = a^{(p+q)}a^r$$

According to the rule for the grouping of factors in a product, we can take the parentheses out of the left-hand side of the above equation and get

$$a^p a^q a^r = a^{(p+q)}a^r$$

The left-hand side of the equation now contains the expression we want to evaluate. Let's consider $(p+q)$ to be a single quantity. We can then use the GAOE rule on the right-hand side of this equation, getting

$$a^p a^q a^r = a^{[(p+q)+r]}$$

Again, we invoke the privilege of ungrouping, this time to the addends in the exponent on the right-hand side. This gives us

$$a^p a^q a^r = a^{(p+q+r)}$$

Q.E.D. Mission accomplished!

Table A-11. Solution to Prob. 7 in Chap. 8. This shows that the multiplication-of-exponents rule applies to a "power of a power of a power." As you read down the left-hand column, each statement is equal to all the statements above it.

Statements	Reasons
$(a^p)^q = a^{pq}$	GMOE rule as given in Chap. 8 text, where a is any number except 0, and p and q are rational numbers
$[(a^p)^q]^r = (a^{pq})^r$	Take rth power of both sides, where r is a rational number
$[(a^p)^q]^r = a^{(pq)r}$	Consider (pq) as a single quantity and then use GMOE rule on right-hand side
$[(a^p)^q]^r = a^{pqr}$	Ungrouping of products in exponent on right-hand side
Q.E.D.	Mission accomplished

7. This S/R proof is shown in Table A-11. We can "legally" take the rth power of both sides in line 2 of the proof even if r is a reciprocal power. Remember, if there is any positive-negative ambiguity when taking a reciprocal power, the positive value is the "default."

8. We can start by stating the generalized multiplication-of-exponents (GMOE) rule as it appears in the chapter text. The exponent names are changed to keep us out of a "rote-memorization rut," and also to conform to the way the problem is stated. For any number x except 0, and for any rational numbers r and s,

$$(x^r)^s = x^{rs}$$

Applying the commutative law for multiplication to the entire exponent on the right-hand side of this equation, we get

$$(x^r)^s = x^{sr}$$

Finally, we can invoke the GMOE rule "in reverse" to the right-hand side, obtaining

$$(x^r)^s = (x^s)^r$$

Q.E.D. Mission accomplished!

9. Let's suppose that x is a positive number. It can by any number we want, as long as it is larger than 0. We take the 4th root or 1/4 power of x, and then square the result. That gives us

$$(x^{1/4})^2$$

According to the GMOE rule, that is the same as

$$x^{(1/4)\times 2}$$

which simplifies to $x^{2/4}$. The fraction 2/4 can be reduced to 1/2. That means we actually have x raised to the 1/2 power, or the square root of x.

10. Imagine that y is a positive number. We take the 6th power of y, and then take the cube root or 1/3 power of the result. That gives us

$$(y^6)^{1/3}$$

According to the GMOE rule, that is the same as

$$y^{6\times(1/3)}$$

which simplifies to $y^{6/3}$ and then reduces to y^2.

Chapter 9

1. We can suspect that quantity (d), $27^{1/2}$, is irrational. It is not a natural number. When we use a calculator to evaluate it, we get 5.196... followed by an apparently random jumble of digits. That suggests its decimal expansion is endless and nonrepeating. The other three quantities can be evaluated and found rational:

   (a) $16^{3/4} = (16^{1/4})^3 = 2^3 = 8$

   (b) $(1/4)^{1/2} = 1/(4^{1/2}) = 1/2$

   (c) $(-27)^{-1/3} = 1/(-27^{1/3}) = 1/(-3) = -1/3$

2. If we have an irrational number expanded into endless, nonrepeating decimal form, we can multiply it by any natural-number power of 10 and always get the same string of digits. The only difference will be that the decimal point moves to the right by one place for each power of 10. As an example, consider the square root of 7, or $7^{1/2}$. Using a calculator with a large display, we see that this expands to

$$7^{1/2} = 2.6457513106459059...$$

As we multiply by increasing natural-number powers of 10, we get this sequence of numbers, each one 10 times as large as the one above it:

$$10 \times 7^{1/2} = 26.457513106459059...$$
$$100 \times 7^{1/2} = 264.57513106459059...$$
$$1{,}000 \times 7^{1/2} = 2{,}645.7513106459059...$$
$$10{,}000 \times 7^{1/2} = 26{,}457.513106459059...$$
$$\downarrow$$

and so on, as long as we want

These are all endless non-repeating decimals, so they're all irrational numbers. This will happen for any endless non-repeating string of digits.

3. All of these sets are infinite, and the elements of each set can be completely defined by means of an "implied list." Therefore, the cardinality of every one of these sets is $\aleph_0$. We can pair off any of these sets one-to-one with the set N of naturals. As an optional exercise, you might want to show how this can be done. Here's a hint: Multiply the naturals all by 2, 10, 100, or any whole-number power of 10 to create "implied lists" for the sets.

4. The original equation is

$$36x + 48y = 216$$

When we divide through by 12 on either side, we get

$$(36x + 48y)/12 = 216/12$$

which can be morphed to

$$(36/12)x + (48/12)y = 18$$

and finally to

$$3x + 4y = 18$$

That's as simple as we can get it.

5. To figure this out, note that $18 = 9 \times 2$. Therefore, according to the power of product rule, we have

$$\begin{aligned} 18^{1/2} &= (9 \times 2)^{1/2} \\ &= 9^{1/2} \times 2^{1/2} \\ &= 3 \times 2^{1/2} \end{aligned}$$

This is a product of a natural number and an irrational number. The nonnegative square roots of large numbers can often be resolved in this way.

6. The number 83 is prime. If we want to factor this into a product of two natural numbers and no remainder, the best we can do is 83×1. Therefore, we can't resolve the nonnegative square root of 83 into anything simpler than $83^{1/2}$. We can also tell that it is in the most simplified form because it has no factors that are perfect squares.

7. The ratio of $50^{1/2}$ to $2^{1/2}$ is the same thing as the quotient of these two numbers. Note that both 50 and 2 are taken to the same real-number power, that is, 1/2. According to the power of quotient rule, then, we have

$$\begin{aligned} 50^{1/2}/2^{1/2} &= (50/2)^{1/2} \\ &= 25^{1/2} \\ &= 5 \end{aligned}$$

The ratio of the nonnegative square root of 50 to the nonnegative square root of 2 is exactly equal to 5, even though the dividend and the divisor are both irrational.

8. Here's the sum-of-quotients rule again:

$$w/x + y/z = (wz + xy)/(xz)$$

In this example, we can let $w=7$, $x=11$, $y=-5$, and $z=17$. None of these are equal to 0, so we can be sure the rule will work properly. Now it's simply a matter of doing the arithmetic:

$$\begin{aligned} 7/11 + (-5/17) &= \{(7 \times 17) + [11 \times (-5)]\} / (11 \times 17) \\ &= [119 + (-55)] / 187 \\ &= 64 / 187 \end{aligned}$$

This can't be reduced to lower terms because the denominator, 187, is the product of two primes, 11 and 17. Neither of these factors "goes into" the numerator, 64, without leaving a remainder.

9. To solve this, we can rewrite $(x - y)$ as $[x + (-y)]$. Then, using the product of sums rule, we have

$$(x + y)[x + (-y)] = xx + x(-y) + yx + y(-y)$$

We can use familiar arithmetic rules to write this as

$$= x^2 - xy + xy - y^2$$

The addends $-xy$ and xy cancel out here, so we get the final result

$$= x^2 - y^2$$

10. Let u, v, w, x, y, and z be real numbers. We're given a product of sums of three variables, and we're told to multiply it out using the product of sums rule. That rule, as stated in the chapter text, only allows us to use two variables in each addend. But we can "cheat" by renaming certain sums! Here is the original expression:

$$(u + v + w)(x + y + z)$$

Let's rename $(v + w)$ as r, and $(y + z)$ as s. Then we have

$$(u + r)(x + s)$$

Using the product of sums rule, we can multiply this out to

$$ux + us + rx + rs$$

Now let's substitute the original values for r and s back into the expression. This gives us

$$ux + u(y + z) + (v + w)x + (v + w)(y + z)$$

The first addend in this expression is ready to go! We can use the distributive laws on the second and third terms to get

$$ux + uy + uz + vx + wx + (v + w)(y + z)$$

The final term can be multiplied out using the product of sums rule from the chapter text. This gives us

$$ux + uy + uz + vx + wx + vy + vz + wy + wz$$

We can use the commutative law of addition, in its generalized form, to rearrange these terms so the whole sum is easier to remember:

$$ux + uy + uz + vx + vy + vz + wx + wy + wz$$

APPENDIX

B

Worked-Out Solutions to Exercises: Chapters 11 to 19

These worked-out solutions do not necessarily represent the only way a problem can be figured out. If you think you can solve a particular problem in a quicker or better way than you see here, by all means go ahead! But always check your work to be sure your "alternative" answer is correct.

Chapter 11

1. We can take a brute-force approach and multiply all the denominators together, getting $2 \times 4 \times 6 = 48$, and then multiply the entire equation through by that number. We get:

$$(7/2) \times 48 = (14/4) \times 48 = (21/6) \times 48$$

which simplifies to

$$168 = 168 = 168$$

A more elegant way (at least in this situation) is to multiply the original equation through by 2, getting

$$14/2 = 28/4 = 42/6$$

Then, dividing the fractions out, we get

$$7 = 7 = 7$$

2. We can do the same thing as we did in the previous solution, and then multiply the three-way equations through by −1, getting

$$-168 = -168 = -168$$

or

$$-7 = -7 = -7$$

3. If we multiply an equation through by the number 0, we will always get a statement to the effect that 0 equals itself. That's true, but it's trivial and isn't good for much of anything. We might multiply an equation through by a variable or expression that's equal to 0, even though we aren't aware of it at the time. That's likely to make the equation more complicated, but it won't make it false. Consider this:

$$x = 2$$

Let's multiply this through by $(2 - x)$. We get

$$x(2 - x) = 2(2 - x)$$

which expands to

$$2x - x^2 = 4 - 2x$$

In this case, our manipulation does us no good. But it does no real harm either, as the inadvertent division by 0 can do. Occasionally, a manipulation like this can put a complicated equation into a form that's easier to work with.

4. Let's call the set of negative integers Z_-. Remember the standard name for the set of natural numbers; it's N. We have

$$Z_- = \{..., -5, -4, -3, -2, -1\}$$

and

$$N = \{0, 1, 2, 3, 4, 5, ...\}$$

From these statements, we can see that any negative integer we choose will be smaller than any natural number we choose. Therefore, if x is an element of Z_- and y is an element of N, x is smaller than y. In logical form along with set notation, we can write this as

$$[(x \in Z_-) \;\&\; (y \in N)] \Rightarrow x < y$$

5. Let's call the set of nonpositive reals R_{0-} (for "0 and all the negative reals") and the set of nonnegative reals R_{0+} (for "0 and all the positive reals"). Both of these sets include 0, but that's the only element they share. Therefore, any nonpositive real number we choose must be smaller than or equal to any nonnegative real number we choose. In other words, if x is an element of R_{0-} and y is an element of R_{0+}, then x is smaller than or equal to y. In logical form along with set notation, we can write this as

$$[(x \in R_{0-}) \;\&\; (y \in R_{0+})] \Rightarrow x \leq y$$

6. We have standard names for the sets of rational and irrational numbers: Q and S, respectively. These sets are disjoint; they have no elements in common. If x is an element of Q and y is an element of S, then x is never equal to y. In logical form along with set notation, we can write

$$[(x \in Q) \,\&\, (y \in S)] \Rightarrow x \neq y$$

7. We can write the statement as "mathematical verse" by reading it out loud and taking careful note of each symbol. Here's the logical statement again, for reference.

$$(\forall\ a, b, c) : [(a \geq b) \,\&\, (b \leq c)] \Rightarrow (a = c)$$

When we break up the statement into parts and write them down on separate lines, we come up with the following:

For all *a*, *b*, and *c*:
If
a is larger than or equal to *b*,
and
b is smaller than or equal to *c*,
then
a is equal to *c*.

This little poem might be cute, but it doesn't state a valid mathematical law. Suppose that $a = 5$, $b = 3$, and $c = 7$. In that case, a is larger than or equal to b and b is smaller than or equal to c. However, a is not equal to c.

8. Our task is to simplify the equation to a form where x appears all by itself on the left side of the equality symbol, and a plain numeral appears all by itself on the right. Here's the equation again, for reference:

$$x + 4 = 2x$$

We can subtract x from both sides, getting

$$x + 4 - x = 2x - x$$

which simplifies to

$$4 = x$$

We can reverse the order to get the solution in its proper form:

$$x = 4$$

The original equation holds true only when x is equal to 4.

9. We must simplify the inequality so that y appears all by itself on the left side of the "not equal" symbol, and a plain numeral appears all by itself on the right. Here's the inequality again, for reference:

$$y/2 \neq 4y + 7$$

First, let's multiply through by 2. That gives us

$$(y/2) \times 2 \neq (4y + 7) \times 2$$

which multiplies out to

$$y \neq 8y + 14$$

Now, let's subtract $8y$ from each side. That gives us

$$y - 8y \neq 8y + 14 - 8y$$

which simplifies to

$$-7y \neq 14$$

We can divide this through by −7 to get

$$(-7y)/(-7) \neq 14/(-7)$$

which simplifies to

$$y \neq -2$$

The original inequality holds true for all values of y except −2.

10. We must simplify the inequality so that z appears all by itself on the left side of the "smaller than or equal" symbol, and a plain numeral appears all by itself on the right. Here's the inequality again, for reference:

$$z/(-3) \leq 6z + 6$$

Let's multiply through by −3, remembering that we must reverse the sense of the inequality whenever we multiply through by a negative. That gives us

$$[z/(-3)] \times (-3) \geq (6z + 6) \times (-3)$$

which simplifies to

$$z \geq -18z - 18$$

If we add $18z$ to each side, we get

$$z + 18z \geq -18z - 18 + 18z$$

which simplifies to

$$19z \geq -18$$

We finish by dividing each side by 19. That leaves us with

$$z \geq -18/19$$

The original inequality holds true for all values of z larger than or equal to $-18/19$.

Chapter 12

1. See Table B-1.
2. See Table B-2.
3. See Table B-3.

Table B-1. Solution to Prob. 1 in Chap. 12.

Statements	Reasons
$4x + 4 = 2x - 2$	This is the equation we are given
$2x + 4 = -2$	Subtract $2x$ from each side
$2x + 6 = 0$	Add 2 to each side

Table B-2. Solution to Prob. 2 in Chap. 12.

Statements	Reasons
$x/3 = 6x + 2$	This is the equation we are given
$x = 3(6x + 2)$	Multiply through by 3
$x = 18x + 6$	Distributive law applied to the right side
$-17x = 6$	Subtract $18x$ from each side
$-17x - 6 = 0$	Subtract 6 from each side
$17x + 6 = 0$	Multiply through by -1 and apply the distributive law to the right side, obtaining a more elegant equation

Table B-3. Solution to Prob. 3 in Chap. 12.

Statements	Reasons
$x - 7 = 7x + x/7$	This is the equation we are given
$7(x - 7) = 7(7x + x/7)$	Multiply through by 7
$7x - 49 = 49x + x$	Apply distributive law to each side
$7x - 49 = 50x$	Simplify the right side
$-43x - 49 = 0$	Subtract $50x$ from each side
$43x + 49 = 0$	Multiply through by -1 and apply the distributive law to the right side, obtaining a more elegant equation

4. We are given the equation

$$x/3 + x/6 = 12$$

Multiplying through by 6, we get

$$6(x/3 + x/6) = 72$$

When we apply the distributive law to the left side and then simplify the addends, we obtain

$$2x + x = 72$$

Simplifying the left side further, we get

$$3x = 72$$

Subtracting 72 from each side yields

$$3x - 72 = 0$$

Finally, we can divide through by 3 to obtain lowest terms:

$$x - 24 = 0$$

5. When we have a first-degree equation in standard form, the solution process never takes more than two steps. Let's solve the results of Probs. 1 through 4

For Prob. 1 (Table B-1). The standard-form equation we got was

$$2x + 6 = 0$$

We subtract 6 from each side, getting

$$2x = -6$$

Then we divide through by 2, getting

$$x = -3$$

For Prob. 2 (Table B-2). The standard-form equation we got was

$$17x + 6 = 0$$

We subtract 6 from each side, getting

$$17x = -6$$

Then we divide through by 17, getting

$$x = -6/17$$

For Prob. 3 (Table B-3). The standard-form equation we got was

$$43x + 49 = 0$$

We subtract 49 from each side, getting

$$43x = -49$$

Then we divide through by 43, getting

$$x = -49/43$$

For Prob. 4. The standard-form equation we got was

$$x - 24 = 0$$

We add 24 to each side, getting

$$x = 24$$

In this case, we need not multiply or divide through by anything, because x is already by itself on the left side of the equation.

6. If we call the unknown number x, then we have the following equation to solve:

$$(2x + 8)/4 = -1$$

Multiplying through by 4, we get

$$2x + 8 = -4$$

Subtracting 8 from each side gives us

$$2x = -12$$

Dividing through by 2 produces

$$x = -6$$

7. If we call the unknown number x, then we have the following equation to solve:

$$(x - x/10)/2 = 135$$

Multiplying through by 2 gives us

$$x - x/10 = 270$$

Note that $x - x/10$ is the same as $x - (1/10)x$, or $(9/10)x$. So we have

$$(9/10)x = 270$$

Multiplying through by 10/9 gives us

$$(10/9)(9/10)x = 10/9 \times 270$$

which simplifies to

$$x = 300$$

8. Let's call Bonnie's weight, in kilograms (kg), the unknown x. Then Bruce's weight in kilograms is $x + 5$, and Bill's weight in kilograms is $(x + 5) + 10$, or $x + 15$. The total weight is 200 kg. We can now write the equation

$$x + (x + 5) + (x + 15) = 200$$

Applying the commutative law and adding the constants on the left side gives us

$$x + x + x + 20 = 200$$

which simplifies to

$$3x + 20 = 200$$

Subtracting 20 from each side gives us

$$3x = 180$$

Dividing through by 3, we get

$$x = 60$$

That means Bonnie weighs 60 kg. Bruce weighs 5 kg more than Bonnie, or 65 kg. Bill weighs 10 kg more than Bruce, or 75 kg.

9. We can work through this without resorting to equations. It's first-degree algebra in a single variable, but we don't have to express it symbolically! Let's start by figuring out the land speed of the boat as it traveled upstream. The distance from our cabin to our cousin's cabin is 18 miles (mi), and it took 1 h 12 min for the boat to travel that far. Because 12 min = 1/5 h, the trip took 6/5 h. The speed of the boat relative to the land was therefore 18 mi per 6/5 h, or

$$\begin{aligned} 18/(6/5) &= 18 \times 5/6 \\ &= 15 \text{ mi/h} \end{aligned}$$

If there were no current, the boat would have traveled at 18 mi/h relative to the water and relative to the land. The water speed was indeed 18 mi/h, but the land speed was only 15 mi/h. The current slowed our boat down by 18 − 15, or 3 mi/h. The river must therefore have been flowing at 3 mi/h.

10. When we go downstream, the river current will add 3 mi/h to our boat's land speed. Again, if there were no current, the land and water speeds would both be 18 mi/h. That means the downstream speed of the boat will be 18 + 3, or 21 mi/h. We must travel a land distance of 18 mi. Time equals distance divided by speed. Therefore, if we let t represent the time in hours, we have

$$\begin{aligned} t &= 18/21 \\ &= 6/7 \text{ h} \end{aligned}$$

Because 1 h = 60 min, 6/7 hour = 6/7 × 60 min, or, approximately, 51.43 min.

Chapter 13

1. For every boy there is exactly one girl dance partner and vice-versa, so the mapping is an injection. It's also onto the entire set of girls. No girl has to sit out the dance without a partner, so it's a surjection. Any mapping that's both an injection and a surjection is, by definition, a bijection.

2. This mapping is not one-to-one, so it's not an injection. That means it can't be a bijection. However, it's a surjection, because it's onto the entire set of girls.

3. This mapping, like the one in the Prob. 2, is not one-to-one, so it can't be an injection or a bijection. But it's a surjection, because it's onto the entire set of boys.

4. Our mapping was not an injection, because it was not one-to-one. If it wasn't an injection, it couldn't have been a bijection. It was a surjection from A to B, because we mapped our message onto the entire set B, the set of all 175,000 subscribers to Internet Network Beta. The maximal domain was A, the set of all 60,000 subscribers to Internet Network Alpha (before we were kicked out). The essential domain was the set containing only you and me. The co-domain and the range were both the whole set B.

5. This relation is one-to-one. For every integer q, there is exactly one even integer r, which we can get by doubling q. Conversely, for every even integer r, there is exactly one integer q, which we get when we divide y by 2. Our relation is therefore an injection from Q to R. It is not onto R, however, because there are plenty of real numbers that are not even integers. That means we do not have a surjection from Q onto R. If it's not a surjection, then it can't be a bijection.

6. This relation is not one-to-one. For example, $q = 3/7$ and $q = 3/11$ both map into $z = 3$. That means it's not an injection, so it cannot be a bijection, either. It is a surjection, because it's onto the entire set Z of integers. We can choose any integer z, place it into the numerator of a fraction, choose a denominator such that the fraction is in lowest terms, and have a rational number q that maps to z.

7. This relation is not quite one-to-one. For every nonzero integer z, there is exactly one rational number $q = 1/z$. But if $z = 0$, there is no q such that $q = 1/z$. The relation is therefore not injective nor bijective. It is not surjective, either. There are plenty of rational numbers that are not reciprocals of integers; 3/7 is an example.

8. This relation is one-to-one. We "patched the hole" in the relation defined in Prob. 7. We declared that if $z = 0$, then $q = 0$. There is no nonzero integer z such that $1/z = 0$, so we don't get a "dupe" by making this declaration. We have an injective relation now! But as with the relation in Prob. 7, we do not have a surjection. That means the relation is not bijective.

9. This relation is a function. For any value of x we choose, there is exactly one value of y such that $y = x^4$. It's not one-to-one between the set of reals and the set of nonnegative reals, but two-to-one except when $x = 0$. That means the function is not injective, so it can't be bijective, either. It's surjective, because it's onto the entire set of nonnegative reals. For any nonnegative real number y, we can find a real number x such that $y = x^4$.

10. When we transpose the values of the independent and dependent variables while leaving their names the same, we get

$$x = y^4$$

which is equivalent to

$$y = \pm(x^{1/4})$$

The plus-or-minus symbol indicates that for every nonzero x, there are two values of y, one positive and the other negative. This relation is one-to-two except when $x = 0$, so it is not a function. It is not an injection because it's not one-to-one. That means it cannot be a bijection. The relation does map onto its entire range (the set of all reals), so it's a surjection.

Chapter 14

1. If we multiply x by -1 and leave y the same, the point will move to the other side of the y axis, but it will stay on the same side of the x axis. If it starts out in the first quadrant, it will move to the second. If it starts out in the second quadrant, it will move to the first. If it starts out in the third quadrant, it will move to the fourth. If it starts out in the fourth quadrant, it will move to the third. The y axis will act as a "point reflector."

2. If we multiply y by -1 and leave x the same, the point will move to the other side of the x axis, but it will stay on the same side of the y axis. If it starts out in the first quadrant, it will move to the fourth. If it starts out in the second quadrant, it will move to the third. If it starts out in the third quadrant, it will move to the second. If it starts out in the fourth quadrant, it will move to the first. The x axis will act as a "point reflector."

3. The point for $(6x, 6y)$ will be in the same quadrant as the point for (x, y), but 6 times as far from the origin. The point for $(x/4, y/4)$ will be in the same quadrant as the point for (x, y), but 1/4 of the distance from the origin. The origin, the point for (x, y), the point for $(6x, 6y)$, and the point for $(x/4, y/4)$ will all lie along a single straight line.

4. If the vertical test line intersects the graph (once for a function, and once or more for a relation), then the point where the test line intersects the independent-variable (horizontal) axis represents a numerical value in the domain. If the test line does not intersect the graph, then the point where the test line intersects the horizontal axis does not represent a value in the domain.

5. This process works just like the process for determining whether or not a point is in the domain, except that everything is rotated by 90°! If the horizontal test line intersects the graph, then the point where the test line intersects the dependent-variable (vertical) axis represents a numerical value in the range. If the test line does not intersect the graph, then the point where the test line intersects the vertical axis does not represent a value in the range.

6. We can plot several specific points for $y = |x|$, and then we can determine the graph on the basis of those points. We can deduce, using some common sense, that the lines are straight. See Fig. B-1. The vertical-line test tells us that this is a function of x.

7. We can plot several specific points for $y = |x + 1|$, and then we can determine the graph on that basis. Again, we can deduce, using some common sense, that the lines are straight. See Fig. B-2. The vertical-line test reveals that this is a function of x.

8. Figure B-3 is a graph of the inverse of $y = x + 1$. If we apply the "point reflector" method to Fig. 14-11, we don't have to mathematically derive an equation for the inverse to figure out what its graph looks like.

9. Figure B-4 is a graph of the inverse of $w = v^2$. This is the result of using the "point reflector" scheme to modify Fig. 14-12. Again, it is not necessary to derive an equation for the inverse.

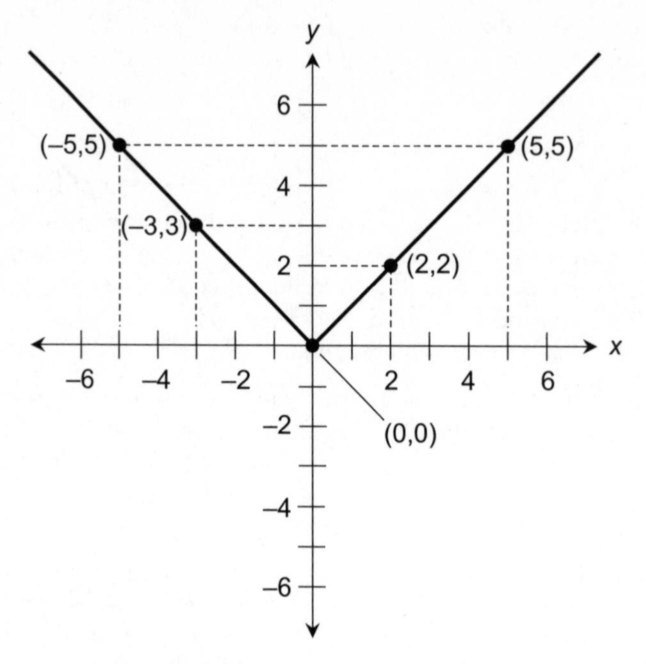

Figure B-1 Illustration for the solution to Prob. 6 in Chap. 14.

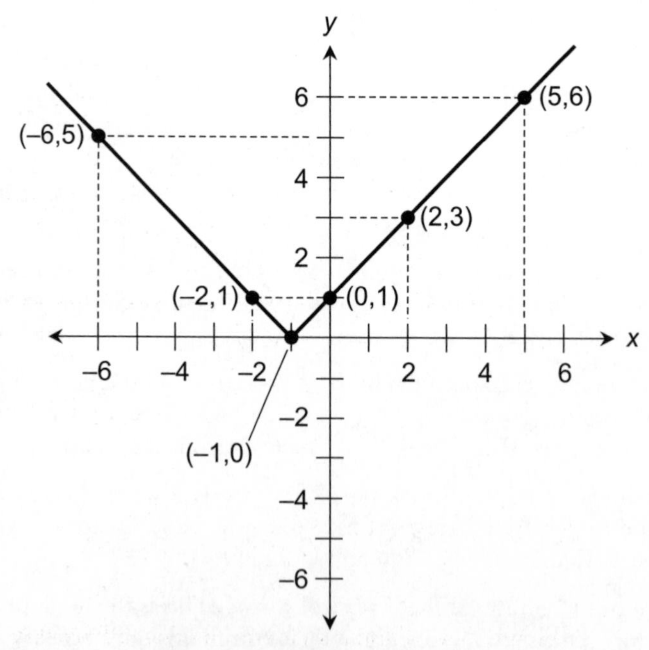

Figure B-2 Illustration for the solution to Prob. 7 in Chap. 14.

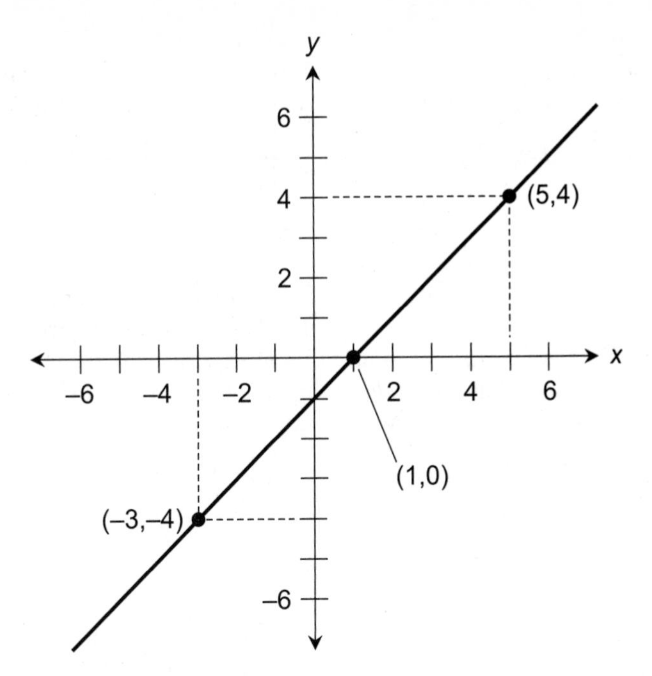

Figure B-3 Illustration for the solution to Prob. 8 in Chap. 14.

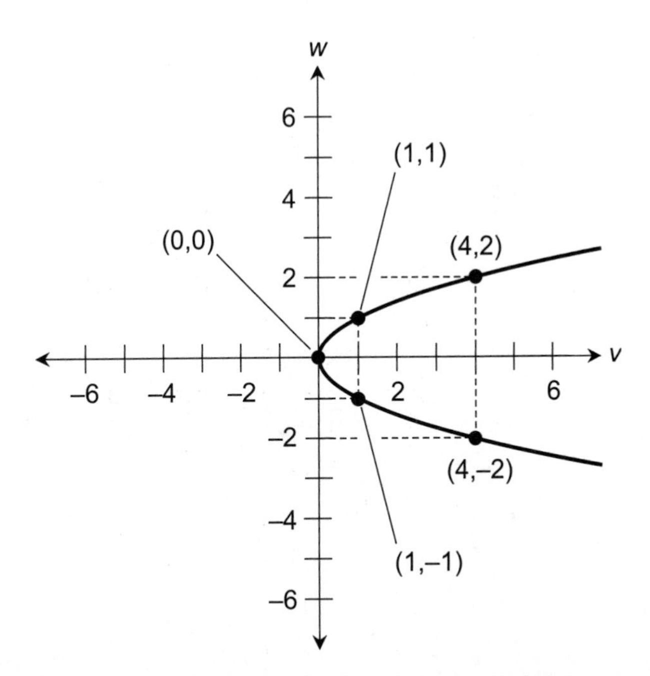

Figure B-4 Illustration for the solution to Prob. 9 in Chap. 14.

10. When the scale increments in a graph are not the same (as is the case in Fig. 14-13), we cannot use the "point reflector" scheme directly to see what the inverse graph looks like. It's better to derive an equation for the inverse relation, and then plot its graph on the basis of that equation. We want to derive the inverse of $u = t^3$. We begin by transposing the values of the variables without changing their names, getting

$$t = u^3$$

We can take the cube root of both sides of the equation here, and we don't run any risk of ambiguity. That's because the original function is a bijection. Remember what that means: Every value in the domain has exactly one "mate" in the range, and vice-versa. There's no chance for confusion or duplicity as there would be if we were dealing with an even-numbered power of *u*. When we take the cube root of both sides, we get

$$\pm t^{1/3} = u$$

Reversing the sense gives us

$$u = \pm t^{1/3}$$

This is the inverse of $u = t^3$. Figure B-5 is a graph of this relation. Note that the increments of the scales have been transposed, as well as the points in the graph, so the graph will fit neatly into the available space.

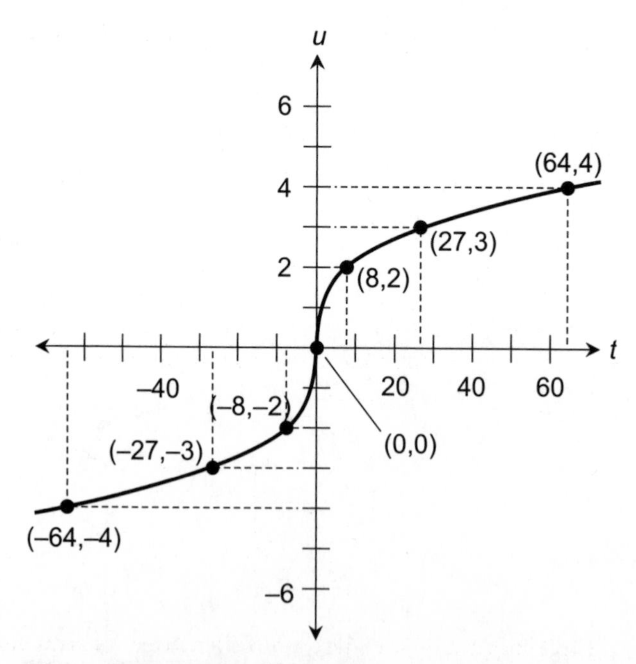

Figure B-5 Illustration for the solution to Prob. 10 in Chap. 14.

Chapter 15

1. We have these two ordered pairs defining the points P and Q, respectively:

$$P = (u_1, v_1) = (-1, -6)$$

and

$$Q = (u_2, v_2) = (2, 2)$$

The slope m is equal to the difference in the dependent-variable coordinates divided by the difference in the independent-variable coordinates, or $\Delta v/\Delta u$. If we move along the line from P to Q, we find the slope on the basis of the ratio between the differences in the point values with the "destination" values listed first:

$$m = (v_2 - v_1) / (u_2 - u_1)$$

Plugging in the values $v_2 = 2$, $v_1 = -6$, $u_2 = 2$, and $u_1 = -1$, we get

$$\begin{aligned} m &= [2 - (-6)] / [2 - (-1)] \\ &= (2 + 6) / (2 + 1) \\ &= 8/3 \end{aligned}$$

2. We still have the same two points, defined by the same two ordered pairs. Points P and Q, respectively, are still defined by

$$P = (u_1, v_1) = (-1, -6)$$

and

$$Q = (u_2, v_2) = (2, 2)$$

If we want to go from Q to P rather than from P to Q, we must reverse the order of v_1 and v_2 in the numerator of the slope equation, and we must also reverse the order of u_1 and u_2 in the denominator. When we do that, we get

$$\begin{aligned} m &= (v_1 - v_2) / (u_1 - u_2) \\ &= (-6 - 2) / (-1 - 2) \\ &= -8 / (-3) \\ &= 8/3 \end{aligned}$$

The slope in either direction is equal to the difference in the v values divided by the difference in the u values, or $\Delta v/\Delta u$. Reversing the direction in which we move along the line simply multiplies both Δv and Δu by -1. The ratio turns out the same either way.

3. We know the coordinates of at least one point (two, actually) and we also know the slope. Let's use the point (2, 2) as the starting basis. We've determined that the slope is 8/3. The general PS equation, using u and v as the variable names rather than the familiar x and y, is

$$v - v_0 = m(u - u_0)$$

Plugging in 2 for v_0, 8/3 for m, and 2 for u_0, we have

$$v - 2 = (8/3)(u - 2)$$

That's the PS form of the equation.

4. To get the equation in SI form, we can manipulate the PS equation we obtained in the previous solution. That equation, again, is

$$v - 2 = (8/3)(u - 2)$$

Using the distributive law for multiplication over subtraction on the right-hand side, we obtain

$$v - 2 = (8/3)u - 16/3$$

Adding 2 to the left side, and adding 6/3 (which is equal to 2) to the right side, we get

$$v = (8/3)u - 16/3 + 6/3$$

which simplifies to

$$v = (8/3)u - 10/3$$

That's the SI form of the equation.

5. The simplest possible way to graph this equation is to plot the two points we were originally given, and then draw a straight line through them. This is done in Fig. B-6. The slope, m, is 8/3 as we derived it. The v-intercept, b, is −10/3 as we derived it.

6. The first equation is in SI form. We can tell, by looking at this equation, that the slope of the line will be 1 when we graph it. The second equation can be put into SI form by considering the subtraction of s as the addition of its negative, getting

$$t = 5 + (-s)$$

We can apply the commutative law on the right side to get

$$t = -s + 5$$

Now we can see that the slope of this line will be −1 when we graph it.

In a Cartesian plane where both axes are graduated in increments of the same size, a slope of 1 corresponds to a ramp angle of 45°, and a slope of −1 corresponds to a ramp

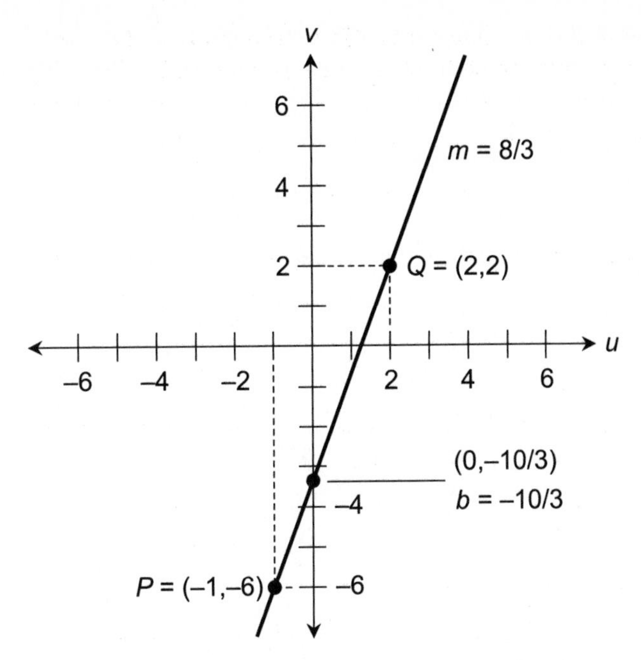

Figure B-6 Illustration for the solution to Prob. 5 in Chap. 15.

angle of –45°. That means the first line will go "uphill" at 45° as we go to the right, and the second line will go "downhill" at 45° as we go to the right. The angle between the two lines will therefore be 45 + 45, or 90°.

If the increments on the s axis are not the same size as those on the t axis, then slopes of 1 and –1 will not appear as "uphill" and "downhill" 45° angles. Depending on which axis has the larger increments, both lines will be either steeper or less steep. Our advisor, who claimed that the lines would intersect at a 90° angle, will be mistaken if we draw the lines on a coordinate system having axes graduated in unequal increments.

7. To determine the point where the two lines intersect, we must find an ordered pair of the form (s, t) that satisfies both equations. Look at the SI forms of the two equations again:

$$t = s + 5$$

and

$$t = -s + 5$$

If we add the left sides of these equations, we get $t + t$, which is equal to $2t$. If we add the right sides, we get $s + 5 + (-s) + 5$, which is equal to 10. That means the sum of the two equations is

$$2t = 10$$

Dividing through by 2, we get $t = 5$. That's the t value of the intersection point. We can plug 5 in for t in either of the original equations to solve for s. Let's use the first one. We then get

$$5 = s + 5$$

It's not too difficult to tell from this equation that $s = 0$. Now we know that $s = 0$ and $t = 5$, so $(s, t) = (0, 5)$ defines the point where the two lines intersect. That point lies on the t axis, because the s coordinate is equal to 0.

8. Figure B-7 shows the graphs of the two lines, based on their known slopes and t-intercepts. The intersection point is, coincidentally, at the t-intercept for both lines. The lines intersect at a 90° angle because the axes are graduated in equal increments.

9. For reference, here's the general two-point equation we derived for a line in Cartesian coordinates:

$$y - y_1 = (x - x_1)(y_2 - y_1) / (x_2 - x_1)$$

where points are represented by ordered pairs (x_1, y_1) and (x_2, y_2). We are told that (2, 8) and (0, −4) both lie on the graph. Let's assign $x_1 = 2$, $x_2 = 0$, $y_1 = 8$, and $y_2 = -4$. When we plug these numbers into the above equation, we get

$$y - 8 = (x - 2)(-4 - 8) / (0 - 2)$$

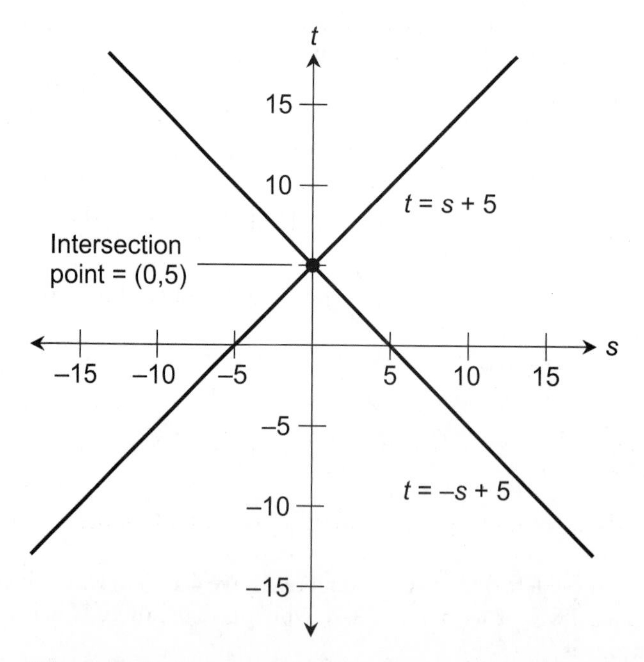

Figure B-7 Illustration for the solution to Prob. 8 in Chap. 15.

which simplifies to

$$y - 8 = (x - 2) \times (-12) / (-2)$$

and further to

$$y - 8 = 6(x - 2)$$

That's the PS form of the equation.

10. Once again, here's the general two-point equation:

$$y - y_1 = (x - x_1)(y_2 - y_1) / (x_2 - x_1)$$

This time, we're told that (−6, −10) and (6, −12) lie on the graph. Let's assign $x_1 = -6$, $x_2 = 6$, $y_1 = -10$, and $y_2 = -12$. When we plug in these numbers, we get

$$y - (-10) = [x - (-6)][-12 - (-10)] / [6 - (-6)]$$

This "nightmare of negatives" simplifies to

$$y + 10 = (x + 6) \times (-2) / 12$$

and further to

$$y + 10 = (-1/6)(x + 6)$$

We want the SI form, so we have a little more manipulation to do. Applying the distributive law of multiplication over addition to the right side, we get

$$y + 10 = (-1/6)x - 1$$

Subtracting 10 from each side produces the desired result:

$$y = (-1/6)x - 11$$

That's the SI form of the equation.

Chapter 16

1. Let's call the numbers x and y. We're told that both of the following facts are true:

$$x + y = 44$$

and

$$x - y = 10$$

Let's get the equations into SI form. In the first equation, we can subtract x from each side to get

$$y = -x + 44$$

In the second equation, we can subtract x from each side and then multiply through by -1 to obtain

$$y = x - 10$$

Mixing the right sides of these two SI equations produces this:

$$-x + 44 = x - 10$$

Adding 10 to each side gives us

$$-x + 54 = x$$

Adding x to each side, we get

$$54 = 2x$$

Dividing through by 2, we determine that $x = 27$. We can plug this into either of the SI equations to solve for y. Let's use the second one. We have

$$\begin{aligned} y &= x - 10 \\ &= 27 - 10 \\ &= 17 \end{aligned}$$

The two numbers are 27 and 17.

2. Again, let's call the numbers x and y. We are told that these two facts are true:

$$x + y = 100$$

and

$$y = 6x$$

Actually, we could just as well say that $x = 6y$; it doesn't matter. Let's stick with the equations above. The first equation can be put into SI form by subtracting x from each side. That gives us

$$y = -x + 100$$

The second equation is already in SI form (the y-intercept is 0). Mixing the right-hand sides, we obtain

$$-x + 100 = 6x$$

Adding x to each side produces

$$100 = 7x$$

Dividing through by 7, we find that $x = 100/7$. We can plug this into the second original equation to get

$$\begin{aligned} y &= 6 \times 100/7 \\ &= 600/7 \end{aligned}$$

The two numbers are 100/7 and 600/7. We can also express them in whole-number-and-fraction form as 14-2/7 and 85-5/7.

3. The process for solving this problem is rather long and a little tricky as well! Let x be the speed of the ball relative to the car. Let y be the speed of the car relative to the pavement.

   When you throw the first baseball straight out in front of the car, the ball's speed adds to the car's speed, so the ball moves at a speed of $x + y$ relative to the pavement. That's simple enough! When you throw the second ball straight backward, the ball's speed subtracts from the car's speed, so the ball moves at a speed of $y - x$ relative to the pavement.

   When the second ball hits the pavement, it's moving backward, opposite to the motion of the car. Therefore, we must consider the direction of the motions relative to the pavement. Let's define forward motion relative to the pavement (that is, in the direction of the car) as *positive speed*, and backward motion relative to the pavement (opposite to the car's motion) as *negative speed*. Keep in mind that these definitions apply only to motions that are observed with respect to the pavement.

   The equations describing the movement of the ball relative to the pavement can be written out:

$$x + y = 135$$

   when for the ball you throw straight out in front of the car, and

$$y - x = -15$$

   for the ball you throw straight out behind the car. The speed in the second case is negative because the ball hits the pavement moving backward. When we morph these two equations into SI form, we obtain

$$y = -x + 135$$

   and

$$y = x - 15$$

   When we mix the right sides, we get

$$-x + 135 = x - 15$$

Adding 15 to each side gives us

$$-x + 150 = x$$

Adding x to each side, we get

$$150 = 2x$$

Therefore, $x = 150/2 = 75$ mi/h. That's the speed of each ball relative to the car. (You have a pretty good throwing arm, considering you're sitting in a car seat and throwing balls out of an open window!) When we plug this value for x into the second SI equation, we get

$$\begin{aligned} y &= x - 15 \\ &= 75 - 15 \\ &= 60 \end{aligned}$$

That means the car is moving at 60 mi/h relative to the pavement—in a forward direction, of course.

4. Let's call the numbers x and y. We are told that both of the following facts are true:

$$x + y = -83$$

and

$$x - y = 13$$

These equations are in the same form, so we're ready to go. We can multiply the first equation through by −1, getting

$$-x - y = 83$$

We add this to the second original equation:

$$\begin{array}{r} -x - y = 83 \\ x - y = 13 \\ \hline -2y = 96 \end{array}$$

This tells us that $y = 96/(-2) = -48$. Now let's add the two original equations directly:

$$\begin{array}{r} x + y = -83 \\ x - y = 13 \\ \hline 2x = -70 \end{array}$$

This tells us that $x = -70/2 = -35$.

5. Let's state the two equations again for reference, and then try to solve them using double elimination:

$$2x + y = 3$$

and

$$6x + 3y = 12$$

Let's eliminate x. We can multiply the first equation through by -3 to get

$$-6x - 3y = -9$$

Here's what happens when we add this to the second original equation:

$$\begin{array}{r} -6x - 3y = -9 \\ 6x + 3y = 12 \\ \hline 0 = 3 \end{array}$$

That's nonsense! No matter what other method we use in an attempt to solve this system, we'll arrive at some sort of contradiction. When this happens with a two-by-two linear system, the system is said to be *inconsistent.* (Most two-by-two linear systems are *consistent,* meaning that they have a single solution that can be expressed as an ordered pair.) Nothing is technically wrong with either equation here. They simply don't get along together. Inconsistent linear systems have no solutions.

6. Let's put the two equations from Prob. 5 into SI form, and see if that tells us anything about what their graphs look like. First, this:

$$2x + y = 3$$

When we subtract $2x$ from each side, we get

$$y = -2x + 3$$

This indicates that the slope of the graph, which is a straight line, is -2. The y-intercept is 3. Now for the second equation:

$$6x + 3y = 12$$

When we subtract $6x$ from each side, we get

$$3y = -6x + 12$$

We can divide through by 3 to obtain

$$y = -2x + 4$$

The slope of this graph is −2, the same as the slope of the graph of the other equation. But the y-intercept is 4, and that's different. If we plot the graphs, we get two parallel lines. On the Cartesian plane, two lines have the same slope but different y-intercepts if and only if they're parallel. Now remember from plane geometry: parallel lines do not intersect. That means they have no point in common. When two parallel lines appear on the Cartesian plane, no ordered pair (x,y) can give us a point that falls on both lines, so no ordered pair (x,y) can satisfy both equations.

7. Let's put the two equations from Probs. 5 and 6 into the format we used to solve the challenge in the section "Double Elimination." Here again are those general equations:

$$ax + by = c$$

and

$$dx + ey = f$$

where x and y are the variables, and a through f are constants. Here are the two equations from Probs. 5 and 6 that we could not solve as a linear system:

$$2x + y = 3$$

and

$$6x + 3y = 12$$

We have $a = 2$, $b = 1$, $d = 6$, and $e = 3$. Therefore,

$$\begin{aligned} ae &= 2 \times 3 \\ &= 6 \end{aligned}$$

and

$$\begin{aligned} bd &= 1 \times 6 \\ &= 6 \end{aligned}$$

Now remember that in the general derivation, we were not allowed to let $ae = bd$, because that would cause us to divide by 0 in the course of trying to solve the system. You've already seen some of the bad things that can happen when we divide by 0 directly or indirectly, knowingly or unknowingly. Let this serve as another example!

8. Let's call the numbers x and y, as we did in Prob. 1. The substitution process is similar to the morph-and-mix process, at least for this system. Here are the equations again:

$$x + y = 44$$

and

$$x - y = 10$$

In the first equation, we can subtract x from each side to get

$$y = -x + 44$$

We can substitute the quantity $(-x + 44)$ for y in the second original equation, getting

$$x - (-x + 44) = 10$$

This can be rewritten as

$$x + [-1(-x + 44)] = 10$$

and simplified to

$$x + x - 44 = 10$$

When we add 44 to each side and note that $x + x = 2x$, we obtain

$$2x = 54$$

This tells us that $x = 27$. Now we can plug this into the SI equation and solve for y, as follows:

$$\begin{aligned} y &= -x + 44 \\ &= -27 + 44 \\ &= 17 \end{aligned}$$

When we check back and compare this with solution to Prob. 1, we see that the answers are the same: $x = 27$ and $y = 17$.

9. Let's call the numbers x and y, as we did in Prob. 4. Here again are the equations that we must solve as a two-by-two linear system:

$$x + y = -83$$

and

$$x - y = 13$$

The first equation can be put into SI form if we subtract x from each side. That gives us

$$y = -x - 83$$

When we substitute $(-x - 83)$ for y in the second equation, we get

$$x - (-x - 83) = 13$$

This can be rewritten as

$$x + [-1(-x - 83)] = 13$$

and simplified to

$$x + x + 83 = 13$$

When we subtract 83 from each side and note that $x + x = 2x$, we get

$$2x = -70$$

Therefore, $x = -70/2 = -35$. We can replace x with -35 in the SI equation above to solve for y, getting

$$\begin{aligned} y &= -x - 83 \\ &= -(-35) - 83 \\ &= 35 - 83 \\ &= -48 \end{aligned}$$

When we check back and compare this with solution to Prob.4, we see that the answers are the same: $x = -35$ and $y = -48$.

10. Here are the equations again, for reference:

$$s = 2r - 3$$

and

$$-10r + 5s + 15 = 0$$

This pair of equations appears well suited to a solution by substitution. The first equation gives us s directly in terms of r. Let's replace s by $(2r - 3)$ in the second equation. We get

$$-10r + 5(2r - 3) + 15 = 0$$

The distributive law allows us to morph the left side of this equation into a straight sum:

$$-10r + 10r - 15 + 15 = 0$$

which simplifies to

$$0 = 0$$

This statement is true, so we can't claim a contradiction. But it's useless for solving this system. (If we try any other method to solve it, we'll encounter a similar barrier.) The trouble becomes clear if we solve the second original equation for s directly in terms of r. We start with

$$-10r + 5s + 15 = 0$$

Subtracting 15 from each side gives us

$$-10r + 5s = -15$$

When we add $10r$ to each side, we get

$$5s = 10r - 15$$

Finally, we can divide through by 5, obtaining

$$s = 2r - 3$$

That's identical to the first original equation! There are infinitely many solutions here; an infinite number of ordered pairs (r,s) will make the equation true. When a linear system consists of two equations that look different but can both be morphed to get a single equation, the system is said to be *redundant.* Some texts use the word *dependent* instead. A redundant linear system always has infinitely many solutions.

Chapter 17

1. In Fig. 17-9, line L passes through the points $(-3, 0)$ and $(0, 4)$. The y-intercept is equal to 4. We can travel along L from $(-3, 0)$ to $(0, 4)$ if we move to the right by $\Delta x = 3$ units and upward by $\Delta y = 4$ units. The slope is therefore

$$m = \Delta y / \Delta x$$
$$= 4/3$$

Now that we know the slope and the y-intercept for line L, we can write its SI equation as

$$y = (4/3)x + 4$$

2. In Fig. 17-9, line M passes through $(-3, 0)$ and $(0, -2)$. The y-intercept is -2. We can travel from $(-3, 0)$ to $(0, -2)$ if we move along the line to the right by $\Delta x = 3$ units and upward by $\Delta y = -2$ units (the equivalent of going downward by 2 units). The slope is therefore

$$m = \Delta y / \Delta x$$
$$= -2/3$$

We now have the slope and the y-intercept for M, so we can write its SI equation as

$$y = (-2/3)x - 2$$

3. We have the equations for lines L and M from Fig. 17-9 in SI form. Together, they constitute a two-by-two linear system:

$$y = (4/3)x + 4$$

and

$$y = (-2/3)x - 2$$

There's no morphing for us to do. So let's go ahead and mix the right sides of these two equations. We get

$$(4/3)x + 4 = (-2/3)x - 2$$

We can multiply this equation through by 3 to obtain

$$4x + 12 = -2x - 6$$

Adding $2x$ to each side gives us

$$6x + 12 = -6$$

Subtracting 12 from each side, we get

$$6x = -18$$

Dividing through by 6 tells us that $x = -18/6 = -3$. We can plug this value for x into either of the SI equations to solve for y. Let's use the first equation. We have

$$\begin{aligned} y &= (4/3)x + 4 \\ &= (4/3) \times (-3) + 4 \\ &= -12/3 + 4 \\ &= -4 + 4 \\ &= 0 \end{aligned}$$

Having found the solution $x = -3$ and $y = 0$, we can state it as the ordered pair $(-3, 0)$. This represents the point where lines L and M intersect in Fig. 17-9.

4. Figure B-8 shows the transformation process, one step at a time, exactly as I did it using the rotate and mirror functions in my computer graphics program. At A, we see the original graph, identical to Fig. 17-9. At B, the entire coordinate grid, the lines, and the points have been rotated counterclockwise, all together, by 90°. Even the label characters have been rotated! At C, the graph from B has been mirrored. Even the label characters have been reversed! (That looks strange, but it can help us see what is taking place.) At D, everything has been relabeled to conform to the new coordinate system. The transposed lines are called L^* and M^*. The numbers in the ordered pairs have been transposed, because y is now the independent variable and x is the dependent variable.
5. In part D of Fig. B-8, line L^* passes through $(0, -3)$ and $(4, 0)$. The x-intercept is -3. Remember that x is now the dependent variable, so it's the x intercept, not the y-intercept, that concerns us. We can travel along L^* from $(0, -3)$ to $(4, 0)$

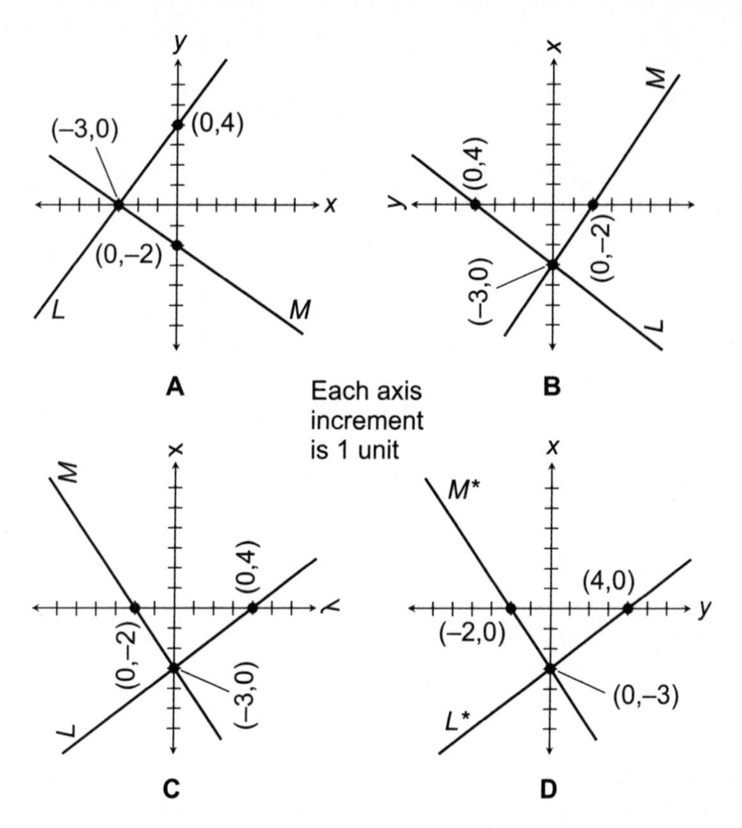

Figure B-8 Illustration for the solutions to Probs. 4 through 7 in Chap. 17. (The rotated and reversed characters are not typos! The text explains this.)

by moving to the right by $\Delta y = 4$ units and upward by $\Delta x = 3$ units. The slope is therefore

$$m = \Delta x / \Delta y$$
$$= 3/4$$

Remember that the slope of a line is the ratio of a change in the dependent variable to the change in the independent variable. That means the slope is now $\Delta x/\Delta y$, not $\Delta y/\Delta x$. We have determined the slope and the *x*-intercept for line L^*, so we can write its SI equation as

$$x = (3/4)y - 3$$

6. In part D of Fig. B-8, line M^* passes through (–2, 0) and (0, –3). The *x*-intercept is –3. When we go from (–2, 0) to (0, –3), we move to the right by $\Delta y = 2$ units and upward by $\Delta x = -3$ units (the equivalent of downward by 3 units). The slope is therefore

$$m = \Delta x / \Delta y$$
$$= -3/2$$

Now that we know the slope and the x-intercept for line M^*, we can write its SI equation as

$$x = (-3/2)y - 3$$

7. We have the equations for lines L^* and M^* from part D of Fig. B-8 in SI form. Together, they constitute a two-by-two linear system:

$$x = (3/4)y - 3$$

and

$$x = (-3/2)y - 3$$

There's no morphing to do here; these equations are ready to mix. When we combine the right sides, we get

$$(3/4)y - 3 = (-3/2)y - 3$$

We can multiply this through by 4 to obtain

$$3y - 12 = -6y - 12$$

Adding 12 to each side gives us

$$3y = -6y$$

When we add $6y$ to each side, we get

$$9y = 0$$

Dividing through by 9 tells us that $y = 0$. We can plug this value for y into either of the SI equations to solve for x. Let's use the first equation. We have

$$\begin{aligned} x &= (3/4)y - 3 \\ &= (3/4) \times 0 - 3 \\ &= 0 - 3 \\ &= -3 \end{aligned}$$

Having found the solution $y = 0$ and $x = -3$, we can state it as the ordered pair $(0,-3)$. This represents the point where lines L^* and M^* intersect in part D of Fig. B-8. Remember that this ordered pair is of the form (y,x), not (x,y). That's because in this situation, y is the independent variable and x is the dependent variable.

8. If a linear function has a graph with a slope of 0, then the inverse relation is not a function. That's because the graph of the inverse relation is a line parallel to the dependent-variable axis. The domain of that inverse relation is a single number, and

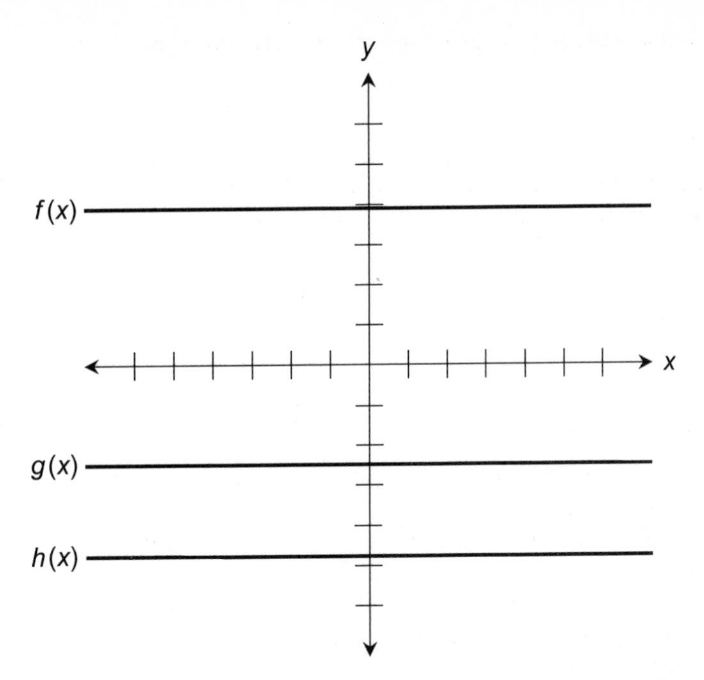

Figure B-9 Illustration for the solution to Prob. 9 in Chap. 17.

the range is the entire set of real numbers. That means there is more than one value in the range for the single value in the domain, causing the inverse to fail the "function test."

9. Figure B-9 shows graphs of three linear functions whose inverses are relations but not functions. We've given the original functions the arbitrary names $f(x)$, $g(x)$, and $h(x)$. (There are infinitely many other examples, of course.) The equations always take the form

$$y = 0x + b$$

where b is the y-intercept. The slope is always 0, and the y-intercept can be any real number. In these examples, x is the independent variable and y is the dependent variable. Axis increments are not indicated, because it doesn't matter what they are! When the variables are transposed to obtain the inverse relations, the lines become parallel to the dependent-variable axis. The inverse relations always fail the "function test," because there is more than one value in the range for the single value in the domain.

10. Let's start with the general SI form of a linear equation, as the hint suggests. The function can then be stated as

$$y = mx + b$$

with the understanding that $y = f(x)$. We want to morph this into SI form, treating y as the independent variable and x as the dependent variable. Let's begin by subtracting b from each side. That gives us

$$y - b = mx$$

Now let's switch the left and right sides of the equation to get

$$mx = y - b$$

We can divide through by m, provided $m \neq 0$, and then use the right-hand distributive law of division over subtraction to obtain

$$x = y/m - b/m$$

If we want it in strict SI form, we can rewrite it as

$$x = (1/m)y - b/m$$

Because $x = f^{-1}(y)$, we have

$$f^{-1}(y) = (1/m)y - b/m$$

The slope of the inverse function is $1/m$, and the x-intercept is $-b/m$. These values have meaning only when $m \neq 0$. But if $m \neq 0$, f^{-1} is always a function. A straight line with defined slope (that is, a nonvertical line) in Cartesian coordinates can *never* produce more than one value of the dependent variable for any single value of the independent variable. Draw some sample graphs, and you'll see why this is true. If you're really ambitious, you might try to formally prove it!

Chapter 18

1. Stated again for reference, the first and third revised equations are

$$-4x + 2y - 3z = 5$$

and

$$3x + 6y - 7z = 0$$

We can multiply the top equation through by 7 to get

$$-28x + 14y - 21z = 35$$

We can multiply the bottom equation through by -3 to get

$$-9x - 18y + 21z = 0$$

When we add these two new equations in their entirety, we obtain the sum

$$\begin{array}{r} -28x + 14y - 21z = 35 \\ -9x - 18y + 21z = 0 \\ \hline -37x - 4y = 35 \end{array}$$

2. The second and third revised equations we obtained in the section "Eliminate One Variable" were

$$2x - 5y - z = -1$$

and

$$3x + 6y - 7z = 0$$

The two-variable equation in x and y that we derived from these, as a result of eliminating the variable z, was

$$-11x + 41y = 7$$

If we take this equation together with the solution to Prob. 1, we have the two-by-two linear system

$$-37x - 4y = 35$$

and

$$-11x + 41y = 7$$

3. First, let's get rid of x. We can multiply the top equation through by -11 and the bottom equation through by 37. When we do these maneuvers and then add the resulting equations, we get

$$\begin{array}{r} 407x + 44y = -385 \\ -407x + 1{,}517y = 259 \\ \hline 1{,}561y = -126 \end{array}$$

Dividing through by 1,561 tells us that $y = -126\ /\ 1{,}561$, which reduces to $-18/223$. Now, let's get rid of y. We can multiply the top equation through by 41 and the bottom equation through by 4. When we do that and then add the results, we get

$$\begin{array}{r} -1{,}517x - 164y = 1{,}435 \\ -44x + 164y = 28 \\ \hline -1{,}561x = 1{,}463 \end{array}$$

We divide through by $-1{,}561$ to get $x = 1{,}463\ /\ (-1{,}561)$, which reduces to $-209/223$. The solution to this two-by-two linear system is therefore

$$x = -209/223$$

and

$$y = -18/223$$

These are the same solutions we obtained in the chapter text.

4. For reference, here again is the first equation stated in Prob. 1:

$$-4x + 2y - 3z = 5$$

The solutions for x and y can be plugged into this equation, and it can then be solved for z in steps, as follows:

$$-4 \times (-209/223) + 2 \times (-18/223) - 3z = 5$$
$$836/223 - 36/223 - 3z = 5$$
$$800/223 - 3z = 5$$
$$-3z = 5 - 800/223$$
$$-3z = 1{,}115 \,/\, 223 - 800/223$$
$$-3z = 315/223$$
$$z = (315/223) \,/\, (-3)$$
$$z = -105/223$$

This is the same solution we obtained in the chapter text.

5. The first and second revised equations we obtained in the section "Eliminate One Variable" were

$$-4x + 2y - 3z = 5$$

and

$$2x - 5y - z = -1$$

The two-variable equation in x and y that we derived from these, as a result of eliminating the variable z, was

$$-10x + 17y = 8$$

If we take this equation together with the solution to Prob. 1, we have the two-by-two linear system

$$-37x - 4y = 35$$

and

$$-10x + 17y = 8$$

6. To use the morph-and-mix method, we must get the equations into SI form. We're told to treat x as the dependent variable, so we must isolate x on the left sides of the equals signs. The equations morph like this:

$$-37x - 4y = 35$$
$$-37x = 4y + 35$$
$$x = (-4/37)y - 35/37$$

and

$$-10x + 17y = 8$$
$$-10x = -17y + 8$$
$$x = (17/10)y - 8/10$$

Now we mix the right sides to get

$$(-4/37)y - 35/37 = (17/10)y - 8/10$$

It will simplify things if we can get a common denominator. Let's multiply the numerators and denominators on the left side of this equation by 10, and multiply the numerators and denominators on the right side by 37. That gives us

$$(-40/370)y - 350/370 = (629/370)y - 296/370$$

Multiplying this entire equation through by 370, we obtain an equation without fractions, which we can solve in steps as follows:

$$-40y - 350 = 629y - 296$$
$$-40y = 629y + 54$$
$$-669y = 54$$
$$y = -54/669$$
$$= -18/223$$

This agrees with the solution we obtained in the chapter text. Now we can plug this into either of the SI equations we derived earlier. Let's use the second one. We get

$$x = (17/10)y - 8/10$$
$$= (17/10) \times (-18/223) - 8/10$$
$$= -306 / 2{,}230 - 8/10$$
$$= -306 / 2{,}230 - (8 \times 223) / (10 \times 223)$$
$$= -306 / 2{,}230 - 1{,}784 / 2{,}230$$
$$= -2{,}090 / 2{,}230$$
$$= -209/223$$

This, too, agrees with the result we obtained in the chapter text.

7. For reference, here again is the second equation stated in Prob. 1:

$$2x - 5y - z = -1$$

The solutions for x and y can be plugged into this equation, and it can then be solved for z in steps, as follows:

$$2 \times (-209/223) + 5 \times (-18/223) - z = -1$$
$$-418/223 + 90/223 - z = -1$$
$$-328/223 - z = -1$$
$$-z = -1 + 328/223$$
$$-z = 105/223$$
$$z = -105/223$$

This is the same solution we obtained in the chapter text.

8. All of these equations are in SI form. We know that their graphs must all be straight lines, because they are linear equations. All the slopes are different, but all the y-intercepts are equal to 1. Therefore, although no two of the lines coincide, all four of them pass through the point (0, 1). We can conclude that the system therefore has a unique solution: $x = 0$ and $y = 1$.

9. Let's go through each equation, plugging in $x = 0$ and $y = 1$, and then grinding out the arithmetic. Here's the first equation:

$$y = -x + 1$$
$$1 = -0 + 1$$
$$1 = 0 + 1$$
$$1 = 1$$

Check! Now the second equation:

$$y = -2x + 1$$
$$1 = -2 \times 0 + 1$$
$$1 = 0 + 1$$
$$1 = 1$$

Check! Now the third:

$$y = 3x + 1$$
$$1 = 3 \times 0 + 1$$
$$1 = 0 + 1$$
$$1 = 1$$

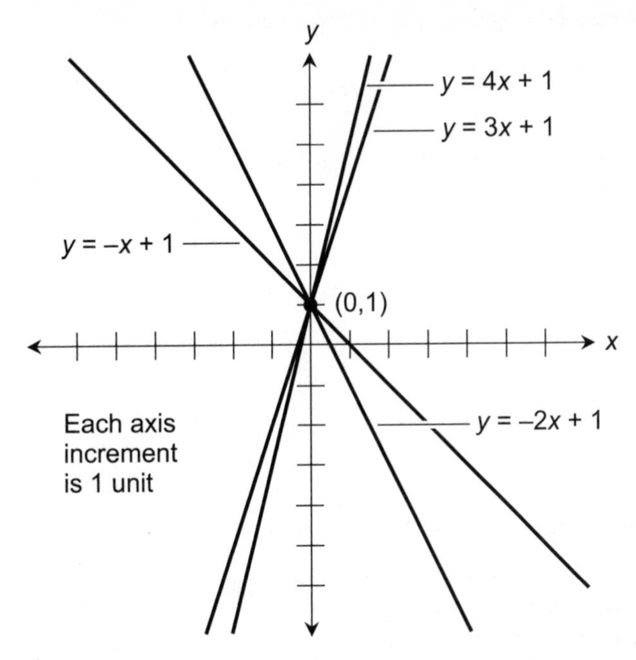

Figure B-10 Illustration for the solution to Prob. 10 in Chap. 18.

Check! Finally the fourth:

$$y = 4x + 1$$
$$1 = 4 \times 0 + 1$$
$$1 = 0 + 1$$
$$1 = 1$$

We can now be confident that the solution to the system is indeed $x = 0$ and $y = 1$.

10. Figure B-10 shows graphs of all four equations on the Cartesian plane. It's visually apparent that if we choose any pair or triplet of these lines, they intersect at the point (0, 1) and nowhere else. Therefore, any pair or triplet of the equations, taken as a two-by-two or three-by-two linear system, has the unique solution $x = 0$ and $y = 1$.

Chapter 19

1. Here are the original three equations, stated again for convenience, followed by the step-by-step processes that get the equations into form for conversion to a matrix:

$$x = y - z - 7$$
$$y = 2x + 2z + 2$$
$$z = 3x - 5y + 4$$

We morph the first equation as follows:

$$x = y - z - 7$$
$$x - y = -z - 7$$
$$x - y + z = -7$$

Next, the second equation:

$$y = 2x + 2z + 2$$
$$-2x + y = 2z + 2$$
$$-2x + y - 2z = 2$$

Finally, the third:

$$z = 3x - 5y + 4$$
$$-3x + z = -5y + 4$$
$$-3x + 5y + z = 4$$

Now we have this set of equations representing our three-by-three linear system:

$$x - y + z = -7$$
$$-2x + y - 2z = 2$$
$$-3x + 5y + z = 4$$

2. Before we write down the matrix, we must be sure we have the correct signs for the coefficients. Subtraction of a positive is equivalent to addition of a negative. With that in mind, we can "pigeonhole" the coefficients into the matrix form:

1	−1	1	−7
−2	1	−2	2
−3	5	1	4

3. It's easy to convert a matrix into a set of linear equations, but we must pay attention to the signs. Here's the matrix:

0	4	−1	−2
5	−3/2	8	1
1	1	1	1

Here are the equations, derived directly from the coefficients in the matrix:

$$0x + 4y + (-z) = -2$$
$$5x + (-3/2)y + 8z = 1$$
$$x + y + z = 1$$

Here are the equations with the negative additions converted to subtractions, and the term $0x$ eliminated from the first equation for simplicity:

$$4y - z = -2$$
$$5x - (3/2)y + 8z = 1$$
$$x + y + z = 1$$

4. Here's the matrix again, for reference:

0	4	−1	−2
5	−3/2	8	1
1	1	1	1

Let's interchange the first and second rows. Then we get

5	−3/2	8	1
0	4	−1	−2
1	1	1	1

If we multiply the first row by 2 to get rid of the fractional expression, we have

10	−3	16	2
0	4	−1	−2
1	1	1	1

We can multiply the bottom row by −10, getting

10	−3	16	2
0	4	−1	−2
−10	−10	−10	−10

Adding the first row to the third and then replacing the third row with the sum, we obtain

10	−3	16	2
0	4	−1	−2
0	−13	6	−8

Now let's multiply the second row by 13 and the third row by 4. That gives us

10	−3	16	2
0	52	−13	−26
0	−52	24	−32

Adding the second row to the third and then replacing the third row with the sum, we get

10	−3	16	2
0	52	−13	−26
0	0	11	−58

This matrix is in echelon form.

5. We want to put the matrix from solution to Prob. 4 into diagonal form. We can multiply the second row by 11 and the third row by 13 to obtain

10	−3	16	2
0	572	−143	−286
0	0	143	−754

Adding the second row to the third and then replacing the second row with the sum, we get

10	−3	16	2
0	572	0	−1,040
0	0	143	−754

Now let's multiply the first row by 572 and the second row by 3. That produces

5,720	−1,716	9,152	1,144
0	1,716	0	−3,120
0	0	143	−754

Adding the first row to the second and then replacing the first row with the sum gives us

5,720	0	9,152	−1,976
0	1,716	0	−3,120
0	0	143	−754

We need to turn the 9,152 in the first row into 0. We'll have to work with the third row to make it happen. Suppose 9,152 divides cleanly by 143? Let's give it a try. A calculator tells us that 9,152 / 143 = 64. Let's multiply the third row by –64 to get the matrix

5,720	0	9,152	–1,976
0	1,716	0	–3,120
0	0	–9,152	48,256

Adding the first row to the third and then replacing the first row with the sum, we have

5,720	0	0	46,280
0	1,716	0	–3,120
0	0	–9,152	48,256

This matrix is unwieldy, but it's in diagonal form.

6. We want to get the absolute values of the numbers in solution to Prob. 5 as small as possible and still have all integers. That means we must find *largest common divisors* for each row. This process is something like reducing fractions to lowest form. Let's divide the first row by 520, the second row by 156, and the third row by –832. That gives us

11	0	0	89
0	11	0	–20
0	0	11	–58

That's as far as we can reduce the matrix, but it's quite an improvement! This is not a coincidence, even though it may appear that way at first. In the matrix morphing process, we did a lot of multiplying. Those multiples "went along for the ride," inflating the numbers. They have served their purpose. It's good to be rid of them.

7. To reduce the matrix in solution to Prob. 6 to unit diagonal form, we divide each row by 11. That gives us

1	0	0	89/11
0	1	0	–20/11
0	0	1	–58/11

The solution to the original linear system, stated at the end of solution to Prob. 3, is apparently

$$x = 89/11$$
$$y = -20/11$$
$$z = -58/11$$

8. Here are the equations from solution to Prob. 3, which we hope we have solved as a linear system:

$$4y - z = -2$$
$$5x - (3/2)y + 8z = 1$$
$$x + y + z = 1$$

Let's check the first equation with the values from the solution to Prob. 7:

$$4y - z = -2$$
$$4 \times (-20/11) - (-58/11) = -2$$
$$-80/11 + 58/11 = -2$$
$$-22/11 = -2$$
$$-2 = -2$$

Checking in the second equation:

$$5x - (3/2)y + 8z = 1$$
$$5 \times 89/11 - [(3/2) \times (-20/11)] + 8 \times (-58/11) = 1$$
$$445/11 - (-30/11) - 464/11 = 1$$
$$445/11 + 30/11 - 464/11 = 1$$
$$11/11 = 1$$
$$1 = 1$$

Checking in the third equation:

$$x + y + z = 1$$
$$89/11 + (-20/11) + (-58/11) = 1$$
$$89/11 - 20/11 - 58/11 = 1$$
$$11/11 = 1$$
$$1 = 1$$

We can now be confident that the solution to the three-by-three linear system is

$$x = 89/11$$
$$y = -20/11$$
$$z = -58/11$$

9. Here's the three-by-three linear system we have been told to describe as a matrix:

$$x + y + z = 1$$
$$x + y + z = 2$$
$$x + y + z = 3$$

These equations are all in ideal form for conversion to the matrix

1	1	1	1
1	1	1	2
1	1	1	3

Now let's try to get this into unit diagonal form. The first step along the way is to seek the echelon form. We can start by doing one of three things:

- Make the first entry in the second row vanish
- Make the first entry in the third row vanish
- Make the second entry in the third row vanish

This is easy—too easy! Suppose we want to make the first entry in the third row vanish. We can multiply the first row through by -1, getting

−1	−1	−1	−1
1	1	1	2
1	1	1	3

Adding the first row to the third row and then replacing the third row with the sum gives us

−1	−1	−1	−1
1	1	1	2
0	0	0	2

That takes care of not only one, but two of the elements we wanted to turn into 0. But there's a problem starting to take shape. We want the third entry in the third row to end up as a *nonzero* element. We won't be able to do that without making both the first and the second elements in that row nonzero as well. We can go further and add the first two rows in the above matrix together, replacing the first row with the sum. Then we get

0	0	0	1
1	1	1	2
0	0	0	2

This in effect states the following three equations:

$$0x + 0y + 0z = 1$$
$$x + y + z = 2$$
$$0x + 0y + 0z = 2$$

There are no real numbers x, y, or z such that, when they are each multiplied by 0, the result is 1 or 2. No matter how we approach this problem, we'll get a statement that, in effect, says that 0 is equal to some nonzero real number. That's absurd! The reason for this "hangup" is that the original three-by-three linear system is inconsistent. If we could draw a graph of this system in Cartesian three-space, we'd get three parallel planes, no two of which would intersect anywhere.

10. Here's the three-by-three linear system we've been told to describe as a matrix:

$$x + y + z = 1$$
$$2x + 2y + 2z = 2$$
$$3x + 3y + 3z = 3$$

These equations are all in ideal form for conversion to

1	1	1	1
2	2	2	2
3	3	3	3

Dividing the second row by 2 and the third row by 3, we get

1	1	1	1
1	1	1	1
1	1	1	1

We won't be able to make any of these elements vanish without making a whole row vanish, giving us the equation

$$0x + 0y + 0z = 0$$

which is utterly useless. However, the above matrix tells us that

$$x + y + z = 1$$

An infinite number of ordered triples (x,y,z) satisfy this equation. Our original three-by-three linear system is actually one equation stated three different ways. It's redundant, so a single solution does not exist.

APPENDIX

C

Worked-Out Solutions to Exercises: Chapters 21 to 29

These worked-out solutions do not necessarily represent the only way a problem can be figured out. If you think you can solve a particular problem in a quicker or better way than you see here, by all means go ahead! But always check your work to be sure your "alternative" answer is correct.

Chapter 21

1. The 0th power of j is equal to 1. We know that $j^2 = -1$, so according to the difference of powers law,

$$\begin{aligned} j^0 &= j^{2-2} \\ &= j^2/j^2 \\ &= -1/(-1) \\ &= 1 \end{aligned}$$

2. We determined in this chapter that $j^2 = -1$, $j^4 = 1$, $j^6 = -1$, $j^8 = 1$, and so on for increasing even-integer powers of j. Based on this knowledge, we can use the negative powers rule to determine the following facts:

$$\begin{aligned} j^{-2} &= 1/(j^2) \\ &= 1/(-1) \\ &= -1 \\ j^{-4} &= 1/(j^4) \\ &= 1/1 \\ &= 1 \end{aligned}$$

$$
\begin{aligned}
j^{-6} &= 1/(j^6) \\
&= 1/(-1) \\
&= -1 \\
j^{-8} &= 1/(j^8) \\
&= 1/1 \\
&= 1 \\
&\downarrow
\end{aligned}
$$

and so on, forever

3. To solve this problem, we need a little intuition. First, let's apply the difference of powers law. Note that

$$
\begin{aligned}
j^{-1} &= j^{3-4} \\
&= j^3/j^4
\end{aligned}
$$

We have already determined that $j^3 = -j$ and $j^4 = 1$. Therefore,

$$
\begin{aligned}
j^3/j^4 &= (-j)/1 \\
&= -j
\end{aligned}
$$

We can conclude that $j^{-1} = -j$. Now let's use cross multiplication and see if we get the same result. Note that j^{-1} is the reciprocal of j, or $1/j$. If we let this quantity equal an unknown z, we can formulate this equation:

$$1/j = z/1$$

According to the law of cross multiplication, the above expression is equivalent to

$$1 \times 1 = jz$$

which tells us that $jz = 1$. Let's make an educated guess as to what z might be. It's easy enough to see that z can't be equal to $1, -1$, or j. How about $-j$? When we multiply j by $-j$, we get

$$
\begin{aligned}
j \times (-j) &= j \times (-1 \times j) \\
&= j \times j \times (-1) \\
&= j^2 \times (-1) \\
&= -1 \times (-1) \\
&= 1
\end{aligned}
$$

It works! This tells us that $z = -j$, and therefore that $1/j = -j$. We've now shown, in two different ways, that the reciprocal of the unit imaginary number is the same as its negative. If you want to use technical language, the additive inverse of j is the same as its multiplicative inverse. No real number behaves like that!

4. To figure out the value of j^{-3} using the difference of powers law, note that

$$j^{-3} = j^{1-4}$$
$$= j/j^4$$

We have determined that $j^4 = 1$. Therefore,

$$j/j^4 = j/1$$
$$= j$$

We can conclude that $j^{-3} = j$. Now let's determine j^{-5}. Again using the difference of powers law, we can say that

$$j^{-5} = j^{-1-4}$$
$$= j^{-1}/j^4$$

We have found that $j^{-1} = -j$, and also that $j^4 = 1$. Therefore

$$j^{-1}/j^4 = (-j)/1$$
$$= -j$$

Now we know that $j^{-5} = -j$. Finally, let's figure out the value of j^{-7}. Once again choosing numbers and applying the difference of powers law, we can say that

$$j^{-7} = j^{-3-4}$$
$$= j^{-3}/j^4$$

We have found that $j^{-3} = j$, and also that $j^4 = 1$. Therefore

$$j^{-3}/j^4 = j/1$$
$$= j$$

This tells us that $j^{-7} = j$. By now, it is apparent that we'll alternate between $-j$ and j as we raise j to ever-decreasing negative odd integer powers of -9, -11, -13, and so on.

5. Refer to Table C-1. The four-way cycle of values goes on forever in both directions, that is, for positive and negative integer powers of j.

6. The answers, along with explanations, are as follows.

   (a) To find the sum $(4 + j5) + (3 - j8)$, we add the real parts and the imaginary parts separately. This gives us

$$(4 + j5) + (3 - j8) = (4 + 3) + j(5 - 8)$$
$$= 7 + j(-3)$$
$$= 7 - j3$$

Table C-1. Solution to Prob. 5 in Chap. 21.

Expression	Value
↑	↑
j^8	1
j^7	$-j$
j^6	-1
j^5	j
j^4	1
j^3	$-j$
j^2	-1
j^1	j
j^0	1
j^{-1}	$-j$
j^{-2}	-1
j^{-3}	j
j^{-4}	1
j^{-5}	$-j$
j^{-6}	-1
j^{-7}	j
j^{-8}	1
↓	↓

(b) To find the difference $(4 + j5) - (3 - j8)$, we multiply the second complex number through by -1, and then add the real parts and the imaginary parts separately, getting

$$\begin{aligned}(4 + j5) - (3 - j8) &= (4 + j5) + [-1(3 - j8)]\\ &= (4 + j5) + (-3 + j8)\\ &= (4 - 3) + j(5 + 8)\\ &= 1 + j13\end{aligned}$$

(c) To find the product $(4 + j5)(3 - j8)$, we use the product of sums rule. This gives us

$$\begin{aligned}(4 + j5)(3 - j8) &= 4 \times 3 + 4 \times (-j8) + j5 \times 3 + j5 \times (-j8)\\ &= 12 + (-j32) + j15 + j \times j \times (-40)\\ &= 12 + (-j17) + (-1) \times (-40)\\ &= (12 + 40) + (-j17)\\ &= 52 - j17\end{aligned}$$

(d) To find the quotient $(4 + j5) / (3 - j8)$, we use the quotient formula from the text. If a, b, c, and d are real numbers, and as long as c and d aren't both equal to 0, then

$$\begin{aligned}&(a + jb) / (c + jd)\\ &= (ac + bd) / (c^2 + d^2) + j(bc - ad) / (c^2 + d^2)\end{aligned}$$

If we let $a = 4$, $b = 5$, $c = 3$, and $d = -8$, then we have

$$c^2 + d^2 = 3^2 + (-8)^2$$
$$= 9 + 64$$
$$= 73$$

and therefore

$$(4 + j5) / (3 - j8)$$
$$= [4 \times 3 + 5 \times (-8)] / 73 + j\{5 \times 3 - [4 \times (-8)]\} / 73$$
$$= (12 - 40) / 73 + j(15 + 32) / 73$$
$$= -28/73 + j(47/73)$$

7. To find $(a + jb) - (a - jb)$, we multiply the second complex number through by -1, and then add the real parts and the imaginary parts separately, getting

$$(a + jb) - (a - jb) = (a + jb) + [-1(a - jb)]$$
$$= (a + jb) + (-a + jb)$$
$$= a + (-a) + jb + jb$$
$$= j2b$$

To find $(a - jb) - (a + jb)$, we again multiply the second complex number through by -1, and then add the real parts and the imaginary parts separately, getting

$$(a - jb) - (a + jb) = (a - jb) + [-1(a + jb)]$$
$$= (a - jb) + (-a - jb)$$
$$= a + (-a) + (-jb) + (-jb)$$
$$= -j2b$$

Note that in these answers, the numerals 2 are not exponents! We have j times the quantity $2b$ in the first case, and $-j$ times the quantity $2b$ in the second case.

8. To find $(a + jb) / (a - jb)$, let's first change the subtraction in the denominator to negative addition. That will give us the expression

$$(a + jb) / [a + j(-b)]$$

Now we can use the quotient formula for complex numbers. Let's state it again for reference. If a, b, c, and d are real numbers, and as long as c and d aren't both equal to 0, then

$$(a + jb) / (c + jd)$$
$$= (ac + bd) / (c^2 + d^2) + j(bc - ad) / (c^2 + d^2)$$

Now we can make these substitutions:

- Let a from the formula equal a in our problem
- Let b from the formula equal b in our problem
- Let c from the formula equal a in our problem
- Let d from the formula equal $-b$ in our problem

The signs will be tricky, now! The quotient formula looks like this:

$$(a + jb) / [a + j(-b)]$$
$$= (aa + b \times (-b)] / [a^2 + (-b)^2] + j[ba - a \times (-b)] / [a^2 + (-b)^2]$$

For any real number b, $(-b)^2 = b^2$. Knowing that, and simplifying the above expression as much as possible, we get

$$(aa + b \times (-b)] / [a^2 + (-b)^2] + j[ba - a \times (-b)] / [a^2 + (-b)^2]$$
$$= (a^2 - b^2) / (a^2 + b^2) + j[2ab / (a^2 + b^2)]$$

9. To find $(a - jb) / (a + jb)$, let's first change the subtraction in the numerator to negative addition. That will give us the expression

$$[a + j(-b)] / (a + jb)$$

Now we can again use the quotient formula for complex numbers. This time, let's make these substitutions:

- Let a from the formula equal a in our problem
- Let b from the formula equal $-b$ in our problem
- Let c from the formula equal a in our problem
- Let d from the formula equal b in our problem

Once again, we must pay close attention to the signs. The quotient formula now looks like this:

$$[a + j(-b)] / (a + jb)$$
$$= (aa + (-b) \times b] / (a^2 + b^2) + j(-ba - ab) / (a^2 + b^2)$$

Simplifying the above expression as much as possible, we get

$$(aa + b \times (-b)] / [a^2 + (-b)^2] + j[ba - a \times (-b)] / [a^2 + (-b)^2]$$
$$= (a^2 - b^2) / (a^2 + b^2) + j[-2ab / (a^2 + b^2)]$$
$$= (a^2 - b^2) / (a^2 + b^2) - j[2ab / (a^2 + b^2)]$$

This is the complex conjugate of the result we got in Prob. 8.

10. If k is a positive real number, then two pure real numbers have absolute values equal to k. These numbers are k and $-k$. Two pure imaginary numbers, jk and $-jk$, also have

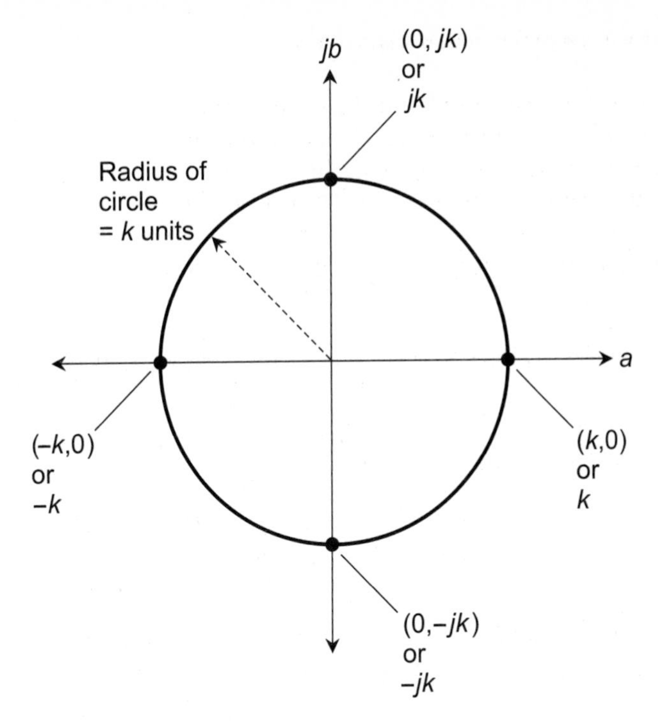

Figure C-1 Illustration for the solution to Prob. 10 in Chap. 21.

absolute values equal to k. These are shown as points in Fig. C-1. There are infinitely many complex numbers with absolute values equal to k. This fact can be shown in the complex-number plane by plotting the set of all points at a distance of k units from the origin. This set of points forms a circle with radius k.

Chapter 22

1. To multiply out this equation, we must apply the product of sums rule on the left side of the equals sign. To keep the signs right, let's change the subtraction into a negative addition before we start multiplying, and then change the negative additions back to subtractions when we're done. Here are the steps:

$$(-7x - 5)(-2x + 9) = 0$$
$$[-7x + (-5)](-2x + 9) = 0$$
$$(-7x) \times (-2x) + (-7x) \times 9 + (-5) \times (-2x) + (-5) \times 9 = 0$$
$$14x^2 + (-63x) + 10x + (-45) = 0$$
$$14x^2 + (-53x) + (-45) = 0$$
$$14x^2 - 53x - 45 = 0$$

2. We've been told to factor the following quadratic, and we've been assured that the coefficients and constants are all integers.

$$x^2 + 10x + 25 = 0$$

Because the coefficient of x^2 is equal to 1, we know that the coefficients of x in both binomials must be equal to 1. That means the factored equation looks like this:

$$(x + \#)(x + \#) = 0$$

where # represents an integer (not necessarily the same one in each case). The sum of these integers is 10, and their product is 25. A good guess will tell us that the numbers are both 5. Let's see what happens if we use those numbers and multiply out:

$$(x + 5)(x + 5) = 0$$
$$x^2 + 5x + 5x + 25 = 0$$
$$x^2 + 10x + 25 = 0$$

That works, so the factored form is

$$(x + 5)(x + 5) = 0$$

which can also be written as

$$(x + 5)^2 = 0$$

To solve this, we can take the square root of both sides. There is no "plus-or-minus" ambiguity. We get

$$x + 5 = 0$$

There is only one root here, and it is -5. The solution set is therefore $\{-5\}$.

3. We want to factor the following quadratic, and we've been told that the coefficients and constants in the binomials are all integers.

$$2x^2 + 8x - 10 = 0$$

The coefficient of x^2 is equal to 2. Therefore, the general form of the equation in binomial factor form will be

$$(x + \#)(2x + \#) = 0$$

where # represents an integer (not necessarily the same one in each case). The product of these unknown integers is -10. If we plug in integers whose product is -10 and multiply the resulting products of binomials out, we'll eventually come up with

$$(x + 5)(2x - 2) = 0$$

Multiplying to confirm, we get

$$(x + 5)(2x - 2) = 0$$
$$2x^2 - 2x + 10x - 10 = 0$$
$$2x^2 + 8x - 10 = 0$$

The roots are found by solving these two first-degree equations:

$$x + 5 = 0$$

and

$$2x - 2 = 0$$

The solution in the first case is easily seen to be $x = -5$, and in the second case the solution is $x = 1$. The roots of the quadratic are -5 and 1. The solution set is $\{-5, 1\}$.

4. We want to morph the left side of the following quadratic into a product of binomials whose coefficients and constants in the binomials are all integers.

$$12x^2 + 7x - 10 = 0$$

In this case, the coefficient of x^2 is equal to 12. That means the general binomial factor form will look like one of these:

$$(x + \#)(12x + \#) = 0$$
$$(2x + \#)(6x + \#) = 0$$
$$(4x + \#)(3x + \#) = 0$$

where # represents an integer (not necessarily the same one in each case). The product of these unknown integers is -10. As before, we can start plugging in integers whose product is -10, multiply the resulting product of binomials out on every attempt, and see what we get. There are lots of choices here, and the process could take time. Eventually we'll come up with

$$(4x + 5)(3x - 2) = 0$$

When we multiply this out, we find

$$(4x + 5)(3x - 2) = 0$$
$$12x^2 + (-8x) + 15x + (-10) = 0$$
$$12x^2 + 7x - 10 = 0$$

To find the roots, we must solve

$$4x + 5 = 0$$

and

$$3x - 2 = 0$$

The solution to the first of these equations is derived like this:

$$\begin{aligned} 4x + 5 &= 0 \\ 4x &= -5 \\ x &= -5/4 \end{aligned}$$

In the second case, the process is similar:

$$\begin{aligned} 3x - 2 &= 0 \\ 3x &= 2 \\ x &= 2/3 \end{aligned}$$

The roots of the quadratic are $-5/4$ and $2/3$, and the solution set is $\{-5/4, 2/3\}$.

5. We want to morph the following quadratic so the left side becomes a product of a binomial with itself:

$$16x^2 - 40x + 25 = 0$$

The coefficient of x^2 is equal to 16, and we know it has to be the square of the first term in the binomial. That means the first term must be $4x$ or $-4x$. The constant in the polynomial is equal to 25, and it must be the square of the constant in the binomial. That means the constant in the binomial must be 5 or -5. The coefficient of x in the polynomial is negative, telling us that the coefficient and the constant in the binomial must have opposite signs. As things work out, we get

$$(4x - 5)^2 = 0$$

Checking to be sure this is right, we can multiply it out:

$$\begin{aligned} (4x - 5)(4x - 5) &= 0 \\ 16x^2 + (-20x) + (-20x) + 25 &= 0 \\ 16x^2 - 40x + 25 &= 0 \end{aligned}$$

It works! We can also use

$$(-4x + 5)^2 = 0$$

This equation is equivalent to the other one. To show this, we can derive one squared binomial from the other:

$$\begin{aligned}(4x-5)^2 &= (4x-5)(4x-5)\\ &= (-1)^2(4x-5)(4x-5)\\ &= (-1)(4x-5)(-1)(4x-5)\\ &= (-4x+5)(-4x+5)\\ &= (-4x+5)^2\end{aligned}$$

This duplicity occurs with all squared binomials. It simply comes out of the fact that $(-1)^2 = 1$.

6. To find the root of the quadratic, we can start with either of the binomial factor equations we found. Let's use the first one:

$$(4x-5)^2 = 0$$

Taking the square root of both sides, we obtain

$$4x-5=0$$

We can add 5 to each side, getting

$$4x=5$$

Dividing through by 4 gives us the root $x = 5/4$. The solution set is $\{5/4\}$. Let's plug the root into the original quadratic to be sure that it works:

$$\begin{gathered}16x^2 - 40x + 25 = 0\\ 16 \times (5/4)^2 - 40 \times (5/4) + 25 = 0\\ 16 \times 25/16 - 50 + 25 = 0\\ 25 - 50 + 25 = 0\\ -25 + 25 = 0\\ 0 = 0\end{gathered}$$

7. We want to morph the following quadratic so we can get the left side into a product of a binomial with itself.

$$x^2 + 6x - 7 = 0$$

We can add 16 to each side to get

$$x^2 + 6x + 9 = 16$$

The left side can now be factored into a square of a binomial:

$$(x+3)^2 = 16$$

8. Let's take the square root of both sides of the binomial factor equation from solution to Prob. 7, remembering to include both the negative and positive results:

$$[(x+3)^2]^{1/2} = \pm(16^{1/2})$$

This simplifies to

$$x+3 = \pm 4$$

which can be stated as the following pair of first degree equations:

$$x+3=4 \qquad \text{or} \qquad x+3=-4$$

The first of these solves to $x = 1$, and the second solves to $x = -7$. The roots of the quadratic are therefore $x = 1$ or $x = -7$, so the solution set is $\{1, -7\}$. Let's check these roots in the original quadratic. When $x = 1$, we get

$$\begin{aligned} x^2 + 6x - 7 &= 0 \\ 1^2 + 6 \times 1 - 7 &= 0 \\ 1 + 6 - 7 &= 0 \\ 7 - 7 &= 0 \\ 0 &= 0 \end{aligned}$$

When $x = -7$, we get

$$\begin{aligned} x^2 + 6x - 7 &= 0 \\ (-7)^2 + 6 \times (-7) - 7 &= 0 \\ 49 + (-42) - 7 &= 0 \\ 49 - 42 - 7 &= 0 \\ 7 - 7 &= 0 \\ 0 &= 0 \end{aligned}$$

9. To determine how many real roots a quadratic has, we can calculate the discriminant. For a quadratic of the form

$$ax^2 + bx + c = 0$$

the discriminant is $b^2 - 4ac$. The equation of interest is

$$-2x^2 + 3x + 35 = 0$$

Here, $a = -2$, $b = 3$, and $c = 35$. Therefore

$$\begin{aligned} b^2 - 4ac &= 3^2 - 4 \times (-2) \times 35 \\ &= 9 - (-280) \\ &= 9 + 280 \\ &= 289 \end{aligned}$$

The fact that the discriminant is positive tells us that this quadratic has two distinct real roots. To find the roots, we can use the quadratic formula:

$$x = [-b \pm (b^2 - 4ac)^{1/2}] \,/\, (2a)$$

We already know the discriminant, so we can plug it in directly along with the values for a, b, and c, getting

$$\begin{aligned} x &= [-3 \pm 289^{1/2}] \,/\, [2 \times (-2)] \\ &= (-3 \pm 17) \,/\, (-4) \\ &= (-3 + 17) \,/\, (-4) \text{ or } (-3 - 17) \,/\, (-4) \\ &= 14 \,/\, (-4) \text{ or } (-20) \,/\, (-4) \\ &= -7/2 \text{ or } 5 \end{aligned}$$

The real roots are $x = -7/2$ or $x = 5$, and the real-number solution set is $\{-7/2, 5\}$. For complementary credit, you can check these roots by plugging them into the original quadratic to be sure that they work.

10. Once again, we can calculate the discriminant to find out how many real roots there are. The equation of interest is

$$4x^2 + x + 3 = 0$$

Here, $a = 4$, $b = 1$, and $c = 3$. Therefore

$$\begin{aligned} b^2 - 4ac &= 1^2 - 4 \times 4 \times 3 \\ &= 1 - 48 \\ &= -47 \end{aligned}$$

Because this is negative, we can conclude that the quadratic has no real roots. The real-number solution set is therefore $\varnothing$, the empty set. But this does *not* mean that there are no roots at all! In the next chapter, we'll learn about the roots of quadratics that have negative discriminants.

Chapter 23

1. We've been told to find the roots of this quadratic:

$$(x - j7)(x + j7) = 0$$

Because this equation is in binomial factor form, the roots can be found by solving these two first-degree equations:

$$x - j7 = 0$$

and

$$x + j7 = 0$$

In the top equation, we can add $j7$ to each side, getting the root $x = j7$. In the bottom equation, we can subtract $j7$ from each side, getting the root $x = -j7$. The roots can be formally expressed this way:

$$x = j7 \qquad \text{or} \qquad x = -j7$$

The solution set is

$$X = \{j7, -j7\}$$

2. To obtain the polynomial standard form of the equation stated in Prob. 1, we must multiply out the left side using the product of sums rule. We can minimize the risk of getting confused by the signs if we convert the first binomial factor to a sum. We proceed as follows:

$$[x + (-j7)](x + j7) = 0$$
$$x^2 + xj7 + (-j7x) + (-j7)(j7) = 0$$
$$x^2 + j7x + (-j7x) + (-j \times j) \times 7 \times 7 = 0$$
$$x^2 + 49 = 0$$

3. Remember the general polynomial standard form of a quadratic equation:

$$ax^2 + bx + c = 0$$

In the equation we derived in solution to Prob. 2, we have $a = 1$, $b = 0$, and $c = 49$. Plugging these values into the quadratic formula and then working out the arithmetic, we get

$$\begin{aligned} x &= [-b \pm (b^2 - 4ac)^{1/2}] \;/\; (2a) \\ &= [-0 \pm (0^2 - 4 \times 1 \times 49)^{1/2}] \;/\; (2 \times 1) \\ &= [\pm(-196)^{1/2}] \;/\; 2 \\ &= \pm j14/2 \\ &= \pm j7 \end{aligned}$$

This agrees with the results we got when we solved Prob. 1.

4. Here are the roots again, for reference:

$$x = j7 \qquad \text{or} \qquad x = -j3$$

Both of these statements are equations. In the first one, we can subtract $j7$ from each side. In the second one, we can add $j3$ to each side. This gives us

$$x - j7 = 0$$

and

$$x + j3 = 0$$

The binomial factor form of a quadratic with the above mentioned roots is therefore

$$(x - j7)(x + j3) = 0$$

5. To find the polynomial standard form of the equation we just found, we must multiply out the left side using the product of sums rule. Let's convert the first binomial to a sum to avoid sign confusion. Then we can take it from there:

$$[x + (-j7)](x + j3) = 0$$
$$x^2 + xj3 + (-j7x) + (-j7)(j3) = 0$$
$$x^2 + j3x + (-j7x) + (-j \times j) \times 7 \times 3 = 0$$
$$x^2 - j4x + 21 = 0$$

This is a twist we haven't seen yet! One of the coefficients is imaginary.

6. Once again, for reference, let's state the general polynomial standard form of a quadratic equation:

$$ax^2 + bx + c = 0$$

In the equation we derived in solution to Prob. 5, we have $a = 1$, $b = -j4$, and $c = 21$. Plugging these values into the quadratic formula and then working out the arithmetic, we get

$$x = [-b \pm (b^2 - 4ac)^{1/2}] \,/\, (2a)$$
$$= \{-(-j4) \pm [(-j4)^2 - 4 \times 1 \times 21]^{1/2}\} \,/\, (2 \times 1)$$

Now let's be careful with $(-j4)^2$. This is the sort of expression that can easily cause us to make a mistake. We can break it down like this, and then solve, being careful with the signs:

$$(-j4)^2 = (-1 \times j \times 4)^2$$
$$= (-1)^2 \times j^2 \times 4^2$$
$$= 1 \times (-1) \times 16$$
$$= -16$$

Now let's substitute back in where we left off. That gives us an expression that's still tricky. But it can be simplified like this:

$$
\begin{aligned}
x &= [-(-j4) \pm (-16 - 4 \times 1 \times 21)^{1/2}] \;/\; (2 \times 1) \\
&= [j4 \pm (-100)^{1/2}] \;/\; 2 \\
&= (j4 \pm j10) \;/\; 2
\end{aligned}
$$

This breaks down to

$$x = (j4 + j10)/2 \qquad \text{or} \qquad x = (j4 - j10)/2$$

which simplifies to

$$x = j14/2 \qquad \text{or} \qquad x = j(-6)/2$$

and further to

$$x = j7 \qquad \text{or} \qquad x = -j3$$

These are the roots we chose in Prob. 4 to "manufacture" the quadratic.

7. We've been told to find the roots of this quadratic:

$$(x + 2 + j3)(x - 2 - j3) = 0$$

We can convert this pair of trinomial factors to a pair of binomial factors by changing the first subtraction in the second factor to addition, and also by grouping the terms within the factors, as follows:

$$[x + (2 + j3)][x + (-1)(2 + j3)] = 0$$

which can be rewritten as

$$[x + (2 + j3)][x - (2 + j3)] = 0$$

Now we have an equation in binomial factor form, where the factors both consist of the variable x plus or minus a numerical constant. The roots can therefore be found by solving the following two first-degree equations:

$$x + (2 + j3) = 0$$

and

$$x - (2 + j3) = 0$$

In the top equation, we can subtract the quantity $(2 + j3)$ from each side, getting

$$
\begin{aligned}
x &= -(2 + j3) \\
&= -2 - j3
\end{aligned}
$$

In the bottom equation, we can add the quantity $(2 + j3)$ to each side, getting

$$x = 2 + j3$$

The roots can be formally expressed this way:

$$x = -2 - j3 \qquad \text{or} \qquad x = 2 + j3$$

The solution set is

$$X = \{(-2 - j3), (2 + j3)\}$$

8. To get the polynomial form of the quadratic stated in Prob. 7, we can multiply out the trinomial factors. Here's the original equation again:

$$(x + 2 + j3)(x - 2 - j3) = 0$$

Let's convert the subtractions in the second factor to negative additions individually to minimize the risk of getting the signs mixed up when we expand the equation into polynomial form. That gives us

$$(x + 2 + j3)[x + (-2) + (-j3)] = 0$$

Now we can multiply out, obtaining

$$\begin{gathered} x^2 + (-2x) + (-j3x) \\ + 2x + (-4) + (-j6) \\ + j3x + (-j6) + 9 \\ = 0 \end{gathered}$$

which simplifies to

$$x^2 + (-j12) + 5 = 0$$

and further to

$$x^2 + (5 - j12) = 0$$

9. Here's the polynomial equation we derived. It's interesting, because the coefficient of x is equal to 0, while the stand-alone constant is complex.

$$x^2 + (5 - j12) = 0$$

Here are the roots we found:

$$x = -2 - j3 \qquad \text{or} \qquad x = 2 + j3$$

Plugging in the first root and converting the subtractions to additions, we can proceed like this, refining the equation step-by-step:

$$[-2 + (-j3)]^2 + [(5 + (-j12)] = 0$$
$$[-2 + (-j3)][-2 + (-j3)] + [(5 + (-j12)] = 0$$
$$4 + j6 + j6 + (-j)^2 9 + 5 + (-j12) = 0$$

Keeping in mind that $(-j)^2 = -1$, we can simplify to

$$4 + j6 + j6 + (-9) + 5 + (-j12) = 0$$

When we add all the terms on the left side, we get $0 = 0$. The first root checks! Now we'll plug in the second root and convert the subtraction to negative addition. We can proceed like this, step-by-step:

$$(2 + j3)^2 + [5 + (-j12)] = 0$$
$$(2 + j3)(2 + j3) + [5 + (-j12)] = 0$$
$$4 + j6 + j6 + j^2 9 + 5 + (-j12) = 0$$
$$4 + j6 + j6 + (-9) + 5 + (-j12) = 0$$

That's the same equation we got when we plugged in the other root. When we add all the terms on the left side, we get $0 = 0$. The second root checks!

10. We've been told to convert the following equation, which consists of two trinomial factors, into the polynomial standard form for a quadratic:

$$(j2x + 2 + j3)(-j5x + 4 - j5) = 0$$

This might look daunting at first, but our assigned task is merely a matter of multiplying things out, taking care with the signs, and not rushing it! Let's begin by converting the subtraction in the second factor to addition. That gives us

$$(j2x + 2 + j3)[(-j5x + 4 + (-j5)] = 0$$

We can use the product of sums rule, remembering that $-j \times j = 1$. It goes like this:

$$(j2x)(-j5x) + j8x + (j2x)(-j5)$$
$$+ (-j10x) + 8 + (-j10)$$
$$+ (j3)(-j5x) + j12 + (j3)(-j5)$$
$$= 0$$

Simplifying the individual terms, we get

$$10x^2 + j8x + 10x$$
$$+ (-j10x) + 8 + (-j10)$$
$$+ 15x + j12 + 15$$
$$= 0$$

Let's use the commutative law for addition to rearrange the terms according to powers of x with each power of x on its own line:

$$\begin{gathered} 10x^2 \\ + j8x + 10x + (-j10x) + 15x \\ + 8 + (-j10) + j12 + 15 \\ = 0 \end{gathered}$$

We can rewrite this as

$$\begin{gathered} 10x^2 \\ + [j8 + 10 + 15 + (-j10)]x \\ + 8 + 15 + (-j10) + j12 \\ = 0 \end{gathered}$$

This simplifies to

$$10x^2 + (25 - j2)x + (23 + j2) = 0$$

Chapter 24

1. This function is stated in binomial factor form. To figure out whether its graph opens upward or downward, we must morph it into polynomial standard form by multiplying out the factors:

$$\begin{aligned} y &= (x - 3)(4x - 1) \\ &= 4x^2 + (-x) + (-12x) + 3 \\ &= 4x^2 - 13x + 3 \end{aligned}$$

The coefficient of x^2 is positive. That means the parabola opens upward.

2. The real zeros of the quadratic function stated in Prob. 1 can be found from the factors in the original version. We must solve these two first-degree equations:

$$x - 3 = 0$$

and

$$4x - 1 = 0$$

In the top equation, we can add 3 to each side, getting $x = 3$. In the bottom equation, we can add 1 to each side and then divide through by 4, obtaining $x = 1/4$. The real zeros of the quadratic function are therefore $x = 3$ or $x = 1/4$.

3. Because the parabola opens upward, we know that its vertex is an absolute minimum. To find the x-coordinate of this point, we average the two zeros:

$$\begin{aligned} x_{min} &= (3 + 1/4)/2 \\ &= (12/4 + 1/4)/2 \\ &= (13/4)/2 \\ &= 13/8 \end{aligned}$$

We can calculate the y-coordinate of this point by plugging in 13/8 for x in the polynomial form of the function:

$$\begin{aligned} y_{min} &= 4x_{min}^2 - 13x_{min} + 3 \\ &= 4 \times (13/8)^2 - 13 \times 13/8 + 3 \\ &= 4 \times 169/64 - 169/8 + 3 \\ &= 676/64 - 1{,}352/64 + 192/64 \\ &= (676 - 1{,}352 + 192) / 64 \\ &= -484/64 \\ &= -121/16 \end{aligned}$$

The coordinates of the absolute minimum are (13/8, −121/16).

4. We know the two points where the curve crosses the x axis (representing the zeros of the function). The left-hand x-intercept point is (1/4, 0). The right-hand x-intercept point is (3, 0). We also know that the absolute minimum point is (13/8, −121/16). The graph, based on these three known points, is shown in Fig. C-2. On both axes, each increment represents 1 unit.

5. Remember the polynomial standard form for a quadratic function of a variable x, where we have coefficients a and b, and a stand-alone constant c. If we let the dependent variable be called y, then

$$y = ax^2 + bx + c$$

The function we're interested in is

$$y = 7x^2 + 5x + 2$$

In the polynomial, we have $a > 0$, so we know that the graph of the function is a parabola that opens upward. When we examine the discriminant d, we find that

$$\begin{aligned} d &= b^2 - 4ac \\ &= 5^2 - 4 \times 7 \times 2 \\ &= 25 - 56 \\ &= -31 \end{aligned}$$

The fact that d is negative tells us that the quadratic equation

$$7x^2 + 5x + 2 = 0$$

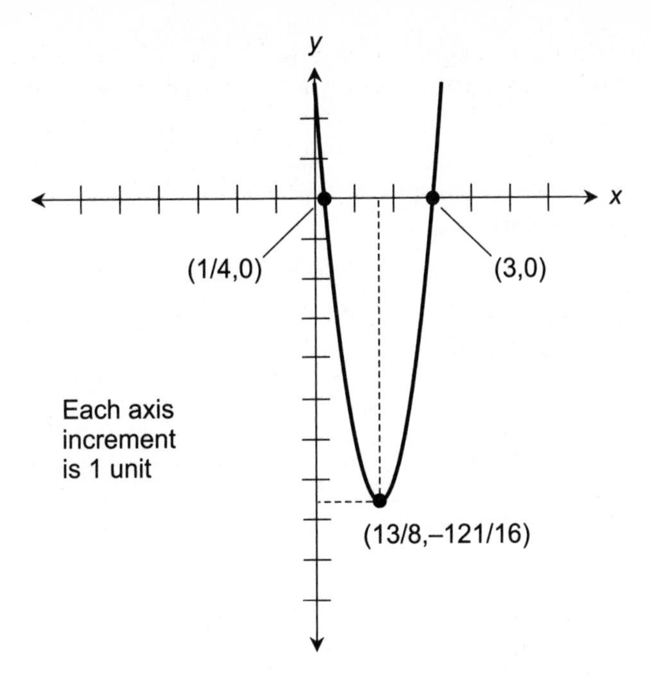

Figure C-2 Illustration for the solution to Prob. 4 in Chap. 24.

has no real roots, so the quadratic function has no real zeros. That means the graph does not cross the *x* axis anywhere. If a parabola opens upward and fails to cross the *x* axis, then that parabola must lie entirely above the *x* axis.

6. In the polynomial, we have $a = 7$ and $b = 5$. The *x*-value of the absolute minimum point, x_{min}, is therefore

$$\begin{aligned} x_{min} &= -b/(2a) \\ &= -5 / (2 \times 7) \\ &= -5/14 \end{aligned}$$

We can find the *y*-value of the absolute minimum point, y_{min}, by plugging in x_{min} to the function and doing the arithmetic:

$$\begin{aligned} y_{min} &= 7{x_{min}}^2 + 5x_{min} + 2 \\ &= 7 \times (-5/14)^2 + 5 \times (-5/14) + 2 \\ &= 7 \times 25/196 - 25/14 + 2 \\ &= 175/196 - 350/196 + 392/196 \\ &= -175/196 + 392/196 \\ &= 217/196 \\ &= 31/28 \end{aligned}$$

The coordinates of the vertex point on the parabola are (–5/14, 31/28).

7. Here's the quadratic function again, for reference:

$$y = -2x^2 + 2x - 5$$

The coefficient of x^2 is negative. Therefore, when we graph the function, we get a parabola that opens downward.

8. To find the real zeros (if any), we can calculate the discriminant based on the general standard polynomial equation for a quadratic:

$$y = ax^2 + bx + c$$

In this situation, we have $a = -2$, $b = 2$, and $c = -5$. The discriminant, d, is

$$\begin{aligned} d &= b^2 - 4ac \\ &= 2^2 - 4 \times (-2) \times (-5) \\ &= 4 - 40 \\ &= -36 \end{aligned}$$

Because d is negative, we know that this function has no real zeros.

9. The x-coordinate of the vertex point, which represents an absolute maximum because the parabola opens downward, is

$$\begin{aligned} x_{max} &= -b/(2a) \\ &= -2 / [2 \times (-2)] \\ &= -2 / (-4) \\ &= 2/4 \\ &= 1/2 \end{aligned}$$

The y-coordinate can be found by plugging in x_{max} and calculating from the function:

$$\begin{aligned} y_{max} &= -2x_{max}^2 + 2x_{max} - 5 \\ &= -2 \times (1/2)^2 + 2 \times 1/2 - 5 \\ &= -2 \times 1/4 + 1 - 5 \\ &= -1/2 + 1 - 5 \\ &= 1/2 - 5 \\ &= -9/2 \end{aligned}$$

The coordinates of the vertex are $(1/2, -9/2)$.

10. We need more points for reference. We can find two more points on the graph by plugging in values of x smaller and larger than x_{max}. Let's try $x = -1$. Then

$$\begin{aligned} y &= -2x^2 + 2x - 5 \\ &= -2 \times (-1)^2 + 2 \times (-1) - 5 \\ &= -2 - 2 - 5 \\ &= -4 - 5 \\ &= -9 \end{aligned}$$

Now we know that $(-1, -9)$ is on the parabola. The x-value that we chose to find that point, -1, happens to be 3/2 units smaller than x_{max}. Let's choose a value of x that's 3/2 units larger than x_{max}. That would be $x = 2$. Then

$$\begin{aligned} y &= -2x^2 + 2x - 5 \\ &= -2 \times 2^2 + 2 \times 2 - 5 \\ &= -8 + 4 - 5 \\ &= -4 - 5 \\ &= -9 \end{aligned}$$

This gives us $(2, -9)$ as a third point on the graph. Now that we know three points on the curve, we can draw an approximation of the graph. Figure C-3 illustrates this parabola. On both axes, each increment represents 1 unit.

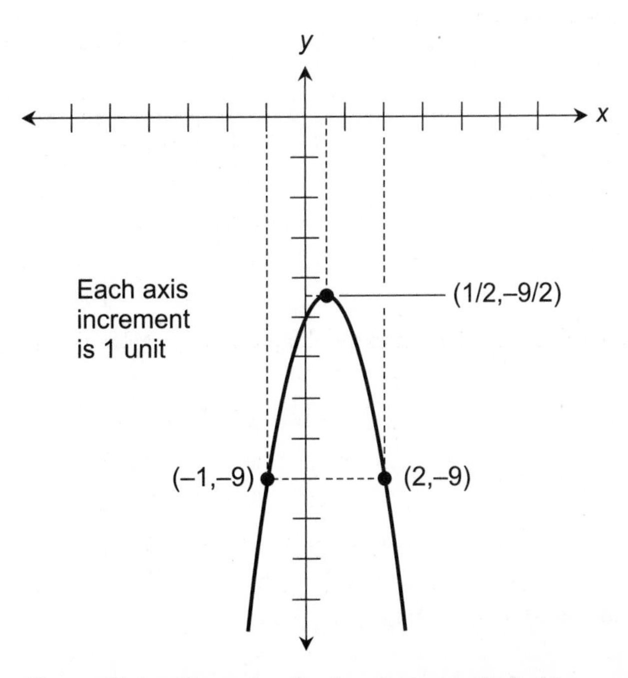

Figure C-3 Illustration for the solution to Prob. 10 in Chap. 24.

Chapter 25

1. Here's the binomial-cubed equation that we've been told to multiply out:

$$(a + b)^3 = 0$$

Let's start by rewriting the equation as a product of three identical binomials:

$$(ax + b)(ax + b)(ax + b) = 0$$

We can multiply the second two factors and then simplify to get

$$(ax + b)(a^2x^2 + 2abx + b^2) = 0$$

When we multiply these two factors and then simplify, we obtain

$$a^3x^3 + 3a^2bx^2 + 3ab^2x + b^3 = 0$$

2. Here's the binomial-cubed equation that we've been told to multiply out:

$$(3^{1/2}x - 12^{1/2})^3 = 0$$

We can rewrite this as

$$(3^{1/2}x - 12^{1/2})(3^{1/2}x - 12^{1/2})(3^{1/2}x - 12^{1/2}) = 0$$

Let's multiply out the second two factors. We must pay attention to the signs! The expression evolves as follows:

$$(3^{1/2}x - 12^{1/2})(3^{1/2}x - 12^{1/2})$$
$$(3^{1/2})^2x^2 - (3^{1/2} \times 12^{1/2})x - (12^{1/2} \times 3^{1/2})x + (12^{1/2})^2$$
$$3x^2 - (3 \times 12)^{1/2}x - (12 \times 3)^{1/2}x + 12$$
$$3x^2 - 36^{1/2}x - 36^{1/2}x + 12$$
$$3x^2 - 6x - 6x + 12$$
$$3x^2 - 12x + 12$$

Now, for the cubic, we have

$$(3^{1/2}x - 12^{1/2})(3x^2 - 12x + 12) = 0$$

When we multiply these factors, we get

$$(3^{1/2} \times 3)x^3 - (3^{1/2} \times 12)x^2 + (3^{1/2} \times 12)x$$
$$- (12^{1/2} \times 3)x^2 + (12^{1/2} \times 12)x - 12^{1/2} \times 12$$

We can morph the mixed products to get

$$(3^{1/2} \times 9^{1/2})x^3 - (3^{1/2} \times 144^{1/2})x^2 + (3^{1/2} \times 144^{1/2})x$$
$$- (12^{1/2} \times 9^{1/2})x^2 + (12^{1/2} \times 144^{1/2})x - 12^{1/2} \times 144^{1/2}$$

Applying the product of powers rule to each of these terms, we obtain

$$(3 \times 9)^{1/2}x^3 - (3 \times 144)^{1/2}x^2 + (3 \times 144)^{1/2}x$$
$$- (12 \times 9)^{1/2}x^2 + (12 \times 144)^{1/2}x - (12 \times 144)^{1/2}$$

This simplifies to

$$27^{1/2}x^3 - 432^{1/2}x^2 + 432^{1/2}x$$
$$- 108^{1/2}x^2 + 1{,}728^{1/2}x - 1{,}728^{1/2}$$

Consolidating the terms for each power of x and then returning the expression to the left side of the complete equation, we get

$$27^{1/2}x^3 - (432^{1/2} + 108^{1/2})x^2$$
$$+ (432^{1/2} + 1{,}728^{1/2})x - 1{,}728^{1/2} = 0$$

That's as "simple" as we can get this cubic in polynomial standard form. All of the coefficients, and the stand-alone constant, are irrational. This is the same equation that we solved in the first "challenge" in the chapter text, finding a single real root of 2. For extra credit, find a calculator that displays a lot of digits (such as the scientific calculator program in a personal computer), substitute 2 for x in the result we just got, and verify that it works out!

3. Here again, for reference, is the binomial factor form of the equation that we have been told to multiply out:

$$(3x + 2)(5x + 6)(-7x - 1) = 0$$

Let's multiply the second two factors first. We get

$$(3x + 2)(-35x^2 - 5x - 42x - 6) = 0$$

which consolidates to

$$(3x + 2)(-35x^2 - 47x - 6) = 0$$

Now we can multiply the binomial by the trinomial, obtaining

$$-105x^3 - 141x^2 - 18x - 70x^2 - 94x - 12 = 0$$

which consolidates to

$$-105x^3 - 211x^2 - 112x - 12 = 0$$

We can multiply through by −1 to get

$$105x^3 + 211x^2 + 112x + 12 = 0$$

Either of the last two equations is a legitimate expression of the polynomial standard form for the cubic. The second one more "sign-friendly."

4. Here are the real roots we found in the chapter text for the equation stated back at the beginning of Prob. 3:

$$x = -2/3 \qquad \text{or} \qquad x = -6/5 \qquad \text{or} \qquad x = -1/7$$

When we plug $x = -2/3$ into the final polynomial equation we found in solution to Prob. 3 and then grind through the arithmetic, the process goes like this, step-by-step:

$$105x^3 + 211x^2 + 112x + 12 = 0$$
$$105 \times (-2/3)^3 + 211 \times (-2/3)^2 + 112 \times (-2/3) + 12 = 0$$
$$105 \times (-8/27) + 211 \times (4/9) + 112 \times (-2/3) + 12 = 0$$
$$-840/27 + 844/9 - 224/3 + 12 = 0$$
$$-840/27 + 2{,}532/27 - 2{,}016/27 + 324/27 = 0$$
$$(-840 + 2{,}532 - 2{,}016 + 324) \ / \ 27 = 0$$
$$-840 + 2{,}532 - 2{,}016 + 324 = 0$$
$$0 = 0$$

To check the root $x = -6/5$, we go through this sequence of calculations:

$$105x^3 + 211x^2 + 112x + 12 = 0$$
$$105 \times (-6/5)^3 + 211 \times (-6/5)^2 + 112 \times (-6/5) + 12 = 0$$
$$105 \times (-216/125) + 211 \times (36/25) + 112 \times (-6/5) + 12 = 0$$
$$-22{,}680/125 + 7{,}596/25 - 672/5 + 12 = 0$$
$$-22{,}680/125 + 37{,}980/125 - 16{,}800/125 + 1{,}500/125 = 0$$
$$(-22{,}680 + 37{,}980 - 16{,}800 + 1{,}500) \ / \ 125 = 0$$
$$-22{,}680 + 37{,}980 - 16{,}800 + 1{,}500 = 0$$
$$0 = 0$$

To check the root $x = -1/7$, we go through this arithmetic, step-by-step:

$$105x^3 + 211x^2 + 112x + 12 = 0$$
$$105 \times (-1/7)^3 + 211 \times (-1/7)^2 + 112 \times (-1/7) + 12 = 0$$
$$105 \times (-1/343) + 211 \times (1/49) + 112 \times (-1/7) + 12 = 0$$
$$-105/343 + 211/49 - 112/7 + 12 = 0$$
$$-105/343 + 1{,}477/343 - 5{,}488/343 + 4{,}116/343 = 0$$
$$(-105 + 1{,}477 - 5{,}488 + 4{,}116) \ / \ 343 = 0$$
$$-105 + 1{,}477 - 5{,}488 + 4{,}116 = 0$$
$$0 = 0$$

5. Here's the general binomial-trinomial equation once again:

$$(a_1x + b_1)(a_2x^2 + b_2x + c) = 0$$

Using the product of sums rule, we can rewrite this as

$$a_1a_2x^3 + a_1b_2x^2 + a_1cx + b_1a_2x^2 + b_1b_2x + b_1c = 0$$

When we bring the terms for x^2 next to each other, and then do the same thing with the terms for x, we get

$$a_1a_2x^3 + a_1b_2x^2 + b_1a_2x^2 + a_1cx + b_1b_2x + b_1c = 0$$

Let's use the commutative law for multiplication on the third term to get "*a* before *b*" in the interest of elegance! That gives us

$$a_1a_2x^3 + a_1b_2x^2 + a_2b_1x^2 + a_1cx + b_1b_2x + b_1c = 0$$

Finally, we can use the distributive law for multiplication over addition to consolidate the coefficients for x^2 and x, giving us the equation in true polynomial standard form:

$$a_1a_2x^3 + (a_1b_2 + a_2b_1)x^2 + (a_1c + b_1b_2)x + b_1c = 0$$

6. If the coefficients and constants are real numbers, and if a_1 and a_2 are both nonzero, then the cubic

$$(a_1x + b_1)(a_2x^2 + b_2x + c) = 0$$

has at least one real root, which is $x = -b_1/a_1$. That's the root that we get when we create a first-degree equation from the binomial term by setting it equal to 0. The cubic might have no more real roots (therefore one real root in total), one more (two in total), or two more (three in total). To find out which of these situations is the true case, we can look at the discriminant d for the quadratic

$$a_2x^2 + b_2x + c = 0$$

In this notation,

$$d = b_2^2 - 4a_2c$$

If $d > 0$, then the quadratic has two real roots, so the original cubic has three. If $d = 0$, then the quadratic has one real root with multiplicity 2, so the original cubic has two real roots, one of which has multiplicity 2. If $d < 0$, then the quadratic has no real roots, so the original cubic has only one.

7. For reference, here's the equation again:

$$(3x + 5)(16x^2 - 56x + 49) = 0$$

We found the real roots

$$x = -5/3 \qquad \text{or} \qquad x = 7/4$$

Let's put −5/3 in place of x, and then carry out the calculations. We get

$$[3 \times (-5/3) + 5][16 \times (-5/3)^2 - 56 \times (-5/3) + 49] = 0$$

The first term in square brackets is equal to 0. Let's check:

$$\begin{aligned} &3 \times (-5/3) + 5 \\ &= -15/3 + 5 \\ &= -5 + 5 \\ &= 0 \end{aligned}$$

This means the entire expression must be 0; it doesn't matter what the second term is. Now let's insert 7/4 for x. We have only to work with the trinomial term this time, and it comes out equal to 0. Let's try it:

$$\begin{aligned} &16 \times (7/4)^2 - 56 \times (7/4) + 49 \\ &= 16 \times 49/16 - 392/4 + 49 \\ &= 49 - 98 + 49 \\ &= -49 + 49 \\ &= 0 \end{aligned}$$

The whole expression must equal 0 because the second factor is 0.

8. We begin by setting up the synthetic division array with the "test root," $x = 5$, and the coefficients in the top row, like this:

5	−9	21	104	80
		#	#	#
	#	#	#	#

Subsequent steps proceed as follows.

5	−9	21	104	80
		#	#	#
	−9	#	#	#

5	−9	21	104	80
		−45	#	#
	−9	#	#	#

5	−9	21	104	80
		−45	#	#
	−9	−24	#	#

5	−9	21	104	80
		−45	−120	#
	−9	−24	#	#

5	−9	21	104	80
		−45	−120	#
	−9	−24	−16	#

5	−9	21	104	80
		−45	−120	−80
	−9	−24	−16	#

5	−9	21	104	80
		−45	−120	−80
	−9	−24	−16	0

We get a remainder of 0, so we know that $x = 5$ is a real root of the original cubic.

9. We can write the cubic presented in Prob. 8 in binomial-trinomial form on the basis of the results of the synthetic division. Because $x = 5$ is a real root of the cubic, we know that it has a binomial factor of $(x - 5)$. In the trinomial factor, the coefficient of x^2 is -9, the coefficient of x is -24, and the stand-alone constant is -16, because those three numbers appear, in that order, in the bottom row before the remainder 0. We can now write down the entire binomial-trinomial cubic equation:

$$(x - 5)(-9x^2 - 24x - 16) = 0$$

To figure out whether there are any other roots besides $x = 5$, we must find the discriminant of the trinomial factor. If we let $a_2 = -9$, $b_2 = -24$, and $c = -16$, we can find the discriminant, d, as follows:

$$\begin{aligned} d &= b_2{}^2 - 4a_2c \\ &= (-24)^2 - 4 \times (-9) \times (-16) \\ &= 576 - 576 \\ &= 0 \end{aligned}$$

Because $d = 0$, we know that the quadratic we get by setting the trinomial equal to 0 has one real root with multiplicity 2. That means the original cubic has one more real root besides $x = 5$, and that root has multiplicity 2. To find it, we can use the quadratic formula with the coefficients and constant named according to the above scheme:

$$x = [-b_2 \pm (b_2^2 - 4a_2c)^{1/2}] / (2a_2)$$

Because the discriminant is equal to 0, we can simplify this to

$$x = -b_2 / (2a_2)$$

Substituting in the values $a_2 = -9$ and $b_2 = -24$, we get

$$\begin{aligned} x &= -(-24) / [2 \times (-9)] \\ &= 24/(-18) \\ &= -24/18 \\ &= -4/3 \end{aligned}$$

The roots of the cubic are therefore $x = 5$ or $x = -4/3$. The root $x = -4/3$ occurs with multiplicity 2. The solution set is $X = \{5, -4/3\}$.

10. The new cubic, written in binomial-trinomial form, looks like this:

$$(x + 3/2)(6x^2 - 4x + 2) = 0$$

Let's examine the discriminant d of the trinomial. Setting $a_2 = 6$, $b_2 = -4$, and $c = 2$, we get

$$\begin{aligned} d &= b_2^2 - 4a_2c \\ &= (-4)^2 - 4 \times 6 \times 2 \\ &= 16 - 48 \\ &= -32 \end{aligned}$$

Because $d < 0$, the new cubic has only one real root, $x = -3/2$, exactly as the original cubic did. (The two complex roots, however, differ in this equation compared with those in the final "challenge" in the chapter text. For extra credit, you can verify this fact.)

Chapter 26

1. In each of these situations, the trinomial can be factored into the square of a binomial. Then that squared binomial is raised to the indicated power.

   (a) Here is the original equation:

$$(x^2 + 6x + 9)^2 = 0$$

In the trinomial, the coefficient of x^2 is 1, the coefficient of x is 6, and the stand-alone constant is 9. We must find a number, such that adding it to itself yields 6 while squaring it yields 9. That number is 3. The binomial is therefore $(x + 3)$, and we have

$$[(x+3)^2]^2 = 0$$

which simplifies to

$$(x+3)^4 = 0$$

(b) Here is the original equation:

$$(x^2 - 4x + 4)^3 = 0$$

In the trinomial, the coefficient of x^2 is 1, the coefficient of x is -4, and the constant is 4. We must find a number, such that adding it to itself yields -4 while squaring it yields 4. That number is -2. The binomial is therefore $(x - 2)$, and we have

$$[(x-2)^2]^3 = 0$$

which simplifies to

$$(x-2)^6 = 0$$

(c) Here is the original equation:

$$(16x^2 - 24x + 9)^4 = 0$$

This trinomial is the square of $(4x - 3)$. Therefore, the original equation is equivalent to

$$[(4x-3)^2]^4 = 0$$

which simplifies to

$$(4x-3)^8 = 0$$

2. In each case, we can remove the exponent from the binomial and then set it equal to 0, obtaining a first-degree equation. The real root of the higher-degree equation is equal to the solution of the first-degree equation. The multiplicity of the root is the value of the exponent n to which the binomial is raised.

(a) The real root is found by solving

$$x + 3 = 0$$

That root is $x = -3$. Because the binomial is raised to the fourth power, this single real root has multiplicity 4.

(b) The real root is found by solving

$$x - 2 = 0$$

That root is $x = 2$. Because the binomial is raised to the sixth power, this single real root has multiplicity 6.

(c) The real root is found by solving

$$4x - 3 = 0$$

We can add 3 to each side and then divide through by 4, obtaining the root $x = 3/4$. Because the binomial is raised to the eighth power, this single real root has multiplicity 8.

3. In each of these situations, the trinomial can be factored into the product of two different binomials. Then that product is raised to the indicated power.

(a) Here is the original equation:

$$(x^2 - 3x + 2)^2 = 0$$

In the trinomial, the coefficient of x^2 is 1, the coefficient of x is -3, and the stand-alone constant is 2. This trinomial factors into the product of $(x - 1)$ and $(x - 2)$. Therefore, the original equation can be rewritten as

$$[(x - 1)(x - 2)]^2 = 0$$

which can be further broken down to

$$(x - 1)^2(x - 2)^2 = 0$$

(b) Here is the original equation:

$$(-3x^2 - 5x + 2)^5 = 0$$

In the trinomial, the coefficient of x^2 is -3, the coefficient of x is -5, and the constant is 4. This trinomial factors into the product of $(x + 2)$ and $(-3x + 1)$. Therefore, we can rewrite the original equation as

$$[(x + 2)(-3x + 1)]^5 = 0$$

and break it down to

$$(x + 2)^5(-3x + 1)^5 = 0$$

(c) Here is the original equation:

$$(4x^2 - 9)^3 = 0$$

Here, the coefficient of x^2 is 4, the coefficient of x is 0, and the constant is −9. This trinomial factors into the product of $(2x + 3)$ and $(2x - 3)$. Therefore, we can rewrite the original equation as

$$[(2x + 3)(2x - 3)]^3 = 0$$

and break it down to

$$(2x + 3)^3(2x - 3)^3 = 0$$

4. In each case, we can remove the exponents from the binomials, and then consider each binomial separately as a first-degree equation when it is set equal to 0. The real roots of the higher-degree equation are equal to the solutions of the first-degree equations. The multiplicity of each root is the power to which its associated binomial is raised.

(a) The real roots are found by solving

$$x - 1 = 0$$

and

$$x - 2 = 0$$

Those roots are $x = 1$ or $x = 2$. Because each binomial is squared, each of these roots has multiplicity 2.

(b) The real roots are found by solving

$$x + 2 = 0$$

and

$$-3x + 1 = 0$$

Those roots are $x = -2$ or $x = 1/3$. Because each binomial is raised to the fifth power, each of these roots has multiplicity 5.

(c) The real roots are found by solving

$$2x + 3 = 0$$

and

$$2x - 3 = 0$$

Those roots are $x = -3/2$ or $x = 3/2$. Because each binomial is cubed, each of these roots has multiplicity 3.

5. Here's the original binomial factor equation, which we have been told to solve and scrutinize:

$$(x - 3/2)^2(2x - 7)^2(7x)^3(-3x + 5)^5 = 0$$

Let's set each binomial equal to 0, and then solve the resulting first-degree equations. Those equations are

$$x - 3/2 = 0$$
$$2x - 7 = 0$$
$$7x = 0$$
$$-3x + 5 = 0$$

The solutions to these first-degree equations, and therefore the real roots of the higher-degree equation, are

$$x = 3/2 \quad \text{or} \quad x = 7/2 \quad \text{or} \quad x = 0 \quad \text{or} \quad x = 5/3$$

The solution set is $X = \{3/2, 7/2, 0, 5/3\}$. The multiplicity of each root is the same as the power to which its binomial is raised in the original equation. Therefore, the root $x = 3/2$ has multiplicity 2, the root $x = 7/2$ has multiplicity 2, the root $x = 0$ has multiplicity 3, and the root $x = 5/3$ has multiplicity 5. The degree of the original equation is the sum of the exponents attached to the factors, which is $2 + 2 + 3 + 5 = 12$.

6. Here's the original binomial factor equation once again, for reference:

$$(x + 4)(2x - 8)^2(x/3 + 12)^3 = 0$$

Let's set each binomial equal to 0, and then solve the resulting first-degree equations. Those equations are

$$x + 4 = 0$$
$$2x - 8 = 0$$
$$x/3 + 12 = 0$$

The solution to the first of these is $x = -4$. The solution to the second is $x = 4$. To solve the third equation, we can subtract 12 from each side and then multiply through by 3, obtaining $x = -36$. The roots of the original equation are therefore

$$x = -4 \quad \text{or} \quad x = 4 \quad \text{or} \quad x = -36$$

The solution set is $X = \{-4, 4, -36\}$. The multiplicity of each root is the same as the power to which its binomial is raised in the original equation. Therefore, the root $x = -4$ has multiplicity 1, the root $x = 4$ has multiplicity 2, and the root $x = -36$ has multiplicity 3. The degree of the original equation is the sum of the exponents attached to the factors, which is $1 + 2 + 3 = 6$.

7. For reference, the polynomial equation is

$$2x^5 - 3x^3 - 2x + 2 = 0$$

We have many options! The largest absolute value of any coefficient or constant is 3, so we can try 3 for the upper bound and −3 for the lower bound. If either or both of these fail, we can try values farther from 0. The coefficients and constant, in order of decreasing powers of x, are 2, 0, −3, 0, −2, and 2. (The coefficients of x^4 and x^2 are both equal to 0.) Here's the synthetic division array for the "test root" 3:

3	2	0	−3	0	−2	2
		#	#	#	#	#
	#	#	#	#	#	#

When we go through the synthetic division process, we end up with

3	2	0	−3	0	−2	2
		6	18	45	135	399
	2	6	15	45	133	401

None of the numbers in the bottom row are negative. This tells us that 3 is an upper bound for the real roots. To try the "test root" −3, we set up the array

3	2	0	−3	0	−2	2
		#	#	#	#	#
	#	#	#	#	#	#

The synthetic division process leads us to

−3	2	0	−3	0	−2	2
		−6	−18	−45	135	−399
	2	−6	15	−45	133	−397

The numbers in the bottom row alternate in sign. This indicates that −3 is a lower bound for the real roots.

8. Here is an outline of the process for finding the rational roots of the equation

$$2x^5 - 3x^3 - 2x + 2 = 0$$

- All the coefficients, as well as the stand-alone constant, are integers, so we don't have to multiply the equation through by anything.
- The integer factors m of the stand-alone constant are 2, 1, −2, and −1.
- The integer factors n of the leading coefficient are 2, 1, −2, and −1.
- All the possible ratios $r = m/n$ are 2, 1, 1/2, −2, −1, and −1/2.
- We input rational numbers r of 2, 1, 1/2, −2, −1, and −1/2 to synthetic division arrays, and see if we get a remainder of 0 for any of them.
- We don't get a remainder of 0 when we input any of the above values of r to the synthetic division array. Therefore, the equation has no rational roots.

9. For reference, the polynomial equation is

$$3x^5 - 3x^2 + 2x - 2 = 0$$

Let's use the same method as we did in solution to Prob. 7. The largest absolute value of any coefficient or constant is 3, so we can try 3 for the upper bound and −3 for the lower bound. If either or both of these fail, we can try values farther from 0. The coefficients and constant, in order of decreasing powers of x, are 3, 0, 0, −3, 2, and −2. (The coefficients of x^4 and x^3 are both equal to 0.) Here's the synthetic division array for the "test root" 3:

3	3	0	0	−3	2	−2
		#	#	#	#	#
	#	#	#	#	#	#

When we go through the synthetic division process, we end up with

3	3	0	0	−3	2	−2
		9	27	81	234	708
	3	9	27	78	236	706

None of the numbers in the bottom row are negative. This tells us that 3 is an upper bound for the real roots. To try the "test root" −3, we set up the array

3	3	0	0	−3	2	−2
		#	#	#	#	#
	#	#	#	#	#	#

The synthetic division process leads us to

−3	3	0	0	−3	2	−2
		−9	27	−81	252	−762
	3	−9	27	−84	254	−764

The numbers in the bottom row alternate in sign. This indicates that −3 is a lower bound for the real roots.

10. Here is an outline of the process for finding the rational roots of the equation

$$3x^5 - 3x^2 + 2x - 2 = 0$$

- All the coefficients, as well as the stand-alone constant, are integers, so we don't have to multiply the equation through by anything.
- The integer factors m of the stand-alone constant are 2, 1, −2, and −1.
- The integer factors n of the leading coefficient are 3, 1, −3, and −1.
- All the possible ratios $r = m/n$ are 2, 1, 2/3, 1/3, −2, −1, −2/3, and −1/3.
- We input rational numbers r of 2, 1, 2/3, 1/3, −2, −1, −2/3, and −1/3 to synthetic division arrays, and see if we get a remainder of 0 for any of them.
- We get a remainder of 0 only when $r = 1$. Therefore, $x = 1$ is the only rational root of the equation.

Chapter 27

1. Here are the two equations in their original forms:

$$3x + y - 1 = 0$$

and

$$2x^2 - y + 1 = 0$$

We can morph these to obtain the following functions of x:

$$y = -3x + 1$$

and

$$y = 2x^2 + 1$$

When we mix the right sides of these equations, we obtain

$$-3x + 1 = 2x^2 + 1$$

which morphs into the quadratic equation

$$2x^2 + 3x = 0$$

Let a be the coefficient of x^2, let b be the coefficient of x, and let c be the stand-alone constant. Then $a = 2$, $b = 3$, and $c = 0$. The quadratic formula tells us that

$$\begin{aligned} x &= [-b \pm (b^2 - 4ac)^{1/2}] / (2a) \\ &= [-3 \pm (3^2 - 4 \times 2 \times 0)^{1/2}] / (2 \times 2) \\ &= (-3 \pm 3) / 4 \\ &= 0/4 \quad \text{or} \quad -6/4 \\ &= 0 \quad \text{or} \quad -3/2 \end{aligned}$$

The roots are $x = 0$ or $x = -3/2$. We can plug these into either of the original functions to get the y-values. The linear function is easier. When $x = 0$, we have

$$\begin{aligned} y &= -3x + 1 \\ &= -3 \times 0 + 1 \\ &= 0 + 1 \\ &= 1 \end{aligned}$$

When $x = -3/2$, we have

$$\begin{aligned} y &= -3x + 1 \\ &= -3 \times (-3/2) + 1 \\ &= 9/2 + 1 \\ &= 9/2 + 2/2 \\ &= 11/2 \end{aligned}$$

The solutions to the system are $(x,y) = (0,1)$ and $(x,y) = (-3/2,11/2)$.

2. First, let's check (0,1) in the original linear equation:

$$\begin{gathered} 3x + y - 1 = 0 \\ 3 \times 0 + 1 - 1 = 0 \\ 0 + 1 - 1 = 0 \\ 1 - 1 = 0 \\ 0 = 0 \end{gathered}$$

Next, let's check (−3/2,11/2) in that same equation:

$$\begin{gathered} 3x + y - 1 = 0 \\ 3 \times (-3/2) + 11/2 - 1 = 0 \\ -9/2 + 11/2 - 1 = 0 \\ 2/2 - 1 = 0 \\ 1 - 1 = 0 \\ 0 = 0 \end{gathered}$$

Next, let's check (0,1) in the original two-variable quadratic equation:

$$2x^2 - y + 1 = 0$$
$$2 \times 0^2 - 1 + 1 = 0$$
$$2 \times 0 - 1 + 1 = 0$$
$$0 - 1 + 1 = 0$$
$$-1 + 1 = 0$$
$$0 = 0$$

Finally, let's check (−3/2,11/2) in that same equation:

$$2x^2 - y + 1 = 0$$
$$2 \times (-3/2)^2 - 11/2 + 1 = 0$$
$$2 \times 9/4 - 11/2 + 1 = 0$$
$$9/2 - 11/2 + 1 = 0$$
$$-2/2 + 1 = 0$$
$$-1 + 1 = 0$$
$$0 = 0$$

3. Here are the two equations in their original forms:

$$3x + y - 1 = 0$$

and

$$2x^2 - 3x - y + 3 = 0$$

We can manipulate these to obtain the following functions of x:

$$y = -3x + 1$$

and

$$y = 2x^2 - 3x + 3$$

When we mix the right sides of these equations, we obtain

$$-3x + 1 = 2x^2 - 3x + 3$$

which can be rewritten in standard form as the quadratic equation

$$2x^2 + 2 = 0$$

Let a be the coefficient of x^2, let b be the coefficient of x, and let c be the stand-alone constant. Then $a = 2$, $b = 0$, and $c = 2$. The quadratic formula tells us that

$$\begin{aligned} x &= [-b \pm (b^2 - 4ac)^{1/2}] \,/\, (2a) \\ &= [0 \pm (0^2 - 4 \times 2 \times 2)^{1/2}] \,/\, (2 \times 2) \\ &= \pm (-16)^{1/2} \,/\, 4 \\ &= \pm j4 \,/\, 4 \\ &= \pm j \end{aligned}$$

The roots are $x = j$ or $x = -j$. We can plug these into either of the original functions to get the y-values. This time, let's use the quadratic function. When $x = j$, we have

$$\begin{aligned} y &= 2j^2 - 3j + 3 \\ &= 2 \times (-1) - j3 + 3 \\ &= -2 - j3 + 3 \\ &= 1 - j3 \end{aligned}$$

When $x = -j$, we have

$$\begin{aligned} y &= 2(-j)^2 - 3(-j) + 3 \\ &= 2 \times (-1) + j3 + 3 \\ &= -2 + j3 + 3 \\ &= 1 + j3 \end{aligned}$$

The solutions to the system are $(x,y) = [j,(1 - j3)]$ and $(x,y) = [-j,(1 + j3)]$.

4. After we plug in the values, let's convert subtractions to negative additions to be sure we keep the signs in order. First, let's check $[j,(1 - j3)]$ in the original linear equation:

$$\begin{aligned} 3x + y - 1 &= 0 \\ 3j + (1 - j3) - 1 &= 0 \\ j3 + 1 + (-j3) + (-1) &= 0 \\ j3 + (-j3) + 1 + (-1) &= 0 \\ 0 &= 0 \end{aligned}$$

Next, let's check $[-j,(1 + j3)]$ in that equation:

$$\begin{aligned} 3x + y - 1 &= 0 \\ 3(-j) + (1 + j3) - 1 &= 0 \\ -j3 + 1 + j3 + (-1) &= 0 \\ -j3 + j3 + 1 + (-1) &= 0 \\ 0 &= 0 \end{aligned}$$

Next, let's check $[j,(1-j3)]$ in the original quadratic equation:

$$2x^2 - 3x - y + 3 = 0$$
$$2j^2 - 3j - (1 - j3) + 3 = 0$$
$$2 \times (-1) + (-j3) + [-(1 - j3)] + 3 = 0$$
$$-2 + (-j3) + (-1) + j3 + 3 = 0$$
$$-j3 + j3 + (-2) + (-1) + 3 = 0$$
$$0 = 0$$

Finally, let's check $[-j,(1+j3)]$ in that equation:

$$2x^2 - 3x - y + 3 = 0$$
$$2(-j)^2 - 3(-j) - (1 + j3) + 3 = 0$$
$$2 \times (-1) + j3 + [-(1 + j3)] + 3 = 0$$
$$-2 + j3 + (-1) + (-j3) + 3 = 0$$
$$j3 + (-j3) + (-2) + (-1) + 3 = 0$$
$$0 = 0$$

5. Here are the two equations in their original forms:

$$x^2 + x - y = -1$$

and

$$x^2 - 2x - y = 2$$

We can morph these into functions of x, obtaining

$$y = x^2 + x + 1$$

and

$$y = x^2 - 2x - 2$$

When we mix the right sides of these equations, we obtain

$$x^2 + x + 1 = x^2 - 2x - 2$$

Adding the quantity $(-x^2 + 2x + 2)$ to each side gives us

$$3x + 3 = 0$$

which resolves to $x = -1$. That's the only root of the equation we got by morphing and mixing. To obtain the y-value for the solution to the two-by-two system, we can plug the

x-value into either of the original quadratic functions. Let's use the first one:

$$\begin{aligned} y &= x^2 + x + 1 \\ &= (-1)^2 + (-1) + 1 \\ &= 1 + (-1) + 1 \\ &= 1 \end{aligned}$$

The system has the single solution $(x,y) = (-1,1)$.

6. First, let's check $(-1,1)$ in the first original quadratic:

$$x^2 + x - y = -1$$
$$(-1)^2 + (-1) - 1 = -1$$
$$1 + (-1) - 1 = -1$$
$$0 - 1 = -1$$
$$-1 = -1$$

Next, let's check $(-1,1)$ in the second original quadratic:

$$x^2 - 2x - y = 2$$
$$(-1)^2 - 2 \times (-1) - 1 = 2$$
$$1 - (-2) - 1 = 2$$
$$1 + 2 - 1 = 2$$
$$3 - 1 = 2$$
$$2 = 2$$

7. Here are the two equations in their original forms:

$$x^2 + y = 0$$

and

$$2x^3 - y = 0$$

We can morph these into functions of x, getting

$$y = -x^2$$

and

$$y = 2x^3$$

Mixing the right sides of these equations gives us

$$-x^2 = 2x^3$$

Adding x^2 to each side and then transposing the left and right sides gives us a cubic equation in polynomial standard form:

$$2x^3 + x^2 = 0$$

It's tempting to divide this equation through by x^2. But if $x = 0$ happens to be a root, dividing through by x^2 will blind us to the existence of that root (and might cause other problems, too). We can use a two-step trick to avoid that trouble. First, let's check to see if $x = 0$ is a root by plugging it in and doing the arithmetic. We get

$$2 \times 0^3 + 0^2 = 0$$
$$0 + 0 = 0$$
$$0 = 0$$

This cubic does have the root $x = 0$! Now that we're aware of this fact, the second step in our trick is to see if the equation has any other roots. Let's impose a temporary restriction on x: It can have any value *except* 0. That makes it "legal" to divide through by x^2, obtaining the equation

$$(2x^3 + x^2) / x^2 = 0/x^2$$

We can rewrite this as

$$2x^3/x^2 + x^2/x^2 = 0$$

which simplifies to

$$2x + 1 = 0$$

Subtracting 1 from each side and then dividing through by 2 tells us that $x = -1/2$. We can now remove the temporary restriction on the value of x, making sure we include the root $x = 0$. The cubic equation we got by morphing and mixing therefore has two roots:

$$x = 0 \qquad \text{or} \qquad x = -1/2$$

We can plug these into either of the original functions to get the y-values. Let's use the second one. When $x = 0$, we have

$$\begin{aligned} y &= 2x^3 \\ &= 2 \times 0^3 \\ &= 2 \times 0 \\ &= 0 \end{aligned}$$

When $x = -1/2$, we have

$$\begin{aligned} y &= 2x^3 \\ &= 2 \times (-1/2)^3 \\ &= 2 \times (-1/8) \\ &= -1/4 \end{aligned}$$

The solutions to the system are $(x,y) = (0,0)$ and $(x,y) = (-1/2,-1/4)$. The solution (0,0) has multiplicity 2. Consider again the cubic that we got by mixing:

$$2x^3 + x^2 = 0$$

This factors into

$$(x)(x)(2x + 1) = 0$$

The root $x = 0$ occurs once for each factor of x, or twice in total. That means the corresponding solution (0,0) has multiplicity 2 in the two-by-two system.

8. First, let's check (0,0) in the original two-variable quadratic equation:

$$\begin{gathered} x^2 + y = 0 \\ 0^2 + 0 = 0 \\ 0 + 0 = 0 \\ 0 = 0 \end{gathered}$$

Next, let's check $(-1/2,-1/4)$ in that equation:

$$\begin{gathered} x^2 + y = 0 \\ (-1/2)^2 + (-1/4) = 0 \\ 1/4 + (-1/4) = 0 \\ 0 = 0 \end{gathered}$$

Next, let's check (0,0) in the original two-variable cubic equation:

$$\begin{gathered} 2x^3 - y = 0 \\ 2 \times 0^3 - 0 = 0 \\ 2 \times 0 - 0 = 0 \\ 0 - 0 = 0 \\ 0 = 0 \end{gathered}$$

Finally, let's check (−1/2,−1/4) in that equation:

$$2x^3 - y = 0$$
$$2 \times (-1/2)^3 - (-1/4) = 0$$
$$2 \times (-1/8) + 1/4 = 0$$
$$-1/4 + 1/4 = 0$$
$$0 = 0$$

9. Here are the two equations in their original forms:

$$4x^3 + 2x^2 + 2x - 2y - 8 = 0$$

and

$$3x^3 - 2x^2 + 4x - y - 5 = 0$$

The first of these can be simplified if we divide through by 2. That gives us

$$2x^3 + x^2 + x - y - 4 = 0$$

We can morph this into a function of x by adding y to each side, getting

$$y = 2x^3 + x^2 + x - 4$$

The second original equation can also be modified by adding y to each side, obtaining the function

$$y = 3x^3 - 2x^2 + 4x - 5$$

Mixing the right sides of these two cubic functions, we get

$$2x^3 + x^2 + x - 4 = 3x^3 - 2x^2 + 4x - 5$$

Now let's add the quantity $(-2x^3 - x^2 - x + 4)$ to each side and then transpose the equation left-to-right. That gives us

$$x^3 - 3x^2 + 3x - 1 = 0$$

The coefficients and constant in this equation show a certain symmetry. Whenever we see a pattern of this sort in a cubic or higher-degree equation, it suggests that the equation can be factored. After a few trials and errors, we discover that we have a binomial cubed:

$$(x - 1)^3 = 0$$

Which can be written out fully as

$$(x - 1)(x - 1)(x - 1) = 0$$

This equation has the single root $x = 1$, which occurs with multiplicity 3. We can plug $x = 1$ into either of the original functions to get the y-value. Let's use the first function. That gives us

$$\begin{aligned} y &= 2x^3 + x^2 + x - 4 \\ &= 2 \times 1^3 + 1^2 + 1 - 4 \\ &= 2 \times 1 + 1 + 1 - 4 \\ &= 2 + 1 + 1 - 4 \\ &= 0 \end{aligned}$$

The solution to the system is therefore $(x,y) = (1,0)$, with multiplicity 3.

10. First, we check (1,0) in the first original two-variable cubic:

$$4x^3 + 2x^2 + 2x - 2y - 8 = 0$$
$$4 \times 1^3 + 2 \times 1^2 + 2 \times 1 - 2 \times 0 - 8 = 0$$
$$4 \times 1 + 2 \times 1 + 2 - 0 - 8 = 0$$
$$4 + 2 + 2 - 8 = 0$$
$$8 - 8 = 0$$
$$0 = 0$$

Next, we check (1,0) in the second original cubic:

$$3x^3 - 2x^2 + 4x - y - 5 = 0$$
$$3 \times 1^3 - 2 \times 1^2 + 4 \times 1 - 0 - 5 = 0$$
$$3 \times 1 - 2 \times 1 + 4 - 5 = 0$$
$$3 - 2 + 4 - 5 = 0$$
$$1 + 4 - 5 = 0$$
$$5 - 5 = 0$$
$$0 = 0$$

Chapter 28

1. See Table C-2, which shows selected values for the functions

$$y = -3x + 1$$

and

$$y = 2x^2 + 1$$

Bold numerals indicate the real solutions we found when we worked out Prob. 1 in Chap. 27. The fractional values are also shown in decimal form so they'll be easier to graph.

2. See Fig. C-4. Each horizontal increment is 1/2 unit. Each vertical increment is 3 units.

Table C-2. Solution to Prob. 1 in Chap. 28.

x	$-3x + 1$	$2x^2 + 1$
−3	10	19
−2	7	9
−3/2	**11/2**	**11/2**
or −1.5	**or 5.5**	**or 5.5**
−1	4	3
−1/2	5/2	3/2
or −0.5	or 2.5	or 1.5
0	**1**	**1**
1	−2	3
2	−5	9

3. See Table C-3, which shows selected values for the functions

$$y = -3x + 1$$

and

$$y = 2x^2 - 3x + 3$$

There are no real solutions to this system, as we discovered when we worked out Prob. 3 in Chap. 27.

4. See Fig. C-5. Each horizontal increment is 1 unit. Each vertical increment is 5 units.

5. See Table C-4, which shows selected values for the functions

$$y = x^2 + x + 1$$

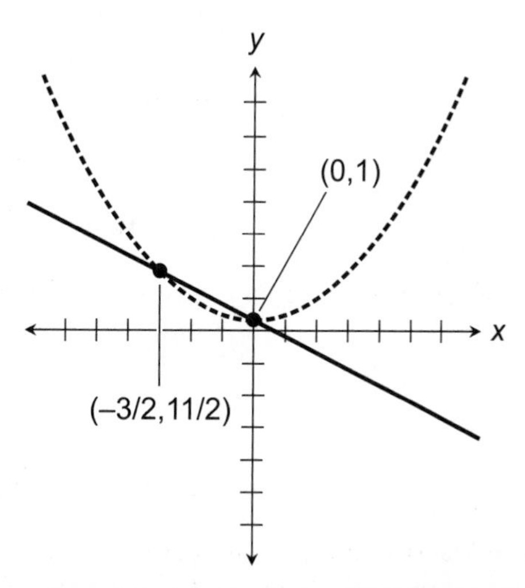

Figure C-4 Illustration for the solution to Prob. 2 in Chap. 28.

Table C-3. Solution to Prob. 3 in Chap. 28.

x	$-3x+1$	$2x^2-3x+3$
−3	10	30
−2	7	17
−1	4	8
0	1	3
1	−2	2
2	−5	5
3	−8	12
4	−11	23

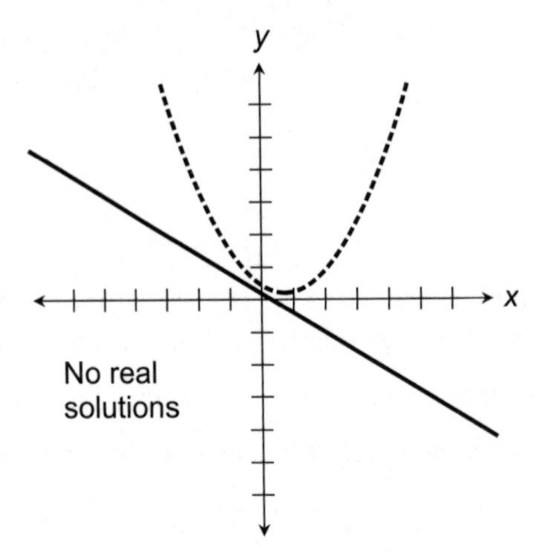

Figure C-5 Illustration for the solution to Prob. 4 in Chap. 28.

Table C-4. Solution to Prob. 5 in Chap. 28.

x	x^2+x+1	x^2-2x-2
−4	13	22
−3	7	13
−2	3	6
−1	**1**	**1**
0	1	−2
1	3	−3
2	7	−2
3	13	1
4	21	6

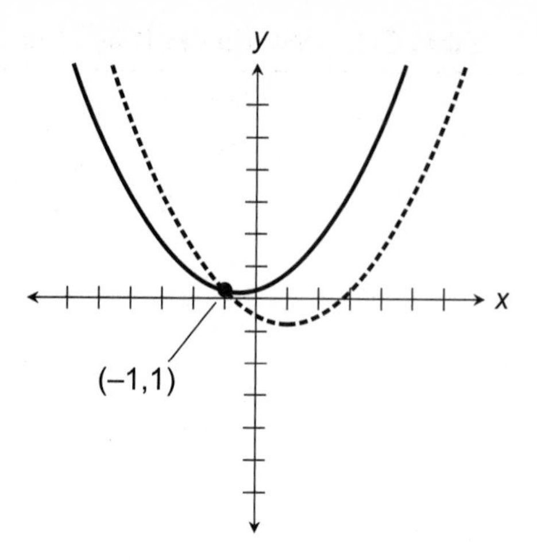

Figure C-6 Illustration for the solution to Prob. 6 in Chap. 28.

and

$$y = x^2 - 2x - 2$$

Bold numerals indicate the real solution we found when we worked out Prob. 5 in Chap. 27.

6. See Fig. C-6. Each horizontal increment is 1 unit. Each vertical increment is 4 units.
7. See Table C-5, which shows selected values for the functions

$$y = -x^2$$

and

$$y = 2x^3$$

Bold numerals indicate the real solution we found when we worked out Prob. 7 in Chap. 27.

Table C-5. Solution to Prob. 7 in Chap. 28.

x	$-x^2$	$2x^3$
−3	−9	−54
−2	−4	−16
−1	−1	−2
0	**0**	**0**
1	−1	2
2	−4	16
3	−9	54

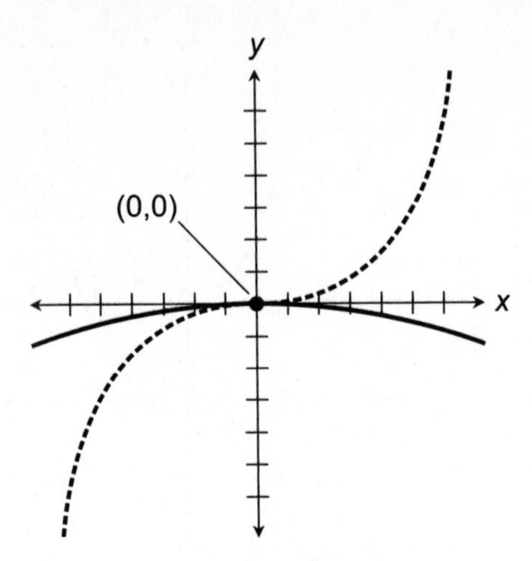

Figure C-7 Illustration for the solution to Prob. 8 in Chap. 28.

8. See Fig. C-7. Each horizontal increment is 1/2 unit. Each vertical increment is 10 units.
9. See Table C-6, which shows selected values for the functions

$$y = 2x^3 + x^2 + x - 4$$

and

$$y = 3x^3 - 2x^2 + 4x - 5$$

Bold numerals indicate the real solution we found when we worked out Prob. 9 in Chap. 27.

10. See Fig. C-8. Each horizontal increment is 1 unit. Each vertical increment is 20 units.

Table C-6. Solution to Prob. 9 in Chap. 28.

x	$2x^3 + x^2 + x - 4$	$3x^3 - 2x^2 + 4x - 5$
−3	−52	−116
−2	−18	−45
−1	−6	−14
0	−4	−5
1	**0**	**0**
2	18	19
3	62	70
4	144	171

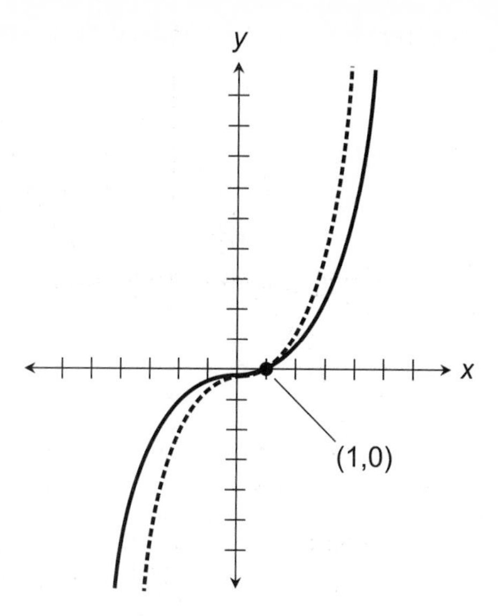

Figure C-8 Illustration for the solution to Prob. 10 in Chap. 28.

Chapter 29

1. We can use the property of common logarithms that converts a product into a sum:

$$\log_{10} xy = \log_{10} x + \log_{10} y$$

In this case, $x = 2.3713018568$ and $y = 0.902780337$. If we use the calculator found in the Windows computer operating system:

$$\begin{aligned} &\log_{10} (2.3713018568 \times 0.902780337) \\ &= \log_{10} 2.3713018568 + \log_{10} 0.902780337 \\ &\approx 0.37498684137 + (-0.0444179086) \\ &\approx 0.37498684137 - 0.0444179086 \\ &\approx 0.3305689328 \end{aligned}$$

This is the common logarithm of the product we wish to find. If we find the common antilogarithm of this, we'll get the desired result. Inputting this to a calculator and then rounding to three decimal places:

$$\text{antilog}_{10} (0.3305689328) \approx 2.141$$

2. We can use the property of natural logarithms that converts a product into a sum:

$$\ln xy = \ln x + \ln y$$

In this case, $x = 2.3713018568$ and $y = 0.902780337$. Therefore:

$$\begin{aligned}
&\ln (2.3713018568 \times 0.902780337) \\
&= \ln 2.3713018568 + \ln 0.902780337 \\
&\approx 0.86343911100 + (-0.102276014) \\
&\approx 0.86343911100 - 0.102276014 \\
&\approx 0.761163097
\end{aligned}$$

This is the natural logarithm of the product we wish to find. If we find the natural antilogarithm of this, we'll get the desired result. Inputting this to a calculator and then rounding to three decimal places:

$$\text{antiln } (0.761163097) \approx 2.141$$

3. In this situation, $P_{out} = 23.7$ and $P_{in} = 0.535$. We can plug these numbers into the formula for gain G in decibels (dB), and then round off as follows:

$$\begin{aligned}
G &= 10 \log_{10} (P_{out}/P_{in}) \\
&= 10 \log_{10} (23.7 / 0.535) \\
&= 10 \log_{10} 44.299 \\
&\approx 10 \times 1.6464 \\
&\approx 16.5 \text{ dB}
\end{aligned}$$

4. In this situation, $P_{out} = 19.3$ and $P_{in} = 23.7$. We're interested in the power gain of the speaker wire, not the power gain of the amplifier. We can plug these numbers into the formula for gain G in decibels, and then round off as follows:

$$\begin{aligned}
G &= 10 \log_{10} (P_{out}/P_{in}) \\
&= 10 \log_{10} (19.3 / 23.7) \\
&= 10 \log_{10} (0.81435) \\
&\approx 10 \times (-0.089189) \\
&\approx -0.892 \text{ dB}
\end{aligned}$$

5. If a positive real number increases by a factor of exactly 10, then its common logarithm increases (it becomes more positive or less negative) by exactly 1.

6. Let x be the original number, and let y be the final number. We're told that $y = 10x$. Taking the common logarithm of each side of this equation gives us

$$\log y = \log_{10} (10x)$$

From the formula for the common logarithm of a product, we can rewrite this as

$$\log_{10} y = \log_{10} 10 + \log_{10} x$$

But $\log_{10} 10 = 1$. Therefore

$$\log_{10} y = 1 + \log_{10} x$$

7. If a positive real number decreases by a factor of exactly 100 (it becomes 1/100 as great), then its common logarithm decreases by exactly 2.
8. Let x be the original number, and let y be the final number. We are told that $y = x/100$. Taking the common logarithm of each side of this equation gives us

$$\log_{10} y = \log_{10} (x/100)$$

From the formula for the common logarithm of a product, we can rewrite this as

$$\log_{10} y = \log_{10} x - \log_{10} 100$$

But $\log_{10} 100 = 2$. Therefore

$$\log_{10} y = (\log_{10} x) - 2$$

9. If a positive real number decreases by a factor of 357, then its natural logarithm decreases by ln 357 or, approximately, 5.88.
10. Let x be the original number, and let y be the final number. We are told that $y = x/357$. Taking the natural logarithm of each side of this equation gives us

$$\ln y = \ln (x/357)$$

From the formula for the natural logarithm of a ratio, we can rewrite this as

$$\ln y = \ln x - \ln 357$$

Using a calculator and rounding to two decimal places, we get $\ln 357 \approx 5.88$, so

$$\ln y \approx (\ln x) - 5.88$$

APPENDIX

D

Answers to Final Exam Questions

1. e	2. e	3. b	4. c	5. b
6. a	7. d	8. d	9. c	10. a
11. b	12. a	13. b	14. b	15. e
16. b	17. d	18. c	19. e	20. b
21. d	22. c	23. c	24. b	25. a
26. c	27. a	28. a	29. d	30. b
31. c	32. c	33. d	34. e	35. b
36. d	37. e	38. a	39. a	40. b
41. b	42. d	43. a	44. a	45. c
46. c	47. b	48. e	49. d	50. e
51. a	52. d	53. b	54. c	55. d
56. a	57. e	58. c	59. d	60. e
61. c	62. a	63. c	64. e	65. a
66. a	67. b	68. d	69. e	70. b
71. e	72. a	73. b	74. e	75. b
76. a	77. e	78. c	79. b	80. a
81. a	82. a	83. d	84. e	85. b
86. c	87. e	88. c	89. a	90. c
91. d	92. a	93. c	94. e	95. b
96. b	97. a	98. a	99. b	100. a
101. c	102. a	103. e	104. b	105. c
106. c	107. e	108. b	109. b	110. d
111. a	112. d	113. b	114. d	115. a
116. d	117. c	118. a	119. e	120. a
121. b	122. e	123. c	124. c	125. d
126. d	127. a	128. d	129. a	130. d
131. e	132. a	133. e	134. e	135. c
136. a	137. b	138. a	139. a	140. c
141. d	142. c	143. e	144. a	145. b
146. b	147. a	148. d	149. c	150. a

APPENDIX

E

Special Characters in Order of Appearance

Symbol	First use	Meaning
I	Chapter 1	Roman numeral for 1
V	Chapter 1	Roman numeral for 5
X	Chapter 1	Roman numeral for 10
L	Chapter 1	Roman numeral for 50
C	Chapter 1	Roman numeral for 100
D	Chapter 1	Roman numeral for 500
M	Chapter 1	Roman numeral for 1,000
K	Chapter 1	Alternative Roman numeral for 1,000
∞	Chapter 1	Lemniscate symbol for infinity
ω	Chapter 1	Lowercase Greek omega symbol for infinity
$\aleph$	Chapter 1	Uppercase Hebrew aleph symbol for infinity
$\aleph_0$	Chapter 1	Aleph-null, the number of whole numbers
...	Chapter 1	Ellipsis, indicating repetition of a sequence or pattern
$=$	Chapter 1	Conventional symbol for numerical equality
$+$	Chapter 1	Conventional symbol for addition
$\in$	Chapter 2	Set symbol meaning "is an element of"
$\notin$	Chapter 2	Set symbol meaning "is not an element of"
$\{\}$	Chapter 2	Braces for enclosing list of set elements
$\varnothing$	Chapter 2	Symbol for the null (empty) set
$/$	Chapter 2	Conventional symbol for division, fraction, or ratio
$\subseteq$	Chapter 2	Set symbol meaning "is a subset of"
$\subset$	Chapter 2	Set symbol meaning "is a proper subset of"
$=$	Chapter 2	Set symbol meaning "is congruent to"
$\equiv$	Chapter 2	Alternative set symbol meaning "is congruent to"
$\cap$	Chapter 2	Set symbol meaning "intersect"
$\cup$	Chapter 2	Set symbol meaning "union"
$\times$	Chapter 3	Conventional symbol for multiplication
$-$	Chapter 3	Conventional symbol for negative numerical value
$()$	Chapter 4	Parentheses for grouping of quantities

Symbol	First use	Meaning
$\| \|$	Chapter 4	Symbol for absolute value (of quantity between bars)
$-$	Chapter 4	Conventional symbol for subtraction
[ ]	Chapter 4	Brackets for grouping of quantities
$\cdot$	Chapter 5	Alternative symbol for multiplication
$\div$	Chapter 5	Alternative symbol for division
$\neq$	Chapter 5	Symbol meaning "is not equal to"
:	Chapter 6	Symbol for a ratio or proportion
$>$	Chapter 6	Inequality symbol meaning "is strictly larger than"
$<$	Chapter 6	Inequality symbol meaning "is strictly smaller than"
$\geq$	Chapter 6	Symbol meaning "is larger than or equal to"
$\leq$	Chapter 6	Symbol meaning "is smaller than or equal to"
{ }	Chapter 6	Braces for grouping of quantities
.	Chapter 7	Decimal point
Σ	Chapter 7	Uppercase Greek sigma symbol for sum
$\pm$	Chapter 8	Plus-or-minus sign
$\surd$	Chapter 8	Surd symbol for square root
$\Rightarrow$	Chapter 11	Symbol meaning "implies" or "if/then"
$\Leftrightarrow$	Chapter 11	Symbol meaning "if and only if"
Δ	Chapter 15	Uppercase Greek delta symbol for difference
f^{-1}	Chapter 17	Notation for inverse of a function f

Suggested Additional Reading

- Bluman, A. *Math Word Problems Demystified.* New York: McGraw-Hill, 2005.
- Bluman, A., *Pre-Algebra Demystified.* New York: McGraw-Hill, 2004.
- Gibilisco, S. *Everyday Math Demystified.* New York: McGraw-Hill, 2004.
- Gibilisco, S. *Technical Math Demystified.* New York: McGraw-Hill, 2006.
- Huettenmueller, R. *Algebra Demystified.* New York: McGraw-Hill, 2003.
- Huettenmueller, R. *College Algebra Demystified.* New York: McGraw-Hill, 2004.
- Huettenmueller, R. *Pre-Calculus Demystified.* New York: McGraw-Hill, 2005.
- Olive, J. *Maths: A Student's Survival Guide,* 2d ed. Cambridge, England: Cambridge University Press, 2003.
- Prindle, A. *Math the Easy Way,* 3d ed. Hauppauge, N.Y.: Barron's Educational Series, 1996.
- Shankar, R. *Basic Training in Mathematics: A Fitness Program for Science Students.* New York: Plenum Publishing Corporation, 1995.

Index

D

E

F

J

L

M

N

S

T

U

V

W

X

Y

Z